Lecture Notes in Bioinformatics 8394

Edited by S. Istrail, P. Pevzner, and M. Waterman

Editorial Board: A. Apostolico S. Brunak M. Gelfand

T. Lengauer S. Miyano G. Myers M.-F. Sagot D. Sankoff

R. Shamir T. Speed M. Vingron W. Wong

Subseries of Lecture Notes in Computer Science

W0234535

Roded Sharan (Ed.)

Research in Computational Molecular Biology

18th Annual International Conference, RECOMB 2014
Pittsburgh, PA, USA, April 2-5, 2014
Proceedings

 Springer

Volume Editor

Roded Sharan
Tel-Aviv University
School of Computer Science
Tel-Aviv 69978, Israel
E-mail: roded@post.tau.ac.il

ISSN 0302-9743   e-ISSN 1611-3349
ISBN 978-3-319-05268-7   e-ISBN 978-3-319-05269-4
DOI 10.1007/978-3-319-05269-4
Springer Cham Heidelberg New York Dordrecht London

Library of Congress Control Number: 2014932543

LNCS Sublibrary: SL 8 – Bioinformatics

© Springer International Publishing Switzerland 2014
This work is subject to copyright. All rights are reserved by the Publisher, whether the whole or part of
the material is concerned, specifically the rights of translation, reprinting, reuse of illustrations, recitation,
broadcasting, reproduction on microfilms or in any other physical way, and transmission or information
storage and retrieval, electronic adaptation, computer software, or by similar or dissimilar methodology
now known or hereafter developed. Exempted from this legal reservation are brief excerpts in connection
with reviews or scholarly analysis or material supplied specifically for the purpose of being entered and
executed on a computer system, for exclusive use by the purchaser of the work. Duplication of this publication
or parts thereof is permitted only under the provisions of the Copyright Law of the Publisher's location,
in its current version, and permission for use must always be obtained from Springer. Permissions for use
may be obtained through RightsLink at the Copyright Clearance Center. Violations are liable to prosecution
under the respective Copyright Law.
The use of general descriptive names, registered names, trademarks, service marks, etc. in this publication
does not imply, even in the absence of a specific statement, that such names are exempt from the relevant
protective laws and regulations and therefore free for general use.
While the advice and information in this book are believed to be true and accurate at the date of publication,
neither the authors nor the editors nor the publisher can accept any legal responsibility for any errors or
omissions that may be made. The publisher makes no warranty, express or implied, with respect to the
material contained herein.

Typesetting: Camera-ready by author, data conversion by Scientific Publishing Services, Chennai, India

Printed on acid-free paper

Springer is part of Springer Science+Business Media (www.springer.com)

Preface

The RECOMB conference series – with the full name of the Annual International Conference on Research in Computational Molecular Biology – was started in 1997 by Sorin Istrail, Pavel Pevzner, and Michael Waterman. The 18th RECOMB conference, RECOMB 2014, was held in Pittsburgh, PA, USA during April 2–5, 2014. It was hosted by Carnegie Mellon University and the University of Pittsburgh.

This volume contains the 35 extended abstracts selected for oral presentation at RECOMB 2014 by the Program Committee (PC) out of 154 submissions. Each submission was assigned to at least three PC members and reviewed with the help of many external reviewers. Following the initial reviews, there was an extensive discussion of the submissions among the members of the PC, leading to the final decisions.

This year RECOMB teamed up with *PLoS Computational Biology* to allow parallel submissions to the proceedings and the journal in a coordinated manner. I would like to thank Thomas Lengauer, deputy editor of *PLoS Computational Biology*, for the countless hours he spent on coordinating the review process and ensuring the success of this partnership. In total, 55 papers were submitted for consideration by RECOMB and *PLoS Computational Biology*. Eighteen of those papers passed an initial pre-screening for the journal and underwent a full review process. Five papers were subsequently accepted to both venues and appear as two-page abstracts in this volume; a few additional papers were accepted to only one of the venues. The five papers were published in full in a special RECOMB 2014 section of *PLoS Computational Biology*. All other papers that were accepted to RECOMB 2014 were invited for submission of an edited journal version to a special issue of the *Journal of Computational Biology*.

In addition to the paper presentations, RECOMB 2014 featured six invited keynote talks by leading scientists world-wide. The keynote speakers were Ian T. Baldwin (Max Planck Institute for Chemical Ecology), Atul Butte (Stanford University), James J. Collins (Harvard University), Trey Ideker (University of California at San Diego), Tom Mitchell (Carnegie Mellon University), and Sarah A. Tishkoff (University of Pennsylvania). Following the tradition started at RECOMB 2010, RECOMB 2014 also featured a special highlights track containing computational biology papers that were published in journals during the last 18 months. There were 48 submissions, eight of which were selected for oral presentation.

The success of RECOMB depends on the effort, dedication, and devotion of many colleagues who contributed to the organization of the conference. We thank the PC members and the external reviewers for the timely review of the assigned papers despite their busy schedules; Teresa Przytycka for chairing the highlights track; Carl Kingsford for chairing the posters track; the Steering Committee and its chair, Martin Vingron, for many excellent suggestions on the organization of the conference; the local Organizing Committee members, especially Thom Gulish and the Co-chairs Russell Schwartz and Panayiotis (Takis) Benos, for hosting the conference and providing the administrative, logistic, and financial support; and the authors of the papers, highlights, and posters and all the attendees for their enthusiastic participation in the conference. We also thank our generous sponsors, including the International Society of Computational Biology (ISCB), the US National Science Foundation, Biomed Central GigaScience, Carnegie Mellon's Lane Center for Computational biology, and University of Pittsburgh's Department of Computational and Systems Biology. Finally, I would like to thank Yehuda Afek (Tel Aviv University), Ron Shamir, and Benny Chor (Program Chairs of RECOMB 2000 and 2012, respectively) for their support and advice.

February 2014   Roded Sharan

Conference Organization

Program Committee

Tatsuya Akutsu	Kyoto University, Japan
Max Alekseyev	University of South Carolina, USA
Joel Bader	Johns Hopkins University, USA
Vineet Bafna	University of California at San Diego, USA
Nuno Bandeira	University of California at San Diego, USA
Mukul S. Bansal	Massachusetts Institute of Technology, USA
Ziv Bar-Joseph	Carnegie Mellon University, USA
Serafim Batzoglou	Stanford University, USA
Asa Ben-Hur	Colorado State University, USA
Bonnie Berger	Massachusetts Institute of Technology, USA
Michael Brudno	University of Toronto, Canada
Sebastian Böcker	Friedrich Schiller University Jena, Germany
Brian Chen	Lehigh University, USA
Ting Chen	University of Southern California, USA
Yi-Ping Phoebe Chen	La Trobe University, USA
Chakra Chennubhotla	University of Pittsburgh, USA
Benny Chor	Tel Aviv University, Israel
Lenore Cowen	Tufts University, USA
Vincent Danos	Centre National de la Recherche Scientifique, France
Minghua Deng	Peking University, China
Colin Dewey	University of Wisconsin-Madison, USA
Nadia El-Mabrouk	University of Montreal, Canada
Sarel Fleishman	Weizmann Institute, Israel
Irit Gat-Viks	Tel Aviv University, Israel
Mikhail Gelfand	Institute for Information Transmission Problems, Russia
David Gifford	Massachusetts Institute of Technology, USA
Assaf Gottlieb	Stanford University, USA
Bjarni Halldorsson	Reykjavik University, Iceland
Wenlian Hsu	Academia Sinica, Taiwan
Sorin Istrail	Brown University, USA
Rui Jiang	Tsinghua University, China
Tao Jiang	University of California at Riverside, USA

Igor Jurisica   Ontario Cancer Institute, Canada
Tamer Kahveci   University of Florida, USA
Simon Kasif   Boston University, USA
Seyoung Kim   Carnegie Mellon University, USA
Carl Kingsford   Carnegie Mellon University, USA
                                     (Posters Chair)
Mehmet Koyuturk   Case Western Reserve University, USA
Rui Kuang   University of Minnesota Twin Cities, USA
Jens Lagergren   KTH Royal Institute of Technology, Sweden
Hyunju Lee   Gwangju Institute of Science and Technology,
                                     South Korea
Michal Linial   The Hebrew University of Jerusalem, Israel
Jinze Liu   University of Kentucky, USA
Stefano Lonardi   University of California at Riverside, USA
Tijana Milenkovic   University of Notre Dame, USA
Satoru Miyano   University of Tokyo, Japan
Bernard Moret   Ecole Polytechnique Federale de Lausanne,
                                     Switzerland
T. M. Murali   Virginia Polytechnic Institute and State
                                     University, USA
William Stafford Noble   University of Washington, USA
Arlindo Oliveira   IST/INESC-ID and Cadence Research
                                     Laboratories, Portugal
Laxmi Parida   IBM T. J. Watson Research Center, USA
Itsik Pe'er   Columbia University, USA
Teresa Przytycka   National Institute of Health, USA
                                     (Highlights Chair)
Ben Raphael   Brown University, USA
Knut Reinert   Free University Berlin, Germany
Marie-France Sagot   Inria Grenoble Rhône-Alpes and Université
                                     de Lyon 1, France
S. Cenk Sahinalp   Simon Fraser University, Canada
Dina Schneidman-Duhovny   University of California at San Francisco, USA
Denise Scholtens   Northwestern University, USA
Russell Schwartz   Carnegie Mellon University, USA
Roded Sharan   Tel Aviv University, Israel (Chair)
Mona Singh   Princeton University, USA
Andrew Smith   University of Southern California, USA
Berend Snel   Utrecht University, The Netherlands
Kai Tan   University of Iowa, USA

Alfonso Valencia	Spanish National Cancer Research Centre, Spain
Fabio Vandin	University of Southern Denmark, Denmark
Jean-Philippe Vert	Ecole des Mines de Paris, France
Martin Vingron	Max Planck Institute for Molecular Genetics, Germany
Lusheng Wang	City University of Hong Kong, Hong Kong
Tandy Warnow	University of Texas at Austin, USA
Haim J. Wolfson	Tel Aviv University, Israel
Limsoon Wong	National University of Singapore, Singapore
Xiaohui Xie	University of California at Irvine, USA
Dong Xu	University of Missouri-Columbia, USA
Jinbo Xu	Toyota Technology Institute at Chicago, USA
Yuzhen Ye	Indiana University, USA
Esti Yeger-Lotem	Ben-Gurion University, Israel
Sungroh Yoon	Seoul National University, South Korea
Louxin Zhang	National University of Singapore, Singapore
Xuegong Zhang	Tsinghua University, China
Hongyu Zhao	Yale University, USA

Steering Committee

Vineet Bafna	University of California at San Diego, USA
Serafim Batzoglou	Stanford University, USA
Bonnie Berger	Massachusetts Institute of Technology, USA
Sorin Istrail	Brown University, USA
Michal Linial	The Hebrew University of Jerusalem, Israel
Martin Vingron	Max Planck Institute for Molecular Genetics, Germany (Chair)

Organizing Committee

Russell Schwartz	Carnegie Mellon University, USA (Co-chair)
Panayiotis (Takis) Benos	University of Pittsburgh, USA (Co-chair)
Thom Gulish	Carnegie Mellon University, USA (Local Arrangements Coordinator)
Carl Kingsford	Carnegie Mellon University, USA
Darya Filippova	Carnegie Mellon University, USA
Nichole Merrit	Carnegie Mellon University, USA

Previous RECOMB Meetings

Dates	Hosting Institution	Program Chair	Conference Chair
January 20-23, 1997 Santa Fe, NM, USA	Sandia National Lab	Michael Waterman	Sorin Istrail
March 22-25, 1998 New York, NY, USA	Mt. Sinai School of Medicine	Pavel Pevzner	Gary Benson
April 22-25, 1999 Lyon, France	Inria	Sorin Istrail	Mireille Regnier
April 8-11, 2000 Tokyo, Japan	University of Tokyo	Ron Shamir	Satoru Miyano
April 22-25, 2001 Montreal, Canada	Université de Montreal	Thomas Lengauer	David Sankoff
April 18-21, 2002 Washington, DC, USA	Celera	Gene Myers	Sridhar Hannenhalli
April 10-13, 2003 Berlin, Germany	German Federal Ministry for Education and Research	Webb Miller	Martin Vingron
March 27-31, 2004 San Diego, USA	University of California, San Diego	Dan Gusfield	Philip E. Bourne
May 14-18, 2005 Boston, MA, USA	Broad Institute of MIT and Harvard	Satoru Miyano	Jill P. Mesirov and Simon Kasif
April 2-5, 2006 Venice, Italy	University of Padova	Alberto Apostolico	Concettina Guerra
April 21-25, 2007 San Francisco, CA, USA	QB3	Terry Speed	Sandrine Dudoit
March 30-April 2, 2008 Singapore, Singapore	National University of Singapore	Martin Vingron	Limsoon Wong
May 18-21, 2009 Tucson, AZ, USA	University of Arizona	Serafim Batzoglou	John Kececioglu
August 12-15, 2010 Lisbon, Portugal	INESC-ID and Instituto Superior Técnico	Bonnie Berger	Arlindo Oliveira
March 28-31, 2011 Vancouver, Canada	Lab for Computational Biology, Simon Fraser University	Vineet Bafna	S. Cenk Sahinalp
April 21-24, 2012 Barcelona, Spain	Centre for Genomic Regulation	Benny Chor	Roderic Guigó
April 7-10, 2013 Beijing, China	Tsinghua University	Fengzhu Sun	Xuegong Zhang
April 2-5, 2014 Pittsburgh, PA, USA	Carnegie Mellon University and University of Pittsburgh	Roded Sharan	Russell Schwartz and Panayiotis (Takis) Benos

External Reviewers

Aganezov, Sergey
Aguiar, Derek
Ali, Raja Hashim
Antonov, Ivan
Arvestad, Lars
Atias, Nir
Ayati, Marzieh
Bandyopadhyay, Nirmalya
Bebek, Gurkan
Behnam, Ehsan
Beissbarth, Tim
Ben-Bassat, Ilan
Bienkowska, Jadwiga
Biesinger, Jacob
Botzman, Maya

Brodt, Avital
Bucher, Philipp
Carmi, Shai
Chen, Caster
Chiu, Jimmy Ka Ho
Cho, Dongyeon
Chowdhury, Salim
Chung, Ho-Ryun
Cicek, Ercument
Csürös, Miklós
Daley, Timothy
Daniels, Noah
Dao, Phuong
Dolzhenko, Egor
Donmez, Nilgun

Dror, Ron
Duma, Denise
Edwards, Matt
Emde, Anne-Katrin
Eskin, Eleazar
Faisal, Fazle
Farahani, Hossein S.
Frishberg, Amit
Frånberg, Mattias
Fujita, Andre
Gabr, Haitham
Gao, Long
Ghersi, Dario
Gitter, Anthony
Greenman, Christopher
Guo, Yuchun
Guthals, Adrian
Hach, Faraz
Haiminen, Niina
Hajirasouliha, Iman
Hasan, Mahmudul
Haws, David
He, Bing
He, Dan
He, Xin
He, Zhiquan
Hildebrandt, Andreas
Hodzic, Ermin
Hufsky, Franziska
Hui, Ken
Hulovatyy, Yuriy
Jiang, Shuai
Jiang, Yue
Jónsson, Björn
Kalinina, Anastasiya
Kashef, Dorna
Keles, Sunduz
Kim, Yoo-Ah
Korobeynikov, Anton
Kreimer, Anat
Kuang, Rui
Kuleshov, Volodymyr
Kwon, Sunyoung
Kyriazopoulou-Panagiotopoulou, Sofia
Lafond, Manuel

Lavenier, Dominique
Leiserson, Mark
Li, Cong
Li, Ming
Li, Wei
Li, Yi
Libbrecht, Max
Lin, Yen Yi
Lynn, Ke-Shiuan
Ma, Jianzhu
Ma, Wenxiu
Ma, Xiaoke
Makeev, Vsevolod
Mariadassou, Mahendra
McPherson, Andrew
Melsted, Páll
Meusel, Marvin
Mirebrahim, Seyed
Mohammadi, Shahin
Morris, Quaid
Na, Seungjin
Nachshon, Aharon
Navlakha, Saket
Newburger, Daniel
Niida, Atsushi
Nikolenko, Sergey
Numanagic, Ibrahim
O'Donnell, Charles W.
Oesper, Layla
Ounit, Rachid
Palmer, Cameron
Park, Heewon
Park, Seunghyun
Pauwels, Edouard
Platt, Daniel
Polishko, Anton
Popic, Victoria
Pritykin, Yuri
Prjibelski, Andrey D.
Przytycki, Pawel
Qu, Jenny
Quang, Daniel
Rajasekaran, Rajalakshmi
Richard, Hugues
Ritz, Anna

Robin, Stéphane
Rubinov, Anatoly
Ruffalo, Matthew
Röst, Hannes
Sacomoto, Gustavo
Sadeh, Mohammad
Sahlin, Kristoffer
Salari, Raheleh
Savel, Daniel
Scheubert, Kerstin
Schloissnig, Siegfried
Schulz, Marcel
Sefer, Emre
Shen, Yang
Sheridan, Paul
Shiraishi, Yuichi
Song, Jimin
Srivastava, Gyan
Steuerman, Yael
Tegge, Allison
Teng, Li
Todor, Andrei
Ullah, Ikram
Uren, Philip

Utro, Filippo
Vinogradov, Dmitry
Wagner, Allon
Wang, Jian
Wang, Mingxun
Wang, Qingguo
Wang, Sheng
Wang, Zheng
Wang, Zhiyong
White, Tim
Wilenzik, Roni
Wise, Aaron
Wu, Hsin-Ta
Wu, Yu-Wei
Wójtowicz, Damian
Yang, Can
Yang, Ei-Wen
Yao, Qiuming
Yiu, Siuming
Yu, Seunghak
Yu, Zhaoxia
Zhang, Kui
Zhao, Feng
Zhong, Shan

Table of Contents

Tractatus: An Exact and Subquadratic Algorithm for Inferring Identical-by-Descent Multi-shared Haplotype Tracts

Derek Aguiar[1], Eric Morrow[2], and Sorin Istrail[1,*]

[1] Department of Computer Science and Center for Computational Biology,
Brown University, Providence, Rhode Island 02912, USA
{Derek_Aguiar,Sorin_Istrail}@brown.edu
[2] Departments of Molecular Biology, Cell Biology & Biochemistry and Psychiatry
& Human Behavior, Brown University, Providence, Rhode Island 02912, USA
Eric_Morrow@brown.edu

Abstract. In this work we present graph theoretic algorithms for the identification of all identical-by-descent (IBD) multi-shared haplotype tracts for an $m \times n$ haplotype matrix. We introduce Tractatus, an exact algorithm for computing all IBD haplotype tracts in time linear in the size of the input, $O(mn)$. Tractatus resolves a long standing open problem, breaking optimally the (worst-case) quadratic time barrier of $O(m^2n)$ of previous methods often cited as a bottleneck in haplotype analysis of genome-wide association study-sized data. This advance in algorithm efficiency makes an impact in a number of areas of population genomics rooted in the seminal Li-Stephens framework for modeling multi-loci linkage disequilibrium (LD) patterns with applications to the estimation of recombination rates, imputation, haplotype-based LD mapping, and haplotype phasing. We extend the Tractatus algorithm to include computation of haplotype tracts with allele mismatches and shared homozygous haplotypes in a set of genotypes. Lastly, we present a comparison of algorithmic runtime, power to infer IBD tracts, and false positive rates for simulated data and computations of homozygous haplotypes in genome-wide association study data of autism. The Tractatus algorithm is available for download at http://www.brown.edu/Research/Istrail_Lab/.

Keywords: haplotypes, haplotype tracts, graph theory, identity-by-descent.

1   Introduction

1.1   Li-Stephens PAC-Likelihood Model and the $O(m^2n)$ Time Bound

Understanding and interpreting patters of linkage disequilibrium (LD) among multiple variants in a genome-wide population sample is a major technical challenge in population genomics. A large body of research literature is devoted to

---

* Corresponding author.

R. Sharan (Ed.): RECOMB 2014, LNBI 8394, pp. 1–17, 2014.
© Springer International Publishing Switzerland 2014

the topic including the computational framework presented in the seminal work of Li and Stephens[1]. Building on the work by Stephens *et al.* 2001[2], Hudson[3], and Fearnhead and Donnelly[4], the Li-Stephens framework led the way towards major advances in the understanding and modeling of linkage disequilibrium patterns and recombination.

The difficulties associated with modeling LD patterns at multiple loci include a number of long standing analytical obstacles. Among existing bottlenecks is the notorious (1) *curse of the pairwise*, as all the popular LD measures in the literature are pairwise measures, and the (2) *haplotype block-free* approach to avoid *ad hoc* haplotype block definitions and "fake blocks" due to recombination rate heterogeneity. Current methods for computing haplotype blocks result in the definition of *ad hoc* boundaries that sometimes present less LD within blocks than between blocks due to different patterns of recombination. This phenomenon leads to spurious block-like clusters. The Li-Stephens statistical model for LD, named the *Product of Approximate Conditionals* (PAC), is based on a generalization of coalescent theory to include recombination [3,5].

The optimization problem introduces the PAC likelihood $L_{PAC}(\rho)$

$$L_{PAC}(\rho) = \tilde{\pi}(h_1 \mid \rho)\tilde{\pi}(h_2 \mid h_1, \rho)...\tilde{\pi}(h_m \mid h_1, ..., h_{m-1}, \rho)$$

where $h_1, ..., h_m$ are the m sampled haplotypes, ρ denotes the recombination parameter, and $\tilde{\pi}$ represents an approximation of the corresponding conditional probabilities. Li and Stephens propose a number of such approximations for approximate likelihood functions[1]. $L_{PAC}(\rho)$ represents the unknown distribution

$$Prob(h_1, ..., h_m \mid \rho) = Prob(h_1 \mid \rho)Prob(h_2 \mid h_1, \rho)...Prob(h_m \mid h_1, ..., h_{m-1}, \rho)$$

The choice of $\tilde{\pi}$ gives the form of the likelihood objective function.

The PAC likelihood is based on expanding the modeling to capture realistic genomic structure while generalizing Ewens' sampling formula and coalescent theory. The framework iteratively samples the m haplotypes; if the first k haplotypes have been sampled $h_1, ..., h_k$, then the conditional distribution for the next sampled haplotype is $Prob(h_{k+1} \mid h_1, ..., h_k)$. $\tilde{\pi}$ approximates this distribution and is constructed to satisfy the following axioms:

1. h_{k+1} is more likely to match a haplotype from $h_1, ..., h_k$ that has been observed many times rather than a haplotype that has been observed less frequently.
2. The probability of observing a novel haplotype decreases as k increases.
3. The probability of observing a novel haplotype increases as $\theta = 4N\mu$ increases, where N is the population size and μ is the mutation rate.
4. If the next haplotype is not identical to a previously observed haplotype, it will tend to differ by a small number of mutations from an existing haplotype (as in the Ewens' sampling formula model).
5. Due to recombination, h_{k+1} will resemble haplotypes $h_1, ..., h_k$ over contiguous genomic regions; the average physical length of these regions should be larger in genomic regions where the local rate of recombination is low.

Intuitively, the next haplotype h_{k+1} should be an imperfect *mosaic* of the first k haplotypes, with the size of the mosaic fragments being smaller for higher values of the recombination rate. Although the proposed model ($\tilde{\pi}_A$ in the notation of [1]) satisfies the above axioms and has the desirable property of being efficiently computable, it has a serious disadvantage. As is stated in their article, this "unwelcome" feature of the PAC likelihoods corresponding to the choices for $\tilde{\pi}$ is *order dependence*, that is, the choices are dependent on the order of the haplotypes sampled. Other methods used in the literature, notable, Stephens *et al.* 2001[2] and Fearnhead and Donnelly[4], present the same problem of order dependence. Different haplotype sampling permutations correspond to different distributions; these probability distributions *do not satisfy the property of exchangeability* that we would expect to be satisfied by the true but unknown distribution.

1.2   Identical-by-Descent Haplotype Tracts

Haplotype tracts, or contiguous segments of haplotypes, are identical-by-descent (IBD) if they are inherited from a common ancestor [6]. Tracts of haplotypes shared IBD are disrupted by recombination so the expected lengths of the IDB tracts depends on the pedigree structure of the sample and the number of generations till the least common ancestor at that haplotype region. The computation of IBD is fundamental to genetic mapping and can be inferred using the PAC likelihood model.

To model the effects of recombination, a hidden Markov model (HMM) is defined to achieve a mosaic construction. At every variant, it is possible to transition to any of the haplotypes generated so far with a given probability. Thus, a path through the chain starts with a segment from one haplotype and continues with a segment from another haplotype and so on. To enforce the mosaic segments to resemble haplotype tracts, the probability of continuing in the same haplotype without jumping is defined exponentially in terms of the physical distance (assumed known) between the markers; that is, if sites j and $j+1$ are at a small genetic distance apart, then they are highly likely to exist on the same haplotype. The computation of the L_{PAC} is linear in the number of variants (n) and quadratic in the number of haplotypes (m) in the sample, hence the $O(m^2 n)$ time bound.

In this work we present results that remove the pairwise quadratic dependence by computing multi-shared haplotype tracts. Multi-shared haplotype tracts are maximally shared contiguous segments of haplotypes starting and ending at the same genomic position that cannot be extended by adding more haplotypes in the sample. Because we represent the pairwise sharing in sets of haplotypes, no more than $O(mn)$ multi-shared haplotype tracts may exist.

1.3   Prior Work

Building on the PAC model, the IMPUTE2 [7] and MaCH [8] algorithms employ HMMs to model a sample set of haplotypes as an imperfect mosaic of reference haplotypes. The usage of the forward-backward HMM algorithm brings these methods in the same $O(m^2n)$ time bound class. The phasing program SHAPEIT (segmented haplotype estimation and imputation tool) also builds on the PAC model by decomposing the haplotype matrix uniformly into a number of segments and creating linear time mosaics within each such *ad hoc* segmented structure[9]. The dependence on the number of segments is not considered in the time complexity.

PLINK [10], FastIBD [11], DASH [12], and IBD-Groupon [13] are algorithms based on HMMs or graph theory clustering methods that consider pairs of haplotypes to compute IBD tracts. Iterating over all such pairs takes time $O(m^2n)$ and is intractable for large samples; this intractability is best described in the recent work of Gusev *et al.* 2011.

> "Although the HMM schemes offer high resolution of detection [of IBD], the implementations require examining all pairs of samples and are intractable for GWAS-sized cohorts. ... In aggregate, these identical-by-descent segments can represent the totality of detectable recent haplotype sharing and could thus serve as refined proxies for recent variants that are generally rare and difficult to detect otherwise." Gusev *et al.* 2011 [12]

Gusev *et al.* 2009 describes the computationally efficient algorithm GERMLINE which employs a dictionary hashing approach[14]. The input haplotype matrix is divided into discrete slices or windows and haplotype words that hash to the same value are identified as shared. Due to this dependence on windows, the algorithm is inherently inexact. While the identification of small haplotype tracts within error-free windows can be performed in linear time, GERMLINE's method for handling base call errors is worst case quadratic. However, GERMLINE's runtime has been shown to be near linear time in practice [6].

In what follows, we describe the Tractatus algorithm for computing IBD multi-shared haplotype tracts from a sample of haplotypes and the Tractatus-HH algorithm for computing homozygous haplotypes in a sample of genotypes. Section 2 introduces the computational model and algorithms. Section 3 compares the runtime of Tractatus to a generic pairwise algorithm, compares false positive rates and power with GERMLINE, and provides an example computation of homozygous haplotype regions in genome-wide association study data of autism. Finally, sections 4 and 5 discuss implications of this algorithm, conclusions, and future directions.

2   Methods

Our work presented here addresses the lack of exchangeability in the sampling methods of the Li-Stephens model and provides a rigorous result that gives a basis for sampling with the assured exchangeability property. We also present a data structure that speeds up the HMM and the graph clustering models for the detection of identical-by-descent haplotype tracts. Informally, a *haplotype tract* or simply *tract* is a contiguous segment of a haplotype – defined by start and end variant indices – that is shared (identical) by two or more haplotypes in a given sample of haplotypes. One can then view each of the haplotypes in the set as a mosaic concatenation of tracts. Such a haplotype tract decomposition is unique and a global property of the sample. Our Tractatus algorithm computes the *Tract tree* of all the tracts of the haplotype sample in linear time in the size of the sample. The Tract tree, related to a suffix tree, represents each haplotype tract in a single root-to-internal-node path. Repeated substrings in distinct haplotypes are compressed and represented only once in the Tract tree.

2.1   The Tractatus Model

Suffix trees are graph theoretic data structures for compressing the suffixes of a character string. Several algorithms exist for suffix tree construction including the notable McCreight and Ukkonen algorithms that achieve linear time and space constructions for $O(1)$ alphabets [15,16]. In 1997, Farach introduced the first suffix-tree construction algorithm that is linear time and space for integer alphabets [17]. Extensions to suffix-trees, commonly known as generalized suffix trees, allow for suffix-tree construction of multiple strings.

The input to the problem of IBD tract inference is m haplotypes which are encoded as n-length strings of 0's and 1's corresponding to the major and minor alleles of genomic variants $v_1, ..., v_n$. Because we are interested in IBD relationships which are by definition interhaplotype, naive application of suffix-tree construction algorithms to the set of haplotypes would poorly model IBD by including intrahaplotype relationships. Let haplotype i be denoted h_i and the allele of h_i at position j be denoted $h_{i,j}$. Then, we model each haplotype $h_i = h_{i,1}, h_{i,2}, ..., h_{i,n}$ with a new string $d_i = (h_{i,1}, 1), (h_{i,2}, 2), ..., (h_{i,n}, n)$ for $1 \leq i \leq m$. Computationally, the position-allele pairs can be modeled as integers $\in [0, 1, 2, ..., 2n - 1]$ where $(h_{i,j}, j)$ is $2 * j + h_{i,j}$ where $h_{i,j} \in 0, 1$. The transformed haplotype strings are termed *tractized*.

2.2   The Tractatus Algorithm without Errors

The Tractatus algorithm incorporates elements from integer alphabet suffix trees with auxiliary data structures and algorithms for computing IBD haplotype tracts. Firstly, a suffix tree is built from the set of m tractized haplotypes each of length n. To represent the tractized haplotypes, the alphabet size is $O(n)$, so Farach's algorithm may be used to construct a suffix tree in linear time[17]. The suffix tree built from the tractized haplotypes is termed the *Tract tree*. After the Tract tree is built, an $O(mn)$ depth first post-order traversal (DFS) is computed to label each vertex with the number of haplotype descendants. These pointers enable the computation of groups of individuals sharing a tract in linear time.

Substrings of haplotypes are compressed if they are identical and contain the same start and end positions in two or more haplotypes. We consider a path from the root to a node with k descendants as maximal if it is not contained within any other path in the Tract tree. The maximal paths can be computed using a depth first search of the Tract tree, starting with suffixes beginning at 0 and ending at suffixes beginning with $2n - 1$. Of course, if a tract is shared by $k \geq 2$ haplotypes, it is represented only once in the Tract tree. Figure 1 shows the construction of the Tract tree and computation of IBD tracts.

The internal nodes of the Tract tree also have an interpretation in regards to Fisher junctions. A Fisher junction is a position in DNA between two variants such that the DNA segments that meet in this virtual point were ancestrally on different chromosomes. Fisher junctions are represented in the Tract tree where maximal tracts branch.

After maximal tracts are computed, they are quantified as IBD or IBS. Tractatus implements two methods for calling maximally shared tracts IBD or IBS. A simple tract calling method thresholds the length L (number of variants) or area (variants × haplotypes) of the tract in terms of the haplotype matrix input. A more complex method considers the probability of a region being shared IBD or identical-by-state (IBS). If two individuals are k^{th} degree cousins, the probability they share a haplotype tract IBD is 2^{-2k} due to the number of meioses between them[18]. Let the frequency of a variant at position i be f_i. Then, the probability of IBD and IBS can be combined to define the probability that a shared haplotype tract of length L and starting at position s for k^{th} degree cousins is IBD (Equation 1) [19]

$$P(IBD|L) = \frac{2^{-2k}}{2^{-2k} + \prod_{i=s}^{s+L} \left(f_i^2 + (1 - f_i)^2\right)} \tag{1}$$

The value of k can be approximated if the population structure is known. Tractatus without errors is presented in Algorithm 1. Because the suffix tree is computable in $O(mn)$ time with $O(mn)$ nodes, the tree traversals can be computed in $O(mn)$ time thus giving Theorem 1.

Theorem 1. *Given a set of m haplotypes each of length n, Algorithm 1 computes the Tract tree and the set of IBD tracts in $O(mn)$ time and space.*

2.3   The Tractatus Algorithm with Errors and Allele Mismatches

Incorporating base call errors and additional variability gained after differentiation from the least common ancestor requires additional computations on the Tract tree and a statistical modeling of haplotype allele mismatches. The Tractatus algorithm with errors is parameterized by an estimated probability of error or mismatched alleles p_t, a p-value threshold corresponding to a test for the number of errors in a tract p_h, a minimum length partial IBD tract l, and a minimum length of calling a full IBD tract L (or alternatively P(IBD) as defined in Equation 1). We will, in turn, explain the significance of each parameter.

The algorithm proceeds similarly to the error-free case. We build the tractized haplotypes, Tract tree and populate necessary data structures with a DFS. Because errors and additional variation now exist which can break up tracts (and therefore paths in the Tract tree), we compute partial tracts as evidence of IBD.

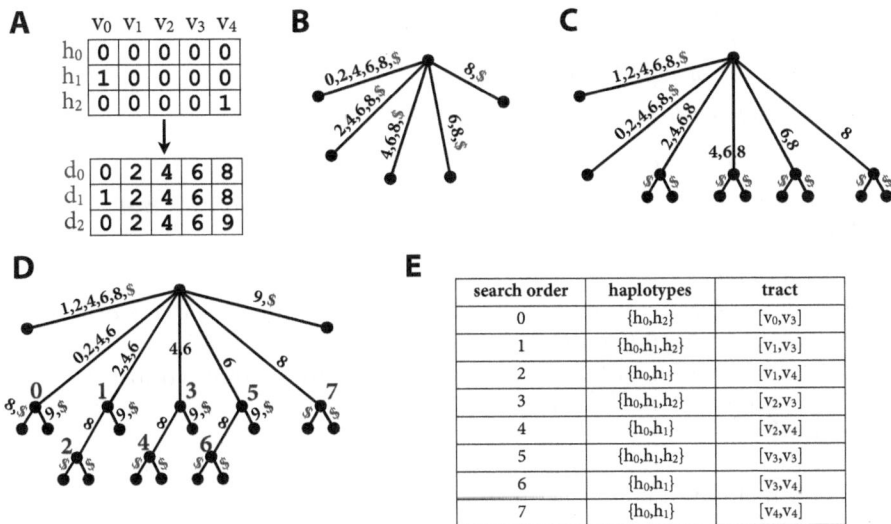

Fig. 1. Construction of the Tract tree and running Tractatus on example input without errors or allele mismatches. Terminator characters $ are colored to match tractized haplotypes and the empty string (simply the terminator character) is omitted in this example. (A) The haplotype matrix is encoded by an integer alphabet representing position-allele pairs. (B) Tractized haplotype d_0 is inserted in the Tract tree. The first tractized haplotype inserts $O(n)$ nodes into the Tract tree. (C) Tractized haplotype d_1 is inserted in the Tract tree. The suffix of d_1 starting at v_0 requires generation of a new node in the Tract tree but subsequent suffixes can be compressed along paths from the root. (D) Tractized haplotype d_2 is inserted in the Tract tree and the algorithmic search order is given in brown integers adjacent to internal nodes. Leaf nodes have exactly one terminating character (haplotype) and therefore do not have to be visited during the search. (E) The largest IBD tracts are found at search numbers 0, 1, and 2. Saving references to these tracts enables the determination that subsequent tracts are contained within already processed tracts.

We compute a DFS from the root, and a maximal partial tract is saved when the algorithm arrives at a node with path length at least l and at least 2 haplotype descendants. If we find a partial tract in a subsequent traversal, we can check in $O(1)$ time is it is contained in a maximal partial tract already computed. Figure 2 shows an example of the Tract tree construction with a single allele mismatch.

Because we computed the partial tracts using a DFS, the tracts are ordered by starting position. For each tract, tracts starting at a position prior and including a subset of the same haplotypes are combined if the extension is statistically probable. To determine the scan distance, we can compute a probability of observing a partially shared tract of length l given a window distance w (or this can be user defined). Assuming the generation of errors is independent and the probability of generating an error is p_e, we model the probability of generating at least k errors in an interval of l_i in t haplotypes as a Poisson process with $\lambda = p_e l_i t$. For each extension we calculate the probability of observing at least

input : m haplotypes each of length n, minimum length L or IBD probability p
output: set of IBD tracts

$H \leftarrow$ *tractized haplotypes*

$T(H) \leftarrow$ *Tract tree of H*

Post-order DFS of $T(H)$ to compute descendant haplotypes from each node

DFS of $T(H)$:
if *path in DFS is longer than L or $P(IBD) > p$ and node has at least 2 descendant haplotypes* **then**
| **if** *tract is not contained in previously computed tract* **then**
| | *report as an IBD tract*
| **end**
else
| *push children nodes on stack*
end

Algorithm 1. Tractatus (error free)

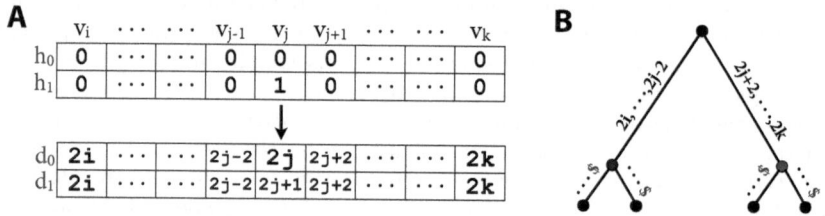

Fig. 2. Construction of the Tract tree and running Tractatus on example input with allele mismatches. (A) h_0 and h_1 share a tract IBD in the interval $[v_i, v_k]$ with a single allele mismatch at v_j. (B) By the construction of the Tract tree, there must be some path (here shown as a single edge but it may be split into a path by other haplotypes) from root to internal node that includes both $[v_i, v_j)$ and $(v_j, v_k]$.

k mismatches and accept the extension if the probability is greater than p_h. The parameter p_t is used as an approximation of p_e. The haplotype consensus sequence of the tract is taken by majority rule at each variant position.

Pseudocode is given in Algorithm 2. While the algorithm is parameterized with five parameters, they are optional and default values are suitable in most cases.

Construction of the Tract-tree takes $O(mn)$ time and space. $O(mn)$ time is needed to prepare data structures and compute maximally shared partial tracts (DFS). A tract can be checked if it is contained in a previously processed tract in $O(1)$ time. It takes $O(sw)$ to merge partial IBD tracts in the worst case when we have to extend many tracts covering a large portion of the matrix, thus yielding Theorem 2.

Theorem 2. *Given a set of m haplotypes each of length n, a scan distance w and a set of partial haplotype tracts s, Algorithm 2 computes the Tract tree and set of IBD tracts in time and space $O(mn + sw)$.*

2.4   Extensions for Homozygous Haplotypes

A particular class of identical-by-descent relationships are long regions of extended homozygosity in genotypes. The two dominant concepts of extended regions of allelic homozygosity are the homozygous haplotype (HH) concept introduced by Miyazawa *et al.* 2007 and the well-known region or run of homozygosity (ROH)[20]. A HH is defined as a genotype after the removal of heterozygous variants such that only homozygous variants remain. Miyazawa *et al.* 2007 compared every pair of HH in a small cohort and reported regions of consecutive matches over a threshold. ROHs are defined as extended genomic regions of homozygous variants allowing for a small number of heterozygous variants contained within. We can rigorously capture both concepts using Tractatus.

input   : m haplotypes each of length n, partial tract length l,
          minimum length L or IBD probability p, p-value threshold p_h,
          estimated probability of error p_t, length of scan w
output: set of IBD tracts

$H \leftarrow$ *tractized haplotypes*

$T(H) \leftarrow$ *Tract tree of H*

Post-order DFS of T(H) to compute descendant haplotypes from each node

DFS of T(H):
if *path in DFS is longer than l, node has at least 2 descendant haplotypes, and is maximal* **then**
| add partial IBD tract to set of tracts S
else
| push children nodes on stack
end
for *tract $s \in S$* **do**
| *Check for extension in previously processed tracts within scan region w*
| *Compute probability according to number of errors in extension, p_t, the length of the extension, and the number of individuals*
| *If probability $> p_h$, merge tracts*
end
for *tract $s \in S$* **do**
| *If length of s is greater than L or $P(IBD) > p$, report as IBD tract*
end

Algorithm 2. Tractatus (with errors)

A naive model for computing HH would consider each heterozygous site as a wildcard allowing for either the 0 or 1 allele. A haplotype with k heterozygous sites would require insertion of 2^k haplotypes into the Tract tree. This immediately suggests a fixed-parameter tractable algorithm using the same machinery as Tractatus. However, we can remove the dependence on k using a key insight regarding the structure of the Tract tree and tractized haplotypes.

Errors split tracts in the Tract tree such that the shared tract fragments are on different paths from the root. Instead of encoding all 2^k possible haplotypes, we simply remove the heterozygous alleles from the tractized string. Because the position is inherently encoded in the tractized string, the removal of the heterozygous alleles would have the same effect as an error. Therefore, if we encode genotypes by simply removing heterozygous variants in the tractized string, we can run Algorithm 2 to produce all the homozygous haplotypes for a set of genotypes in linear time and space.

3   Results

The principle advantages of Tractatus over existing methods are the theoretically guaranteed subquadratic runtime and exact results in the error-free case which translate to improved results in the case with errors and allele mismatches. We evaluate the runtime of Tractatus against a generic algorithm that processes individuals in pairs using phased HapMap haplotypes from several populations. We then compare the power and false positive rates of both Tractatus and GERMLINE which is a leading method for IBD inference[14]. Finally, we show an application of Tractatus-HH by inferring homozygous haplotypes in a previously known homozygous region in the Simons Simplex Collection genome-wide association study data[21].

3.1   Tractatus vs. Pairwise Algorithm Runtimes

To evaluate the runtime of Tractatus versus pairwise methods, we implemented the pairwise equivalent algorithm which iterates through pairs of individuals and reports tracts of variants occurring in both individuals over some threshold length of variants. The data consist of phased haplotypes from HapMap Phase III Release 2 in the ASW, CEU, CHB, CHD, GIH, JPT, LWK, MEX, MKK, TSI, and YRI populations[22]. Figure 3 left shows the independence between chromosome and computation time for the Tractatus suffix tree and the pairwise algorithm. Because the runtime of each algorithm does not depend on the chromosome, we varied the population sizes while keeping the number of variants constant for chromosome 22. Figure 3 right shows the quadratic computation time growth for the pairwise algorithm while Tractatus tree construction remains linear in the number of individuals.

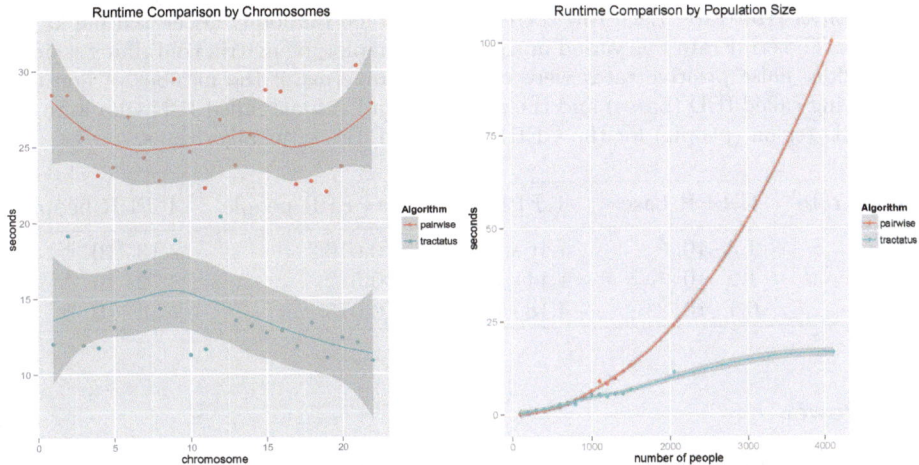

Fig. 3. Left: Tractatus and the pairwise algorithm were run on haplotypes from each chromosome of all HapMap populations for a minimum tract length of 100, and a randomly selected interval of 1000 variants. The experiment was repeated 100 times for each chromosome and elapsed time was averaged. Right: Tractatus and the pairwise algorithm were run on a randomly selected interval of 1000 variants from chromosome 22. The population size varied from 100 to twice the actual population size by resampling haplotypes with a 0.05 allele switch rate (per base).

3.2 False Positive Rates

Because it is difficult to construct a gold-standard baseline of true IBD regions in real data, our false positive rate and power calculations are performed on simulated data. To estimate the false positive rates for GERMLINE and Tractatus we simulated haplotypes at random and generated a single IBD region defined as having identical haplotype alleles in the region of IBD. We generated 100 haplotype matrices where $m = n = 500$ for all possible combinations of the number of individuals sharing a segment IBD $\in [3, 5, 10]$, the number of variants in the IBD region $\in [50, 60, 70, 80, 90, 100, 150, 200]$ and the single base substitution error rates $\in [0.0, 0.01, 0.05]$. In total, we generated 7200 haplotype matrices but aggregated the data across the number of individuals and variants in the IBD region because the false positive rates did not vary over these dimensions.

Table 1 shows that both algorithms have very low false positive rates in terms of the number of bases incorrectly called in an IBD region. However, Tractatus incorrectly calls less individuals in IBD regions than GERMLINE. In this experiment, IBD regions were generated in block sizes and GERMLINE benefits from calling IBD regions in terms of blocks or windows. GERMLINE and Tractatus call a similar amount of bases IBD because Tractatus can over-estimate the ends of blocks. However, when individuals are compared, Table 1 shows that Tractatus computes a significantly smaller number of false positive IBD regions.

Table 1. False positive rates for the GERMLINE (G) and Tractatus (T) algorithms as a function of error rate. Each row corresponds to 2400 randomly generated haplotype matrices. The error rate was varied in a simulated haplotype matrix containing a single IBD region. False positive rates were calculated in terms of the number of non-IBD bases being called IBD (bases) and the number of individuals called IBD who were not in an IBD region (people) for the GERMLINE and Tractatus algorithms.

error rate	G FPR bases	T FPR bases	G FPR people	T FPR people
0.0	$1.3 \cdot 10^{-4}$	$1.16 \cdot 10^{-4}$	0.016	$2.13 \cdot 10^{-3}$
0.01	$1.2 \cdot 10^{-4}$	$1.11 \cdot 10^{-4}$	0.012	$8.72 \cdot 10^{-3}$
0.05	$6.1 \cdot 10^{-5}$	$4.18 \cdot 10^{-5}$	0.015	$7.43 \cdot 10^{-3}$

3.3   Power

We apply Tractatus and GERMLINE to the simulated data from Section 3.2 and estimate power by computing the number of times GERMLINE and Tractatus correctly call the IBD region in terms of variants and individuals. We considered an individual being called correctly if an IBD region was called and overlapped anywhere in the interval of the true IBD tract. We set the -bits and min_m options of GERMLINE to 20 and 40 respectively which sets the slice size for exact matches to 20 consecutive variants and the minimum length of a match to be 40 MB (which corresponds to 40 variants in our simulated data). For a valid comparison, we set Tractatus to accept partial tract sizes of 20 variants and a minimum length of an IBD region to 40 variants.

Figure 4 shows the power of GERMLINE and Tractatus to infer IBD as a function of IBD region length, number of haplotypes sharing the region, and the probability of base call error. Figure 4 right displays a *jagged* curve for GERMLINE which is likely due to the algorithmic dependence on window size. Both algorithms perform relatively well for shorter IBD tracts but Tractatus is clearly more powerful when the number of haplotypes sharing the tract increases or the base call error rates are low. Additionally, the minimum partial tract length for Tractatus could be lowered to increase the power to find smaller IBD tracts (at a cost of higher false positive rates). Another interesting observation is that both GERMLINE and Tractatus are able to perfectly infer all individuals sharing the IBD region in the perfect data case, but, GERMLINE is unable to compute the entire IBD interval in some data.

3.4   Homozygous Haplotypes in Autism GWAS Data

As a proof of concept for Tractatus-HH, we extracted a $250kb$ genomic region identified as having a strong homozygosity signal in the Simons Simplex Collection[23]. The families analyzed include 1,159 simplex families each with at least one child affected with autism and genotyped on the Illumina 1Mv3 Duo microarray. Gamsiz et al. 2013 approached the problem by treating a homozygous region as a marker and testing for association or burden for the region as

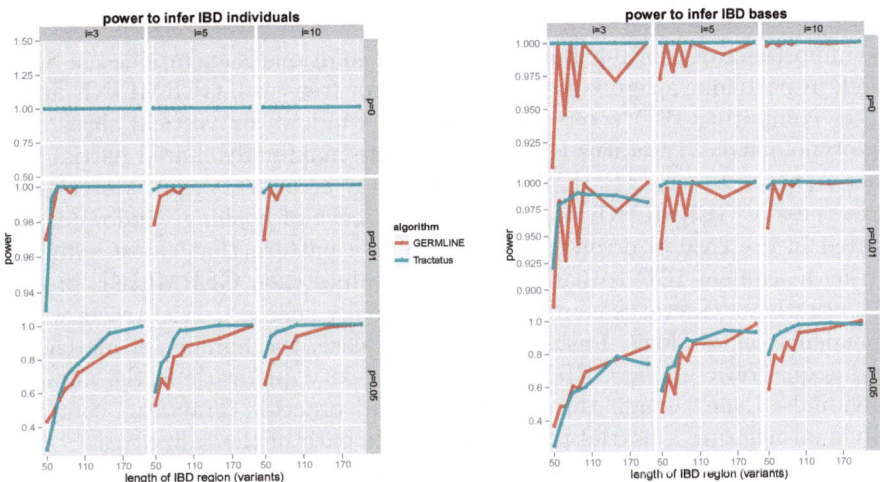

Fig. 4. The power to infer IBD by individual haplotypes (left) and variant bases (right) as a function of the length of the IBD region in variants (x-axis), the probability of base call error (p), and the number of individual haplotypes sharing the IBD segment (i).

a whole[23]. Our analysis shows that regions of homozygosity are more complex than previously assumed and there can be multiple regions overlapping and sharing some segments of homozygous haplotypes but largely different in other segments (Table 2). We found more individuals possessing a homozygous haplotype than Gamsiz *et al.* 2013 because the probability of generating an error or heterozygous site was set to a large value (0.1) but in general this parameter can be adjusted to be more conservative.

Table 2. Analysis of a 250*kb* region of homozygosity in the Simons Simplex Collection. The homozygous interval is defined as a region start and end in terms of variants in the genomic interval, a number of individuals (size), and the number of individuals unique to the particular homozygous haplotype group (unique). One region is dominant and contains most of the individuals, but there are smaller regions with some overlap that contain unique individuals not sharing a homozygous haplotype with the larger region.

region start	region end	size	unique
0	111	20	10
0	109	20	12
0	109	252	238

4   Discussion

The importance of provable bounds and exact solutions is exemplified in Section 3 and, in particular, Figure 4. Even in the error free case, GERMLINE approximates computing IBD tracts by processing windows or vertical slices of the haplotype matrix. Tractatus is able to compute maximally shared partial tracts exactly (which are exactly the IBD tracts in the error-free case). Moreover, the inexactness of GERMLINE due to the dependence of hashing windows is exacerbated in the case of errors. If errors fall in a pattern that cause individuals sharing a segment IBD to hash to different values then GERMLINE produces false negatives. Tractatus computes all maximally shared partial tracts without dependence on windows. Lastly, in the worst case, the number of matches per word is quadratic giving GERMLINE a complexity quadratic in the number of individuals. Even though this is unrealistic in practice, Tractatus compresses individuals sharing a partial tract into a single path of the suffix tree.

The Tract tree in itself is an interesting data structure with many possible applications. Once the Tract tree is computed for a set of haplotypes, the statistics of constructing the mosaic of tract combinations can be done rigorously and completely such that sampling can be implemented in an order independent manner satisfying the exchangeability property. For the HMM constructions, the availability of the complete set of tracts would provide a rigorous basis for defining the transition probabilities and overall linear time construction. For the graph clustering methods, the Tract tree represents tracts occurring multiple times together and thus this construction will maximize the power in association studies.

Unfortunately, the issue of acquiring the haplotypes remains. Almost exclusively, algorithms for computing IBD require haplotypes due, in part, to the higher power to infer a more subtle IBD sharing than in genotype data. However, this is not a major roadblock considering haplotype phasing algorithms can be highly parallelized or made more efficient using reference panels. Additionally, haplotype assembly algorithms are very efficient and can extend genome-wide [24].

A related and important unanswered problem is to compute IBD regions in genotypes faster than the naive quadratic allele sharing algorithm. Haplotype-based IBD inference algorithms have difficulties modeling genotypes predominately because the heterozygous site introduces ambiguity in haplotype phase. We believe an approach exploiting the Tract tree may infer IBD in genotypes in subquadratic time perhaps with a direct application of the Tractatus-HH algorithm. However, the number of heterozygous variants is usually very high, so additional computation would be required to handle the large quantity of ambiguous sites.

Our analysis of the autism genome-wide association study data shows that homozygous regions cannot simply be treated as a biallelic markers. Distinct homozygous haplotypes, while having a similar signature of homozygosity, can be composed of entirely different alleles. These finding suggest that homozygous regions are complex, multi-allelic markers.

Finally, we note that a similar linear time construction could be used for constructing a Tract tree for a set of haplotypes where there is known genetic information about the distance between variants as in the Li-Stephens PAC model[1]. The genetic distance can be modeled approximately as an integer and used in a similar encoding to compress "identical" tracts.

5 Conclusions

In this work, we described the Tractatus algorithm for computing IBD tracts with and without errors and homozygous haplotypes. Tractatus represents the first provably exact algorithm for finding multi-shared IBD tracts given a set of haplotypes as input; it computes all subsets of individuals that share tracts and the corresponding shared tracts in time linear in the size of the input. By starting from an exact and rigorous algorithmic baseline, we are able to modify downstream decisions based on the global IBD tract decomposition. We compare the runtimes of Tractatus and a generic pairwise algorithm that process individuals in pairs using phased HapMap haplotypes from several populations and show decreased runtimes. Also, we exhibit superior statistical power to infer IBD tracts with less false positives than GERMLINE. Finally, with a conceptual change to the interpretation of genotypes, we showed that homozygous haplotype inference in genotypes can be modeled in the same Tractatus framework and demonstrated Tractatus-HH in a previously known homozygous region of the Simons Simplex Collection autism data.

Acknowledgements. This work was supported by the National Science Foundation [1048831 and 1321000 to S.I.] and NIGMS-NIH (P20GM103645). We are grateful to all of the families at the participating Simons Simplex Collection (SSC) sites, the Simons Foundation Autism Research Initiative, and the principal investigators (A.L. Beaudet, R. Bernier, J. Constantino, E.H.C., Jr., E. Fombonne, D.H.G., E. Hanson, D.E. Grice, A. Klin, R. Kochel, D. Ledbetter, C. Lord, C. Martin, D.M. Martin, R. Maxim, J. Miles, O. Ousley, K. Pelphrey, B. Peterson, J. Piggot, C. Saulnier, M.W.S., W. Stone, J.S. Sutcliffe, C.A. Walsh, Z. Warren, and E. Wijsman). We appreciate obtaining access to phenotypic data in SFARI Base.

References

1. Li, N., Stephens, M.: Modeling linkage disequilibrium and identifying recombination hotspots using single-nucleotide polymorphism data. Genetics 165(4), 2213–2233 (2003)
2. Stephens, M., Smith, N.J., Donnelly, P.: A new statistical method for haplotype reconstruction from population data. American Journal of Human Genetics 68(4), 978–989 (2001)
3. Hudson, R.R.: Gene genealogies and the coalescent process. Oxford Survey in Evolutionary Biology 7, 1–44 (1991)

4. Fearnhead, P., Donnelly, P.: Estimating recombination rates from population genetic data. Genetics 159(3), 1299–1318 (2001)
5. Kingman, J.F.C.: On the Genealogy of Large Populations. Journal of Applied Probability 19, 27–43 (1982)
6. Browning, S.R., Browning, B.L.: Identity by descent between distant relatives: Detection and applications. Annual Review of Genetics 46(1), 617–633 (2012)
7. Howie, B.N., Donnelly, P., Marchini, J.: A flexible and accurate genotype imputation method for the next generation of genome-wide association studies. PLoS Genet. 5(6), e1000529 (2009)
8. Li, Y., Willer, C.J., Ding, J., Scheet, P., Abecasis, G.R.: Mach: using sequence and genotype data to estimate haplotypes and unobserved genotypes. Genetic Epidemiology 34(8), 816–834 (2010)
9. Delaneau, O., Marchini, J., Zagury, J.F.: A linear complexity phasing method for thousands of genomes. Nat. Meth. 9(2), 179–181 (2011)
10. Purcell, S., Neale, B., Todd-Brown, K., Thomas, L., Ferreira, M.A., Bender, D., Maller, J., Sklar, P., de Bakker, P.I., Daly, M.J., Sham, P.C.: PLINK: a tool set for whole-genome association and population-based linkage analyses. American Journal of Human Genetics 81(3), 559–575 (2007)
11. Browning, B.L., Browning, S.R.: A fast, powerful method for detecting identity by descent. American Journal of Human Genetics 88(2), 173–182 (2011)
12. Gusev, A., Kenny, E.E., Lowe, J.K., Salit, J., Saxena, R., Kathiresan, S., Altshuler, D.M., Friedman, J.M., Breslow, J.L., Pe'er, I.: DASH: A Method for Identical-by-Descent Haplotype Mapping Uncovers Association with Recent Variation. Am. J. Hum. Genet. 88(6), 706–717 (2011)
13. He, D.: IBD-Groupon: an efficient method for detecting group-wise identity-by-descent regions simultaneously in multiple individuals based on pairwise IBD relationships. Bioinformatics 29(13), 162–170 (2013)
14. Gusev, A., Lowe, J.K., Stoffel, M., Daly, M.J., Altshuler, D., Breslow, J.L., Friedman, J.M., Pe'er, I.: Whole population, genome-wide mapping of hidden relatedness. Genome Research 19(2), 318–326 (2009)
15. McCreight, E.M.: A space-economical suffix tree construction algorithm. J. ACM 23(2), 262–272 (1976)
16. Ukkonen, E.: On-line construction of suffix trees. Algorithmica 14(3), 249–260 (1995)
17. Farach, M.: Optimal suffix tree construction with large alphabets. In: Proceedings of the 38th Annual Symposium on Foundations of Computer Science, FOCS 1997, pp. 137–143. IEEE Computer Society, Washington, DC (1997)
18. Kong, A., Masson, G., Frigge, M.L., Gylfason, A., Zusmanovich, P., Thorleifsson, G., Olason, P.I., Ingason, A., Steinberg, S., Rafnar, T., et al.: Detection of sharing by descent, long-range phasing and haplotype imputation. Nature Genetics 40(9), 1068–1075 (2008)
19. Halldorsson, B.V., Aguiar, D., Tarpine, R., Istrail, S.: The Clark Phaseable sample size problem: long-range phasing and loss of heterozygosity in GWAS. Journal of Computational Biology 18(3), 323–333 (2011)
20. Miyazawa, H., Kato, M., Awata, T., Kohda, M., Iwasa, H., Koyama, N., Tanaka, T., Huqu, N., Kyo, S., Okazaki, Y.: Homozygosity Haplotype Allows a Genomewide Search for the Autosomal Segments Shared among Patients. The American Journal of Human Genetics 80(6), 1090–1102 (2007)
21. Fischbach, G.D., Lord, C.: The Simons Simplex Collection: A Resource for Identification of Autism Genetic Risk Factors. Neuron 68(2), 192–195 (2010)

22. International HapMap Consortium: The International HapMap Project. Nature 426(6968), 789–796 (December 2003)
23. Gamsiz, E., Viscidi, E., Frederick, A., Nagpal, S., Sanders, S., Murtha, M., Schmidt, M., Triche, E., Geschwind, D., State, M., Istrail, S., Cook Jr., E., Devlin, B., Morrow, E.: Intellectual disability is associated with increased runs of homozygosity in simplex autism. The American Journal of Human Genetics 93(1), 103–109 (2013)
24. Aguiar, D., Istrail, S.: Hapcompass: A fast cycle basis algorithm for accurate haplotype assembly of sequence data. Journal of Computational Biology (2012)

HapTree: A Novel Bayesian Framework for Single Individual Polyplotyping Using NGS Data

Emily Berger[1,2,3], Deniz Yorukoglu[2,*], Jian Peng[1,2], and Bonnie Berger[1,2,*]

[1] Department of Mathematics, MIT, Cambridge, MA, USA
[2] Computer Science & Artificial Intelligence Laboratory, MIT, Cambridge, MA, USA
[3] Department of Mathematics, UC Berkeley, Berkeley, CA, USA
{bab,denizy}@mit.edu

1 Background

Using standard genotype calling tools, it is possible to accurately identify the number of "wild type" and "mutant" alleles (A, C, G, or T) for each single-nucleotide polymorphism (SNP) site. In the case of two heterozygous SNP sites however, genotype calling tools cannot determine whether "mutant" alleles from different SNP loci are on the same or different chromosomes. While in many cases the former would be healthy, the latter can cause loss of function; it is therefore important to identify the phase—the copies of a chromosome on which the mutant alleles occur—in addition to the genotype. This need necessitates efficient algorithms to obtain an accurate and comprehensive haplotype reconstruction (the phase of heterozygous SNPs in the genome) directly from the next-generation sequencing (NGS) read data. Nearly all previous haplotype reconstruction studies have focused on diploid genomes and are rarely scalable to genomes of higher ploidy; however, computational investigations into polyploid genomes carry great importance, impacting plant, yeast and fish genomics, as well as studies into the evolution of modern-day eukaryotes and (epi)genetic interactions between copies of genes.

2 Method

We propose a novel maximum-likelihood estimation framework, HapTree, for polyploid haplotype assembly of an individual genome using NGS read datasets. The HapTree pipeline is designed to perform the haplotype reconstruction of a single genome. The key component of HapTree is a relative likelihood function which measures the concordance between the aligned read data and a given haplotype under a probabilistic model that also accounts for possible sequencing errors. To identify a haplotype of maximal likelihood, HapTree finds a collection of high-likelihood haplotype partial solutions, which are restricted to the first m SNP loci, and extends those to high-likelihood partial solutions on the first

---

* Corresponding authors.

R. Sharan (Ed.): RECOMB 2014, LNBI 8394, pp. 18–19, 2014.
© Springer International Publishing Switzerland 2014

$m + 1$ SNP loci, for each incremental m. In each step, HapTree maintains only the set of sufficiently likely partial solutions to be extended in the next steps. The goal is to find a full haplotype of maximal likelihood.

3   Results

We evaluate the performance of HapTree on simulated polyploid sequencing read data modeled after Illumina and 454 sequencing technologies. For triploid and higher ploidy genomes, we demonstrate that HapTree substantially improves haplotype assembly accuracy and efficiency over the state-of-the-art [1,2] for varying read depth coverage and length of haplotype block; moreover, HapTree is the first scalable polyplotyping method for higher ploidy. To evaluate Hap-Tree's performance, we consider the probability that HapTree finds the exact solution, as well as compute the *vector error* of a proposed solution, a scoring mechanism for when the exact solution is known, which we newly define to generalize the commonly-used switch error to genomes of higher ploidy. In addition, for triploid genomes, we demonstrate that our relative likelihood measure significantly outperforms the commonly used minimum error correction (MEC) score [4,5]; this outperformance becomes even greater as the ploidy increases. Finally, as a proof of concept, we test our method on real diploid sequencing data from NA12878 (1000 Genomes Project) and evaluate the quality of assembled haplotypes with respect to trio-based diplotype annotation as the ground truth. The results indicate that even for diploid genomes, HapTree improves the switch error accuracy within phased haplotype blocks as compared to existing haplotype assembly methods [3], while producing comparable MEC values.

Availability: An implementation of our method, HapTree, is available for download at: `http://groups.csail.mit.edu/cb/haptree/`

References

1. Aguiar, D., Istrail, S.: Hapcompass: a fast cycle basis algorithm for accurate haplotype assembly of sequence data. Journal of Computational Biology 19(6), 577–590 (2012)
2. Aguiar, D., Istrail, S.: Haplotype assembly in polyploid genomes and identical by descent shared tracts. Bioinformatics 29(13), i352–i360 (2013)
3. Bansal, V., Bafna, V.: Hapcut: an efficient and accurate algorithm for the haplotype assembly problem. Bioinformatics 24(16), i153–i159 (2008)
4. Lancia, G., Bafna, V., Istrail, S., Lippert, R., Schwartz, R.: SNPs problems, complexity, and algorithms. In: Meyer auf der Heide, F. (ed.) ESA 2001. LNCS, vol. 2161, pp. 182–193. Springer, Heidelberg (2001)
5. Lippert, R., Schwartz, R., Lancia, G., Istrail, S.: Algorithmic strategies for the single nucleotide polymorphism haplotype assembly problem. Briefings in Bioinformatics 3(1), 23–31 (2002)

Changepoint Analysis
for Efficient Variant Calling

Adam Bloniarz[1,*], Ameet Talwalkar[2,*], Jonathan Terhorst[1,*],
Michael I. Jordan[1,2], David Patterson[2], Bin Yu[1], and Yun S. Song[1,2,3,**]

[1] Department of Statistics, University of California, Berkeley
[2] Computer Science Division, University of California, Berkeley
[3] Department of Integrative Biology, University of California, Berkeley
yss@eecs.berkeley.edu

Abstract. We present CAGE, a statistical algorithm which exploits high sequence identity between sampled genomes and a reference assembly to streamline the variant calling process. Using a combination of changepoint detection, classification, and online variant detection, CAGE is able to call simple variants quickly and accurately on the 90-95% of a sampled genome which differs little from the reference, while correctly learning the remaining 5-10% that must be processed using more computationally expensive methods. CAGE runs on a deeply sequenced human whole genome sample in approximately 20 minutes, potentially reducing the burden of variant calling by an order of magnitude after one memory-efficient pass over the data.

Keywords: genome complexity, next-generation sequencing, variant calling, changepoint detection.

1 Introduction

A central goal in computational biology is to accurately reconstruct sampled genomes from next-generation sequencing (NGS) data, a procedure termed *variant calling*. A vast number of algorithms have been developed in pursuit of this goal, and they are notoriously computationally demanding. This is due both to the difficulty of the underlying problem, as well as the sheer size of the data: a whole human genome sequenced to 30× coverage produces roughly 250 GB of sequence information and metadata; thus even one sample cannot be represented in memory on a typical workstation. As a result, variant calling algorithms spend significant time simply transferring and storing the information needed to carry out the analysis.

A potential solution to this problem is to harness inherent similarity in genetic data. Unrelated humans are estimated to share over 99% sequence identity [1], so most sequencer output will be similar to a corresponding region of the human reference sequence. So-called "reference-based" compression techniques

---

* These authors contributed equally.
** Corresponding author.

R. Sharan (Ed.): RECOMB 2014, LNBI 8394, pp. 20–34, 2014.
© Springer International Publishing Switzerland 2014

which exploit this feature have been proposed [2, 3, 4]; however few existing tools can operate natively on reference-compressed genomic data. The standard compressed format for aligned sequence data [BAM; 5] enjoys widespread support, but only achieves roughly 50% compression owing to its use of a generic compression algorithm (zlib).

In lieu of compression, a promising alternative is to use statistical methods to discover the small fraction of the sampled genome believed to harbor interesting variation, and focus further computational resources in these limited regions. We formalize this idea in terms of *complexity*. Regions which are highly mutated, have low coverage and/or were subject to sequencing errors are *complex*: they contain additional signal which cannot be retrieved from the reference genome. Conversely, regions which are similar to the reference, display expected coverage levels, and show low rates of mutation and sequencer error have low complexity. Our goal then becomes to algorithmically classify genomic regions according to their complexity level. Concretely, we propose the following hybrid approach:

1. Perform a first-pass analysis to isolate a small fraction of the sampled genome which is "non-reference" and complex;
2. Pass these high-complexity regions to the computationally intensive variant detection algorithms described above;
3. Process the remaining low-complexity regions using a fast algorithm designed to detect simple variation.

In this work, we explore methods to isolate such regions by exploiting statistical features of the sequencer output, which can be computed quickly and without recourse to fully decoding the underlying genome. Our algorithm, called **C**hangepoint **A**nalysis of **G**enomic r**e**ads (CAGE), is fast and trivially parallelizable across the genome, and hence well-suited to process large amounts of NGS data quickly. Using several benchmark datasets, we demonstrate that our approach maintains state-of-the-art variant calling accuracy while subjecting less than 10% of the sampled genome to computationally intensive analysis. Additionally, we present an extension of our algorithm, called CAGE++, in which we simultaneously perform variant detection and variant calling on low-complexity genomic regions, potentially obviating the need for the third step of the hybrid approach described above. Finally, our approach is very cheap when compared to standard analysis tools [e.g. 5, 6, 7]: CAGE and CAGE++ can process a human whole genome in approximately 20 minutes on a single 32 core machine while consuming less than 16 GB of memory, thus illustrating that our proposed hybrid variant calling pipeline has the potential to significantly speedup the variant calling process.

2   Related Work

High-throughput sequencing has inspired various efforts aimed at reducing the amount of data needed to be stored and analyzed, primarily in the form of compression algorithms. Lossless compression methods include reference-based approaches [3, 4] which store variation relative to a reference sequence, as well as non reference-based methods which specialize certain existing compression techniques to genomic data [2]. Greater compression ratios may be achieved if sequencer quality scores are lossily compressed while still losslessly compressing the actual sequence data [8]. The primary disadvantage of these techniques is that few existing software tools can operate directly on the compressed data, mandating a time- and space-intensive decompression step each time the data are analyzed.

Recent versions of the Genome Analysis Toolkit [GATK, 6] employ a lossy compression tool, ReduceReads, to reduce alignment data before being processed by other variant calling tools. The tool works by discarding data in regions of the genome which contain little variation, and is thus similar in motivation to the algorithm we report here. However, the algorithms differ in several regards. CAGE is based on a statistical model of the observed data (Section 3), and is tuned using intuitive quantities such as read coverage rate, sequencer error rate, and mutation rate. ReduceReads appears to employ several heuristics when creating the compressed output, and it is not necessarily clear how to parameterize these heuristics in order to achieve a desired compression ratio or data fidelity. Additionally, though we are unaware of any formal publication or other effort to benchmark the ReduceReads algorithm, user reports from the GATK support forums indicate that it requires costly preprocessing steps in order to run, and can require a large amount of memory and processing time in order to compress a whole genome sequence.

CAGE uses a changepoint detection method to mark regions of variable complexity as it moves along the genome. A similar idea was used by Shen and Zhang [9] to detect abrupt changes in copy number. One way to view these methods is as an alternative to the hidden Markov model (HMM), which has also been previously used to detect genomic variation [10, 11]. In contrast to the latter methods, which require the number of hidden states to be known *a priori*, changepoint methods allow the number of detected segments to vary in accordance with the data. We leverage this observation, in conjunction with simple rule-based classifier, to learn the number of hidden genomic complexity states in a semiparametric manner.

Various distributional aspects of the data we consider have been previously studied. In a seminal paper, Lander and Waterman [12] showed that read depth in whole-genome shotgun sequencing experiments is well modeled by a Poisson distribution, a fact which we exploit in our model. Evans et al. [13] considered fragment site–insert length pairs embedded into the plane. This construction can be used to derive null distributions of several coverage-related statistics [13, 14]. They also describe an interesting visualization technique which can be used to detect deviations from the null coverage distribution. This approach is similar in spirit to our goal, but here we rely on automated techniques in order to detect these deviations in a high-throughput environment.

3   Methods

Following sequencing and alignment, evidence of genetic variation in NGS data is detectable in several ways. Sites which harbor isolated, single-nucleotide variants can usually be aligned unambiguously to the reference genome, resulting in a characteristic SNP signature common to half or all of the reads (depending on zygosity) in the alignment. Small insertions and deletions ($\approx$2-10bp) are also frequently detected and compensated for by the aligner. In both of these scenarios, coverage and concordance statistics are usually unaffected by the presence of the nearby variant since the aligner is able to "explain away" the variant.

In contrast, larger structural variants produce several noticeable signals in the alignment. Novel insertions are typically flanked by reads with low mapping quality or missing mate pairs, and may also result in a coverage dropoff or decreased insert size near the insertion site. Similarly, deleted regions are evidenced by larger than expected insert sizes and a coverage dropoff. Reads that straddle the boundary of a structural variant often have a high percentage of soft-clipped bases with high Phred-scores. More complicated forms of rearrangement result in other distinctive patterns involving, for example, split mapping and orientation bias [15].

Formally, we define complexity in terms of point processes and their associated rates. At genomic position i, let

- $M_i \in \{0, 1\}$ denote the (unobserved) mutation state, assuming a biallelic mutation model;
- $R_i \in \mathbb{Z}^+$ the number of short-reads whose alignment begins at i;
- $D_i \geq R_i$ be number of sequenced bases ("coverage depth") at i; and
- $\mathbf{E}_i = (e_{i,1}, \ldots, e_{i,D_i}) \in \{0, 1\}^{D_i}$ denote a vector of indicators for whether a sequencing error occurred in each of the D_i bases aligned to i.

Note that we observe the random variables R_i and D_i but not M_i or $\mathbf{E}_i$; the M_i are what we ultimately hope to infer through later variant calling analysis, and we only observe a noisy signal of $\mathbf{E}_i$ through the sequencer quality score.

These random variables generate our data as follows. After sequencing and read mapping, we observe a collection of vectors $\mathbf{B}_1, \ldots, \mathbf{B}_L$, where L is the length of the reference genome ($\approx 3.3 \times 10^9$ in humans) and $\mathbf{B}_i = (B_{i,1}, \ldots, B_{i,D_i})$ is the vector of sequenced bases at site i, with

$$B_{i,j} = \mathbf{1}\{\text{base } j \text{ at location } i \text{ matches the reference}\}$$
$$= M_i(1 - e_{i,j}) + (1 - M_i)e_{i,j}.$$

To compute the likelihood of the data, we make the following distributional assumptions:

- The R_i are independent and Poisson distributed with intensity λ_i [12].
- Conditional on the $R_1, \ldots, R_i$, the coverage depth D_i is deterministic.
- $M_i \sim \text{Bernoulli}(\mu_i)$ has a Bernoulli distribution with success parameter μ_i, the probability that a mutation occurs at i.

- The indicators $e_{i,j}$ have a common Bernoulli(ϵ_i) distribution.
- R_i, $e_{i,j}$ and M_i are mutually independent within and across sites.
- All reference bases at a mutated site are sequencer errors, as are all non-reference bases at a non-mutated site.

These assumptions are not expected to hold for real data, however they lead to a fast, easily estimated model, and moreover they do not appear to greatly affect the quality of our inference. Henceforth we abbreviate the genomic region $\{i, i+1, \ldots, j-1, j\}$ as $i : j$. In a region of uniform genomic complexity we expect that the parameters $\lambda_k, \mu_k, \epsilon_k$ are approximately constant, $(\lambda_k, \mu_k, \epsilon_k) = (\lambda, \mu, \epsilon) \triangleq \theta$ for $k \in i : j$.

The complete log-likelihood of the data, $\ell_\theta(i : j)$, is then

$$\ell_\theta(i, j) \triangleq \ell_\theta(\mathbf{B}_{i:j}, M_{i:j}, \mathbf{E}_{i:j}) = \ell_\theta(R_{i:j}) + \ell_\theta(M_{i:j}) + \ell_\theta(\mathbf{E}_{i:j} \mid M_{i:j}, \mathbf{B}_{i:j}) \quad (1)$$

where:

$$\ell_\theta(R_{i:j}) = \sum_{b=i}^{j} \log \mathcal{P}_\lambda(R_b) \tag{2}$$

$$\ell_\theta(M_{i:j}) = \sum_{b=i}^{j} \left[M_b \log \mu + (1 - M_b) \log(1 - \mu) \right] \tag{3}$$

$$\ell_\theta(\mathbf{E}_{i:j} \mid M_{i:j}, \mathbf{B}_{i:j}) = \sum_{b=i}^{j} \left[(1 - M_b)\bar{B}_b \log \epsilon + M_b(D_i - \bar{B}_b) \log(1 - \epsilon) \right], \tag{4}$$

$\mathcal{P}_\lambda$ is the Poisson likelihood with rate λ and $\bar{B}_b = \sum_k B_{b,k}$.

Let $\ell_{\hat\theta}(i : j) \triangleq \sup_\theta \ell_\theta(i : j)$ denote the maximized log-likelihood. It is clear from the additive form of (2)—(4) that for any $i \leq k \leq j$, we can always increase the likelihood of the data by breaking $i : j$ into two independent segments $i : k$ and $(k + 1) : j$:

$$\ell_{\hat\theta}(i : j) \leq \ell_{\hat\theta}(i : k) + \ell_{\hat\theta}((k + 1) : j).$$

In what follows we use this observation to quickly detect uniformly complex regions using likelihood-based methods.

3.1 Maximum Likelihood Estimation

The simple form of the complete likelihood (1) suggests using the EM algorithm to compute $\ell_{\hat\theta}(i : j)$. To do so, we must evaluate the conditional expectation

$$\mathbb{E}_{M_{i:j}, \mathbf{E}_{i:j} \mid \mathbf{B}_{i:j}, \theta_t} (\ell_\theta(\mathbf{B}_{i:j}, M_{i:j}, \mathbf{E}_{i:j})).$$

Unfortunately, the conditional distribution $M_{i:j}, \mathbf{E}_{i:j} \mid \mathbf{B}_{i:j}, \theta_t$ is intractable because the normalization constant requires integrating over the high-dimensional vectors $M_{i:j}$ and $\mathbf{E}_{i:j}$. Since the main goal of our algorithm is to decrease overall computation time, we instead assume that $M_{i:j}$ is known, either from a public database of mutations [16], or from genotypes estimated on-the-fly using a fast and simple variant calling algorithm. We assume that the sampled genome(s) harbor mutations only at sites contained in this database. This enables us to quickly estimate the sample genotype in a particular region, at the expense of erroneously classifying uncalled sites as sequencer errors. Since these sites are generally segregating at low frequency in the population, or were the result of genotyping error, the overall effect of this assumption on our likelihood calculation should be small. Moreover, by training our algorithm to flag regions with an elevated sequencer error rate, we retain the ability to detect these novel variants downstream.

3.2   Augmented Likelihood

The model described above aims to capture the essential data generating mechanism of an NGS experiment. In practice, we found that augmenting the likelihood with additional terms designed to capture features of coverage, mapping quality, and related statistics improved the accuracy of our algorithm with minimal performance impact. In particular, we assume that, in a region of constant genomic complexity:

1. The mapping quality (MAPQ) distribution of short reads is Bernoulli: with probability $\tau \in [0, 1]$, a read has MAPQ $= 0$; otherwise the read MAPQ > 0. Here we bin MAPQ into two classes, zero and non-zero, since its distribution is unknown, and also because we found that the strongest signal was contained in reads which had zero mapping quality.
2. With probability $\eta \in [0, 1]$, each base pair is inserted or deleted in the sample genome; otherwise, with probability $1 - \eta$ the base is subject to the standard mutational and sequencer error processes described above.

Modern aligners [7, 17] generate MAPQ scores during read mapping procedure, and also call small indels where doing so improves concordance. Hence τ and η can be be estimated with high confidence from the data. Indels which are not detected by the aligner will generate aberrations in the coverage and mismatch signals as described above.

Letting θ denote the vector of all parameters in our model, the augmented likelihood from positions i to j can be written as

$$\ell_{\theta,\tau,\eta}^{\mathrm{aug}}(i,j) \equiv \ell_\theta(\mathbf{B}_{i:j}, M_{i:j}, \mathbf{E}_{i:j}) +$$

$$\sum_{b=i}^{j} I_b \log \eta + Q_b \log \tau + (D_b - I_b) \log(1 - \eta) + (D_b - Q_b) \log(1 - \tau), \quad (5)$$

where I_b and Q_b count the number of inserted/deleted and MAPQ-0 bases at position b. The augmented model no longer has a simple interpretation in the generative sense; and in fact we unrealistically assume that the random variables with which we have augmented the likelihood function are mutually independent and identically distributed. On the other hand, this enables us to quickly estimate these parameters from data, and these parameter estimates in turn enable us to easily detect the signatures of mutational events which mark complex regions of the genome.

3.3   Change Point Detection

CAGE classifies genomic regions using parameter estimates obtained by maximizing (5). We assume that these parameters are piecewise constant within (unobserved) segments of the sample genome, and estimate the segments using a changepoint detection method. Let

$$(\hat{\theta}_{i:j}, \hat{\tau}, \hat{\eta}) = \arg\max_{\theta, \tau, \eta} \ell_{\theta, \tau, \eta}^{\mathrm{aug}}(i, j)$$

$$C(i, j) = -\ell_{\hat{\theta}_{i:j}, \hat{\tau}, \hat{\eta}}^{\mathrm{aug}}(i, j)$$

be defined using the likelihood function given above, and let

$$\boldsymbol{\tau} \triangleq (\tau_0 = 0, \tau_1, \tau_2, \ldots, \tau_{m+1} = s)$$

be a sequence of changepoints dividing the region $0, \ldots, s$. $C(i, j)$ is the negative log-likelihood of segment $i : j$ evaluated at the MLE, and hence

$$\sum_{i=1}^{m+1} [C(\tau_{i-1} + 1, \tau_i) + \beta] \tag{6}$$

is a natural measure of fit for the segmentation $\boldsymbol{\tau}$. Here β is a regularization parameter which penalizes each additional changepoint. In practice, rather than considering all $O(10^9)$ loci in the human genome as potential changepoints, we restrict the τ_i in (6) to integer multiples of some window size w. We typically set $25 \leq w \leq 200$ when evaluating our algorithm. This speeds up the optimization, and also decreases the variance of the maximum likelihood parameter estimates for each segment.

   Exact minimization of (6) over m and $\boldsymbol{\tau}$ can be achieved in quadratic time via a dynamic programming algorithm [18]. For likelihood-based changepoint detection, properties of the likelihood function certify that certain changepoint positions can never be optimal. Killick et al. [19] exploit this property to formulate a pruning algorithm which minimizes (6) in expected linear time. The pruning process enables both computational savings, as well as significant memory savings since the in-memory data structures can be repeatedly purged of all data prior to the earliest possible changepoint. The resulting algorithm consumes only a few gigabytes of memory, even when processing data sets which are tens of gigabytes in size. Thus, multiple chromosomes can be processed simultaneously on a typical workstation.

   Our cost calculations take further advantage of a property of maximum likelihood-based cost functions for parametric families that enable us to avoid calculating the full likelihood. From (2)–(4) and (5) it is seen that our cost function factors as $C(i, j) = g_{\hat{\theta}_{i:j},\hat{\tau},\hat{\eta}}(T(i,j)) + h(i,j)$ where $T(i,j)$ and $h(i,j)$ depend only on the data in region $i : j$ and h does not depend on the parameters. Since our model assumes the data are independent and identically distributed within a segment, we have

$$h(i,j) = \sum_{b=i}^{j} h_1(b)$$

where h_1 is a univariate function. Thus the optimization (6) can be decomposed as

$$\min_{\substack{m,\boldsymbol{\tau} \\ s=\tau_{m+1}}} \sum_{i=1}^{m+1} \left[g_{\hat{\theta},\hat{\tau},\hat{\eta}}(\tau_{i-1}+1,\tau_i) + h(\tau_{i-1}+1,\tau_i) + \beta \right]$$

$$= \min_{\substack{m,\boldsymbol{\tau} \\ s=\tau_{m+1}}} \sum_{i=1}^{m+1} \left[g_{\hat{\theta},\hat{\tau},\hat{\eta}}(\tau_{i-1}+1,\tau_i) + \sum_{b=\tau_{i-1}+1}^{\tau_i} h_1(b) + \beta \right]$$

$$= \sum_{b=1}^{n} h_1(b) + \min_{\substack{m,\boldsymbol{\tau} \\ s=\tau_{m+1}}} \sum_{i=1}^{m+1} \left[g_{\hat{\theta},\hat{\tau},\hat{\eta}}(\tau_{i-1}+1,\tau_i) + \beta \right].$$

We see that it is not necessary to evaluate h at all in order to carry out the optimization. In our setting this function involves a number of log-factorial terms which are relatively expensive to evaluate.

3.4   Identification of High-Complexity Regions

The changepoint detection algorithm described above determines which genomic regions have uniform genomic complexity. Next, we use this information to allocate additional computational resources to complex regions. In this paper we use a binary classification scheme in which a region is labeled as either high or low. For each changepoint region, we compute the features considered in CAGE's augmented likelihood defined in (5), and classify each region as high-complexity if any of these features are outliers, using hand-tuned thresholds for each feature.

3.5   Integrated Variant Calling Algorithm

The hybrid variant calling approach described in Section 1 relies on a fast algorithm to detect variation in low-complexity regions. Additionally, CAGE requires estimates of ground-truth locations of mutations and short indels in its likelihood calculation, as discussed in Section 3.1. We hypothesized that variation in low-complexity regions of the genome should be particularly easy to call, and implemented a simple, rule-based variant calling heuristic to be run alongside the core CAGE algorithm during the initial pass over the data.

We refer to this modified algorithm as CAGE++. It uses a count-based method to identify variants from a pileup, and relies on read depth, strand bias and read quality scores to filter calls. This method has five tuning parameters: CAGE++ ignores all bases with quality scores less than α_1, and calls a variant at a particular pileup location if:

1. The pileup depth is at least α_2;
2. An alternative allele appears in at least α_3 percent of the reads, and no fewer than α_4 actual reads; and
3. The strand bias is less than α_5 percent.

As we show in Section 4, this heuristic is extremely fast while remaining competitive with more sophisticated algorithms in terms of accuracy.

4   Results

We compared our proposed algorithms to two baseline variant calling algorithms from GATK version 2.8. The first is a computationally cheap caller known as UnifiedGenotyper (GATK-ug). The second algorithm, HaplotypeCaller (GATK-ht), is more accurate but relies on computationally demanding local de novo assembly. We compare these two variant callers with two hybrid approaches. In one approach, we first segment the genome by complexity using CAGE, and then use GATK-ug and GATK-ht to process the low- and high-complexity regions, respectively. In the second approach, we use CAGE++ to both segment the genome by complexity and perform variant calling on the low-complexity regions, and then rely on GATK-ht to process the high-complexity regions. To measure the effectiveness of the changepoint detection component of CAGE and CAGE++, we also evaluate a simple alternative hybrid approach, called ALLCHANGE, in which we treat each window as a distinct region and rely solely on our rule-based classifier to determine genomic complexity.[1]

To perform the CAGE and ALLCHANGE hybrid approaches, we first ran GATK-ug on the full genome, and used these predictions as estimates of $M_{i:j}$, as discussed in Section 3.1. Since CAGE++ calls variants directly, the CAGE++ approach did not rely on GATK-ug. For all three hybrid approaches, after identifying regions of high-complexity via binary classification, we then executed GATK-ht on each region, expanding each region by 10% of its length to minimize errors at the boundaries. We then combined the resulting predictions from the low-complexity regions (either from GATK-ug or directly from CAGE++) with the predictions from GATK-ht on the high-complexity regions.

We performed all our experiments on an x86-64 architecture multicore machine with 12 2.4Ghz hyperthreaded cores and 284 GB of main memory. We tuned the changepoint parameters for CAGE and CAGE on a small hold-out set, setting the window size to $w = 100$, the regularization parameter to $\beta = 3.0$. We used the same window size for ALLCHANGE. For CAGE++, we set the variant calling parameters to $(\alpha_1, \alpha_2, \alpha_3, \alpha_4, \alpha_5) = (12, 10, 20, 3, 20)$.

---

[1] ALLCHANGE is similar to GATK's ReduceReads algorithm and is also what we obtain from CAGE with $\beta \equiv 0$.

4.1   Datasets and Evaluation

To evaluate the performance of these algorithms we used SMaSH [20], a recently developed suite of tools for benchmarking variant calling algorithms. Briefly, SMaSH is motivated by the lack of a gold-standard NGS benchmarking dataset which both a) mimics a realistic use-case (i.e. is not generated by simulating sequencer output) and b) includes comprehensive, orthogonal validation data (i.e. is not generating using sequencer output). The datasets comprised by SMaSH present different trade-offs between practical relevance and the quality and breadth of the validation data. We worked with the following SMaSH datasets:

- Venter Chromosome 20 and full genome: generated from Craig Venter's genome [21] and including noise-free validation data and synthetically generated short-reads (30× coverage, mean insert size of 400);
- Mouse Chromosome 19: derived from the mouse reference genome and including noisy validation data and overlapping short-reads generating from a GAIIx sequencer (60× coverage, mean insert size of −34);
- NA12878 Chromosome 20: based on a well studied human subject, including short-reads from a HiSeq2000 sequencer (50× coverage, mean insert size of 300). SMaSH's validation data for this dataset consists primarily of well-studied SNP locations, so we instead rely on a richer set of validated variants (SNPs and indels only) provided by Illumina's Platinum Genomes project [22]. We nonetheless leverage the SMaSH evaluation framework to compute accuracy. Since this validation set is only a sample of the full set of variants, we do not report precision results.

4.2   Accuracy

Table 1 summarizes the accuracy of GATK-ug, AllChange, CAGE/CAGE++ and GATK-ht. Precision and recall are calculated with respect to the validated variants in the SMaSH datasets, except in the case of NA12878 as described above. As expected, GATK-ht generally outperforms GATK-ug; the difference is particularly pronounced for indels. Second, the CAGE approach in which GATK-ht is applied to the high-complexity regions and GATK-ug is applied to the remainder yields comparable accuracy to GATK-ht, with the CAGE approach being slightly better on Mouse, comparable on Venter, and slightly worse on NA12878. Third, AllChange accuracy is comparable to that of CAGE, indicating that the features we consider in our likelihood model are indeed predictive of genome complexity.

As shown in Table 2, the AllChange high-complexity regions are larger than the corresponding CAGE and CAGE++ regions, and by a large margin for the two human datasets, thus highlighting the effectiveness of the changepoint detection algorithm. The table also shows that a large fraction of the structural variants are concentrated in these high-complexity regions. Moreover, when we investigated the handful of remaining structural variants which fell into low-complexity regions, we found that they were difficult to discern even by visually

Table 1. Precision/Recall of various variant calling algorithms

Variant	Dataset	GATK-ug	ALLCHANGE	CAGE	CAGE++	GATK-ht
	Venter	93.6 / 98.7	98.8 / 98.2	98.6 / 98.2	98.9 / 96.5	98.9 / 98.1
SNPs	Mouse	93.3 / 96.4	98.2 / 95.5	98.1 / 95.6	98.6 / 95.4	98.5 / 95.4
	NA12878	- / 99.7	- / 99.6	- / 99.6	- / 99.6	- / 99.6
	Venter	94.3 / 81.3	93.1 / 92.1	93.1 / 91.9	93.1 / 91.8	93.1 / 92.0
Indel-Ins	Mouse	94.2 / 75.1	89.7 / 89.5	89.8 / 89.4	90.0 / 88.3	89.2 / 89.7
	NA12878	- / 62.2	- / 93.0	- / 92.7	- / 92.8	- / 93.4
	Venter	95.0 / 87.2	95.0 / 93.4	95.1 / 93.3	95.0 / 93.6	95.1 / 93.9
Indel-Del	Mouse	92.1 / 90.7	80.1 / 94.4	81.0 / 94.4	85.6 / 94.2	77.7 / 94.6
	NA12878	- / 64.3	- / 94.3	- / 94.3	- / 94.3	- / 94.6

Table 2. Segregation of variants in high-complexity regions for ALLCHANGE and CAGE

Algorithm	Dataset	Size of high	% SNPs	% Indels	% SVs
	Venter	13.1%	28.2%	58.3%	98.4%
ALLCHANGE	Mouse	13.2%	72.5%	79.9%	98.7%
	NA12878	13.5%	-	-	-
	Venter	4.0%	14.3%	53.4%	98.4%
CAGE	Mouse	11.3%	61.3%	77.1%	98.8%
	NA12878	7.1%	-	-	-
	Venter	6.3%	32.1%	99.6%	97.2%
CAGE++	Mouse	9.7%	55.8%	97.3%	98.2%
	NA12878	8.3%	-	-	-

inspecting the raw data. We further note that the basic variant caller in CAGE++ leads to a much higher fraction of indels being placed in high-complexity regions.

4.3   Computational Performance

We evaluated the runtime of CAGE on the full Venter genome, executing it on a single Amazon EC2 cc2.8xlarge instance with 59 GB of main memory and 32 cores. We divided the genome into roughly equal sized subproblems, and CAGE completed in 13 minutes when executing all subproblems in parallel, with peak memory usage of less than 16 GB. Next, we evaluated the performance of CAGE++. Since CAGE is heavily I/O bound, the additional computation required by the variant caller component of CAGE++ has a small impact on overall execution time, increasing runtime relative to CAGE by approximately 50%. Finally, we evaluated the speedup obtained by executing GATK-ht only on CAGE's high-complexity regions, but we observed modest speedups. Indeed, on Venter, where CAGE's high-complexity regions comprise a mere 4% of chromosome 20, we observed a 1.4× speedup. As a baseline comparison, we also executed GATK-ht on randomly selected contiguous regions each consisting of 4% of chromosome 20, and observed an average speedup of 2.8×. The sublinear scaling of GATK-ht suggests that it may not be well suited for a hybrid variant calling approach.

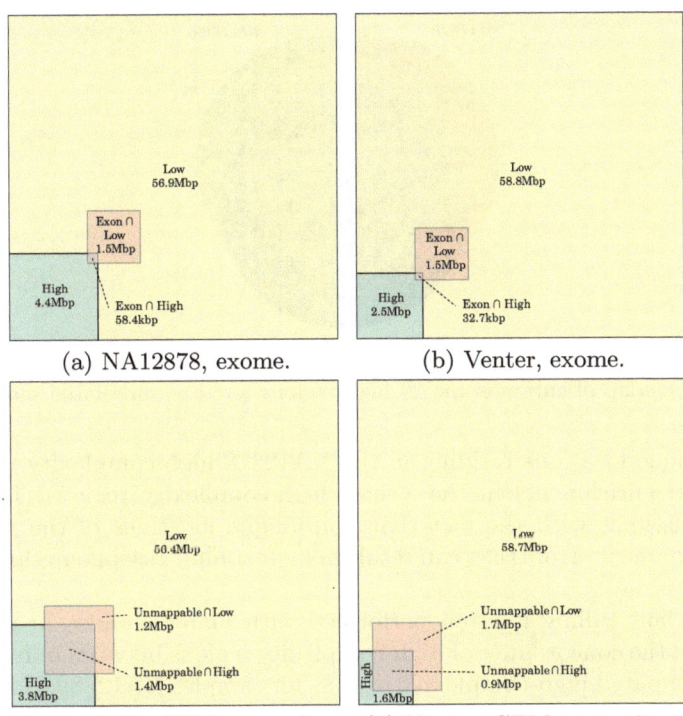

(a) NA12878, exome.　　　　(b) Venter, exome.

(c) NA12878, GEM non-unique. (d) Venter, GEM non-unique.

Fig. 1. Overlap between CAGE regions and known genome annotations. (a,b) Overlap with the exome. (c,d) Overlap with GEM non-unique regions.

4.4 Properties of CAGe Regions

To further validate our approach, we examined how low- and high-complexity regions produced by CAGE interact with various genome annotation tracks, as well as with each other when processing multiple samples at once. Figure 1 (a,b) depicts the overlap between these regions and portions of the sampled genome that are annotated as exons according to the Illumina TruSeq exon capture kit. We found that low-complexity regions are comparatively enriched for exons: 96.3% (97.9%) of the exome falls in low-complexity for NA12878 (Venter). This is expected since exons are under stronger purifying selection than noncoding regions of the genome and hence harbor less variation [23].

We also explored the relationship between genome mappability [24] and CAGE classification. Mappability estimates the uniqueness of each k-mer of the reference genome. Repetitive and duplicated regions have lower mappability scores, while k-mers that are unique among all reference k-mers have a mappability score of 1. In our experiments we set $k = 100$ to match the standard read length of NGS data. Figure 1 (c,d) compares the overlap between high-complexity regions, low-complexity regions, and segments of the genome that have non-unique k-mers (mappability < 1). High-complexity regions are comparatively enriched for segments that are more difficult to map, with 53.3% (7.5× enrichment) of

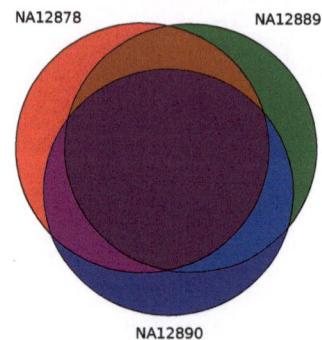

Fig. 2. Overlap of chromosome 20 high regions for three unrelated individuals

the non-unique locations residing in the NA12878 high-complexity regions and 34.1% (8.5× enrichment) in the Venter high-complexity regions. This enrichment is consistent with the fact that non-unique locations of the genome are prone to alignment errors that can result in high-complexity pileups around these locations.

Since variant calling is often performed with many samples in parallel, we next studied the concordance of high-complexity regions between individual samples. We computed high-complexity regions for samples NA12889, NA12890 and NA12878 using data released by the Platinum Genomes project. The individuals are members of the CEPH/UTAH 1463 pedigree but are unrelated (mother and paternal grandparents). We ran CAGE on chromosome 20 of each of these individuals, setting the thresholds of our rule-based classifiers to generate a high-complexity region of consistent size on each of the chromosomes (respectively, 8.1%, 8.2%, 8.3%). The Venn diagram in Figure 2 characterizes the overlap among these regions. The high-complexity regions are fairly consistent among the three individuals; the union of their high-complexity regions consists of 11.6% of the chromosome.

5   Discussion

These experiments illustrate that a hybrid approach has the potential to accurately detect regions of a sampled genome that harbor the majority of complex variation, as well as improve the computational performance of variant calling algorithms. It is possible to partition a genome into high- and low-complexity regions such that:

1. Low-complexity regions comprise a large majority of the genome; and
2. Fast, simple variant calling algorithms work as well as slower, more complex ones on these regions.

This strategy leads to a large increase in throughput compared with traditional variant calling pipelines.

There are several avenues for improving upon this work. Our experiments demonstrate the promise of, at least in the case of deeply sequenced samples, employing a trivial, rule-based variant calling algorithm to process a large fraction of the data with minimal impact on accuracy. More experiments are needed to confirm that this finding translates to larger samples and/or other types of sequencing experiments.

The experimental results presented above used a hand-trained classifier to segment the sampled genomes into high- and low-complexity regions. In order to employ our algorithm on a larger scale it is necessary to automatically train this classifier. Since it is usually straightforward via visual inspection to determine whether a region harbors a complex variant, one potential solution is to build a streamlined program to facilitate the rapid generation of training examples by human supervision. We have implemented a prototype of this software and found that a knowledgeable human subject is capable of generating on the order of 1,000 training examples per hour. More work is needed to integrate this supervised classifier into CAGE.

Another extension would be to the multi-class regime where regions are placed into one of several categories based on whether they are believed to harbor SNPs, indels, various types of structural variants or some combination thereof. The summary statistics generated in the maximum likelihood step of CAGE could be used to send segments to specialized variant calling algorithms designed to handle these respective categories.

Acknowledgments. This research is supported in part by a National Defense Science and Engineering Graduate Fellowship (AB), an NSF award No. 1122732 (AT), NSF award No. 0130526 (JT, AB, BY), an NIH National Research Service Award Trainee appointment on T32-HG00047 (JT), and an NSF CAREER Grant DBI-0846015 (JT, YSS).

References

[1] Tishkoff, S.A., Kidd, K.K.: Implications of biogeography of human populations for 'race' and medicine. Nature Genetics 36, S21–S27 (2004)

[2] Cox, A.J., Bauer, M.J., Jakobi, T., Rosone, G.: Large-scale compression of genomic sequence databases with the Burrows-Wheeler transform. Bioinformatics 28(11), 1415–1419 (2012)

[3] Hsi-Yang, F.M., Leinonen, R., Cochrane, G., Birney, E.: Efficient storage of high throughput DNA sequencing data using reference-based compression. Genome Research 21(5), 734–740 (2011)

[4] Jones, D.C., Ruzzo, W.L., Peng, X., Katze, M.G.: Compression of next-generation sequencing reads aided by highly efficient de novo assembly. Nucleic Acids Research 40(22), e171 (2012)

[5] Li, H., et al.: The sequence alignment/map format and samtools. Bioinformatics 25(16), 2078–2079 (2009)

[6] DePristo, M.A.: et al. A framework for variation discovery and genotyping using next-generation DNA sequencing data. Nature Genetics 43(5), 491–498 (2011)

[7] Zaharia, M., Bolosky, W., Curtis, K., Fox, A., Patterson, P., Shenker, S., Stoica, I., Karp, R., Sittler, T.: Faster and more accurate sequence alignment with SNAP (2011), http://arxiv.org/abs/1111.5572

[8] Popitsch, N., von Haeseler, A.: NGC: lossless and lossy compression of aligned high-throughput sequencing data. Nucleic Acids Research 41(1), e27 (2013)

[9] Shen, J.J., Zhang, N.R.: Change-point model on nonhomogeneous Poisson processes with application in copy number profiling by next-generation DNA sequencing. The Annals of Applied Statistics 40(6), 476–496 (2012)

[10] Shen, Y., Gu, Y., Pe'er, I.: A Hidden Markov Model for Copy Number Variant prediction from whole genome resequencing data. BMC Bioinformatics 12(suppl. 6), S4 (2011)

[11] Wang, K., et al.: PennCNV: An integrated hidden Markov model designed for high-resolution copy number variation detection in whole-genome SNP genotyping data. Genome Research 17(11), 1665–1674 (2007)

[12] Lander, E.S., Waterman, M.S.: Genomic mapping by fingerprinting random clones: a mathematical analysis. Genomics 2(3), 231–239 (1988)

[13] Evans, S.N., Hower, V., Pachter, L.: Coverage statistics for sequence census methods. BMC Bioinformatics 11, 430 (2010)

[14] Hower, V., Starfield, R., Roberts, A., Pachter, L.: Quantifying uniformity of mapped reads. Bioinformatics 28(20), 2680–2682 (2012)

[15] Medvedev, P., Stanciu, M., Brudno, M.: Computational methods for discovering structural variation with next-generation sequencing. Nature Methods 6, S13–S20 (2009)

[16] Sherry, S.T., et al.: dbSNP: the NCBI database of genetic variation. Nucleic Acids Research 29(1), 308–311 (2001)

[17] Li, H., Durbin, R.: Fast and accurate short read alignment with Burrows-Wheeler transform. Bioinformatics 25, 1754–1760 (2009)

[18] Jackson, B., et al.: An algorithm for optimal partitioning of data on an interval. IEEE Signal Processing Letters 12, 105–108 (2005)

[19] Killick, R., Fearnhead, P., Eckley, I.A.: Optimal detection of changepoints with a linear computational cost. Journal of the American Statistical Association 107(500), 1590–1598 (2012)

[20] Talwalkar, A., et al.: SMaSH: A benchmarking toolkit for variant calling (2013), http://arxiv.org/abs/1310.8420

[21] Levy, S., et al.: The diploid genome sequence of an individual human. PLoS Biology 5(10), e254 (2007)

[22] Illumina Corporation. Platinum genomes project (2013), http://www.platinumgenomes.org

[23] Zhao, Z., Fu, Y., Hewett-Emmett, D., Boerwinkle, E.: Investigating single nucleotide polymorphism (SNP) density in the human genome and its implications for molecular evolution. Gene 312, 207–213 (2003)

[24] Derrien, T., et al.: Fast computation and applications of genome mappability. PLoS ONE 7(1), e30377 (2012)

On the Representation of de Bruijn Graphs

Rayan Chikhi[1], Antoine Limasset[2], Shaun Jackman[3],
Jared T. Simpson[4], and Paul Medvedev[1,5,6,*]

[1] Department of Computer Science and Engineering, The Pennsylvania State University, USA
[2] ENS Cachan Brittany, Bruz, France
[3] Canada's Michael Smith Genome Sciences Centre, Canada
[4] Ontario Institute for Cancer Research, Toronto, Canada
[5] Department of Biochemistry and Molecular Biology, The Pennsylvania State University, USA
[6] Genomics Institute of the Huck Institutes of the Life Sciences,
The Pennsylvania State University, USA
pashadag@cse.psu.edu

Abstract. The de Bruijn graph plays an important role in bioinformatics, especially in the context of *de novo* assembly. However, the representation of the de Bruijn graph in memory is a computational bottleneck for many assemblers. Recent papers proposed a navigational data structure approach in order to improve memory usage. We prove several theoretical space lower bounds to show the limitations of these types of approaches. We further design and implement a general data structure (DBGFM) and demonstrate its use on a human whole-genome dataset, achieving space usage of 1.5 GB and a 46% improvement over previous approaches. As part of DBGFM, we develop the notion of frequency-based minimizers and show how it can be used to enumerate all maximal simple paths of the de Bruijn graph using only 43 MB of memory. Finally, we demonstrate that our approach can be integrated into an existing assembler by modifying the ABySS software to use DBGFM.

1 Introduction

De novo assembly continues to be one of the fundamental problems in bioinformatics, with new datasets coming from projects such as the Genome10K [17]. The task is to reconstruct an unknown genome sequence from a set of short sequenced fragments. Most state-of-the-art assemblers (e.g. [14,23,2,41]) start by building a de Bruijn graph (dBG) [28,18], which is a directed graph where each node is a distinct k-mer present in the input fragments, and an edge is present between two k-mers when they share an exact $(k-1)$-overlap. The de Bruijn graph is the basis of many steps in assembly, including path compression, bulge removal, graph simplification, and repeat resolution [26]. In the workflow of most assemblers, the graph must, at least initially, reside in memory; thus, for large genomes, memory is a computational bottleneck. For example, the graph of a human genome consists of nearly three billions nodes and edges and assemblers require computers with hundreds of gigabytes of memory [14,23]. Even these large

---

*   Corresponding author.

R. Sharan (Ed.): RECOMB 2014, LNBI 8394, pp. 35–55, 2014.
© Springer International Publishing Switzerland 2014

resources can be insufficient for many genomes, such as the 20 Gbp white spruce. Recent assembly required a distributed-memory approach and around a hundred large-memory servers, collectively storing a 4.3 TB de Bruijn graph data structure [3].

Several articles have pursued the question of whether smaller data structures could be designed to make large genome assembly more accessible [10,40,27,9,7]. Conway and Bromage [10] gave a lower bound on the number of bits required to encode a de Bruijn graph consisting of n k-mers: $\Omega(n \lg n)$ (assuming $4^k > n$). However, two groups independently observed that assemblers use dBGs in a very narrow manner [9,7] and proposed a data structure that is able to return the set of neighbors of a given node but is not necessarily able to determine if that node is in the graph. We refer to these as *navigational data structures* (NDS). The navigational data structures proposed in [9,7] require $O(n \lg k)$ and $O(n)^1$ bits (respectively), beating the Conway-Bromage lower bound both in theory and in practice [9].

What is the potential of these types of approaches to further reduce memory usage? To answer this question, we first formalize the notion of a navigational data structure and then show that any NDS requires at least $3.24n$ bits. This result leaves a gap with the known upper bounds; however, even if a NDS could be developed to meet this bound, could we hope to do better on inputs that occur in practice? To answer this, we consider a very simple class of inputs: simple paths. We show that on these inputs (called linear dBGs), there are both navigational and general data structures that asymptotically use $2n$ bits and give matching lower bounds. While dBGs occurring in practice are not linear, they can nevertheless be often decomposed into a small collection of long simple paths (where all the internal nodes have in- and out-degree of 1). Could we then take advantage of such a decomposition to develop a data structure that can achieve close to $2n$ bits on practical inputs?

We describe and implement a data structure (DBGFM) to represent de Bruijn graphs in low memory. The first step of the construction uses existing k-mer counting software to transform, in constant memory, the input sequencing dataset to a list of k-mers (i.e. nodes) stored on disk [29]. The second step is a novel low memory algorithm that enumerates all the maximal simple paths without loading the whole graph in memory. We achieve this through the use of non-lexicographic minimizers, ordered based on their frequency in the data. Finally, we use the FM-index [12] to store the simple paths in memory and answer membership and neighborhood queries.

We prove that as the number of simple paths decreases, the space utilization of DBGFM approaches $2n$ bits. In practice, DBGFM uses $4.76n$ bits on a human whole-genome dataset and $3.53n$ bits on a human chr14 dataset, improving the state-of-the-art [33] by 46% and 60%, respectively. We demonstrate the efficiency of frequency-based minimizers by collapsing the dBG of the human whole-genome dataset using only 43 MB of memory. Finally, we show how DBGFM can be integrated into an existing assembler by modifying the ABySS software [38] to use DBGFM instead of a hash table.

---

[1] The paper only showed the number of bits is $O(n \lg n)$. However, the authors recently indicated in a blog post [6] how the dependence on $\lg(n)$ could be removed, though the result has not yet been published.

2   Previous Work

In the last three years, several papers and assemblers have explored novel data structures designed to reduce the space usage of dBGs, and we provide a brief summary of the results here.

ABySS was one of the first genome assemblers capable of representing large dBGs [38]. It uses an open-addressing hash table that stores the k-mers of the graph in the keys. The edges can be inferred from the nodes and do not need to be stored. For every k-mer, ABySS uses $2k$ bits to store the k-mer, plus an additional 43 bits of associated data (stored in 64 bits for ease of implementation). Therefore, in total, the space usage of the dBG data structure in ABySS is $(\ell^{-1}(2k + 64))$ bits per k-mer, where ℓ is the load factor of the hash table (set to 0.8). In the following, we focus on the space needed to store just the dBG, since the type of associated data varies greatly between different assemblers.

Conway and Bromage [10] gave a $\lg \binom{4^k}{n}$ bits lower bound for representing a dBG and demonstrated a sparse bit array data structure that comes close to achieving it. They used an edge-centric definition of the dBG (where edges are all the $(k + 1)$-mers, and nodes are prefixes and suffixes of length k), but their results trivially translate to node-centric dBGs by storing k-mers instead of $(k + 1)$-mers. For a dataset with $k = 27$ and $12 \cdot 10^9$ edges (i.e. $(k + 1)$-mers), their theoretical minimum space is 22 bits per edge while their implementation achieves 28.5 bits per edge.

Later work explored the trade-offs between the amount of information retained from the de Bruijn graph and the space usage of the data structure. Ye $et\ al.$ [40] showed that a graph equivalent to the de Bruijn graph can be stored in a hash table by sub-sampling k-mers. The values of the hash table record sequences that would correspond to paths between k-mers in the de Bruijn graph. The theoretical memory usage of this approach is $\Omega(k/g)$ bits per k-mer, where g is the distance between consecutive sampled k-mers. Pell $et\ al.$ [27] proposed a practical lossy approximation of the de Bruijn graph that stores the nodes in a Bloom filter [4]. They found that a space usage of 4 bits per k-mer provided a reasonable approximation of the de Bruijn graph for their purpose (partitioning and down-sampling DNA sequence datasets). Yet, the structure has not yet been directly applied to $de\ novo$ assembly.

Chikhi and Rizk [9] built upon the structure of Pell $et\ al.$ by additionally storing the set of Bloom filter false positives (false neighbors of true nodes in the graph). In this way, their structure is no longer lossy. They obtained a navigational data structure that allowed the assembler to exactly enumerate the in- and out-neighbors of any graph node in constant time. However, the structure does not support node membership queries, and also does not support storing associated data to k-mers. The theoretical space usage is $(1.44 \lg(\frac{16k}{2.08}) + 2.08)$ bits per k-mer, under certain assumptions about the false positive rate of the Bloom filter. This corresponds to 13.2 bits per k-mer for $k = 27$.

The structure has recently been improved by Salikhov $et\ al.$ with cascading Bloom filters [33], replacing the hash table by a cascade of Bloom filters. In theory, if an infinite number of Bloom filters is used, this scheme would require 7.93 bits per k-mer independently of k. The authors show that using only 4 Bloom filters is satisfactory in practice, yet they do not provide a formula for the theoretical space usage in this case. For $k = 27$ and $2.7 \cdot 10^9$ nodes, they computed that their structure uses 8.4 bits per k-mer. Bowe *et al.* [7] used a tree variant of the Burrows-Wheeler transform [8] to support identical operations. They describe a theoretical navigational data structure for representing the dBG of a set of input sequences that uses a space $4m + M \lg(m) + o(m)$ bits, where M is the number of input strings and m the number of graph edges. Note that the space is independent of k. Another data structure based on a similar principle has been recently proposed [32].

3   Preliminaries

We assume, for the purposes of this paper, that all strings are over the alphabet $\Sigma = \{A, C, G, T\}$. A string of length k is called a k-mer and U is the universe of all k-mers, i.e. $U = \Sigma^k$. The binary relation $u \to v$ between two strings denotes an exact suffix-prefix overlap of length $(k - 1)$ between u and v. For a set of k-mers S, the *de Bruijn graph* of S is a directed graph such that the nodes are exactly the k-mers in S and the edges are given by the $\to$ relation. We define S to be a *linear* dBG if there exists a string x where all the $(k - 1)$-mers of x are distinct and S is the set of k-mers present in x. Equivalently, S is a linear dBG if and only if the graph is a simple path. The de Bruijn graph of a string s is the de Bruijn graph of all the k-mers in s. We adopt the node-centric definition of the de Bruijn graph, where the edges are implicit given the vertices; therefore, we use the terms de Bruijn graph and a set of k-mers interchangeably.

For a node x in the de Bruijn graph, let $\overleftarrow{ext}(x)$ be its four potential in-neighbors (i.e. $\overleftarrow{ext}(x) = \{y : y \in \Sigma^k, y \to x\}$) and $\overrightarrow{ext}(x)$ be its four potential out-neighbors. Let $ext(x) = \overleftarrow{ext}(x) \cup \overrightarrow{ext}(x)$. For a given set of k-mers S, let $ext(S) = \{ext(x), x \in S\}$ (similarly for $\overrightarrow{ext}(S)$ and $\overleftarrow{ext}(S)$).

We will need some notation for working with index sets, which is just a set of integers that is used to select a subset of elements from another set. Define $\text{IDX}(i, j)$ as a set of all index sets that select j out of i elements. Given a set of i elements Y and $X \in \text{IDX}(i, j)$, we then write $Y[X]$ to represent the subset of j elements out of Y, as specified by X. We assume that there is a natural ordering on the elements of the set Y, e.g. if Y is a set of strings, then the ordering might be the lexicographical one.

The families of graphs we will use to construct the lower bounds of Theorems 1 and 2 have k be a polylogarithmic function of $|S|$, i.e. $k = O(\log^c |S|)$ for some c. We note that in some cases, higher lower bounds could be obtained using families of graphs with $k = \Theta(n)$; however, we feel that such values of k are unrealistic given the sequencing technologies. On one hand, the value of k is a bounded from above by the read length, which is fixed and independent of the number of k-mers. On the other hand, k must be at least $\log_4(|S|)$ in order for there to be at least $|S|$ distinct k-mers.

4   Navigational Data Structures

We use the term *membership data structure* to refer to a way of representing a dBG and answering k-mer membership queries. We can view this as a pair of algorithms: (CONST, MEMB). The CONST algorithm takes a set of k-mers S (i.e. a dBG) and outputs a bit string. We call CONST a constructor, since it constructs a representation of a dBG. The MEMB algorithm takes as input a bit string and a k-mer x and outputs true or false. Intuitively, MEMB takes a representation of a dBG created by CONST and outputs whether a given k-mer is present. Formally, we require that for all $x \in \Sigma^k$, MEMB(CONST(S), x) is true if and only if $x \in S$. An example membership data structure, as used in ABySS, is one where the k-mers are put into a hash table (the CONST algorithm) and membership queries are answered by hashing the k-mer to its location in the table (the MEMB algorithm).

Recently, it was observed that most assemblers use the MEMB algorithm in a limited way [9,7]. They do not typically ask for membership of a vertex that is not in $ext(S)$, but, instead, ask for the neighborhood of nodes that it already knows are in the graph. We formalize this idea by introducing the term navigational data structure (NDS), inspired by the similar idea of performing navigational queries on trees [11]. An NDS is a pair of algorithms, CONST and NBR. As before, CONST takes a set of k-mers and outputs a bit string. NBR takes a bit string and a k-mer, and outputs a set of k-mers. The algorithms must satisfy that for every dBG S and a k-mer $x \in S$, NBR(CONST(S), x) = $ext(x)$. Note that if $x \notin S$, then the behavior of NBR(CONST(S), x) is undefined. We observe that a membership data structure immediately implies a NDS because a NBR query can be reduced to eight MEMB queries.

To illustrate how such a data structure can be useful, consider a program that can enumerate nodes using external memory (e.g. a hard drive or a network connection). Using external memory to navigate the graph by testing node membership would be highly inefficient because of long random access times. However, it is acceptable to get a starting node from the device and access the other nodes using the proposed data structure.

There are several important aspects of both a navigational and membership data structures, including the space needed to represent the output of the constructor, the memory usage and running time of the constructor, and the time needed to answer either neighborhood or membership queries. For proving space lower bounds, we make no restriction on the other resources so that our bounds hold more generally. However, adding other constraints (e.g. query time of $\lg n$) may allow us to prove higher lower bounds and is an interesting area for future work.

5   Navigational Data Structure Lower Bound for de Bruijn Graphs

In this section, we prove that a navigational data structure on de Bruijn graphs needs at least 3.24 bits per k-mer to represent the graph:

Theorem 1. *Consider an arbitrary NDS and let* CONST *be its constructor. For any* $0 < \epsilon < 1$, *there exists a* k *and* $x \subseteq \Sigma^k$ *such that* $|\text{CONST}(x)| \geq |x| \cdot (c - \epsilon)$, *where* $c = 8 - 3\lg 3 \approx 3.25$.

Our proof strategy is to construct a family of graphs, for which the number of navigational data structures is at least the size of the family. The full proof of the theorem is in the Appendix, however, we will describe the construction used and the overall outline here. Our first step is to construct a large dBG T and later we will choose subsets

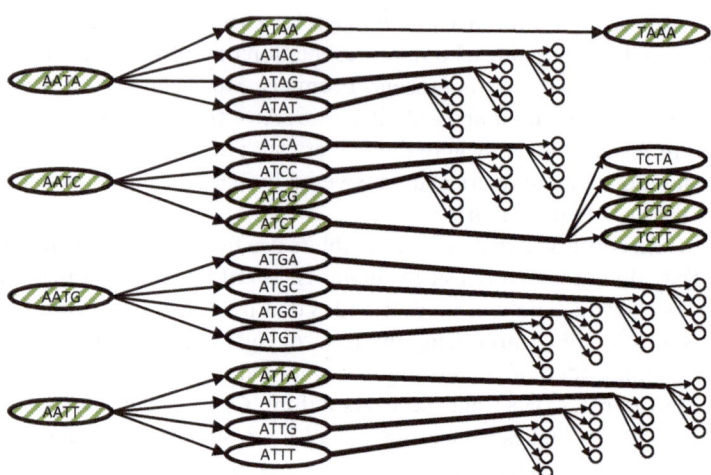

Fig. 1. Example of lower bound construction for $k = 4$. The figure shows T along with some of the node labels. The four nodes on the left form T_0, the 16 nodes in the middle are T_1, and the nodes on the right are T_2. For space purposes, some of the edges from T_1 to T_2 are grouped together. An example of a member from the family is shown with shaded vertices. Note that there are four vertices at each level, and together they form a subforest of T.

as members of our family. Fix an even $k \geq 2$, let $\ell = k/2$, and let $m = 4^{\ell-1}$. T will be defined as the union of $\ell + 1$ levels, $T = \cup_{0 \leq i \leq \ell} T_i$. For $0 \leq i \leq \ell$, we define the ith level as $T_i = \{\text{``}A^{\ell-i}T\alpha\text{''} : \alpha \in \Sigma^{i+\ell-1}\}$. Observe that $T_i = \overrightarrow{ext}(T_{i-1})$, for $1 \leq i \leq \ell$. See Figure 1 for a small example.

We focus on constructing dBGs that are subsets of T because T has some desirable properties. In fact, one can show that the set of k-mers T induces a forest in the dBG of Σ^k (Lemmas 1 and 2 in the Appendix). Each member of our family will be a subforest of T that contains m vertices from every level.

Formally, suppose we are given a sequence of index sets $X = X_1, \ldots, X_\ell$ where every index set is a member of $\text{IDX}(4m, m)$. Each index set will define the subset of vertices we select from a level, and we define $L_0^X = T_0$ and $L_i^X = \overrightarrow{ext}(L_{i-1}^X)[X_i]$, for $1 \leq i \leq \ell$. Note that $L_i^X \subseteq T_i$. In this manner, the index sets define a set of k-mers $s(X) = \cup_{0 \leq i \leq \ell} L_i^X$. Finally, the family of graphs which we will use for our proof is given by:

$$S^k = \{s(X_1, \ldots, X_\ell) : \ell = k/2, m = 4^{\ell-1}, X_i \in \text{IDX}(4m, m)\}$$

To prove Theorem 1, we first show that each of the dBGs of our family have the same amount of k-mers (proof in Appendix):

Property 1. *For all $x \in S^k$, $|x| = 4^{\ell-1}(\ell + 1)$.*

Next, we show that each choice of X leads to a unique graph $s(X)$ (Lemma 3 in the Appendix) and use it to show that the numbers of graphs in our family is large, relative to the number of k-mers in each set (proof in the Appendix):

Property 2. $|S^k| = \binom{4m}{m}^\ell \geq 2^{(c-\epsilon_0)\ell m}$, where $c = 8 - 3\lg 3$ and $\epsilon_0 = \lg(12m)/m$.

Finally, we need to show that for any two graphs in the family, there is at least one k-mer that appears in both graphs but with different neighbors:

Property 3. Let $x = s(X) \in S^k$ and $y = s(Y) \in S^k$ be two distinct elements in S^k. Then, there exists a k-mer $u \in x \cap y$ such that $\overrightarrow{ext}(u) \cap x \neq \overrightarrow{ext}(u) \cap y$.

The proof of Theorem 1 now follows by using the pigeonhole principle to argue that the number of navigational data structures must be at least the size of our family, giving a lower bound on the bits per k-mer.

6 Linear de Bruijn Graphs

In this section, we study data structures to represent linear de Bruijn graphs. Though a linear dBG will never occur in practice, it is an idealized scenario which lets us capture how well a data structure can do in the best case. The bounds obtained here also serve as motivation for our approach in later sections, where we build a membership data structure whose space usage approaches our lower bound from this section the "closer" the graph is to being linear.

We can design a naive membership data structure for linear dBGs. A linear dBG with n k-mers corresponds to a string of length $n + k - 1$. The constructor builds and stores the string from the k-mers, while the membership query simply does a linear scan through the string. The space usage is $2(n + k - 1)$ bits. The query time is prohibitively slow, and we show in Section 7 how to achieve a much faster solution at the cost of a space increase.

We now prove that a NDS on linear de Bruijn graphs needs at least $2n$ bits to represent the graph, meaning one cannot do much better than the naive representation above. In general, representing all strings of length $n + k - 1$ would take $2(n + k - 1)$ bits, however, not all strings of this length correspond to linear dBGs. Fortunately, we can adapt a probabilistic result of Gagie [13] to quantify the number of strings of this length that have no duplicate k-mers (Lemma 5 in Appendix). Our strategy is similar to that of Section 5. We construct a large family of linear dBGs such that for any pair of members, there is always a k-mer that belongs to both but whose neighborhoods are different. We build the family by taking the set of all strings without duplicate $(k - 1)$-mers and identifying a large subset having the same starting k-mer. We then show that by increasing the length of the strings and k, we can create a family of size arbitrarily close to 4^n (Lemma 6 in Appendix). Finally, we show that because each string in the family starts with the same k-mer, there always exists a distinguishing k-mer for any pair of strings. Using the pigeonhole principle, this implies that number of navigational data structures must be at least the number of distinct strings (proof in Appendix):

Theorem 2. Consider an arbitrary NDS for linear de Bruijn graphs and let CONST be its constructor. Then, for any $0 < \epsilon < 1$, there exists (n, k) and a set of k-mers S of cardinality n, such that $|\text{CONST}(S)| \geq 2n(1 - \epsilon)$.

Note that our naive membership data structure of $2(n + k - 1)$ bits immediately implies a NDS of the same size. Similarly, Theorem 2's lower bound of $2n$ bits on a NDS immediately implies the same bound for membership data structures. In practice, k is orders of magnitude less than n, and we view these results as saying that the space usage of membership and navigational data structures on linear dBGs is essentially $2n$ and cannot be improved.

These results, together with Theorem 1, suggested that the potential of navigational data structures may be dampened when the dBG is linear-like in structure. Intuitively, the advantage of a linear dBG is that all the k-mers of a path collapse together onto one string and require storing only one nucleotide per k-mer, except for the overhead of the first k-mer. If the dBG is not linear but can still be decomposed into a few paths, then we could still take advantage of each path while paying an overhead of only a single k-mer per path. This in fact forms the basis of our algorithm in the next section.

7 Data Structure for Representing a de Bruijn Graph in Small Space (DBGFM)

Recall that a simple path is a path where all the internal nodes have in- and out-degree of 1. A *maximal* simple path is one that cannot be extended in either direction. It can be shown that there exists a unique set of edge-disjoint maximal simple paths that completely covers the dBG, and each path p with $|p|$ nodes can be represented compactly as a string of length $k + |p| - 1$. We can thus represent a dBG S containing n k-mers as a set of strings corresponding to the maximal simple paths, denoted by $sp_k(S)$. Let $c_k(S) = |sp_k(S)|$ be the number of maximal simple paths, and let s to be the concatenation of all members of $sp_k(S)$ in arbitrary order, separating each element by a symbol not in Σ (e.g. \$). Using the same idea as in Section 6, we can represent a dBG using s in $2|s| = \sum_{p \in sp_k(S)} 2(|p| + k) \leq 2(n + (k + 2)c_k(S))$ bits. However, this representation requires an inefficient linear scan in order to answer a membership query.

We propose the use of the FM-index of s to speed up query time at the cost of more space. The FM-index [12] is a full-text index which is based on the Burrows-Wheeler transform [8,1] developed for text compression. It has previously been used to map reads to a reference genome [22,20,24], perform *de novo* assembly [36,37,21], and represent the dBG for the purpose of exploring genome characteristics prior to *de novo* assembly [35].

The implementation of the FM-index stores the Huffman-coded Burrows-Wheeler transform of s along with two associated arrays and some $o(1)$ space overhead. Our software, called DBGFM[2], follows the implementation of [12], and we refer the reader there for a more thorough description of how the FM-index works. Here, we will only state its most relevant properties. It allows us to count the number of occurrences of an arbitrary pattern q in s in $O(|q|)$ time. In the context of dBGs, we can test for the membership of a k-mer in S in time $O(k)$. Two sampling parameters (r_1 and r_2) trade-off the size of the associated arrays with the query time. For constructing the FM-index, there are external memory algorithms that do not use more intermediate memory than

---

[2] Source code available at http://github.com/jts/dbgfm

the size of the final output [12]. The space usage of DBGFM is $|s|(H_0(s) + 96r_1^{-1} + 384r_2^{-1}) + o(1) \leq n(H_0(s) + 96r_1^{-1} + 384r_2^{-1})(1 + \frac{k+2}{n}c_k(S)) + o(1)$ bits. H_0 is the zeroth order entropy [1]: $H_0(s) = -\sum_{c \in \Sigma \cup \{\$\}} f_c \lg f_c$, where f_c is the frequency of character c in s. Note that for our five character alphabet H_0 is at most $\lg 5$.

As the value of $c_k(S)$ approaches one, $f_\$$ approaches 0 and hence the upper bound on H_0 approaches 2. If we further set the sampling parameters to be inversely proportional to n, the space utilization approaches at most $2n$ bits. However, this results in impractical query time and, more realistically, typical values for the sampling parameters are $r_1 = 256$ and $r_2 = 16384$, resulting in at most $2.32n$ bits as $c_k(S)$ approaches 1. For the error-free human genome with $k = 55$, there are $c_{55}(S) = 12.9 \cdot 10^6$ maximal simple paths and $n = 2.7 \cdot 10^9$ k-mers. The resulting $H_0(S)$ is at most 2.03, and the space utilization is at most 2.43 bits per k-mer.

An additional benefit of the FM-index is that it allows constant-time access to the in-neighbors of nodes — every edge is part of a simple path, so we can query the FM-index for the symbols preceding a k-mer x. Thus, DBGFM is a membership data structure but supports faster in-neighbor queries. However we note that this is not always the case when reverse complementarity is taken into account.

We wanted to demonstrate how the DBGFM data structure could be incorporated into an existing assembler. We chose ABySS, a popular *de novo* sequence assembly tool used in large-scale genomic projects [38]. In modifying ABySS to look up k-mer membership using DBGFM, we replace its hash table with a simple array. DBGFM associates each k-mer with an integer called a suffix array index (SAI), which could be used to index the simple array. However, some of the k-mers of the DBGFM string include a simple path separator symbol, $\$$, and, hence, not every SAI corresponds to a node in the dBG. We therefore use a rank/select data structure [15] to translate the SAIs into a contiguous enumeration of the nodes, which we then use to index our simple array. We also modified the graph traversal strategy in order to maximize the number of in-neighborhood queries, which are more efficient than out-neighborhood or membership queries.

8 Algorithm to Enumerate the Maximal Simple Paths of a de Bruijn Graph in Low Memory (BCALM)

The DBGFM data structure of Section 7 can construct and represent a dBG in low space from the set of maximal simple paths ($sp_k(S)$). However, constructing the paths (called compaction) generally requires loading the k-mers into memory, which would require large intermediate memory usage. Because our goal is a data structure that is low-memory during both construction and the final output, we developed an algorithm for compacting de Bruijn graphs in low-memory (BCALM[3]).

Our algorithm is based on the idea of minimizers, first introduced by [30,31] and later used by [40,25]. The ℓ-minimizer of a string s is the smallest ℓ-mer that is a substring of s (we assume there is a total ordering of the strings, e.g. lexicographical). We define Lmin(u) (respectively, Rmin(u)) to be the ℓ-minimizer of the $(k-1)$-prefix

---

[3] Source code available at http://github.com/Malfoy/bcalm

(respectively suffix) of u. We refer to these as the left and right minimizers of u, respectively. We use minimizers because of the following observation:

Observation 1. *For two strings u and v, if $u \rightarrow v$, then $Rmin(u) = Lmin(v)$.*

We will first need some definitions. Given a set of strings S, we say that $(u, v) \in S^2$ are *compactable* in a set $V \subseteq S$ if $u \rightarrow v$ and, $\forall w \in V$, if $w \rightarrow v$ then $w = u$ and if $u \rightarrow w$ then $w = v$. The compaction operation is defined on a pair of compactable strings u, v in S. It replaces u and v by a single string $w = u \cdot v[k + 1, |v|]$ where '$\cdot$' is the string concatenation operator. Two strings (u, v) are *m-compactable* in V if they are compactable in V and if $m = Rmin(u) = Lmin(v)$. The m-compaction of a set V is obtained from V by applying the compaction operation as much as possible in any order to all pairs of strings that are m-compactable in V. It is easy to show that the order in which strings are compacted does not lead to different m-compactions. Compaction is a useful notion because a simple way to obtain the simple paths is to greedily perform compaction as long as possible. In the following analysis, we identify a string u with the path $p = u_1 \rightarrow u_2 \rightarrow \ldots \rightarrow u_{|u|-k+1}$ of all the k-mers of u in consecutive order.

We now give a high-level overview of Algorithm 1. The input is a set of k-mers S and a parameter $\ell < k$ which is the minimizer size. For each $m \in \Sigma^\ell$, we maintain a file F_m in external memory. Each file contains a set of strings, and we will later prove that at any point during the execution, each string is a sub-path of a simple path (Lemma 9 in the Appendix). Moreover, we show that at any point of execution, the multi-set of k-mers appearing in the strings and in the output does not change and is always S (Property 4 in the Appendix).

At line 5, we partition the k-mers into the files, according to their ℓ-minimizers. Next, each of the files is processed, starting from the file of the smallest minimizer in increasing order (line 6). For each file, we load the strings into memory and m-compact them (line 7), with the idea being that the size of each of the files is kept small enough

---

Algorithm 1. BCALM: Enumeration of all maximal simple paths in the dBG

1: **Input:** Set of k-mers S, minimizer size $\ell < k$
2: **Output:** Sequences of all simple paths in the de Bruijn graph of S
3: Perform a linear scan of S to get the frequency of all ℓ-mers (in memory)
4: Define the ordering of the minimizers, given by their frequency in S
5: Partition S into files F_m based on the minimizer m of each k-mer
6: **for** each file F_m in increasing order of m **do**
7:   $C_m \leftarrow m$-compaction of F_m (performed in memory)
8:   **for** each string u of C_m **do**
9:     $B_{min} \leftarrow \min(Lmin\,(u), Rmin\,(u))$
10:    $B_{max} \leftarrow \max(Lmin\,(u), Rmin\,(u))$
11:    **if** $B_{min} \leq m$ and $B_{max} \leq m$ **then**
12:      Output u
13:    **else if** $B_{min} \leq m$ and $B_{max} > m$ **then**
14:      Write u to $F_{B_{max}}$
15:    **else if** $B_{min} > m$ and $B_{max} > m$ **then**
16:      Write u to $F_{B_{min}}$
17: Delete F_m

so that memory usage is low. The result of the compaction is a new set of strings, each of which is then either written to one of the files that has not been yet processed or output as a simple path.

The rule of choosing which file to write to is based on the left and right minimizers of the string. If both minimizers are no more than m, then the string is output as a simple path (line 12). Otherwise, we identify m', the smallest of the two minimizers that is bigger than m, and write the string to the file $F_{m'}$. Finally, the file F_m is discarded, and the next file is processed. The rule for placing the strings into the files ensures that as each file F_m is processed (line 6), it will contain every k-mer that has m as a minimizer (Lemma 8 in the Appendix). We can then use this to prove the correctness of the algorithm (proof in Appendix):

Theorem 3. *The output of Algorithm 1 is the set of maximal simple paths of the de Bruijn graph of S.*

There are several implementation details that make the algorithm practical. First, reverse complements are supported in the natural way by identifying each k-mer with its reverse complement and letting the minimizer be the smallest ℓ-mer in both of them. Second, we avoid creating 4^ℓ files, which may be more than the file system supports. Instead, we use virtual files and group them together into a smaller number of physical files. This allowed us to use $\ell = 10$ in our experiments. Third, when we load a file from disk (line 7) we only load the first and last k-mer of each string, since the middle part is never used by the compaction algorithm. We store the middle part in an auxiliary file and use a pointer to keep track of it within the strings in the F_m files.

Consequently, the algorithm memory usage depends on the number of strings in each file F_m, but not on the total size of those files. For a fixed input S, the number of strings in a file F_m depends on the minimizer length ℓ and the ordering of minimizers. When ℓ increases, the number of $(k - 1)$-mers in S that have the same minimizer decreases. Thus, increasing ℓ yields less strings per file, which decreases the memory usage. We also realized that, when highly-repeated ℓ-mers are less likely to be chosen as minimizers, the sequences are more evenly distributed among files. We therefore perform in-memory ℓ-mer counting (line 3) to obtain a sorted frequency table of all ℓ-mers. This step requires an array of $64|\Sigma|^\ell$ bits to store the count of each ℓ-mer in 64 bits, which is negligible memory overhead for small values of ℓ (8 MB for $\ell = 10$). Each ℓ-mer is then mapped to its rank in the frequency array, to create a total ordering of minimizers (line 4). Our experiments showed a drastic improvement over lexicographic ordering (results in Section 9).

9   Results

We tested the effectiveness of our algorithms to assemble two sequencing datasets. Experiments in Tables 1, 2 and 3 were run on a single core of a desktop computer equipped with an Intel i7 3.2 GHz processor, 8 GB of memory and a 7200 RPM hard disk drive. Experiments in Tables 4 and 5 were run on a single core of a cluster node with 24 GB of memory and 2.67 GHz cores. In all experiments, at most 300 GB of temporary disk space was used. The first dataset is 36 million 155bp Illumina human

Table 1. Running times (wall-clock) and memory usage of DSK, BCALM and DBGFM construction on the human chromosome 14 and whole human genome datasets ($k = 55$ and $\ell = 10$ for both)

Dataset	DSK	BCALM	DBGFM
Chromosome 14	43 MB	19 MB	38 MB
	25 mins	15 mins	7 mins
Whole human genome	1.1 GB	43 MB	1.5 GB
	5 h	12 h	7 h

Table 2. Memory usage of de Bruijn graph data structures, on the human chromosome 14 and whole human genome datasets ($k = 55$ for both). We did not run the algorithm of Conway and Bromage because our machine does not have sufficient memory for the whole genome. Instead, we report the theoretical size of their data structure, assuming that it would be constructed from the output of DSK. As described in [10], this gives a lower bound on how well their implementation could perform.

	DBGFM	Salikhov *et al.*	Conway & Bromage
chr14	38.0 MB	94.9 MB	> 875 MB
Full human dataset	1462 MB	2702 MB	> 22951 MB

chromosome 14 reads (2.9 GB compressed fastq) from the GAGE benchmark [34]. The second dataset is 1.4 billion Illumina 100bp reads (54 GB compressed fastq) from the NA18507 human genome (SRX016231). We first processed the reads with k-mer counting software, which is the first step of most assembly pipelines. We used a value of $k = 55$ as we found it gives reasonably good results on both datasets. We used DSK [29], a software that is designed specifically for low memory usage and can also filter out low-count k-mers as they are likely due to sequencing errors (we used < 5 for chr14 and < 3 for the whole genome).

First, we ran BCALM on the of k-mers computed by DSK. The output of BCALM was then passed as input to DBGFM, which constructed the FM-index. Table 1 shows the resulting time and memory usage of DSK, BCALM, and DBGFM. For the whole genome dataset, BCALM used only 43 MB of memory to take a set of $2.5 \cdot 10^9$ 55-mers and output 40 million sequences of total length 4.6 Gbp. DBGFM represented these paths in an index of size 1.5 GB, using no more than that memory during construction. The overall construction time, including DSK, was roughly 24 hours. In comparison, a subset of this dataset was used to construct the data structure of Salikhov *et al.* in 30.7 hours [33].

We compared the space utilization of our DBGFM representation with that of other low space data structures, Salikhov *et al.* [33] and Conway and Bomage [10] (Table 2). Another promising approach is that of Bowe *et al.* [7], but they do not have an implementation available. We use 3.53 bits per k-mer (38.0 MB total) for chr14 and 4.76 bits per k-mer (1462 MB total) for the whole-genome. This is a 60% and 46% improvement over the state-of-the art, respectively.

Table 3. Memory usage of BCALM with three different minimizer orderings: lexicographical, uniformly random, and according to increasing ℓ-mer frequencies. The dataset used is the human chromosome 14 with $k = 55$ and $\ell = 8$.

Ordering type	Lexicographical	Uniformly Random	ℓ-mer frequency
Memory usage	804 MB	222 MB	19 MB

Table 4. Memory usage and wall-clock running time of BCALM with increasing values of minimizer sizes ℓ on the chr14 data. By grouping files into virtual files, these values of ℓ require respectively 4, 16, 64, 256 and 1024 physical files on disk. The ordering of minimizers used is the one based on ℓ-mer counts.

Minimizer size ℓ	2	4	6	8	10
Memory usage	9879 MB	3413 MB	248 MB	19 MB	19 MB
Running time	27m19s	22m2s	20m5s	18m39s	21m4s

Table 5. Memory usage and run time (wall clock) of the ABySS hash table implementation (sparsehash) and of the DBGFM implementation, using a single thread to assemble the human chromosome 14 data set. The dBG bytes/k-mer column corresponds to the space taken by encoded k-mers for sparsehash, and the FM-index for DBGFM. The Data bytes/k-mer column corresponds to associated data. The Overhead bytes/k-mer corresponds to the hash table and heap overheads, as well as the rank/select bit array. The run time of the DBGFM row does not include the time to construct the DBGFM representation.

Data structure	Memory usage	Bytes/ k-mer	dBG (B/k-mer)	Data (B/k-mer)	Overhead (B/k-mer)	Run time
sparsehash	2429 MB	29.50	16	8	5.50	14m4s
DBGFM	739 MB	8.98	0.53	8	0.44	21m1s

During algorithm development, we experimented with different ways to order the minimizers and the effect on memory usage (Table 3). Initially, we used the lexicographical ordering, but experiments with the chromosome 14 dataset showed it was a poor choice, resulting in 804 MB memory usage with $\ell = 8$. The lexicographically smallest ℓ-mer is $m_0 = A^\ell$, which is overly abundant in human chromosomes for $\ell \leq 10$, resulting in a large file F_{m_0}. In a second attempt, we created a uniformly random ordering of all the ℓ-mers. While A^ℓ is no longer likely to have a small value, it is still likely that there is a highly repeated ℓ-mer that comes early in the ordering, resulting in 222 MB memory usage. Finally, we ordered the ℓ-mers according to their frequency in the dataset. This gave a memory usage of 19 MB, resulting in a 40-fold improvement over the initial lexicographical ordering. The running times of all three orderings were comparable. We also evaluated the effect that the minimizer size ℓ has on memory usage and running time (Table 4). Larger ℓ will generally lead to smaller memory usage, however we did not see much improvement past $\ell = 8$ on this dataset.

Finally, we evaluated the performance of ABySS using DBGFM compared with that of the hash table implementation (Table 5). Note, however, that only the graph traversal and marking steps were implemented in the DBGFM version, and none of the graph simplifications. The DBGFM version used 70% less memory, albeit the hash version was 33% faster, indicating the time/space trade-off inherent in the FM-index. In addition to storing the graph, ABySS associates data with each k-mer: the count of each k-mer and its reverse complement (two 16 bits counters), the presence or absence of the four possible in- and out-edges (8 bits), three boolean flags indicating whether the k-mer and its reverse complement have been visited in graph traversal (2 bits), and whether they have been removed (1 bit). While in the hash implementation, the graph structure takes 54% of the memory, in the DBGFM version it only used 6% of memory. This indicates that further memory improvements can be made by optimizing the memory usage of the associated data.

10   Conclusion

This paper has focused on pushing the boundaries of memory efficiency of de Bruijn graphs. Because of the speed/memory trade-offs involved, this has come at the cost of slower data structure construction and query times. Our next focus will be on improving these runtimes through optimization and parallelization of our algorithms.

We see several benefits of low-memory de Bruijn graph data structures in genome assembly. First, there are genomes like the 20 Gbp white spruce which are an order of magnitude longer than the human which cannot be assembled by most assemblers, even on machines with a terabyte of RAM. Second, even for human sized genomes, the memory burden poses unnecessary costs to research biology labs. Finally, in assemblers such as ABySS that store the k-mers explicitly, memory constraints can prevent the use of large k values. With DBGFM, the memory usage becomes much less dependent on k, and allows the use of larger k values to improve the quality of the assembly.

Beyond genome assembly, our work is also relevant to many *de novo* sequencing applications where large de Bruijn graphs are used, e.g. assembly of transcriptomes and meta-genomes [16,5], and *de novo* genotyping [19].

Acknowledgements. The authors would like to acknowledge anonymous referees from a previous submission for their helpful suggestions and for pointing us to the paper of Gagie [13].

References

1. Adjeroh, D., Bell, T.C., Mukherjee, A.: The Burrows-Wheeler Transform: Data Compression, Suffix Arrays, and Pattern Matching. Springer (2008)
2. Bankevich, A., Nurk, S., Antipov, D., Gurevich, A.A., Dvorkin, M., Kulikov, A.S., Lesin, V.M., Nikolenko, S.I., Pham, S.K., Prjibelski, A.D., Pyshkin, A., Sirotkin, A., Vyahhi, N., Tesler, G., Alekseyev, M.A., Pevzner, P.A.: SPAdes: A new genome assembly algorithm and its applications to single-cell sequencing. Journal of Computational Biology 19(5), 455–477 (2012)

3. Birol, I., Raymond, A., Jackman, S.D., Pleasance, S., Coope, R., Taylor, G.A., Yuen, M.M.S., Keeling, C.I., Brand, D., Vandervalk, B.P., Kirk, H., Pandoh, P., Moore, R.A., Zhao, Y., Mungall, A.J., Jaquish, B., Yanchuk, A., Ritland, C., Boyle, B., Bousquet, J., Ritland, K., MacKay, J., Bohlmann, J., Jones, S.J.: Assembling the 20 Gb white spruce (Picea glauca) genome from whole-genome shotgun sequencing data. Bioinformatics (2013)
4. Bloom, B.H.: Space/time trade-offs in hash coding with allowable errors. Commun. ACM 13(7), 422–426 (1970)
5. Boisvert, S., Raymond, F., Godzaridis, É., Laviolette, F., Corbeil, J., et al.: Ray Meta: scalable de novo metagenome assembly and profiling. Genome Biology 13(12), R122 (2012)
6. Bowe, A.: Succinct de Bruijn graphs (blog post), http://alexbowe.com/succinct-debruijn-graphs/ (accessed: October 18, 2013)
7. Bowe, A., Onodera, T., Sadakane, K., Shibuya, T.: Succinct de bruijn graphs. In: Raphael, B., Tang, J. (eds.) WABI 2012. LNCS, vol. 7534, pp. 225–235. Springer, Heidelberg (2012)
8. Burrows, M., Wheeler, D.J.: A block sorting lossless data compression algorithm. Technical report 124. Tech. rep. Digital Equipment Corporation, Palo Alto (1994)
9. Chikhi, R., Rizk, G.: Space-efficient and exact de Bruijn graph representation based on a Bloom filter. In: Raphael, B., Tang, J. (eds.) WABI 2012. LNCS, vol. 7534, pp. 236–248. Springer, Heidelberg (2012)
10. Conway, T.C., Bromage, A.J.: Succinct data structures for assembling large genomes. Bioinformatics 27(4), 479 (2011)
11. Ferragina, P., Luccio, F., Manzini, G., Muthukrishnan, S.: Compressing and indexing labeled trees, with applications. J. ACM 57(1) (2009)
12. Ferragina, P., Manzini, G.: Opportunistic data structures with applications. In: Proceedings of the 41st Annual Symposium on Foundations of Computer Science, pp. 390–398. IEEE (2000)
13. Gagie, T.: Bounds from a card trick. Journal of Discrete Algorithms 10, 2–4 (2012)
14. Gnerre, S., MacCallum, I., Przybylski, D., Ribeiro, F.J., Burton, J.N., Walker, B.J., Sharpe, T., Hall, G., Shea, T.P., Sykes, S.: High-quality draft assemblies of mammalian genomes from massively parallel sequence data. Proceedings of the National Academy of Sciences 108(4), 1513 (2011)
15. González, R., Grabowski, S., Mäkinen, V., Navarro, G.: Practical implementation of rank and select queries. In: Poster Proceedings Volume of 4th Workshop on Efficient and Experimental Algorithms (WEA 2005), Greece, pp. 27–38 (2005)
16. Grabherr, M.G., Haas, B.J., Yassour, M., Levin, J.Z., Thompson, D.A., Amit, I., Adiconis, X., Fan, L., Raychowdhury, R., Zeng, Q., et al.: Full-length transcriptome assembly from RNA-Seq data without a reference genome. Nature Biotechnology 29(7), 644–652 (2011)
17. Haussler, D., O'Brien, S.J., Ryder, O.A., Barker, F.K., Clamp, M., Crawford, A.J., Hanner, R., Hanotte, O., Johnson, W.E., McGuire, J.A., et al.: Genome 10K: a proposal to obtain whole-genome sequence for 10,000 vertebrate species. Journal of Heredity 100(6), 659–674 (2008)
18. Idury, R.M., Waterman, M.S.: A new algorithm for DNA sequence assembly. Journal of Computational Biology 2(2), 291–306 (1995)
19. Iqbal, Z., Caccamo, M., Turner, I., Flicek, P., McVean, G.: De novo assembly and genotyping of variants using colored de Bruijn graphs. Nature Genetics 44(2), 226–232 (2012)
20. Langmead, B., Salzberg, S.L.: Fast gapped-read alignment with Bowtie 2. Nature Methods 9(4), 357–359 (2012)
21. Li, H.: Exploring single-sample SNP and INDEL calling with whole-genome de novo assembly. Bioinformatics 28(14), 1838–1844 (2012)
22. Li, H., Durbin, R.: Fast and accurate short read alignment with Burrows–Wheeler transform. Bioinformatics 25(14), 1754–1760 (2009)

23. Li, R., Zhu, H., Ruan, J., Qian, W., Fang, X., Shi, Z., Li, Y., Li, S., Shan, G., Kristiansen, K.: De novo assembly of human genomes with massively parallel short read sequencing. Genome Research 20(2), 265 (2010)
24. Li, R., Yu, C., Li, Y., Lam, T.W., Yiu, S.M., Kristiansen, K., Wang, J.: SOAP2: an improved ultrafast tool for short read alignment. Bioinformatics 25(15), 1966–1967 (2009)
25. Li, Y., Kamousi, P., Han, F., Yang, S., Yan, X., Suri, S.: Memory efficient minimum substring partitioning. In: Proceedings of the 39th International Conference on Very Large Data Bases, pp. 169–180. VLDB Endowment (2013)
26. Miller, J.R., Koren, S., Sutton, G.: Assembly algorithms for next-generation sequencing data. Genomics 95(6), 315–327 (2010)
27. Pell, J., Hintze, A., Canino-Koning, R., Howe, A., Tiedje, J.M., Brown, C.T.: Scaling metagenome sequence assembly with probabilistic de Bruijn graphs. Proceedings of the National Academy of Sciences 109(33), 13272–13277 (2012)
28. Pevzner, P.A.: l-Tuple DNA sequencing: computer analysis. Journal of Biomolecular Structure & Dynamics 7(1), 63–73 (1989)
29. Rizk, G., Lavenier, D., Chikhi, R.: DSK: k-mer counting with very low memory usage. Bioinformatics 29(5), 652–653 (2013)
30. Roberts, M., Hayes, W., Hunt, B.R., Mount, S.M., Yorke, J.A.: Reducing storage requirements for biological sequence comparison. Bioinformatics 20(18), 3363–3369 (2004)
31. Roberts, M., Hunt, B.R., Yorke, J.A., Bolanos, R.A., Delcher, A.L.: A preprocessor for shotgun assembly of large genomes. Journal of Computational Biology 11(4), 734–752 (2004)
32. Rødland, E.A.: Compact representation of k-mer de bruijn graphs for genome read assembly. BMC Bioinformatics 14(1), 313 (2013)
33. Salikhov, K., Sacomoto, G., Kucherov, G.: Using cascading Bloom filters to improve the memory usage for de Brujin graphs. In: Darling, A., Stoye, J. (eds.) WABI 2013. LNCS, vol. 8126, pp. 364–376. Springer, Heidelberg (2013)
34. Salzberg, S.L., Phillippy, A.M., Zimin, A., Puiu, D., Magoc, T., Koren, S., Treangen, T.J., Schatz, M.C., Delcher, A.L., Roberts, M., et al.: GAGE: A critical evaluation of genome assemblies and assembly algorithms. Genome Research 22(3), 557–567 (2012)
35. Simpson, J.T.: Exploring genome characteristics and sequence quality without a reference. arXiv preprint arXiv:1307.8026 (2013)
36. Simpson, J.T., Durbin, R.: Efficient construction of an assembly string graph using the FM-index. Bioinformatics 26(12), 367–373 (2010)
37. Simpson, J.T., Durbin, R.: Efficient de novo assembly of large genomes using compressed data structures. Genome Research 22(3), 549–556 (2012)
38. Simpson, J.T., Wong, K., Jackman, S.D., Schein, J.E., Jones, S.J., Birol, İ.: ABySS: a parallel assembler for short read sequence data. Genome Research 19(6), 1117–1123 (2009)
39. Sondow, J., Stong, R.: Choice bounds: 11132. The American Mathematical Monthly 114(4), 359–360 (2007)
40. Ye, C., Ma, Z.S., Cannon, C.H., Pop, M., Douglas, W.Y.: Exploiting sparseness in de novo genome assembly. BMC Bioinformatics 13(suppl. 6), S1 (2012)
41. Zerbino, D.R., Birney, E.: Velvet: algorithms for de novo short read assembly using de Bruijn graphs. Genome Research 18(5), 821–829 (2008)

11 Appendix

This Appendix contains lemmas and proofs that are omitted in the main text.

11.1 Lower Bound for General de Bruijn Graphs

Lemma 1. *Let $y \in T$. There exists a unique $0 \leq i \leq \ell$ such that $y \in T_i$.*

Proof. Take two arbitrary levels $i_1 < i_2$ and two arbitrary vertices in those levels, $x_1 \in T_{i_1}$ and $x_2 \in T_{i_2}$. Let $z \in \{1, 2\}$. The k-mer represented by x_z is "$A^{\ell - i_z + 1} T \alpha_z$", where α_z is some string. At position $\ell - i_1 + 1$, x_1 has a T, while x_2 has an A. Therefore, $x_1 \neq x_2$ and the lemma follows. □

Lemma 2. *For all distinct x_1 and x_2 in T that are not in the last level (T_ℓ), $\overrightarrow{ext}(x_1) \cap \overrightarrow{ext}(x_2) = \emptyset$.*

Proof. By Lemma 1, there exist unique levels i_1 and i_2 such that $x_1 \in T_{i_1}$ and $x_2 \in T_{i_2}$. We first observe that $\overrightarrow{ext}(x_z) \in T_{i_z + 1}$, for $z \in \{1, 2\}$. If it is the case that $i_1 \neq i_2$, then Lemma 1 applied to the vertices in the extensions prove the lemma. Now suppose that $i_1 = i_2$, and we write $i = i_1$. Then, for $z \in \{1, 2\}$, the k-mer represented by x_z is "$A^{\ell - i} T \alpha_z$", where α_z is a $(\ell + i - 1)$-mer and $\alpha_1 \neq \alpha_2$. We can then write the extensions as $\overrightarrow{ext}(x_z) = \{$"$A^{\ell - i - 1} T \alpha_z \beta$" $: \beta \in \{A, C, G, T\}\}$. Because $\alpha_1 \neq \alpha_2$, the sets $\overrightarrow{ext}(x_1)$ and $\overrightarrow{ext}(x_2)$ share no common elements. □

Property 1. *For all $x \in S^k$, $|x| = 4^{\ell - 1}(\ell + 1)$.*

Proof. Follows directly from Lemmas 1 and 2. □

Lemma 3. *Let $X = X_1, \ldots, X_\ell$ and $Y = Y_1, \ldots, Y_\ell$ be two sequences of index sets. Then $s(X) = s(Y)$ if and only if $X = Y$.*

Proof. Since the construction is fully deterministic and depends only on the index sets, then $X = Y$ immediately implies $s(X) = s(Y)$. For the other direction, suppose that $X \neq Y$. Let $i > 0$ be the smallest index such that $X_i \neq Y_i$. Then there exists a vertex y such that $y \in L_i^X$ but $y \notin L_i^Y$. Since y is in T_i but not in L_i^Y, Lemma 1 implies that $y \notin s(Y)$. □

Lemma 4. *For all $m > 0$, $\binom{4m}{m} \geq 2^{(c - \epsilon_0)m}$, where $c = 8 - 3 \lg 3$ and $\epsilon_0 = \lg(12m)/m$.*

Proof. Follows directly from an inequality of Sondow and Stong [39]: $\binom{rm}{m} > \frac{2^{cm}}{4m(r-1)}$. □

Property 2. $|S^k| = \binom{4m}{m}^\ell \geq 2^{(c - \epsilon_0)\ell m}$, *where $c = 8 - 3 \lg 3$ and $\epsilon_0 = \lg(12m)/m$.*

Proof. Lemma 3 tells us that the size of S^k is the number of possible ways one could choose $X_1, \ldots, X_\ell$ during the construction of each element $s(X_1, \ldots, X_\ell)$. The choice for each X_i is independent, and there are $\binom{4m}{m}$ possibilities. Hence, there are $\binom{4m}{m}^\ell$ total choices. The inequality follows from Lemma 4. □

Property 3. *Let $x = s(X) \in S^k$ and $y = s(Y) \in S^k$ be two distinct elements in S^k. Then, there exists a k-mer $u \in x \cap y$ such that $\overrightarrow{ext}(u) \cap x \neq \overrightarrow{ext}(u) \cap y$.*

Proof. By Lemma 3, $X \neq Y$. Let i be the smallest index such that $X_i \neq Y_i$, and let v be an element in L_i^X but not in L_i^Y. By construction, there exists a vertex $u \in L_{i-1}^X$ (and hence in L_{i-1}^Y) such that $v \in \overrightarrow{ext}(u)$. Lemma 1 tells us that v is not in y and hence u satisfies the condition of the lemma. □

Theorem 1. *Consider an arbitrary NDS and let* CONST *be its constructor. For any $0 < \epsilon < 1$, there exists a k and $x \subseteq \Sigma^k$ such that $|\text{CONST}(x)| \geq |x| \cdot (c - \epsilon)$, where $c = 8 - 3 \lg 3 \approx 3.25$.*

Proof. Assume for the sake of contradiction that for all x, $|\text{CONST}(x)| < |x|(c - \epsilon)$. Let k be a large enough integer such that $k > 2c\epsilon^{-1}$ and $\epsilon_0 < (\epsilon(\ell + 1) - c)/\ell$ holds (with m, ℓ, ϵ_0 as defined above). The second inequality is verified for any large value of k, since $\epsilon_0 = \Theta(\ell/4^\ell)$ converges to 0 and $(\epsilon(\ell + 1) - c)/\ell$ converges to ϵ. Let $n = 4^{k/2-1}(k/2 + 1)$. Consider the outputs of CONST on the elements of S^k. When the input is constrained to be of size n, the output must use less than $(c - \epsilon)n$ bits (by Lemma 1). Hence the range of CONST over the domain S^k has size less than $2^{(c-\epsilon)n}$. At the same time, Lemma 2 states that there are at least $2^{(c-\epsilon_0)\ell m}$ elements in S^k.

From the inequality $\epsilon_0 < (\epsilon(\ell + 1) - c)/\ell$ we derive that $(c - \epsilon_0)\ell > (c - \epsilon)(\ell + 1)$ and thus $2^{(c-\epsilon_0)\ell m} > 2^{(c-\epsilon)n}$. Therefore, there must exist distinct $s_1, s_2 \in S^k$ such that $\text{CONST}(s_1) = \text{CONST}(s_2)$. We can now apply Lemma 3 to obtain an element $y \in s_1 \cap s_2$ that is a valid input to $\text{CONST}(s_1)$ and to $\text{CONST}(s_2)$. Since the two functions are the same, the return value must also the same. However, we know that the out-neighborhoods of y are different in s_1 and in s_2, hence, one of the results of NBR on y must be incorrect. This contradicts the correctness of CONST. □

11.2   Lower Bound for Linear de Bruijn Graphs

Lemma 5. *The number of DNA strings of length m where each k-mer is seen only once is at least $4^m(1 - \binom{m}{2}4^{-k})$.*

Proof. This Lemma was expressed in a probabilistic setting in [13], but we provide a deterministic proof here. We define a set of strings S and show that it contains all strings with at least one repeated k-mer. Let $\bar{s}^k$ be the string obtained by repeating the pattern s as many times as needed to obtain a string of length exactly k, possibly truncating the last occurrence.

$$S = \{s \mid \exists\, (i, j), 1 \leq i < j \leq m, \exists\, t, |t| = (m - k), |s| = m$$
$$s[1 \ldots j] = t[1 \ldots j], s[j+k+1 \ldots m] = t[j+1 \ldots m-k], s[j+1 \ldots j+k] = \overline{s[i \ldots j]}^k\}$$

Let s' be a string which contains at least one repeated k-mer. Without loss of generality, assume that $i < j$ are two starting positions of identical k-mers ($s'[j \ldots j + k - 1] = s'[i \ldots i+k-1]$). Setting t to be the concatenation of $s'[1..j]$ and $s'[j+k+1 \ldots n]$, it is clear that s' is in S. Thus S contains all strings of length n having at least one repeated k-mer. Since there are $\binom{m}{2}$ choices for (i, j) and 4^{m-k} choices for t, the cardinality of S is at most $\binom{m}{2} 4^{m-k}$, which yields the result.   □

Lemma 6. *Given $0 < \epsilon < 1$, let $n = \lceil 3\epsilon^{-1} \rceil$ and $k = \lceil 1 + (2 + \epsilon) \log_4(2n) \rceil$. The number of DNA strings of length $(n + k - 1)$ which start with the same k-mer, and do not contain any repeated $(k - 1)$-mer, is strictly greater than $4^{n(1-\epsilon)}$.*

Proof. Note that $k < n$, thus $k > (1 + (2 + \epsilon) \log_4(n + k - 1))$ and $4^{-k+1} < (n+k-1)^{(-2-\epsilon)}$. Using Lemma 5, there are at least $(4^{n+k-1}(1 - \binom{n+k-1}{2} 4^{-k+1})) > (4^{n+k-1}(1 - \frac{1}{2(n+k-1)^\epsilon}))$ strings of length $(n + k - 1)$ where each $(k - 1)$-mer is unique. Thus, each string has exactly n k-mers that are all distinct. By binning these strings with respect to their first k-mer, there exists a k-mer k_0 such that there are at least $4^{n-1}(1 - \frac{1}{2(n+k-1)^\epsilon})$ strings starting with k_0, which do not contain any repeated $(k - 1)$-mer. The following inequalities hold: $4^{-1} > 4^{-n\epsilon/2}$ and $(1 - \frac{1}{2(n+k-1)^\epsilon}) > \frac{1}{2} > 4^{-n\epsilon/2}$. Thus, $4^{n-1}(1 - \frac{1}{2(n+k-1)^\epsilon}) > 4^{n(1-\epsilon)}$.   □

Lemma 7. *Two different strings of length $(n+k-1)$ starting with the same k-mer and not containing any repeated $(k - 1)$-mer correspond to two different linear de Bruijn graphs.*

Proof. For two different strings s_1 and s_2 of length $(n + k - 1)$, which start with the same k-mer and do not contain any repeated $(k - 1)$-mer, observe that their sets of k-mers cannot be identical. Suppose they were, and consider the smallest integer i such that $s_1[i \ldots i+k-2] = s_2[i \ldots i+k-2]$ and $s_1[i-k+1] \neq s_2[i-k+1]$. The k-mer $s_1[i \ldots i + k - 1]$ appears in s_2, at some position $j \neq i$. Then $s_2[i \ldots i + k - 2]$ and $s_2[j \ldots j + k - 2]$ are identical $(k - 1)$-mers in s_2, which is a contradiction. Thus, s_1 and s_2 correspond to different sets of k-mers, and therefore correspond to two different linear de Bruijn graphs.   □

Theorem 2. *Consider an arbitrary NDS for linear de Bruijn graphs and let* CONST *be its constructor. Then, for any $0 < \epsilon < 1$, there exists (n, k) and a set of k-mers S of cardinality n, such that $|\text{CONST}(S)| \geq 2n(1 - \epsilon)$.*

Proof. Assume for the sake of contradiction that for all linear de Bruijn graphs, the output of CONST requires less than $2(1 - \epsilon)$ bits per k-mer. Thus for a fixed k-mer length, the number of outputs CONST(S) for sets of k-mers S of size n is no more than $2^{2n(1-\epsilon)}$. Lemma 6 provides values (k, n, k_0), for which there are more strings starting with a k-mer k_0 and containing exactly n k-mers with no duplicate $(k-1)$-mers (strictly more than $2^{2n(1-\epsilon)}$) than the number of outputs CONST(S) for n k-mers.

By the pigeonhole principle, there exists a navigational data structure constructor CONST(S) that takes the same values on two different strings s_1 and s_2 that start with the same k-mer k_0 and do not contain repeated $(k-1)$-mer. By Lemma 7, CONST(S) takes the same values on two different sets of k-mers S_1 and S_2 of cardinality n. Let p be the length of longest prefix common to both strings. Let k_1 be the k-mer at position $(p-k+1)$ in s_1. Note that k_1 is also the k-mer that starts at position $(p-k+1)$ in s_2. By construction of s_1 and s_2, k_1 does not appear anywhere else in s_1 or s_2. Moreover, the k-mer at position $(p-k)$ in s_1 is different to the k-mer at position $(p-k)$ in s_2. In a linear de Bruijn graph corresponding to a string where no $(k-1)$-mer is repeated, each node has at most one out-neighbor. Thus, the out-neighbor of k_1 in the de Bruijn graph of S_1 is different to the out-neighbor of k_1 in the de Bruijn graph of S_2, i.e. NBR(CONST$(S_1), k_1$) $\neq$ NBR(CONST$(S_2), k_1$), which is a contradiction. $\square$

11.3   Algorithm to Enumerate the Simple Paths of a de Bruijn Graph in Low Memory

Property 4. *At any point of execution after line 5, the multi set of k-mers present in the files and in the output is S.*

Proof. We prove by induction. It is trivially true after the partition step. In general, note that the compaction operation preserves the multi set of k-mers. Because the only way the strings are ever changed is through compaction, the property follows. $\square$

Lemma 8. *For each minimizer m, for each k-mer u in S such that $Lmin(u) = m$ (resp. $Rmin(u) = m$), u is the left-most (resp. right-most) k-mer of a string in F_m at the time F_m is processed.*

Proof. We prove this by induction on m. Let m_0 be the smallest minimizer. All k-mers that have m_0 as a left or right minimizer are strings in F_{m_0}, thus the base case is true. Let m be a minimizer and u be a k-mer such that $Lmin(u) = m$ or $Rmin(u) = m$, and assume that the induction hypothesis holds for all smaller minimizers. If $\min(Lmin(u), Rmin(u)) = m$, then u is a string in F_m after execution of line 5. Else, without loss of generality, assume that $m = Rmin(u) > Lmin(u)$. Then, after line 5, u is a string in $F_{Lmin(u)}$. Let $F_{m^1}, \ldots, F_{m^t}$ be all the files, in increasing order of the minimizers, which have a simple path containing u before the maximal-length simple path containing u is outputted by the algorithm. Let m^i be the largest of these minimizers strictly smaller than m. By the induction hypothesis and Property 4, u is at the right extremity of a unique string s_u in F_{m^i}. After the m^i-compactions, since $m = Rmin(s_u) > m^i$, s_u does not go into the output. It is thus written to the next larger minimizer. Since $m = Rmin(u) \leq m^{i+1}$, then it must be that $m^{i+1} = m$, and s_u is written to F_m, which completes the induction. $\square$

Lemma 9. *In Algorithm 1, at any point during execution, each string in F_m corresponds to a sub-path of a maximal-length simple path.*

Proof. We say that a string is *correct* if it corresponds to a sub-path of a maximal-length simple path. We prove the following invariant inductively: at the beginning of the loop at line 6, all the files F_m contain correct strings. The base case is trivially true as all files contain only k-mers in the beginning. Assume that the invariant holds before processing F_m. It suffices to show that no wrong compactions are made; i.e. if two strings from F_m are m-compactable, then they are also compactable in S. The contrapositive is proven. Assume, for the sake of obtaining a contradiction, that two strings (u, v) are not compactable in S, yet are m-compactable in F_m at the time it is processed. Without loss of generality, assume that there exists $w \in S$ such that $u \to w$ and $w \neq v$. Since $u \to v$ and $u \to w$, $m = \text{Rmin}(u) = \text{Lmin}(v) = \text{Lmin}(w)$. Hence, by Lemma 8, w is the left-most k-mer of a string in F_m at the time F_m is processed. This contradicts that (u, v) are m-compactable in F_m at the time it is processed. Thus, all compactions of strings in F_m yield correct strings, and the invariant remains true after F_m is processed. □

Theorem 3. *The output of Algorithm 1 is the set of maximal simple paths of the de Bruijn graph of S.*

Proof. By contradiction, assume that there exists a maximal-length simple path p that is not returned by Algorithm 1. Every input k-mer is returned by Algorithm 1 in some string, and by Lemma 9, every returned string corresponds to a sub-path of a maximal-length simple path. Then, without loss of generality, assume that a simple path of p is split into sub-paths $p_1, p_2, \ldots, p_i$ in the output of Algorithm 1. Let u be the last k-mer of p_1 and v be the first k-mer of p_2. Let $m = \text{Rmin}(u) = \text{Lmin}(v)$ (with Observation 1). By Lemma 8, u and v are both present in F_m when it is processed. As u and v are compactable in S (to form p), they are also compactable in F_m and thus the strings that include u and v in F_m are compacted at line 7. This indicates that u and v cannot be extremities of p_1 and p_2, which yields a contradiction. □

Exact Learning of RNA Energy Parameters from Structure

Hamidreza Chitsaz* and Mohammad Aminisharifabad

Department of Computer Science, Wayne State University,
Detroit, MI 48202
{chitsaz,mohammad.aminisharifabad}@wayne.edu
http://compbio.cs.wayne.edu

Abstract. We consider the problem of exact learning of parameters of a linear RNA energy model from secondary structure data. A necessary and sufficient condition for learnability of parameters is derived, which is based on computing the convex hull of union of translated Newton polytopes of input sequences [15]. The set of learned energy parameters is characterized as the convex cone generated by the normal vectors to those facets of the resulting polytope that are incident to the origin. In practice, the sufficient condition may not be satisfied by the entire training data set; hence, computing a maximal subset of training data for which the sufficient condition is satisfied is often desired. We show that problem is NP-hard in general for an arbitrary dimensional feature space. Using a randomized greedy algorithm, we select a subset of RNA STRAND v2.0 database that satisfies the sufficient condition for separate A-U, C-G, G-U base pair counting model. The set of learned energy parameters includes experimentally measured energies of A-U, C-G, and G-U pairs; hence, our parameter set is in agreement with the Turner parameters.

1 Introduction

The discovery of key regulatory roles of RNA in the cell has recently invigorated interest in RNA structure and RNA-RNA interaction determination or prediction [4,6,17,18,30,33,37]. Due to high chemical reactivity of nucleic acids, experimental determination of RNA structure is time-consuming and challenging. In spite of the fact that computational prediction of RNA structure may not be accurate in a significant number of cases, it is the only viable low-cost high-throughput option to date. Furthermore, with the advent of whole genome synthetic biology [16], accurate high-throughput RNA engineering algorithms are required for both *in vivo* and *in vitro* applications [26–28,32,35].

Since the dawn of RNA secondary structure prediction four decades ago [31], increasingly complex models and algorithms have been proposed. Early approaches considered mere base pair counting [23, 34], followed by the Turner

---

* To whom correspondence should be addressed.

R. Sharan (Ed.): RECOMB 2014, LNBI 8394, pp. 56–68, 2014.
© Springer International Publishing Switzerland 2014

thermodynamics model which was a significant leap forward [22, 38]. Recently, massively feature-rich models empowered by parameter estimation algorithms have been proposed. Despite significant progress in the last three decades, made possible by the work of Turner and others [21] on measuring RNA thermodynamic energy parameters and the work of several groups on novel algorithms [3, 5, 8–10, 19, 20, 24] and machine learning approaches [2, 11, 36], the RNA structure prediction accuracy has not reached a satisfactory level yet [25].

Until now, computational convenience, namely the ability to develop dynamic programming algorithms of polynomial running time, and biophysical intuition have played the main role in development of RNA energy models. The first step towards high accuracy RNA structure prediction is to make sure that the energy model is inherently capable of predicting every observed structure, but it is not over-capable, as accurate estimation of the parameters of a highly complex model often requires a myriad of experimental data, and lack of sufficient experimental data causes overfitting. Recently, we gave a systematic method to assess that inherent capability of a given energy model [15]. Our algorithm decides whether the parameters of an energy model are *learnable*. The parameters of an energy model are defined to be *learnable* iff there exists at least one set of such parameters that renders *every* known RNA structure to date, determined through X-ray or NMR, the minimum free energy structure. Equivalently, we say that the parameters of an energy model are learnable iff 100% structure prediction accuracy can be achieved when the training and test sets are identical. Previously, we gave a necessary condition for the learnability and an algorithm to verify it. Note that a successful RNA folding algorithm needs to have generalization power to predict unseen structures. We leave assessment of the generalization power for future work.

In this paper, we give a necessary and sufficient condition for the learnability and characterize the set of learned energy parameters. Also, we show that selecting the maximum number of RNA sequences for which the sufficient condition is satisfied is NP-hard in general in arbitrary dimensions. Using a randomized greedy algorithm, we select a subset of RNA STRAND v2.0 database that satisfies the sufficient condition for separate A-U, C-G, G-U base pair counting model and yields a set of learned energy parameters that includes the experimentally measured energies of A-U, C-G, and G-U pairs.

2   Methods

2.1   Preliminaries

Let the training set $D = \{(x_i, y_i) \mid i = 1, 2, \ldots, n\}$ be a given collection of RNA sequences x and their corresponding experimentally observed structures y. Throughout this paper, we assume the free energy

$$G(x, s, \mathbf{h}) := \langle c(x, s), \mathbf{h} \rangle \tag{1}$$

associated with a sequence-structure pair (x, s) is a linear function of the energy parameters $\mathbf{h} \in \mathbb{R}^k$, in which k is the number of features, s is a structure in $\mathcal{E}(x)$ the ensemble of possible structures of x, and $c(x, s) \in \mathbb{Z}^k$ is the feature vector.

2.2   Learnability of Energy Parameters

The question that we asked before [15] was: does there exist nonzero parameters $\mathbf{h}^\dagger$ such that for every $(x, y) \in D$, $y = \arg\min_s G(x, s, \mathbf{h}^\dagger)$? We ask a slightly relaxed version of that question in this paper: does there exist nonzero parameters $\mathbf{h}^\dagger$ such that for every $(x, y) \in D$, $G(x, y, \mathbf{h}^\dagger) = \min_s G(x, s, \mathbf{h}^\dagger)$? The answer to this question reveals inherent limitations of the energy model, which can be used to design improved models. We provided a necessary condition for the existence of $\mathbf{h}^\dagger$ and a dynamic programming algorithm to verify it through computing the Newton polytope for every x in D [15]. Following our previous notation, let the *feature ensemble* of sequence x be

$$\mathcal{F}(x) := \{c(x, s) \mid s \in \mathcal{E}(x)\} \subset \mathbb{Z}^k, \tag{2}$$

and call the convex hull of $\mathcal{F}(x)$,

$$\mathcal{N}(x) := \operatorname{conv}\mathcal{F}(x) \subset \mathbb{R}^k, \tag{3}$$

the *Newton polytope* of x. We remind the reader that the convex hull of a set, denoted by 'conv' here, is the minimal convex set that fully contains the set. Let $(x, y) \in D$ and $0 \neq \mathbf{h}^\dagger \in \mathbb{R}^k$. We previously showed in [15] that if y minimizes $G(x, s, \mathbf{h}^\dagger)$ as a function of s, then $c(x, y) \in \partial\mathcal{N}(x)$, i.e. the feature vector of (x, y) is on the boundary of the Newton polytope of x.

   In this paper, we provide a necessary and sufficient condition for the existence of $\mathbf{h}^\dagger$. First, we rewrite that necessary condition by introducing a translated copy of the Newton polytope,

$$\mathcal{N}_y(x) := \mathcal{N}(x) \ominus c(x, y) = \operatorname{conv} \left\{ \mathcal{F}(x) \ominus c(x, y) \right\}, \tag{4}$$

in which $\ominus$ is the Minkowski difference. The necessary condition for learnability then becomes $0 \in \partial\mathcal{N}_y(x)$.

2.3   Necessary and Sufficient Condition for Learnability

The following theorem specifies a necessary and sufficient condition for the learnability.

Theorem 1 (Necessary and Sufficient Condition). *There exists $0 \neq \mathbf{h}^\dagger \in \mathbb{R}^k$ such that for all $(x, y) \in D$, y minimizes $G(x, s, \mathbf{h}^\dagger)$ as a function of s iff $0 \in \partial\mathcal{N}(D)$, in which*

$$\mathcal{N}(D) := \operatorname{conv} \left\{ \bigcup_{(x,y) \in D} \mathcal{N}_y(x) \right\} = \operatorname{conv} \left\{ \bigcup_{(x,y) \in D} \mathcal{F}(x) \ominus c(x, y) \right\}. \tag{5}$$

Proof. ($\Rightarrow$) Suppose $0 \neq \mathbf{h}^\dagger \in \mathbb{R}^k$ exists such that for all $(x, y) \in D$, y minimizes $G(x, s, \mathbf{h}^\dagger)$ as a function of s. To the contrary, assume 0 is in the interior of $\mathcal{N}(D)$. Therefore, there is an open ball of radius $\delta > 0$ centered at $0 \in \mathbb{Z}^k$ completely contained in $\mathcal{N}(D)$, i.e.

$$B_\delta(0) \subset \mathcal{N}(D). \tag{6}$$

Let

$$p = -(\delta/2)\frac{\mathbf{h}^\dagger}{\|\mathbf{h}^\dagger\|}.$$

It is clear that $p \in B_\delta(0) \subset \mathcal{N}(D)$ since $\|p\| = \delta/2 < \delta$. Therefore, p can be written as a convex linear combination of the feature vectors in

$$\mathcal{F}(D) := \bigcup_{(x,y) \in D} \{\mathcal{F}(x) \ominus c(x, y)\} = \{v_1, \ldots, v_N\}, \tag{7}$$

i.e.

$$\exists\ \alpha_1, \ldots \alpha_N \geq 0 : \quad \alpha_1 v_1 + \cdots + \alpha_N v_N = p, \tag{8}$$

$$\alpha_1 + \cdots + \alpha_N = 1. \tag{9}$$

Note that

$$\langle p, \mathbf{h}^\dagger \rangle = -(\delta/2)\|\mathbf{h}^\dagger\| < 0. \tag{10}$$

Therefore, there is $1 \leq i \leq N$, such that $\langle v_i, \mathbf{h}^\dagger \rangle < 0$ for otherwise,

$$\langle p, \mathbf{h}^\dagger \rangle = \sum_{i=1}^{N} \alpha_i \langle v_i, \mathbf{h}^\dagger \rangle \geq 0, \tag{11}$$

which would be a contradiction with (10). Since $v_i \in \mathcal{F}(D)$ in (7), there is $(x, y) \in D$ such that $v_i \in \mathcal{F}(x) \ominus c(x, y)$. It is now sufficient to note that $v_i' = v_i + c(x, y) \in \mathcal{F}(x)$ and $\langle v_i', \mathbf{h}^\dagger \rangle < \langle c(x, y), \mathbf{h}^\dagger \rangle = G(x, y, \mathbf{h}^\dagger)$ which is a contradiction with our assumption that y minimizes $G(x, s, \mathbf{h}^\dagger)$ as a function of s.

($\Leftarrow$) Suppose 0 is on the boundary of $\mathcal{N}(D)$. Note that for all $(x, y) \in D$, $0 \in \partial\mathcal{N}_y(x)$. We construct a nonzero $\mathbf{h}^\dagger \in \mathbb{R}^k$ such that for all $(x, y) \in D$, $G(x, y, \mathbf{h}^\dagger) = \min_{s \in \mathcal{E}(x)} G(x, s, \mathbf{h}^\dagger)$. Since $\mathcal{N}(D)$ is convex, it has a supporting hyperplane $\mathcal{H}$, which passes through 0. Let the positive normal to $\mathcal{H}$ be $\mathbf{h}^\dagger$, i.e. $\langle \mathcal{N}(D), \mathbf{h}^\dagger \rangle \subset [0, \infty)$. Therefore, $\min_{p \in \mathcal{N}(D)} \langle p, \mathbf{h}^\dagger \rangle = \langle 0, \mathbf{h}^\dagger \rangle = 0$, which implies that for all $(x, y) \in D$, $\min_{p \in \mathcal{N}_y(x)} \langle p, \mathbf{h}^\dagger \rangle = \langle 0, \mathbf{h}^\dagger \rangle = 0$, or equivalently, $G(x, y, \mathbf{h}^\dagger) = \min_{s \in \mathcal{E}(x)} G(x, s, \mathbf{h}^\dagger)$. $\square$

The proof above is constructive; hence using a similar argument, we characterize all learned energy parameters in the following theorem.

Theorem 2. *Let*

$$H(D) := \left\{ \mathbf{h}^\dagger \in \mathbb{R}^k \mid \mathbf{h}^\dagger \neq 0, G(x, y, \mathbf{h}^\dagger) = \min_{s \in \mathcal{E}(x)} G(x, s, \mathbf{h}^\dagger) \quad \forall (x, y) \in D \right\}.$$
$$\tag{12}$$

In that case, $H(D)$ is the set of vectors $n \in \mathbb{R}^k$ orthogonal to the supporting hyperplanes of $\mathcal{N}(D)$ such that $\langle \mathcal{N}(D), n \rangle \subset [0, \infty)$. Moreover, that is the convex cone generated by the inward normal vectors to those facets of $\mathcal{N}(D)$ that are incident to 0.

2.4   Compatible Training Set

We say that a training set D is *compatible* if the sufficient condition for learnability is satisfied when D is considered. However, the sufficient condition for the entire training set is often not satisfied in practice, for example when the feature vector is low dimensional. We would like to find a compatible subset of D to estimate the energy parameters. A natural quest is to find the maximal compatible subset of D. In the following section, we show that even when the Newton polytopes with polynomial complexity for all of the sequences in D are given, selection of a maximal compatible subset of D is NP-hard in arbitrary dimensions.

2.5   NP-hardness of Maximal Compatible Subset

Problem 1 (Maximal Compatible Subset (MCS)). We are given a collection of convex polytopes $A = \{\mathcal{P}_1, \mathcal{P}_2, \ldots, \mathcal{P}_n\}$ in $\mathbb{R}^k$ such that $0 \in \mathbb{R}^k$ is on the boundary of every $\mathcal{P}_i$, i.e. $0 \in \partial \mathcal{P}_i$ for $i = 1, 2, \ldots, n$. The desired output is a maximal subcollection $B = \{\mathcal{P}_{j_1}, \mathcal{P}_{j_2}, \ldots, \mathcal{P}_{j_m}\} \subseteq A$ with the following property

$$0 \in \partial \ conv \left\{ \bigcup_{\mathcal{P} \in B} \mathcal{P} \right\}. \tag{13}$$

Theorem 3. *MCS is NP-hard.*

Sketch of proof. We use a direct reduction to MAX 3-SAT. Let $\Phi(w_1, w_2, \ldots, w_k) = \bigwedge_{i=1}^n \tau_i$ be a formula in the 3-conjunctive normal form with clauses

$$\tau_i = q_i^1 \ w_{a_i} \lor q_i^2 \ w_{b_i} \lor q_i^3 \ w_{c_i}, \tag{14}$$

where w_j are binary variables, $1 \leq a_i, b_i, c_i \leq k$, and $q_i^1, q_i^2, q_i^3 \subseteq \{\neg\}$. For every clause τ_i, we build 8 convex polytopes $\mathcal{P}_i^{000}, \ldots, \mathcal{P}_i^{111}$ in $\mathbb{R}^k$, where the superscripts are in binary. To achieve that, we first build a *true* $\mathcal{T}_j^1$ and a *false* $\mathcal{T}_j^0$ convex polytope for every variable w_j. Let $\{e_1, e_2, \ldots, e_k\}$ be the standard orthonormal basis for $\mathbb{R}^k$ and define

$$\mathcal{T}_j^1 = conv \{0, Me_j \pm e_1, \ldots, Me_j \pm e_{j-1}, Me_j \pm e_{j+1}, \ldots, Me_j \pm e_k\}, \tag{15}$$

$$\mathcal{T}_j^0 = -\mathcal{T}_j^1, \tag{16}$$

for an arbitrary $1 \ll M \in \mathbb{Z}$ and $j = 1, 2, \ldots, k$. Note that $\mathcal{T}$ is a k-dimensional narrow arrow with polynomial complexity, and its vertices are on the integer

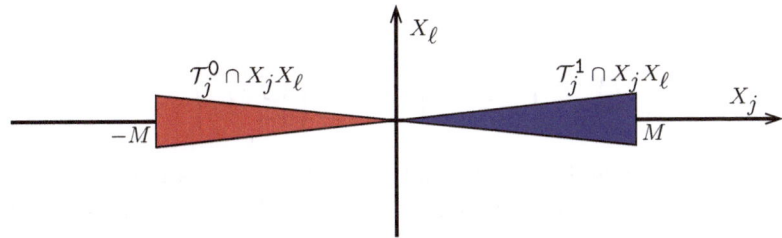

Fig. 1. Intersection of $\mathcal{T}_j^0$ and $\mathcal{T}_j^1$ polytopes with the $X_j X_\ell$ plane in the proof of Theorem 3

lattice and can be computed in polynomial time; see Fig. 1. Moreover, $\{\mathcal{T}_j^0, \mathcal{T}_j^1\}$ is incompatible, i.e. $0 \notin \partial \operatorname{conv}\{\mathcal{T}_j^0 \cup \mathcal{T}_j^1\}$. Let

$$\mathcal{P}_i^{d_1 d_2 d_3} = \operatorname{conv}\left\{\mathcal{T}_{a_i}^{d_1} \cup \mathcal{T}_{b_i}^{d_2} \cup \mathcal{T}_{c_i}^{d_3}\right\}. \tag{17}$$

Note that even though computing the convex hull is NP-hard in general [12], $\mathcal{P}_i^{d_1 d_2 d_3}$ can be computed in polynomial time. Essentially, the vertex set of $\partial \mathcal{P}_i^{d_1 d_2 d_3}$ is the union of vertices of $\partial \mathcal{T}_{a_i}^{d_j}$. Assume $(\sigma_i^1, \sigma_i^2, \sigma_i^3) \in \{0,1\}^3$ is that assignment to variables $(w_{a_i}, w_{b_i}, w_{c_i})$ which makes τ_i false, and let

$$A_i = \left\{\mathcal{P}_i^{000}, \dots, \mathcal{P}_i^{111}\right\} \setminus \left\{\mathcal{P}_i^{\sigma_i^1 \sigma_i^2 \sigma_i^3}\right\}. \tag{18}$$

More precisely,

$$\sigma_i^j = \begin{cases} 0 & \text{if } q_i^j = \emptyset, \\ 1 & \text{otherwise.} \end{cases} \tag{19}$$

Note that for every $\mathcal{P}_i^{d_1 d_2 d_3} \in A_i$, assignment of (d_1, d_2, d_3) to variables $(w_{a_i}, w_{b_i}, w_{c_i})$ makes τ_i true. Finally, define the input to the MCS as

$$A = \bigcup_{i=1}^{n} A_i, \tag{20}$$

and assume B is a maximal subset of A with the property

$$0 \in \partial \operatorname{conv}\left\{\bigcup_{\mathcal{P} \in B} \mathcal{P}\right\}. \tag{21}$$

It is sufficient to show that $|B| = \text{MAX 3-SAT}(\Phi)$. First, we prove that $|B| \geq \text{MAX 3-SAT}(\Phi)$. Suppose $\text{MAX 3-SAT}(\Phi) = m$, and the assignment $(g_1, g_2, \dots, g_k) \in \{0,1\}^k$ renders $\tau_{i_1}, \tau_{i_2}, \dots, \tau_{i_m}$ true. Define

$$B_m = \left\{\mathcal{P}_i^{g_{a_i} g_{b_i} g_{c_i}} \mid i \in \{i_1, i_2, \dots, i_m\}\right\} \subseteq A, \tag{22}$$

and verify that it is compatible, i.e. $0 \in \partial \operatorname{conv} \left\{ \bigcup_{\mathcal{P} \in B_m} \mathcal{P} \right\}$, since

$$\operatorname{conv} \left\{ \bigcup_{\mathcal{P} \in B_m} \mathcal{P} \right\} = \operatorname{conv} \left\{ \bigcup_{j=1}^{k} \mathcal{T}_j^{g_j} \right\} \tag{23}$$

spans only (approximately) one orthant of $\mathbb{R}^k$. Since B is a maximal subset of A with the compatibility property, $|B| \geq |B_m| = m$. Second, we prove that $|B| \leq$ MAX 3-SAT(Φ). We show that for every $1 \leq i \leq n$, at most one $\mathcal{P}_i \in B$. To the contrary, suppose for some $1 \leq i \leq n$, $\mathcal{P}_i^{d_1 d_2 d_3}, \mathcal{P}_i^{d_1' d_2' d_3'} \in B$. Without loss of generality, assume $d_1 \neq d_1'$. In that case,

$$0 \in \operatorname{int} \left\{ \operatorname{conv} \left[\mathcal{T}_{a_i}^0 \cup \mathcal{T}_{a_i}^1 \right] \right\}$$

$$= \operatorname{int} \left\{ \operatorname{conv} \left[\mathcal{T}_{a_i}^{d_1} \cup \mathcal{T}_{a_i}^{d_1'} \right] \right\}$$

$$\subseteq \operatorname{int} \left\{ \operatorname{conv} \left[\mathcal{P}_i^{d_1 d_2 d_3} \cup \mathcal{P}_i^{d_1' d_2' d_3'} \right] \right\} \tag{24}$$

$$\subseteq \operatorname{int} \left[\operatorname{conv} \left\{ \bigcup_{\mathcal{P} \in B} \mathcal{P} \right\} \right],$$

which is a contradiction. Above int denotes the interior. Similarly, B induces a consistent assignment to the variables, that makes $|B|$ clauses true. $\qquad \square$

2.6   Randomized Greedy Algorithm

Since the maximal compatible subset problem is NP-hard in general, we used a randomized greedy algorithm. In the ith iteration, our algorithm starts with a seed B_i which is a random subset of A, the input set of polytopes. In our case, B_i is a single element subset. The algorithm iteratively keeps adding other members of A to B_i as long as 0 remains as a vertex of the convex hull of union of all of the polytopes in B_i. Note that Theorem 1 requires 0 to be on the boundary of the convex hull, not necessarily on a vertex. However in practice, applicable cases have 0 as a vertex. Finally, the B_i with maximum number of elements is returned in the output; see Algorithm 1.

3   Results

We used 2277 unpseudoknotted RNA sequence-structure pairs from RNA STRAND v2.0 database as our data set D. RNA STRAND v2.0 is a convenient source of RNA sequences and structures selected from various Rfam families [7] and contains known RNA secondary structures of any type and organism, particularly with and without pseudoknots [1]. There are 2334 pseudoknot-free RNAs in the RNA STRAND database. We sorted them based on their length and selected the first 2277 ones for computational convenience. We excluded pseudoknotted structures because our current implementation is incapable of

---

Algorithm 1. Randomized Greedy Maximal Compatible Subset

Input: $A = \{\mathcal{P}_1, \mathcal{P}_2, \ldots, \mathcal{P}_n\}$
Output: $B \subseteq A$

$i \leftarrow 1$
$B \leftarrow \emptyset$
while $i < $ MaxIterations **do**   ▷ MaxIterations is a static/dynamic constant
    $A' \leftarrow \{\mathcal{P} \in A \mid 0 \text{ is a vertex of } \partial\mathcal{P}\}$ ▷ remove inapplicable polytopes
    $B_i \leftarrow \emptyset$ ▷ empty subcollection
    $\mathcal{Q} \leftarrow \emptyset$ ▷ empty polytope
    while $A' \neq \emptyset$ **do**
        $\mathcal{P}_r \leftarrow $ Random(A') ▷ pick a random polytope from A'
        $A' \leftarrow A' \setminus \{\mathcal{P}_r\}$
        $C \leftarrow B_i \cup \{\mathcal{P}_r\}$
        $\mathcal{R} \leftarrow \text{conv}\,(\mathcal{Q} \cup \mathcal{P}_r)$
        if 0 is a vertex of $\partial\mathcal{R}$ **then**
            $B_i \leftarrow C$ ▷ greedily expand the subcollection
            $\mathcal{Q} \leftarrow \mathcal{R}$ ▷ update the convex hull of the union
        end if
    end while
    if $|B| < |B_i|$ **then**
        $B \leftarrow B_i$
    end if
    $i \leftarrow i + 1$
end while

---

considering pseudoknots. Some sequences in the data set allow only A-U base pairs (not a single C-G or G-U pair), in which case the Newton polytope degenerates into a line.

We demonstrate the results for the separate A-U, C-G, and G-U base pair counting energy model similar to our previous model [15]. In that case, the feature vector

$$c(x, s) = (c_1(x, s), c_2(x, s), c_3(x, s))$$

is three dimensional: $c_1(x, s)$ is the number of A-U, $c_2(x, s)$ the number of C-G, and $c_3(x, s)$ the number of G-U base pairs in s. First, we computed $c(x, y)$ and used our Newton polytope program to compute $\mathcal{N}(x)$ for each $(x, y) \in D$ [15]. For completeness, we briefly include our dynamic programming algorithm which starts by computing the Newton polytope for all subsequences of unit length, followed by all subsequences of length two and more up to the Newton polytope for the entire sequence x. We denote the Newton polytope of the subsequence $n_i \cdots n_j$ by $\mathcal{N}(i, j)$, i.e.

$$\mathcal{N}(i, j) := \mathcal{N}(n_i \cdots n_j). \tag{25}$$

The following dynamic programming yielded the result

$$
\mathcal{N}(i,j) = \mathrm{conv}\left\{ \bigcup \begin{bmatrix} \mathcal{N}(i,\ell) \oplus \mathcal{N}(\ell+1,j), & i \le \ell \le j-1 \\ \{(1,0,0)\} \oplus \mathcal{N}(i+1,j-1) & \text{if } n_i n_j = \mathrm{AU} \mid \mathrm{UA} \\ \{(0,1,0)\} \oplus \mathcal{N}(i+1,j-1) & \text{if } n_i n_j = \mathrm{CG} \mid \mathrm{GC} \\ \{(0,0,1)\} \oplus \mathcal{N}(i+1,j-1) & \text{if } n_i n_j = \mathrm{GU} \mid \mathrm{UG} \end{bmatrix} \right\},
$$

(26)

with the base case $\mathcal{N}(i,i) = \{(0,0,0)\}$. Above $\oplus$ is the Minkowski sum. We then computed $\mathcal{N}_y(x)$ by translating the Newton polytope so that $c(x,y)$ moves to the origin 0, to obtain $A = \{\mathcal{N}_y(x) \mid (x,y) \in D\}$ as the input to Algorithm 1.

We then removed those polytopes in A that do not have 0 as one of their boundary vertices, to obtain A' in Algorithm 1. It turns out that only 126 sequences out of the initial 2277 remain in A'. Note that our condition here is more stringent than the necessary condition in [15], and that is why fewer sequences satisfy this condition. After 100 iterations (MAXITERATIONS = 100) which took less than a minute, the algorithm returned 3 polytopes that are compatible, i.e. the origin 0 is a vertex of the boundary of convex hull of union of three polytopes. They correspond to sequences PDB_00434 (length: 15nt; bacteriophage HK022 nun-protein-nutboxb-RNA complex [13]), PDB_00200 (length: 21nt; an RNA hairpin derived from the mouse 5' ETS that binds to the two N-terminal RNA binding domains of nucleolin [14]), and PDB_00876 (length: 45nt; solution structure of the HIV-1 frameshift inducing element [29]) which are experimentally verified by NMR or X-ray.

The origin is incident to 5 facets of the resulting convex hull, shown in Fig. 2, the inward normal vector to which are in the rows of

$$
J = \begin{bmatrix} -0.3162 & -0.9487 & 0 \\ -1 & 0 & 0 \\ 0 & 0 & 1 \\ -0.4082 & -0.8165 & -0.4082 \\ -0.5774 & -0.5774 & -0.5774 \end{bmatrix}
$$

(27)

as explained in Theorem 2. The set of those energy parameters that correctly predict the three structures in Fig. 3 is the convex cone generated by these 5 vectors. The Turner model measures the average energies of A-U, C-G, and G-U to be approximately $(-2, -3, -1)$ kcal/mol [21]. To test whether those energy parameters fall into the convex cone generated by J, we solved a convex linear equation. Let

$$
\mathbf{h}^\dagger = \begin{bmatrix} -2 \\ -3 \\ -1 \end{bmatrix}.
$$

We would like to find a positive solution $v \in \mathbb{R}_+^5$ to the linear equation $J^T v = \mathbf{h}^\dagger$. We formulated that as a linear program which was solved using GNU Octave, and here is the answer:

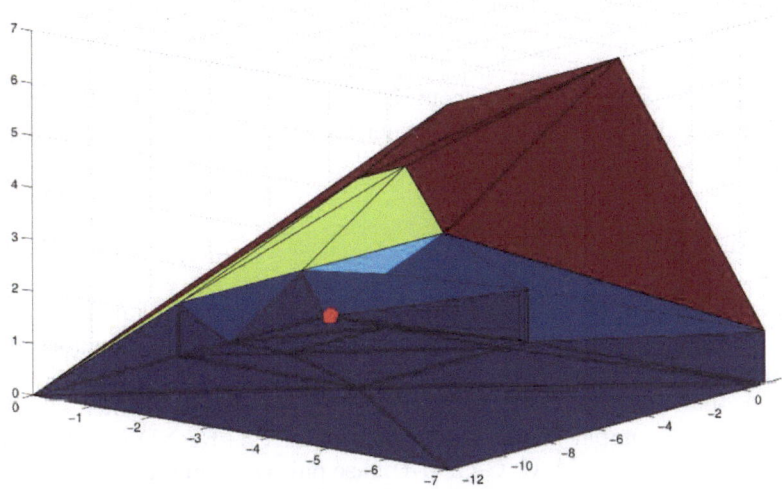

Fig. 2. $\mathcal{N}(D')$ where D' consists of three sequence-structure pairs: PDB_00434, PDB_00200, and PDB_00876. The dot shows the origin 0. Secondary structures in D' are shown in Fig. 3.

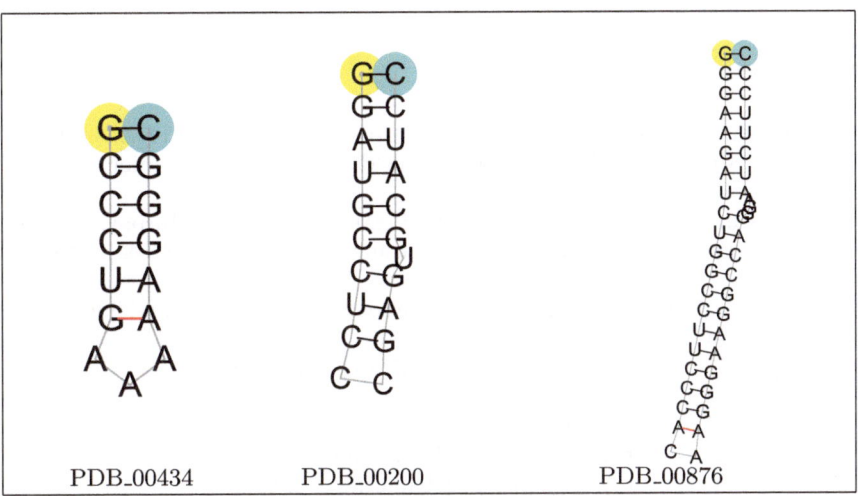

Fig. 3. Structures of PDB_00434, PDB_00200, and PDB_00876 obtained from RNA STRAND v2.0 website [1]

$$v = \begin{bmatrix} 2.10815 \\ 0.33340 \\ 0 \\ 0 \\ 1.73190 \end{bmatrix}.$$

The convex cone generated by J contains the vector $(-2, -3, -1)$. Therefore, our finding is in agreement with the Turner base pairing energies. Note that the three structures are base-pair rich (Fig. 3).

4 Discussion

We further developed the notion of learnability of parameters of an energy model. A necessary and sufficient condition for it was given, and a characterization of the set of energy parameters that realize exact structure prediction followed as a by-product. If an energy model satisfies the sufficient condition, then we say that the training set is compatible. In our case, the RNA STRAND v2.0 training set is not compatible (for the A-U, C-G, G-U base pair counting model). We showed that computing a maximal compatible subset of a set of convex polytopes is NP-hard in general and gave a randomized greedy algorithm for it. The computed set of energy parameters for A-U, C-G, G-U from a maximal compatible subset agreed with the thermodynamic energies. Complexity of the MCS problem is an open and interesting question, particularly if we treat the dimension of the feature space as a constant. Also, assessing the generalization power of an energy model remains for future work.

References

1. Andronescu, M., Bereg, V., Hoos, H.H., Condon, A.: RNA STRAND: the RNA secondary structure and statistical analysis database. BMC Bioinformatics 9, 340 (2008)
2. Andronescu, M., Condon, A., Hoos, H.H., Mathews, D.H., Murphy, K.P.: Computational approaches for RNA energy parameter estimation. RNA 16, 2304–2318 (2010)
3. Backofen, R., Tsur, D., Zakov, S., Ziv-Ukelson, M.: Sparse RNA folding: Time and space efficient algorithms. In: Kucherov, G., Ukkonen, E. (eds.) CPM 2009 Lille. LNCS, vol. 5577, pp. 249–262. Springer, Heidelberg (2009)
4. Bartel, D.P.: MicroRNAs: genomics, biogenesis, mechanism, and function. Cell 116(2), 281–297 (2004)
5. Bernhart, S.H., Tafer, H., Mückstein, U., Flamm, C., Stadler, P.F., Hofacker, I.L.: Partition function and base pairing probabilities of RNA heterodimers. Algorithms Mol. Biol. 1, 3 (2006)
6. Brantl, S.: Antisense-RNA regulation and RNA interference. Bioch. Biophys. Acta 1575(1-3), 15–25 (2002)

7. Burge, S.W., Daub, J., Eberhardt, R., Tate, J., Barquist, L., Nawrocki, E.P., Eddy, S.R., Gardner, P.P., Bateman, A.: Rfam 11.0: 10 years of RNA families. Nucleic Acids Res. 41(database issue), D226–D232 (2013)
8. Chitsaz, H., Backofen, R., Sahinalp, S.C.: biRNA: Fast RNA-RNA binding sites prediction. In: Salzberg, S.L., Warnow, T. (eds.) WABI 2009. LNCS, vol. 5724, pp. 25–36. Springer, Heidelberg (2009)
9. Chitsaz, H., Salari, R., Cenk Sahinalp, S., Backofen, R.: A partition function algorithm for interacting nucleic acid strands. Bioinformatics 25(12), i365–i373 (2009); Also ISMB/ECCB proceedings
10. Dirks, R.M., Pierce, N.A.: A partition function algorithm for nucleic acid secondary structure including pseudoknots. Journal of Computational Chemistry 24(13), 1664–1677 (2003)
11. Do, C.B., Woods, D.A., Batzoglou, S.: CONTRAfold: RNA secondary structure prediction without physics-based models. Bioinformatics 22, 90–98 (2006)
12. Dyer, M.E.: The Complexity of Vertex Enumeration Methods. Mathematics of Operations Research 8(3), 381–402 (1983)
13. Faber, C., Scharpf, M., Becker, T., Sticht, H., Rosch, P.: The structure of the coliphage HK022 Nun protein-lambda-phage boxB RNA complex. Implications for the mechanism of transcription termination. J. Biol. Chem. 276(34), 32064–32070 (2001)
14. Finger, L.D., Trantirek, L., Johansson, C., Feigon, J.: Solution structures of stem-loop RNAs that bind to the two N-terminal RNA-binding domains of nucleolin. Nucleic Acids Res. 31(22), 6461–6472 (2003)
15. Forouzmand, E., Chitsaz, H.: The RNA Newton polytope and learnability of energy parameters. Bioinformatics 29(13), i300–i307 (2013); Also ISMB/ECCB proceedings
16. Gibson, D.G., Glass, J.I., Lartigue, C., Noskov, V.N., Chuang, R.-Y., Algire, M.A., Benders, G.A., Montague, M.G., Ma, L., Moodie, M.M., Merryman, C., Vashee, S., Krishnakumar, R., Assad-Garcia, N., Andrews-Pfannkoch, C., Denisova, E.A., Young, L., Qi, Z.-Q., Segall-Shapiro, T.H., Calvey, C.H., Parmar, P.P., Hutchison, C.A., Smith, H.O., Craig Venter, J.: Creation of a bacterial cell controlled by a chemically synthesized genome. Science 329(5987), 52–56 (2010)
17. Gottesman, S.: Micros for microbes: non-coding regulatory RNAs in bacteria. Trends in Genetics 21(7), 399–404 (2005)
18. Hannon, G.J.: RNA interference. Nature 418(6894), 244–251 (2002)
19. Honer zu Siederdissen, C., Bernhart, S.H., Stadler, P.F., Hofacker, I.L.: A folding algorithm for extended RNA secondary structures. Bioinformatics 27(13), i129–i136 (2011)
20. Huang, F.W.D., Qin, J., Reidys, C.M., Stadler, P.F.: Target prediction and a statistical sampling algorithm for RNA-RNA interaction. Bioinformatics 26(2), 175–181 (2010)
21. Mathews, D.H., Sabina, J., Zuker, M., Turner, D.H.: Expanded sequence dependence of thermodynamic parameters improves prediction of RNA secondary structure. J. Mol. Biol. 288, 911–940 (1999)
22. McCaskill, J.S.: The equilibrium partition function and base pair binding probabilities for RNA secondary structure. Biopolymers 29, 1105–1119 (1990)
23. Nussinov, R., Piecznik, G., Grigg, J.R., Kleitman, D.J.: Algorithms for loop matchings. SIAM Journal on Applied Mathematics 35, 68–82 (1978)

24. Rivas, E., Eddy, S.R.: A dynamic programming algorithm for RNA structure prediction including pseudoknots. J. Mol. Biol. 285(5), 2053–2068 (1999)
25. Rivas, E., Lang, R., Eddy, S.R.: A range of complex probabilistic models for RNA secondary structure prediction that includes the nearest-neighbor model and more. RNA 18(2), 193–212 (2012)
26. Seeman, N.C., Lukeman, P.S.: Nucleic acid nanostructures: bottom-up control of geometry on the nanoscale. Reports on Progress in Physics 68, 237–270 (2005)
27. Seeman, N.C.: From genes to machines: DNA nanomechanical devices. Trends Biochem. Sci. 30, 119–125 (2005)
28. Simmel, F.C., Dittmer, W.U.: DNA nanodevices. Small 1, 284–299 (2005)
29. Staple, D.W., Butcher, S.E.: Solution structure and thermodynamic investigation of the HIV-1 frameshift inducing element. J. Mol. Biol. 349(5), 1011–1023 (2005)
30. Storz, G.: An expanding universe of noncoding RNAs. Science 296(5571), 1260–1263 (2002)
31. Tinoco, I., Borer, P.N., Dengler, B., Levin, M.D., Uhlenbeck, O.C., Crothers, D.M., Bralla, J.: Improved estimation of secondary structure in ribonucleic acids. Nature New Biol. 246(150), 40–41 (1973)
32. Venkataraman, S., Dirks, R.M., Rothemund, P.W., Winfree, E., Pierce, N.A.: An autonomous polymerization motor powered by DNA hybridization. Nat. Nanotechnol. 2, 490–494 (2007)
33. Wagner, E.G., Flardh, K.: Antisense RNAs everywhere? Trends Genet. 18, 223–226 (2002)
34. Waterman, M.S., Smith, T.F.: RNA secondary structure: A complete mathematical analysis. Math. Biosc. 42, 257–266 (1978)
35. Yin, P., Hariadi, R.F., Sahu, S., Choi, H.M., Park, S.H., Labean, T.H., Reif, J.H.: Programming DNA tube circumferences. Science 321, 824–826 (2008)
36. Zakov, S., Goldberg, Y., Elhadad, M., Ziv-Ukelson, M.: Rich parameterization improves RNA structure prediction. In: Bafna, V., Sahinalp, S.C. (eds.) RECOMB 2011. LNCS, vol. 6577, pp. 546–562. Springer, Heidelberg (2011)
37. Zamore, P.D., Haley, B.: Ribo-gnome: the big world of small RNAs. Science 309(5740), 1519–1524 (2005)
38. Zuker, M., Stiegler, P.: Optimal computer folding of large RNA sequences using thermodynamics and auxiliary information. Nucleic Acids Research 9(1), 133–148 (1981)

An Alignment-Free Regression Approach for Estimating Allele-Specific Expression Using RNA-Seq Data

Chen-Ping Fu, Vladimir Jojic, and Leonard McMillan

Department of Computer Science, University of North Carolina, Chapel Hill, NC
{ping,vjojic,mcmillan}@cs.unc.edu

Abstract. RNA-seq technology enables large-scale studies of allele-specific expression (ASE), or the expression difference between maternal and paternal alleles. Here, we study ASE in animals for which parental RNA-seq data are available. While most methods for determining ASE rely on read alignment, read alignment either leads to reference bias or requires knowledge of genomic variants in each parental strain. When RNA-seq data are available for both parental strains of a hybrid animal, it is possible to infer ASE with minimal reference bias and without knowledge of parental genomic variants. Our approach first uses parental RNA-seq reads to discover maternal and paternal versions of transcript sequences. Using these alternative transcript sequences as features, we estimate abundance levels of transcripts in the hybrid animal using a modified lasso linear regression model.

We tested our methods on synthetic data from the mouse transcriptome and compared our results with those of Trinity, a state-of-the-art *de novo* RNA-seq assembler. Our methods achieved high sensitivity and specificity in both identifying expressed transcripts and transcripts exhibiting ASE. We also ran our methods on real RNA-seq mouse data from two F1 samples with wild-derived parental strains and were able to validate known genes exhibiting ASE, as well as confirm the expected maternal contribution ratios in all genes and genes on the X chromosome.

Keywords: Allele-Specific Expression, RNA-seq, Lasso Regression.

1 Introduction

Recent advances in high-throughput RNA-seq technology have enabled the generation of massive amounts of data for investigation of the transcriptome. While this offers exciting potential for studying known gene transcripts and discovering new ones, it also necessitates new bioinformatic tools that can efficiently and accurately analyze such data.

Current RNA-seq techniques generate short reads from RNA sequences at high coverage, and the main challenge in RNA-seq analysis lies in reconstructing transcripts and estimating their relative abundances from millions of short (35-250 bp) read sequences. A common approach is to first map short reads onto

R. Sharan (Ed.): RECOMB 2014, LNBI 8394, pp. 69–84, 2014.
© Springer International Publishing Switzerland 2014

a reference genome, and then estimate the abundance in each annotated gene region. Such reference-alignment methods include TopHat [25], Cufflinks [27] and Scripture [10], which use algorithms such as the Burrows-Wheeler transform [1] to achieve fast read alignment. These methods are well established in the RNA-seq community and there exist many auxiliary tools [25] [26] for downstream analysis.

However, aligning reads to a reference genome has some disadvantages. First, read alignment assumes samples are genetically similar to the reference genome, and as a result, samples that deviate significantly from the reference frequently have a large portion of unmapped reads. This bias favors mapping reads from samples similar to the reference genome and is known as "reference bias." Second, alignment methods typically cannot resolve the origin of reads that map to multiple locations in the genome, resulting in reads being arbitrarily mapped or discarded from analysis. Suggested workarounds to the first problem of reference bias involve creating new genome sequences, typically by incorporating known variants, to use in place of the reference genome for read alignment [23]. However, this requires prior knowledge of genomic variants in the targeted RNA-seq sample, which is sometimes difficult and expensive to obtain.

Another class of methods perform *de novo* assembly of transcriptomes using *De Bruijn* graphs of k-mers from reads [7] [21]. These methods enable reconstruction of the transcriptome in species for which no reference genomic sequence is available. While these methods offer the possibility of novel transcript discovery, their *de novo* nature makes it difficult to map assembled subsequences back to known annotated transcripts. Furthermore, estimation of transcript expression levels in these methods is not straightforward and generally involves alignment of assembled contigs to a reference genome [7] [21], which reintroduces the possibility for reference bias.

Expression level estimation is particularly difficult for outbred diploid organisms, since each expressed transcript may contain two different sets of alleles, one from each parental haplotype. In some transcripts, one parental allele is preferentially expressed over another, resulting in what is known as allele-specific expression (ASE). It is often biologically interesting to identify genes and transcripts exhibiting ASE, as well as estimate the relative expression levels of the maternal and paternal alleles [8] [29]. Prior to the introduction of RNA-seq, ASE studies often relied on microarray technology. Although microarrays are able to identify genes exhibiting ASE, they generally examine a small number of genes, with expression level estimates in highly relative terms [19] [22]. The abundance of data from RNA-seq not only enables large-scale ASE studies incorporating the entire transcriptome, but also provides several means for direct estimation of more accurate expression levels, such as using alignment pile-up heights.

Current RNA-seq-based methods for analyzing ASE rely on reference transcriptome alignment [23] [24], which is again subject to reference bias and requires prior knowledge of genomic variants in the strains of interest. Reference bias is particularly problematic in ASE analysis, since it can falsely enhance relative expression in one parental strain over another.

In the case where RNA-seq data of all three members of a mother-father-child trio are available, we can utilize the RNA-seq data from the parental strains and eliminate the need for prior knowledge of their genomic variants. Here, we examine ASE in F1 mouse strains, which are first-generation offspring of two distinct isogenic parental strains. We separately construct maternal and paternal versions of transcripts using RNA-seq reads from the parental strains and annotated reference transcripts, creating a set of candidate transcript sequences the F1 strain could express. We then estimate the expression level of each candidate transcript in the F1 strain using a modified lasso regression model [11]. Lasso regression has been proposed by Li et al. [17] in the context of RNA-seq isoform expression level estimation, but not in the context of estimating ASE without reference alignment. We choose to use lasso regularization since it drives parameters to zero, enabling us to effectively eliminate non-expressed isoforms that have significant k-mer overlaps with expressed isoforms. We modify the lasso penalty slightly to prefer assigning higher F1 expression levels in transcripts with k-mers that appear frequently in the parental RNA-seq reads, due to the assumption that most highly expressed genes in the parents should also be highly expressed in the F1 strain.

We tested our methods on synthetic RNA-seq data from the wild-derived mouse strains CAST/EiJ and PWK/PhJ, along with F1 offspring CASTxPWK, with CAST/EiJ as the maternal strain and PWK/PhJ as the paternal strain. We also tested on real RNA-seq data from a CAST/EiJ, PWK/PhJ, CASTxPWK trio and a CAST/EiJ, WSB/EiJ, CASTxWSB trio, both using CAST/EiJ as the maternal strain. The CAST/EiJ, PWK/PhJ, and WSB/EiJ mouse strains are isogenic, and all three have well-annotated genomes that differ significantly from each other and from the mouse reference sequence [30], which is largely based on the C57BL/6J strain (NCBI37 [4]). CAST/EiJ and PWK/PhJ each have a high variation rate of approximately one variant per 130 bp with respect to the reference genome, and a slightly higher rate with respect to each other, while WSB/EiJ is more similar to the reference genome with approximately one variant per 375 bp. The genetic distance between these three strains make them ideal candidates for studying ASE, since we expect a large percentage of reads to contain distinguishing variants.

Table 1. Notation

$\mathbf{y}$	F1 k-mer profile. An $n \times 1$ vector where y_i indicates the number of times the ith k-mer appears in the F1 sample
$\mathbf{z}^M, \mathbf{z}^P$	maternal and paternal k-mer profiles
$\mathbf{X}^M$	set of k-mer profiles of candidate transcripts from $\mathbf{z}^M$
$\mathbf{X}^P$	set of k-mer profiles of candidate transcripts from $\mathbf{z}^P$
$\mathbf{X}$	an $n \times m$ matrix equal to $[\mathbf{X}^M \cup \mathbf{X}^P]$, where n is number of k-mers and m is number of transcripts
$\mathbf{x}_j$	k-mer profile of the jth candidate transcript
$x_{i,j}$	number of times the ith k-mer occurs in the jth candidate transcript
θ_j	estimated expression level for the jth candidate transcript

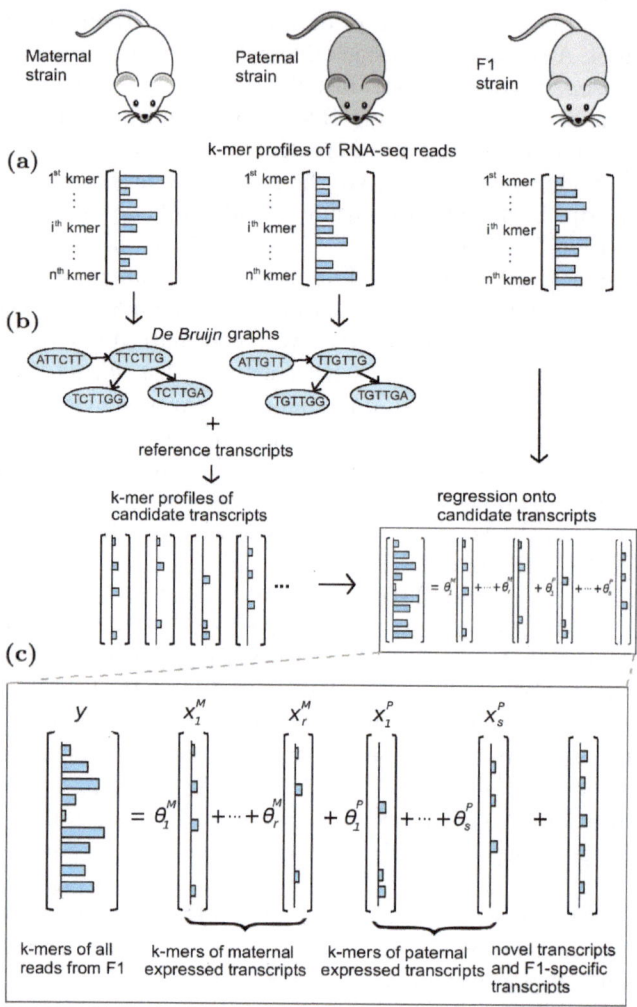

Fig. 1. Our pipeline for estimating allele-specific expression in F1 animals. **(a)** k-mer profiles are created for the maternal, paternal, and F1 strains, using all available RNA-seq reads from one sample of each strain. Each k-mer is also saved as its reverse complement, since we do not know the directionality of the read. **(b)** *De Bruijn* graphs are created for the maternal and paternal samples. Using annotated reference transcripts and the parental *De Bruijn* graphs, we select candidate transcripts which incorporate parental alleles from the *De Bruijn* graphs. **(c)** The k-mer profile of the F1 sample, y, is then regressed onto the candidate parental transcripts, $\{\mathbf{x}_1^M, \mathbf{x}_2^M, ...\mathbf{x}_r^M\} \cup \{\mathbf{x}_1^P, \mathbf{x}_2^P, ...\mathbf{x}_s^P\}$, and we estimate the expression level θ of each candidate transcript.

2 Approach

In this section, we discuss the parameters and assumptions of our proposed model and the underlying optimization problem.

2.1 Notation

Table (1) includes a description of the variables used in this paper. We denote the k-mer profiles of maternal candidate transcripts, $\mathbf{X}^M = \{\mathbf{x}_1^M, \mathbf{x}_2^M, ...\mathbf{x}_r^M\}$, and the k-mer profiles of paternal candidate transcripts, $\mathbf{X}^P = \{\mathbf{x}_1^P, \mathbf{x}_2^P, ...\mathbf{x}_s^P\}$, jointly as $\mathbf{X} = \mathbf{X}^M \cup \mathbf{X}^P$, a matrix representing the k-mer profiles of all candidate transcripts. Each candidate transcript k-mer profile is labeled as originating from the maternal k-mer profile, the paternal k-mer profile, or both if there are no differentiating variants between the parental k-mer profiles.

2.2 Regression Model

We propose a modified lasso penalized regression model for estimating the abundance of each candidate transcript, with the assumption that the F1's k-mer profile y can be expressed as a linear combination of its expressed transcripts $\mathbf{X} = \{\mathbf{x}_1, ...\mathbf{x}_j, ...\mathbf{x}_m\}$ multiplied by their relative expression levels θ_j:

$$\mathbf{y} = \sum_{j=1}^{m} \theta_j \mathbf{x}_j. \tag{1}$$

To filter out non-expressed transcripts and prevent overfitting, each candidate transcript is penalized by an l_1-norm, parameterized by the regularization parameter λ and the inverse of w_j, where

$$w_j = median \begin{cases} \{z_i^M/x_{i,j}, \forall x_{i,j} > 0\}, & \mathbf{x}_j \in \mathbf{X}^M \\ \{z_i^P/x_{i,j}, \forall x_{i,j} > 0\}, & \mathbf{x}_j \in \mathbf{X}^P \\ \{(z_i^M + z_i^P)/x_{i,j}, \forall x_{i,j} > 0\}, & \mathbf{x}_j \in \mathbf{X}^P \cap \mathbf{X}^M \end{cases} \tag{2}$$

Therefore, transcripts that are expressed at a high level in the parental samples are more likely to be expressed at a high level in the F1 sample as well. Our objective function then becomes

$$\operatorname*{argmin}_{\theta} \quad \frac{1}{2} \sum_{i=1}^{n} (y_i - \sum_{j=1}^{m} \theta_j x_{i,j})^2 + \lambda \sum_{j=1}^{m} \frac{\theta_j}{w_j} \tag{3}$$

$$\text{subject to} \quad \theta_j \geq 0, \forall j,$$

with each θ_j constrained to be nonnegative since they represent transcript expression levels.

3   Methods

3.1   Synthetic Data

We used the Flux Simulator [9] to create simulated reads from the CAST/EiJ, PWK/PhJ, and CASTxPWK mouse genomes. We chose these two parental strains because they are well-annotated strains that differ significantly from the reference strain C57BL/6J and from each other. The transcript sequences for CAST/EiJ and PWK/PhJ were created using Cufflinks' gffread utility [27] with genomes from the Wellcome Trust Institute [13] and transcript annotation from the Ensembl Genome Database [3]. The positions from the reference transcript annotation files were updated with positions to the CAST/EiJ and PWK/PhJ genomes using MODtools [12].

We synthesized 10,000,000 100bp paired-end reads from both the CAST/EiJ and the PWK/PhJ genomes to represent reads from a maternal CAST/EiJ genome and a paternal PWK/EiJ genome. We specified the same set of 1000 transcripts with a positive number of expressed RNA molecules in both genomes. In addition, we merged two sets of 5,000,000 separately synthesized reads from both CAST/EiJ and PWK/PhJ to create a simulated F1 fastq file. From the merged CAST/EiJ and PWK/PhJ versions of transcript sequences, the Flux Simulator output 1156 unique transcripts sequences where at least 95% of the sequence is covered by reads, and we define this set of 1156 transcript sequences, representing 626 reference transcripts, as the truly expressed transcripts.

3.2   Real Data

RNA from whole-brain tissues (excluding cerebellum) was extracted from 5 samples (CAST/EiJ female, PWK/PhJ male, WSB/EiJ male, CASTxPWK male and CASTxWSB female) using the Illumina TruSeq RNA Sample Preparation Kit v2. The barcoded cDNA from each sample was multiplexed across four lanes and sequenced on an Illumina HiSeq 2000 to generate 100 bp paired-end reads (2x100). This resulted in $2 \times 71,291,857$ reads for the CAST/EiJ sample, $2 \times 49,877,124$ reads for the PWK/PhJ sample, $2 \times 62,712,206$ reads for the WSB/EiJ sample, $2 \times 77,773,220$ reads for the CASTxPWK hybrid sample, and $2 \times 57,386,133$ reads for the CASTxWSB hybrid sample. Note that the selected samples were not true biological trios, but genetically equivalent. We also used the same female CAST/EiJ sample as the maternal model for both F1 hybrids.

3.3   Selecting Candidate Transcripts

We used a greedy approach for selecting candidate transcript sequences from the *De Bruijn* graphs of each parental k-mer profile. The k-mer size used for this and subsequent analyses was 32 bp. For each of the 93,006 reference transcripts provided by Ensembl [3], we match the reference transcript sequence to a path of k-mers in the *De Bruijn* graph, allowing for a maximum number of 5 mismatches within a sliding window of 25 bp, which is a sensible choice except in the case of unusually dense SNPs or indels. In the case of mismatches, we replace the reference sequence with the sequence in the parental *De Bruijn* graph, thus creating updated candidate transcript sequences which reflect variants in the parental strains. If more than 80% of a transcript's k-mers are found in the *De Bruijn* graph, we consider it a candidate transcript. The k-mer profiles of the selected candidate transcript sequences are then used as features in our regularized regression model.

3.4 Coordinate Descent

To optimize our objective function Eq. (3), we update θ_j using coordinate descent:

$$\theta_j = \frac{max(\sum_{i=1}^{n} y_i^{(-j)} x_{i,j} - \frac{\lambda}{w_j}, 0)}{\|x_j\|_2^2}, \text{where}$$

$$y_i^{(-j)} = y_i - \sum_{k \neq j} \theta_j x_{i,j}.$$

(4)

Due to the high dimensional nature of our data (in real data, the number of k-mers, n, is approximately 5×10^7, and the number of candidate transcripts, m, is approximately 2×10^4), updating each θ_j on every iteration becomes inefficient. We therefore adapt the coordinate descent with a refined sweep algorithm as described by Li and Osher [18], where we greedily select to update only the θ_j that changes the most on every iteration. To save on computation per iteration, we can let $\beta_j = \sum_{i=1}^{n} y_i^{(-j)} x_{i,j}$ and precompute the matrix product $\mathbf{X}^T\mathbf{y}$, so that β can be updated at every iteration using only addition and a scalar-vector multiplication. The algorithm is described in Eq. (5), and proof of its convergence is provided by Li and Osher [18].

Initialize:

$$\theta^0 = \mathbf{0}$$

$$\beta^0 = \mathbf{X}^T\mathbf{y}$$

$$\gamma = diag(\|\mathbf{x}_j\|_2^2) - \mathbf{X}^T\mathbf{X}$$

Iterate until convergence:

$$\theta^* = \frac{max(\beta - \frac{\lambda}{\mathbf{w}}, 0)}{\|\mathbf{x}_j\|_2^2}$$

(5)

$$j = argmax|\theta^* - \theta^k|$$

Updates:

$$\theta_j^{k+1} = \theta_j^*$$

$$\beta^{k+1} = \beta^k + \gamma_{j,:}(\theta_j^* - \theta_j^k)$$

$$\beta_j^{k+1} = \beta_j^k$$

The coordinate descent algorithm terminates when the minimization objective Eq. (3) decreases by less than a threshold of 0.001 per iteration. For computational efficiency, the value of our objective function Eq. (3) is evaluated per τ iterations, where $\tau = 10^4$ initially. We decrease τ as the objective increases, until $\tau = 1$ for the final iterations. This saves significant computation time since the computation of the objective function contains a matrix multiplication and the regular updates do not, and the convergence of the algorithm is not affected as the updates are still being performed per iteration.

The lasso regularization parameter λ is chosen via 4-fold cross validation. It is important to note that the value of λ depends on the mean observed values for w_j, so different values of λ could be chosen for each trio.

4   Results

We analyzed a synthetic data set to ascertain the sensitivity and specificity of our estimation framework. We then applied our technique to two real data sets and evaluated them based on their ability to recapitulate known biological properties.

4.1   Synthetic Data Results

In our synthetic F1 sample, the Flux Simulator generated 1156 unique transcript sequences from both the maternal and paternal haplotypes with positive expression levels, representing 626 reference transcripts. We identified 4517 candidate parental transcript sequences from all reference mouse transcripts annotated by Ensembl, 1055 of which were truly expressed, representing 598 out of 626 truly expressed reference transcripts.

We selected the lasso regularization parameter λ to be 500 using 4-fold cross validataion. We took $\theta_j = 0$ to indicate transcript j was not expressed and calculated the sensitivity and specificity of our method in identifying which transcripts were expressed. For the chosen value of λ, we found the sensitivity to be 0.9553 (598/626) and the specificity to be 0.9880 (91278/92385).

Of the correctly identified expressed transcripts, the true and estimated expression levels had a Pearson correlation coefficient of 0.85, indicating high positive correlation, as shown in Fig. (S1). To allow for comparison of relative expression levels, we normalized both true and predicted expression levels to have a mean value of 1 across all expressed transcripts. The mean absolute error between true and predicted expression levels was 0.3128 for the chosen value of λ. True positive rates, false positive rates, and mean absolute error of predicted expression levels for different values of λ are summarized in Fig. (2).

Among the 598 correctly identified expressed transcripts, 544 had differentiable paternal and maternal candidate sequences. Of these, 141 exhibited ASE, as defined by having a maternal contribution ratio (maternal expression level divided by total expression level) outside the range $[0.4, 0.6]$. Our model correctly

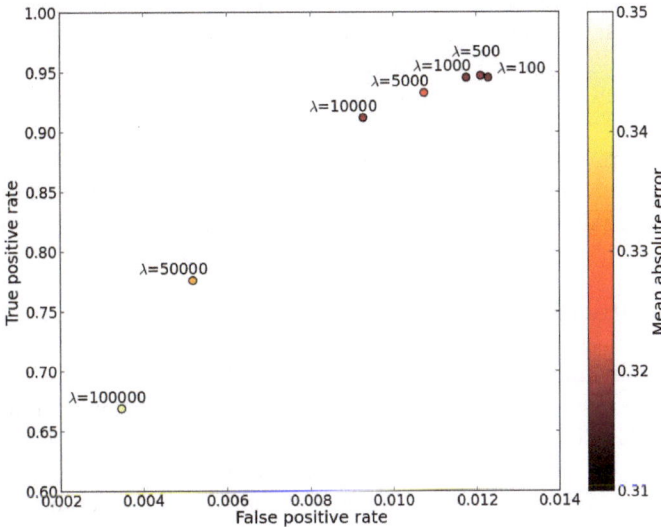

Fig. 2. True positive rate vs. false positive rate for different values of λ. Each point is colored by the mean absolute error between normalized true and estimated expression levels for all transcripts correctly classified as expressed.

identified 109 transcripts exhibiting ASE and correctly rejected 293 transcripts not exhibiting ASE, achieving a sensitivity of 0.77 and specificity of 0.73.

We compared our results with Trinity [7], since its *de novo* assembly methods are able to separate maternal and paternal versions of transcripts better than reference alignment-based methods.

To assemble candidate transcripts from the maternal and paternal strains, we ran Trinity with its default parameters on the synthetic maternal CAST/EiJ and paternal PWK/PhJ samples. Per Trinity's downstream analysis guidelines, we then aligned reads from the synthetic F1 sample to the assembled parental transcript sequences using Bowtie [14] then estimated expression levels using RSEM [16].

Trinity assembled 4215 transcript sequences from both parental strains. Following their guidelines to eliminate false positives, we retained 3336 transcript sequences representing at least 1% of the per-component expression level. We used a criterion of Levenshtein distance less than 10% of the true transcript length to match annotated transcripts to the *de novo* transcripts sequences reported by Trinity. With this criterion, only 110 out of 626 truly expressed transcripts were present in the set of expressed transcripts found by Trinity. In this set, the mean Levenshtein distance from each true transcript sequence to the Trinity sequences was 0.12% of the true transcript length, with the maximum distance being 2.6% of the true transcript length, suggesting our matching criterion of 10% Levenshtein distance was generous.

Out of the 110 assembled transcripts correctly identified, 81 had nonzero expression levels, making the sensitivity for baseline expression detection 0.13. However, of the 81 correctly identified transcripts, the Trinity-Bowtie-RSEM pipeline produced a high correlation of 0.88 between true and estimated expression levels.

Of the 81 expressed transcripts correctly identified by Trinity, 63 originated from reference transcripts with ASE. Trinity correctly identified 20 true positives and 16 true negatives, with a sensitivity of 0.32 and specificity of 0.89.

4.2 Real Data Results

We applied our methods to a male CASTxPWK F1 sample and a female CASTxWSB F1 sample. We first created *De Bruijn* graphs for a CAST/EiJ female, a PWK/EiJ male, and a WSB/EiJ male, representing the parental *De Bruijn* graphs of our two F1 samples. To eliminate erroneous reads in each strain, we filtered k-mers appearing fewer than 5 times. Using Algorithm 2, we selected 15,287 candidate transcripts from the CAST/EiJ *De Bruijn* graph, 9,852 candidate transcripts from the PWK/EiJ graph, and 16,023 candidate transcripts from the WSB/EiJ graph. For each F1 sample, transcript sequences without differentiating variants between the two parental strains were merged into a single candidate transcript. This resulted in 23,585 candidate transcripts for CASTxPWK and 29,155 candidate transcripts for CASTxWSB, representing 7,393 and 8,532 candidate genes, respectively.

The CAST/EiJ, PWK/EiJ and CASTxPWK trio had a merged k-mer profile of 118,100,824 k-mers, 42,688,910 (36.1%) of which appeared in our candidate transcripts. Similarly, the CAST/EiJ, WSB/EiJ and CASTxWSB trio had a merged k-mer profile of 118,383,117 k-mers, 52,715,089 (44.5%) of which appeared in its set of candidate transcripts. We verified most the k-mers in the F1 samples not appearing in candidate transcripts have few occurrences. The k-mers with high profiles which do not appear in candidate transcripts are most likely due to poly(A) tails, transcripts with dense variants in the parental strains, or transcripts expressed by the F1 strain but not the parents, as shown in Fig. (S2)

Table 2. Dimensions and Results from Real Data

	CASTxPWK	CASTxWSB
k-mers in merged trio k-mer profile	118,100,824	118,383,117
k-mers in candidate transcripts	42,688,910	52,715,089
k-mers in estimated expressed transcripts	42,482,315	52,162,586
candidate transcripts	23,585	29,155
estimated expressed transcripts	17,118	20,596
candidate genes	7,393	8,532
estimated expressed genes	7,148	8,242
expressed genes with isoforms from both parents	4,065	5,183

Using the penalty parameter $\lambda = 10^4$ for both F1 samples, our methods found 17,118 non-zero θ values in the CASTxPWK sample and 20,596 non-zero θ values in the CASTxWSB sample, corresponding to as many estimated expressed transcripts. This represented 7,148 of 7,393 and 8,242 of 8,532 estimated expressed genes, respectively. These results are summarized in Table (2). We estimated the expression level of each gene by summing the θ values for all expressed isoforms, both maternal and paternal, of each gene.

To assess our ability to estimate ASE, we looked at the maternal contribution ratio of all expressed genes with candidate isoforms from both parents and differentiating variants between the two parents. Maternal contribution ratio of a gene is defined as the ratio of the expression levels from all maternal isoforms to the expression levels from both paternal and maternal isoforms of the gene. The distribution of maternal contribution ratios for both F1 samples is shown in Fig. (3). The median maternal contribution ratio for both the male CASTxPWK sample and the female CASTxWSB sample is around 0.5, as expected. In the male CASTxPWK sample, a higher number of genes are maternally expressed, which is expected since genes on the X chromosome and mitochondria should be maternally expressed in males. We verified several genes that are known to exhibit ASE [8] [29] as having high maternal contribution ratios if maternally expressed and low maternal contribution ratios if paternally expressed.

In addition, we examined the maternal contribution ratios of all expressed genes on the X chromosome with candidate isoforms from both parents and differentiating variants between the parents. In the male CASTxPWK sample,

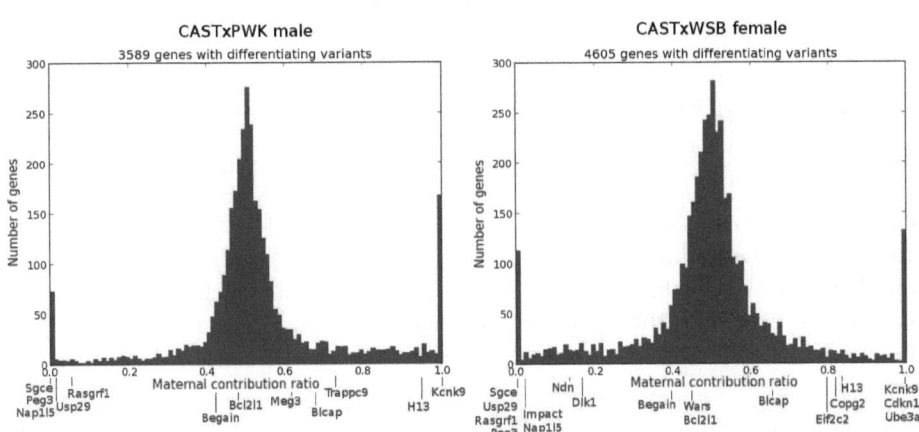

Fig. 3. Histogram of the maternal contribution ratios of all expressed genes with candidate isoforms from both parental strains and containing differentiating variants between the parental strains. On the bottom of each plot, several genes known to be maternally expressed in literature are highlighted in red, and several genes known to be paternally expressed are highlighted in blue.

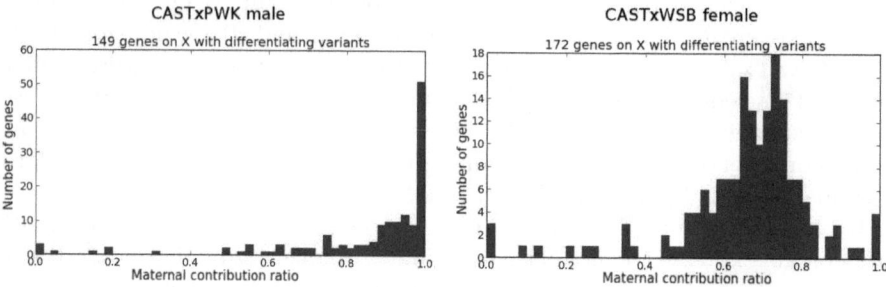

Fig. 4. Histogram of the maternal contribution ratio of all expressed genes on the X chromosome with candidate isoforms from both parental strains and containing differentiating variants between the parental strains. In the male CASTxPWK sample, the median maternal contribution ratio is 0.94. In the female CASTxWSB sample, the median maternal contribution ratio is 0.68. Both are in the expected range of maternal contribution ratio of X-chromosome genes in male and female animals, respectively.

we expect all genes on the X chromosome to be maternally expressed, since its X chromosome is inherited from the maternal strain. In the female CASTxWSB sample, we expect most genes on X to be expressed with a 0.6-0.7 maternal contribution ratio due to a known maternal bias in X inactivation [28] [2]. As expected, we found the median maternal contribution ratio to be 0.94 in the male CASTxPWK sample and 0.68 in the female CASTxWSB sample. The distributions of maternal contribution ratios of genes on the X chromosome are plotted in Fig. (4).

4.3 Speed and Memory

We ran our methods on a single 1600 MHz processor on a machine with 32 GB RAM. The *De Bruijn* graphs of our samples take up around 1GB of disk space. The selection of candidate transcripts takes approximately 2-3 hours per parental strain, and the coordinate descent algorithm converges after approximately 1 to 3 million iterations, which takes around 1-2 hours on our machine. We were able to take advantage of the sparseness of our candidate transcript k-mer profile matrix $\mathbf{X}$ by storing them as sparse matrices using the Scipy.sparse package.

5   Discussion

We have developed methods to estimate expression levels for maternal and paternal versions of transcripts from RNA-seq trio data. Our need for such methods arose when we realized that although we have RNA-seq data of many biological trios and wish to analyze ASE of F1 strains, current methods, both alignment-based and *de novo*, do not include standard pipelines that take advantage of available RNA-seq data from parental strains. Our model is able to exploit the information from the maternal and paternal RNA-seq reads and build candidate transcripts that accurately reflect the F1 strain's transcriptome, and it does so without requiring a database of variants of the parental strains. Our proposed methods still rely on the existence of an annotated reference transcriptome, which is essential for making biologically meaningful observations.

Our methods performed well when compared to a Trinity-Bowtie-RSEM pipeline, which incorporates a state-of-the-art *de novo* assembler and aligner. We were able to achieve high sensitivity and specificity (0.9553 and 0.9883) in detecting baseline expression of transcripts. Of the correctly identified expressed transcripts, we were also able to correctly identify more transcripts exhibiting ASE, with a sensitivity of 0.77, compared to Trinity's low ASE sensitivity of 0.32. The pipeline we used with Trinity also made use of parental RNA-seq data, since we separately assembled transcript sequences from maternal and paternal reads, then aligned the F1 reads to the entire set of assembled transcript sequences. However, Trinity still had a low sensitivity of 0.13 for determining baseline expression, since the main challenge we faced using Trinity was mapping the assembled sequences back to known reference transcripts.

The dimensionality of our data can be large. In our real data, we have approximately 5×10^7 k-mers after filtering and tens of thousands of candidate transcripts. Despite the high dimensionality of our k-mer space and transcripts space, we were able to use a refined coordinate descent algorithm to efficiently perform lasso regression. Although not implemented, we could also decrease our k-mer space without affecting the solution by merging overlapping k-mers into contigs.

Since our candidate transcripts are generated from annotated reference transcripts, our methods do not currently assemble novel transcript sequences. However, it is possible to model the k-mer profiles of all novel transcripts as the residual of our linear regression, and *de novo* assembly of the residual k-mers could then generate sequences of novel transcripts. Another limitation of our model lies in its inability to detect genes exhibiting overdominance, where the expression level is high in the F1 animal but nonexistent in the parental strains. This could be remedied by also selecting candidate transcripts from the F1 *De Bruijn* graph itself to add to our feature space.

The strength of our methods lies in the ability to determine ASE directly from RNA-seq data in diploid trios without prior knowledge of genomic variation in the parental genomes. This straightforward regression approach is tolerant of imbalanced read counts in different samples, as demonstrated by our reasonable maternal contribution ratio distribution in the male CASTxPWK F1 sample (Fig 3), despite the CAST/EiJ read count being nearly 1.5 times as high as the PWK/EiJ read count. Our methods could even be extended to ascertain ASE in any animal that is a hybrid of two or more isogenic ancestral genomes, such as the recombinant inbred strains often used as genetic reference panels.

Acknowledgements. The authors thank Fernando Pardo-Manuel de Villena and the UNC Center for Excellence in Genome Sciences (NIH P50 MH090338) for providing the sequencing data used in this study.

References

1. Burrows, M., Wheeler, D.J.: A block-sorting lossless data compression algorithm, Citeseer (1994)
2. Chadwick, L.H., Pertz, L.M., Broman, K.W., Bartolomei, M.S., Willard, H.F.: Genetic control of x chromosome inactivation in mice: definition of the xce candidate interval. Genetics 173(4), 2103–2110 (2006)
3. Chinwalla, A.T., Cook, L.L., Delehaunty, K.D., Fewell, G.A., Fulton, L.A., Fulton, R.S., Graves, T.A., Hillier, L.D.W., Mardis, E.R., McPherson, J.D., et al.: Initial sequencing and comparative analysis of the mouse genome. Nature 420(6915), 520–562 (2002)
4. Church, D.M., Goodstadt, L., Hillier, L.W., Zody, M.C., Goldstein, S., She, X., Bult, C.J., Agarwala, R., Cherry, J.L., DiCuccio, M., et al.: Lineage-specific biology revealed by a finished genome assembly of the mouse. PLoS Biology 7(5), e1000112 (2009)
5. de Bruijn, N.G., Erdos, P.: A combinatorial problem. Koninklijke Netherlands: Academe Van Wetenschappen 49, 758–764 (1946)
6. Efron, B., Hastie, T., Johnstone, I., Tibshirani, R.: Least angle regression. The Annals of Statistics 32(2), 407–499 (2004)
7. Grabherr, M.G., Haas, B.J., Yassour, M., Levin, J.Z., Thompson, D.A., Amit, I., Adiconis, X., Fan, L., Raychowdhury, R., Zeng, Q., et al.: Full-length transcriptome assembly from rna-seq data without a reference genome. Nature Biotechnology 29(7), 644–652 (2011)
8. Gregg, C., Zhang, J., Weissbourd, B., Luo, S., Schroth, G.P., Haig, D., Dulac, C.: High-resolution analysis of parent-of-origin allelic expression in the mouse brain. Science 329(5992), 643–648 (2010)
9. Griebel, T., Zacher, B., Ribeca, P., Raineri, E., Lacroix, V., Guigó, R., Sammeth, M.: Modelling and simulating generic rna-seq experiments with the flux simulator. Nucleic Acids Research 40(20), 10073–10083 (2012)
10. Guttman, M., Garber, M., Levin, J.Z., Donaghey, J., Robinson, J., Adiconis, X., Fan, L., Koziol, M.J., Gnirke, A., Nusbaum, C., et al.: Ab initio reconstruction of cell type-specific transcriptomes in mouse reveals the conserved multi-exonic structure of lincrnas. Nature Biotechnology 28(5), 503–510 (2010)
11. Hastie, T., Tibshirani, R., Friedman, J., Franklin, J.: The elements of statistical learning: data mining, inference and prediction. The Mathematical Intelligencer 27(2), 83–85 (2005)
12. Huang, S., Kao, C.-Y., McMillan, L., Wang, W.: Transforming genomes using mod files with applications. In: Proceedings of the ACM Conference on Bioinformatics, Computational Biology and Biomedicine. ACM (2013)
13. Keane, T.M., Goodstadt, L., Danecek, P., White, M.A., Wong, K., Yalcin, B., Heger, A., Agam, A., Slater, G., Goodson, M., et al.: Mouse genomic variation and its effect on phenotypes and gene regulation. Nature 477(7364), 289–294 (2011)
14. Langmead, B., Trapnell, C., Pop, M., Salzberg, S.L., et al.: Ultrafast and memory-efficient alignment of short dna sequences to the human genome. Genome Biol. 10(3), R25 (2009)

15. Levenshtein, V.I.: Binary codes capable of correcting deletions, insertions. Technical report, and reversals. Technical Report 8 (1966)
16. Li, B., Dewey, C.N.: Rsem: accurate transcript quantification from rna-seq data with or without a reference genome. BMC Bioinformatics 12(1), 323 (2011)
17. Li, W., Feng, J., Jiang, T.: Isolasso: a lasso regression approach to rna-seq based transcriptome assembly. Journal of Computational Biology 18(11), 1693–1707 (2011)
18. Li, Y., Osher, S.: Coordinate descent optimization for l-1 minimization with application to compressed sensing; a greedy algorithm. Inverse Probl. Imaging 3(3), 487–503 (2009)
19. Liu, R., Maia, A.-T., Russell, R., Caldas, C., Ponder, B.A., Ritchie, M.E.: Allele-specific expression analysis methods for high-density snp microarray data. Bioinformatics 28(8), 1102–1108 (2012)
20. Nesterov, Y.: Efficiency of coordinate descent methods on huge-scale optimization problems. SIAM Journal on Optimization 22(2), 341–362 (2012)
21. Robertson, G., Schein, J., Chiu, R., Corbett, R., Field, M., Jackman, S.D., Mungall, K., Lee, S., Okada, H.M., Qian, J.Q., et al.: De novo assembly and analysis of rna-seq data. Nature Methods 7(11), 909–912 (2010)
22. Ronald, J., Akey, J.M., Whittle, J., Smith, E.N., Yvert, G., Kruglyak, L.: Simultaneous genotyping, gene-expression measurement, and detection of allele-specific expression with oligonucleotide arrays. Genome Research 15(2), 284–291 (2005)
23. Rozowsky, J., Abyzov, A., Wang, J., Alves, P., Raha, D., Harmanci, A., Leng, J., Bjornson, R., Kong, Y., Kitabayashi, N., et al.: Alleleseq: analysis of allele-specific expression and binding in a network framework. Molecular Systems Biology 7(1) (2011)
24. Skelly, D.A., Johansson, M., Madeoy, J., Wakefield, J., Akey, J.M.: A powerful and flexible statistical framework for testing hypotheses of allele-specific gene expression from rna-seq data. Genome Research 21(10), 1728–1737 (2011)
25. Trapnell, C., Pachter, L., Salzberg, S.L.: Tophat: discovering splice junctions with rna-seq. Bioinformatics 25(9), 1105–1111 (2009)
26. Trapnell, C., Roberts, A., Goff, L., Pertea, G., Kim, D., Kelley, D.R., Pimentel, H., Salzberg, S.L., Rinn, J.L., Pachter, L.: Differential gene and transcript expression analysis of rna-seq experiments with tophat and cufflinks. Nature Protocols 7(3), 562–578 (2012)
27. Trapnell, C., Williams, B.A., Pertea, G., Mortazavi, A., Kwan, G., Van Baren, M.J., Salzberg, S.L., Wold, B.J., Pachter, L.: Transcript assembly and quantification by rna-seq reveals unannotated transcripts and isoform switching during cell differentiation. Nature Biotechnology 28(5), 511–515 (2010)
28. Wang, X., Soloway, P.D., Clark, A.G., et al.: Paternally biased x inactivation in mouse neonatal brain. Genome Biol. 11(7), R79 (2010)
29. Wang, X., Sun, Q., McGrath, S.D., Mardis, E.R., Soloway, P.D., Clark, A.G.: Transcriptome-wide identification of novel imprinted genes in neonatal mouse brain. PloS One 3(12), e3839 (2008)
30. Yang, H., Wang, J.R., Didion, J.P., Buus, R.J., Bell, T.A., Welsh, C.E., Bonhomme, F., Yu, A.H.T., Nachman, M.W., Pialek, J., et al.: Subspecific origin and haplotype diversity in the laboratory mouse. Nature Genetics 43(7), 648–655 (2011)

A   Supplemental Figures

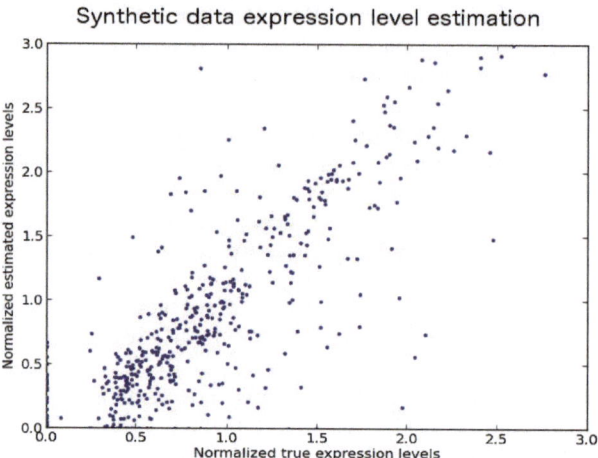

Fig. S1. Predicted versus actual expression levels from synthetic data. Expression levels were normalized to have a mean value of 1.The Pearson correlation coefficient is 0.85 among the 1055 correctly identified expressed transcript sequences.

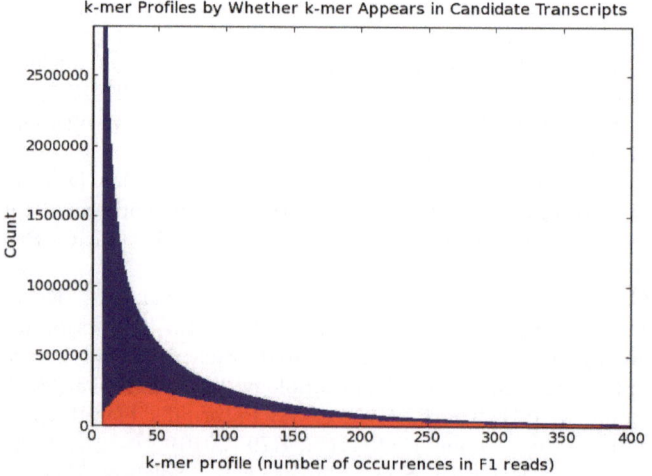

Fig. S2. Stacked histogram of k-mers in the real CASTxPWK k-mer profile, sorted by the number of times each k-mer appears in the F1 reads. K-mers appearing in candidate transcripts are in red, and k-mers not appearing in candidate transcripts are in blue. The majority of k-mers not appearing in candidate transcripts have low number of occurrences, suggesting they are from lowly expressing genes or erroneous reads.

The Generating Function Approach for Peptide Identification in Spectral Networks

Adrian Guthals[1], Christina Boucher[2], and Nuno Bandeira[1,3,*]

[1] Department of Computer Science and Engineering, University of California San Diego,
9500 Gillman Drive, La Jolla, California 92093
[2] Department of Computer Science, Colorado State University, 1873 Campus Delivery,
Fort Collins, CO 80523
[3] Skaggs School of Pharmacy and Pharmaceutical Sciences,
University of California San Diego, 9500 Gillman Drive, La Jolla, California 92093
nbandeira@ucsd.edu

Abstract. Tandem mass (MS/MS) spectrometry has become the method of choice for protein identification and has launched a quest for the identification of every translated protein and peptide. However, computational developments have lagged behind the pace of modern data acquisition protocols and have become a major bottleneck in proteomics analysis of complex samples. As it stands today, attempts to identify MS/MS spectra against large databases (e.g., the human microbiome or 6-frame translation of the human genome) face a search space that is 10-100 times larger than the human proteome where it becomes increasingly challenging to separate between true and false peptide matches. As a result, the sensitivity of current state of the art database search methods drops by nearly 38% to such low identification rates that almost 90% of all MS/MS spectra are left as unidentified. We address this problem by extending the generating function approach to rigorously compute the joint spectral probability of multiple spectra being matched to peptides with overlapping sequences, thus enabling the confident assignment of higher significance to overlapping peptide-spectrum matches (PSMs). We find that these joint spectral probabilities can be several orders of magnitude more significant than individual PSMs, even in the ideal case when perfect separation between signal and noise peaks could be achieved per individual MS/MS spectrum. After benchmarking this approach on a typical lysate MS/MS dataset, we show that the proposed *intersecting spectral probabilities* for spectra from overlapping peptides improve peptide identification by 30-62%.

1   Introduction

The leading method for protein identification by tandem mass spectrometry (MS/MS) involves digesting proteins into peptides, generating an MS/MS spectrum per peptide, and obtaining peptide identifications by individually matching each MS/MS spectrum to putative peptide sequences from a target database. Many computational approaches have been developed for this purpose, such as SEQUEST [1], Mascot [2], Spectrum

---

* Corresponding author.

R. Sharan (Ed.): RECOMB 2014, LNBI 8394, pp. 85–99, 2014.
© Springer International Publishing Switzerland 2014

Mill [3], and more recently MS-GFDB [4], yet they all address the same two problems: Given a MS/MS spectrum S and a collection of possible peptide sequences, i) find the peptide P that most likely produced spectrum S and ii) report the statistical significance of the Peptide-Spectrum Match (P, S) (denoted PSM) while searching many MS/MS spectra against multiple putative peptide sequences from a target database. Problem (i) is typically addressed by maximizing a scoring function proportional to the likelihood that peptide P generated spectrum S while solving problem (ii) involves choosing a score threshold that yields an experiment-wide 1% False-Discovery Rate (FDR [5]), usually based on an estimated distribution of PSM scores for incorrect PSMs [6]. Yet a major limitation comes from ambiguous interpretations of MS/MS fragmentation where the true peptide match for a given spectrum S may only be the 2nd or $100,000^{th}$ highest scoring over all possible PSMs for the same spectrum [7]. We address this issue as it relates to problem (ii) where the probability of false peptides matching S with high score can become common when searching large databases, particularly for meta-proteomics [8] and 6-frame translation [9] searches, thus leading to higher-scoring false matches and stricter significance thresholds resulting in as little as 2% of all spectra being identified [10] since only the highest scoring PSMs become statistically significant even at 5% FDR.

Identifying peptides from a large database is less of a challenge than that of *de novo* sequencing, where the target database contains all possible peptide sequences. Yet, recent advances in *de novo* sequencing have demonstrated 97-99% sequencing accuracy (percent of amino acids in matched peptides that are correct) at nearly the same level of coverage (percent of amino acids in target peptides that were matched) as that of database search for small mixtures of target proteins [11, 12]. At the heart of this approach is the pairing of spectra from *overlapping* peptides (i.e. peptides that have overlapping sequences) to construct *spectral networks* [13, 14] of paired spectra. It is then shown that *de novo* sequences assembled by simultaneous interpretation of multiple spectra from overlapping peptides are much more accurate than individual per-spectrum interpretations [13, 15]. Use of multiple enzyme digestions and SCX [16] fractionation is becoming more common in MS/MS protocols to generate broader coverage of protein sequences and yield wider distributions of overlapping peptides, but current statistical methods still ignore the peptide sequence overlaps and separately compute the significance of individual peptides matched to individual spectra [17].

Given that the set of all possible protein sequences is orders of magnitude larger than the human 6-frame translation (or any other database), application of these *de novo* techniques to database search should substantially improve peptide identification rates, especially for large databases. Since the original generating function approach showed how de novo algorithms can be used to estimate the significance of PSMs for individual spectra, it is expected that advances in de novo sequencing should consequently translate into better estimation of PSM significance. It has already been shown that spectral networks can be used to improve the ranking of database peptides against paired spectra [18], but it is still unclear how to accurately evaluate the statistical significance of peptides matched to multiple overlapping spectra. Intuitively, if it is known that these overlapping spectra yield more accurate de novo sequencing then the probability of observing multiple incorrect high-scoring PSMs with overlapping sequences should be lower than the probability of single incorrect peptides matching single spectra with high scores. To this end we introduce *StarGF*, a novel approach for peptide identification that accurately models the distribution of all peptide sequences against pairs of spectra from overlapping peptides. We demonstrate its performance on a typical lysate mass spectrometry dataset and show that it can improve peptide-level identification by up to 62% compared to a state-of-the-art database search tool.

2   Methods

2.1   Spectral Probabilities and Notation

We describe a method to assess the significance of overlapping peptide-spectrum matches (PSMs) based on the generating function approach for computing the significance of individual PSMs [7]. Although traditional methods for scoring PSMs incorporate prior knowledge of N/C-terminal ions, peak intensities, charges, and mass inaccuracies, these terms are avoided here for simplicity of presentation, and later we describe how these features were considered for real spectra.

Let a peptide P of length n be a string of amino acids $a_1 \dots a_n$ with parent mass $|P| = \sum_i |a_i|$ and each a_i is one of the standard amino acids $a_i \in A$. For clarity of presentation we define acid masses $|a_i|$ to be integer-valued and that each MS/MS spectrum is an integer vector $S = s_1 \dots s_{|S|}$ where $s_i > 0$ if there is a peak at mass i (having intensity s_i), and $s_i = 0$ otherwise (denote $|S|$ as the parent mass of S). Let $Spectrum(P)$ be a spectrum with parent mass $|P|$ such that $s_i = 1$ if i is the mass of a prefix of P. We define the *match score* between spectra $S = s_1 \dots s_{|S|}$ and $S' = s'_1 \dots s'_{|S|}$ as $\sum_{i=1}^{|S|} s_i * s'_i$. Thus, the match score $Score(P,S)$ between a peptide P and a spectrum S is equivalent to the match score between $Spectrum(P)$ and S if both spectra have the same parent mass (otherwise $Score(P,S) = -\infty$). The problem faced by peptide identification algorithms is to find a peptide P from a database of known protein sequences that maximizes $Score(P,S)$, then assess the statistical significance of each top-scoring PSM.

Given a PSM (P,S) with score $Score(P,S) = T$, the *spectral probability* introduced by MSGF [7] computes the significance of the match as the aggregate probability that a random peptide P^* achieves a $Score(P^*,S) \geq T$, otherwise termed as $Prob_T(S)$. The probability of a peptide $P = a_1 \dots a_n$ is defined as the product of probabilities of its amino acids $\prod_{i=1}^{n} prob(a_i)$ where each amino acid $a \in A$ has a fixed probability of occurrence of $1/|A|$ (or could be set to the observed frequencies in a target database). In MSGF, computing $Prob_T(S)$ is done in polynomial time by filling in the dynamic programming matrix $SP(i,t)$, which denotes the aggregate probability that a random peptide P^* with mass $|P^*| = i$ achieves $Score(P^*, S_{1 \to i}) = t$, where $S_{1 \to i} = s_1 \dots s_i$. The SP matrix is initialized to $SP(0,0) = 1$, zero elsewhere, and updated using the following recursion [7].

$$SP(i,t) = \sum_{a \in A: i \geq |a|, t \geq s_i} SP(i - |a|, t - s_i) * prob(a) \qquad (1)$$

$Prob_T(S)$ is calculated from the SP matrix as follows:

$$Prob_T(S) = \sum_{t \geq T} SP(|S|, t) \tag{2}$$

2.2   Pairing of Spectra

A pair of *overlapping* PSMs is defined as a pair (P, S) and (P', S') such that i) both spectra are matched to the same peptide $(P = P')$ *or* ii) the spectra are matched to peptides with *partially-overlapping* sequences: either P' is a substring of P or a prefix of P' matches a suffix of P. As mentioned above, spectral pairs can be detected using spectral alignment without explicitly knowing which peptide sequences produced each spectrum (as described previously [15], [19]). Intersecting spectral probabilities (described below) are calculated for all pairs of spectra with overlapping PSMs. In addition, we use all neighbors of each paired spectrum to calculate the *star probability* for the center nodes in each sub-component defined by S and all of its immediate neighbors.

2.3   Star Probabilities

In the simplest case of a pair of overlapping PSMs (P, S) and (P', S') where $P = P'$, we want to find the aggregate probability that a random peptide matches S with score $\geq T$ and matches S' with score $\geq T'$ (denoted the *intersecting spectral probability* $Prob_{T,T'}(S, S')$). A naïve solution is to simply take the product of $Prob_T(S)$ and $Prob_{T'}(S')$, but this approach fails to capture the dependence between $Prob_{T,T'}(S, S')$ induced by the similarity between S and S'. Intuitively, a high similarity between S and S' should correlate with a high probability that both spectra get matched to the same peptide, regardless of whether it is a correct match.

$Prob_{T,T'}(S, S')$ can be computed efficiently by adding an extra dimension to the dynamic programming recursion SP, yielding a 3-dimensional matrix $ISP_{same}(i, t, t')$ that tracks the aggregate probability that a random peptide P with mass i matches $S_{1 \to i}$ with score t and matches $S'_{1 \to i}$ with score t'. The ISP_{same} matrix is initialized to $ISP_{same}(0,0,0) = 1$, zero elsewhere, and computed as follows.

$$ISP_{same}(i, t, t') = \sum_{a \in A:\ i \geq |a|, t \geq s_i, t' \geq s'_i} ISP_{same}(i - |a|, t - s_i, t' - s'_i) * prob(a) \tag{3}$$

$Prob_{T,T'}(S, S')$ is calculated from the ISP_{same} matrix as follows:

$$Prob_{T,T'}(S, S') = \sum_{t \geq T} \sum_{t' \geq T'} ISP_{same}(|S|, t, t') \tag{4}$$

To generalize intersecting spectral probabilities to include pairs of spectra from partially overlapping peptides, we define $ISP(i, t, t')$ to address the case where S' is shifted in relation to S (see Figure 1) by a given mass shift λ, which may be positive or negative. The shift λ defines an *overlapping mass range* between the spectra; in spectrum S the range starts at mass $b = max(0, \lambda)$ and ends at mass $e = min(|S|, |S'| + \lambda)$ while in spectrum S' the range starts at mass $b' = max(0, -\lambda)$ and ends at mass $e' = min(|S'|, |S| - \lambda)$. Since partially-overlapping spectra may originate from different peptides ($\lambda \neq 0$ or $|S| \neq |S'|$), the probabilities of peptides matching S must be processed differently from those matching S'. If one considers a peptide P matching S, only the portion of P from b to e (denoted as P_{ovlp}) can be matched against $S'_{b' \to e'} = s'_{b'} \ldots s'_{e'}$. For example, in Figure 1, P_{ovlp} is equal to the peptide "PTIDE". First, $ISP(i, t, t')$ is defined to hold the aggregate probability that a random peptide P with mass i achieves $Score(P, S_{1 \to i}) = t$ such that $Score\left(P_{ovlp}, S'_{b' \to min(e', i - \lambda)}\right) = t'$. In cases where i is less than b (i.e. when $\lambda > 0$), P_{ovlp} is empty and is defined to have zero score against S'.

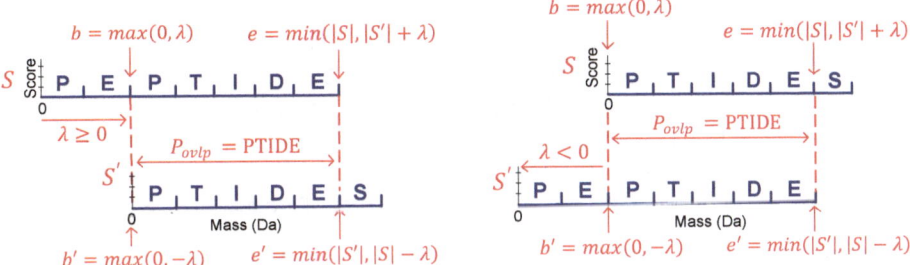

Fig. 1. Illustration of P_{ovlp} and the overlapping mass range between overlapping spectra S and S'

The base case for $ISP(i, t, t')$ is the same as the base case for ISP_{same}, but the recursion must be separated into three separate cases depending on whether $i \le b$, $b < i \le e$, or $i > e$. If $i \le b$, then $ISP(i, t, t')$ is tracking peptides matching $S_{1 \to i}$ with score t, but score 0 against S'.

If $i \le b$ ($t' = 0$):

$$ISP(i, t, 0) = \sum_{a \in A:\ i \ge |a|, t \ge s_i} ISP(i - |a|, t - s_i, 0) * prob(a) \qquad (5)$$

When i is inside the overlapping mass range of S, the matrix tracks peptides matching $S_{1 \to i}$ with score t that contain a suffix matching $S'_{b' \to i - \lambda}$ with score t'.

If $b < i \le e$:

$$ISP(i, t, t') = \sum_{a \in A:\ i \ge |a|, t \ge s_i, t' \ge s'_{i - \lambda}, i - |a| \ge b} ISP(i - |a|, t - s_i, t' - s'_{i - \lambda}) * prob(a) \ (6)$$

When $e < i \le |S|$ and, thus, i is outside the overlapping mass range, $ISP(i, t, t')$ is extending peptides P matching $S_{1 \to i}$ with score t where P_{ovlp} has score t' against $S'_{b' \to e'}$.

If $i > e$:

$$ISP(i, t, t') = \sum_{a \in A:\ i \ge |a|, t \ge s_i, i - |a| \ge e} ISP(i - |a|, t - s_i, t') * prob(a) \qquad (7)$$

If P matches S with score $\ge T$ and P_{ovlp} matches $S'_{b' \to e'}$ with score $\ge T'$, the probability of both events is computed as given below.

$$Prob_{T, T'}\left(S, S'_{b' \to e'}\right) = \sum_{t \ge T} \sum_{t' \ge T'} ISP(|S|, t, t') \qquad (8)$$

Note that since λ may be positive or negative, the intersecting probability of a peptide P matching S' with score $\geq T'$ and P_{ovlp} matching $S_{b\to e}$ with score $\geq T$ is computed by simply setting $\lambda = -\lambda$ before calculating $Prob_{T',T}(S', S_{b\to e})$.

The term *star* is defined as the set of all spectra directly connected with spectrum S in the spectral network [18]. We are interested in the minimum $Prob_{T,T'}(S, S'_{b'\to e'})$ over all S' in the star of S, otherwise termed as the *star probability* of S. Computation of the star probability is more precisely defined in pseudo code below.

```
StarProbability(P,S):
    T := Score(P,S)
    starP := Prob_T(S)
    for all (S,S') in the star of S:
        λ := mass shift of S' in relation to S
        T' := Score(P_ovlp,S'_b'→e')
        if Prob_T,T'(S,S'_b'→e') > 0:
            starP := min(starP, Prob_T,T'(S,S'_b'→e'))

    return starP
```

2.4 Processing Real Spectra

Each MS/MS spectrum was transformed into a PRM spectrum [20] with integer-valued masses and likelihood intensities $s_1 \ldots s_{|S|}$ using the PepNovo$^+$ probabilistic scoring model [21]. PepNovo$^+$ interprets MS/MS fragmentation patterns and converts MS/MS spectra into PRM (prefix residue mass) spectra where peak intensities are replaced with log-likelihood scores and peak masses are replaced by PRMs, or Prefix-Residue Masses (cumulative amino acid masses of putative N-term prefixes of the peptide sequence). PRM scores combine evidence supporting peptide breaks: observed cleavages along the peptide backbone supported by either N- or C-terminal fragments. To minimize rounding errors, floating point peak masses returned by Pep-Novo$^+$ were converted to integer values as in MS-GF [7], where cumulative peak mass rounding errors were reduced by multiplying by 0.9995 before rounding to integers (amino acid masses were also rounded to integer values). High-resolution peak masses could also be supported by using a larger multiplicative constant (e.g., 100.0) prior to rounding. Peak intensities were first normalized so each spectrum contained a maximum total score of $\sigma = 150$, then they were rounded to integers (peaks with score less than 0.5 were effectively removed). With these parameters the time complexity of computing individual and intersecting spectral probabilities is approximately $O(|S|\sigma|A|)$ and $O(|S|\sigma^2|A|)$, respectively.

2.5 Generating Candidate PSMs

A published set of ion-trap CID spectra acquired from the model organism Saccharomyces cerevisiae was used to benchmark this approach [17]. To aid in the acquisition of spectra from overlapping peptides, 12 SCX fractions were obtained for each of five enzyme digests. Three technical replicates were also run for each digest, but only spectra from the second replicate were used here. Thermo RAW files were converted to mzXML using ProteoWizard [22] (version 3.0.3224) with peak-picking enabled and clustered using MSCluster [23] (version 2.0, release 20101018) to merge repeated spectra, yielding 255,561 clusters of one or more spectra.

MS-GFDB [4] (version 7747) was used to match spectra against candidate peptides from target and decoy protein databases. Two sets of target+decoy databases (labeled *small* and *large*) were used to evaluate the performance of individual vs. StarGF spectral probabilities when searching databases of different size. The small target database consisted of all reference Saccharomyces cerevisiae protein sequences downloaded from UniProt [24] (~4 MB on 09/27/2013) while the large database contained all reference fungi UniProt protein sequences (~130 MB on 09/27/2013). The large database (~32 times larger) was used to represent searches against large search spaces, such as meta-proteomics [8] or 6-frame translation [9] searches. Separate small and large decoy databases were generated by randomly shuffling protein sequences from the target database [6].

The 255,561 cluster-consensus spectra were separately searched against the small target, small decoy, large target, and large decoy databases with MS-GFDB [4] configured to report the top 10 PSMs for each spectrum. The "no enzyme" model was selected along with 30ppm parent mass tolerance, "Low-res LCQ/LTQ" instrument ID, one ^{13}C, two allowed non-enzymatic termini, and amino acid probabilities set to 0.05 (the same amino acid probabilities used by StarGF). Target and decoy PSMs were then merged by an in-house program that discarded decoy PSMs whose peptides were also found in the target database (allowing for I/L, Q/K, and M+16/F ambiguities). Although variable post-translational modifications (PTMs) were permitted in each initial search to reproduce typical search parameters (oxidized methionine and deamidated asparagine/glutamine), spectra assigned to modified PSMs were removed from consideration at this stage (the incorporation of PTMs into intersecting spectral probabilities is not considered here). The top-scoring peptide match for each remaining spectrum was then set to the target or decoy PSM with the highest matching score to the PRM spectrum. Each set of unfiltered target+decoy PSMs was evaluated at 1% FDR [5] using star probabilities.

To benchmark StarGF, each set of MS-GFDB results was separately evaluated at 1% FDR using MS-GFDB's spectral probability [7] while allowing MS-GFDB to report the top-scoring PSM per spectrum. X!Tandem [25] Cyclone (2011.12.01.1) was also run on the same set of MS/MS spectra in a separate search against each database and results were filtered at 1% spectrum- and peptide-level FDR using the same target-decoy approach. X!Tandem search parameters consisted of 0.5 Da peak tolerance, 30ppm parent mass tolerance, multiple ^{13}C, and non-specific enzyme cleavage (remaining parameters were set to their default values).

All raw and clustered MS/MS spectra associated with this study have been uploaded to the MassIVE public repository (http://massive.ucsd.edu) and are accessible at ftp://MSV000078538@massive.ucsd.edu with password recomb_ag88 while StarGF can be obtained from http://proteomics.ucsd.edu/Software/StarGF.html.

3   Results

Two sets of pairwise alignments were used to demonstrate the effectiveness of StarGF: *i*) the set of pairs obtained by spectral alignment in the spectral network [18] and *ii*) to simulate the situation when maximal pairwise alignment sensitivity is achieved, pairs were also obtained using sequence-based alignment of the top-scoring peptide matches returned by the MS-GFDB searches. A pair of overlapping PSMs was retained if they shared at least 7 overlapping residues and at least 3 matching theoretical PRM masses from the overlapping sequence. To eliminate the possibility of pairing unique peptides from different proteins, each target PSM pair was also enforced to have at least one target protein containing the full sequence supported by the pair (e.g. the pair (PEPTIDE,PTIDES) must be supported by a protein containing the substring PEPTIDES). Unless otherwise stated, results are reported after applying the sequence-based pairing strategy to 40,926 unmodified target PSMs from the small database (separately identified by MS-GFDB at 1% spectrum-level FDR), yielding 32,777 paired spectra in the network. Using these parameters, less than 1% of pairs contained at least one decoy PSM while 5% of paired PSMs were decoys for the large database set. The significance of each PSM (P, S) was reported as the star probability of S. To evaluate the utility of intersecting probabilities, we separately assessed intersecting spectral probabilities for same-peptide pairs and partially-overlapping pairs: we computed a *same-peptide star probability* (equal to the minimum $Prob_{T,T'}(S, S'_{b'\to e'})$ such that $P = P'$) and a *partially-overlapping star probability* (equal to the minimum $Prob_{T,T'}(S, S'_{b'\to e'})$ such that $P \neq P'$) for each spectrum in the network.

Figure 2 illustrates the substantial separation between individual spectral probabilities, same-peptide star probabilities, and partially-overlapping star probabilities (top panel). Same-peptide star probabilities can be further separated into those where the minimum intersecting probability was selected for a pair of PSMs with equal precursor charge (higher correlation between MS/MS fragmentation patterns [26]), and those where the minimum was selected for a pair with different precursor charge states (less-correlated MS/MS fragmentation). Due to repeated instrument acquisition of multiple spectra from the same peptide and charge state, it was expected that individual spectral probabilities would be approximately the same as intersecting probabilities for most same-peptide/same-charge pairs since duplicate spectra often have high similarity [26]. Nevertheless, star probabilities for same-peptide/same-charge pairs still prove valuable in improving spectral probabilities by an average of ~2 orders of magnitude (Figure 2, bottom left), while same-peptide/different-charge and partially-overlapping pairs enable an even greater improvement in spectral probabilities by an average of ~8 orders of magnitude.

The distributions of decoy spectral probabilities in the bottom right panel of Figure 2 illustrate the effect of star probabilities on paired decoy PSMs. It was rare for decoy PSMs to pair with others in the network (only 919 of 37,522 decoy PSMs were detected in a spectral pair) and those that did had their spectral probabilities improve by an average of ~2 orders of magnitude, which is significantly less than observed for correct PSM pairs. Also shown is the distribution of decoy star probabilities as computed by the product of probabilities $(Prob_{T,T'}(S,S') = Prob_T(S) * Prob_{T'}(S'))$. As expected, the product of spectral probabilities ignores the dependencies between the spectra and severely under-estimates the true intersecting spectral probability by several orders of magnitude. This would likely lead to increased sampling of false-positive PSMs at any given star probability cutoff and thus result in an overall reduced number of identifications by requiring strict probability thresholds to achieve the same 1% FDR. This effect can be explained intuitively for a given pair of PSMs (P,S) and (P',S') where $S = S'$ and $P = P'$: if a random peptide matches S with a high score, then with probability 1 the same random peptide also matches S' with an equally high score. Thus, in this special case, $Prob_{T,T'}(S,S')$ should equal $Prob_T(S) = Prob_{T'}(S')$, not the product of the individual spectral probabilities.

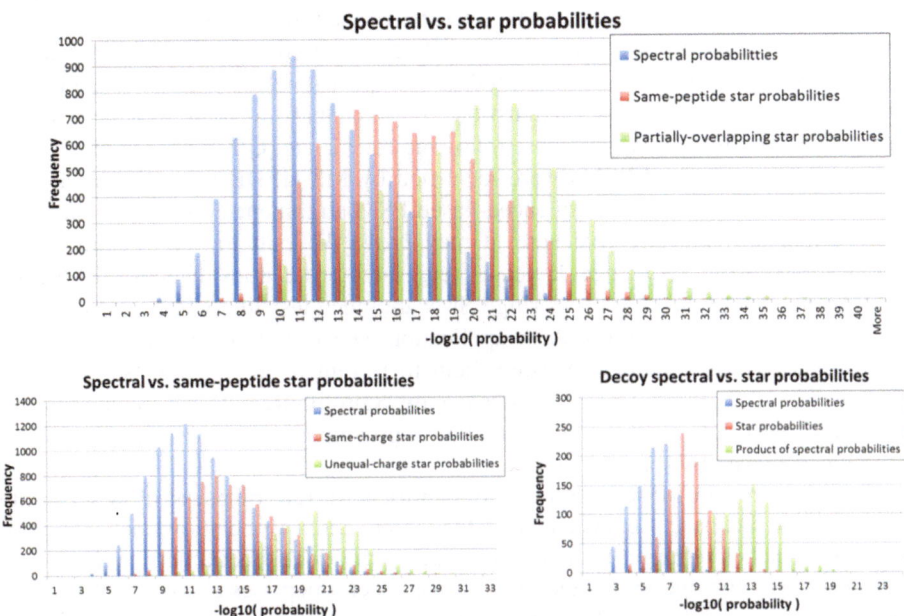

Fig. 2. Spectral and star probability distributions of observed p-values. (**top**) Distribution of the spectral, same-peptide star, and partially-overlapping star probabilities for PSMs with at least one same-peptide pair and at least one partial overlapping pair. (**bottom left**) Distribution of spectral, same-charge star, and unequal-charge star probabilities for PSMs from at least one same-peptide pair. (**bottom right**) Distribution of spectral and star probabilities for all 919 small-database decoy PSMs found in the network where 480 had a same-peptide pair and 450 had a partially-overlapping pair (11 had more than one pair). Also shown is the distribution of the product of individual spectral probabilities for the same decoys (where $Prob_{T,T'}(S,S')$ is computed as $Prob_T(S) * Prob_{T'}(S')$) to illustrate how it would substantially underestimate $Prob_{T,T'}(S,S')$ by ignoring the dependencies between repeated MS/MS spectra acquisitions from the same peptide with the same charge state.

Figure 3 compares every PSM's star probability to its *optimal spectral probability*, which is defined as the spectral probability of the same peptide matched against the subset of peaks from the spectrum that correspond to true PRM masses (i.e., a noise-free version of the spectrum). In general, star probabilities improved the least for spectral probabilities that were already close to optimal. But the vast majority of star probabilities improved past optimal, particularly for stars with same-peptide/unequal-charge and partially-overlapping pairs. Star probabilities can improve past optimal when missing PRMs from one spectrum S are present in the overlapping region of the spectrum S is paired with, thus enforcing that high-scoring peptide matches contain prefix masses that would otherwise be missed. This demonstrates that StarGF probabilities can improve on spectral probabilities by orders of magnitude even if perfect separation between signal and noise peaks could be achieved for any given spectrum.

Star probabilities of unfiltered target+decoy PSMs were evaluated at 1% FDR using both paired and unpaired PSMs (spectral probabilities were computed for unpaired PSMs). Paired PSMs that were identified by StarGF against the large database were verified to have a FDR of 1% (both at the spectrum- and peptide-level) by considering any peptide identified against the fungi database to be a false positive if it was not present in the yeast database (allowing for I/L and Q/K ambiguities). Table 1 shows how many paired PSMs were identified by MS-GFDB [4] and StarGF using either spectral alignments or sequenced-based PSM alignments. Although sequenced-based alignment was effective here, it may prove difficult to pair spectra by top-scoring PSMs from very large databases (e.g. meta-proteomics databases or 6-frame translations) where the highest-scoring PSMs are much less likely to be correct due to the increased search space. For these applications spectral alignment may prove more effective at detecting pairs and using them to re-rank matching PSMs (as done in [18]) before computing PSM significance by StarGF. Results for sequence-based alignments thus indicate the upper bound of improvement when perfect pairwise sensitivity is achieved by spectral alignment.

The 37% drop in MS-GFDB peptide identification rate of paired PSMs from the small to large database is expected since the larger search space allows decoy peptides and false matches to target to randomly match individual spectra with higher scores, thus decreasing the overall number of detected spectra/peptides at a fixed FDR. Using the same set of unfiltered PSMs as MS-GFDB, however, StarGF only lost 20% of paired peptides from the small database as it could identify 36-66% more spectra and 29-62% more peptides by significantly improving the significance of true overlapping PSMs while only marginally increasing the significance of decoy overlapping PSMs (see Table 1). Note that as described here StarGF could not identify any spectra that were matched to decoy peptides, only re-rank them by their star probability. The drop in StarGF identification rate from the small to the large database is explained by this effect; of the 10,648 spectra identified in the small database search but missed in the large database, only 6% were assigned the same peptide from the large database and had their *preferred neighbor* (the paired PSM from which the lowest intersecting probability was selected) matched to the same peptide. The remaining PSMs were either matched to a different peptide (75%) or had their preferred neighbors matched to different peptides (19%). Thus, the majority (94%) of PSMs lost by StarGF from

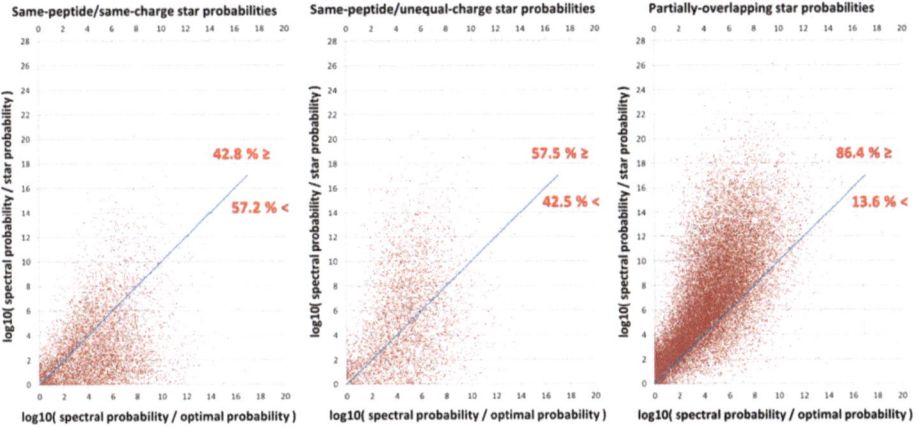

Fig. 3. Reduction of star probability (y-axis) with respect to optimality of starting spectral probability (x-axis). Each red dot denotes either a same-peptide (**left, middle**) or partially-overlapping (**right**) star probability. Values on the x-axis that approach zero indicate a starting spectral probability that approaches optimal while larger values indicate sub-optimal starting spectral probabilities (by orders of magnitude) due to the presence of unexplained PRM masses in the spectrum. Values on the y-axis that approach zero indicate star probabilities that did not improve substantially over the original spectral probabilities while larger values indicate star probabilities that are orders of magnitude smaller than spectral probabilities. The blue line is shown to indicate star probabilities that equal their optimal spectral probability; any data point above the blue line indicates a star probability that is more significant than optimal (see text for a detailed explanation). Red numbers next to the lines indicate the percentage of data points above and below each blue line.

the small to the large database search could potentially be recovered by re-ranking candidate peptides against paired spectra (as done before in spectral networks using de novo sequence tags [18]).

Although the results in Table 1 are over paired PSMs, StarGF still significantly improved spectrum- and peptide-level identification rate for *all* spectra since a large portion (89%) of all PSMs were paired (Table 2). Considering both paired and un-paired (unmodified) PSMs when searching against the small database, MS-GFDB was able to identify 40,926 spectra (34,165 peptides) while StarGF identified 50,310 spectra (35,521 peptides). However, when searching against the large database MS-GFDB could identify only 27,128 spectra (22,782 peptides, 33% loss from the small-database search) while StarGF could identify 40,269 spectra (32,891 peptides, 16% loss from the small-database search) using PSM sequence alignments, an overall im-provement over MS-GFDB of 48% more identified spectra (44% more identified peptides) and revealing StarGF to be nearly as sensitive when searching a 32 times larger database as MS-GFDB is when searching a small database.

Table 1. Spectrum and peptide-level identification rate of paired PSMs at 1% FDR. The "Small Database" column indicates results using the UniProt reference yeast protein database (~4 MB) while results on the right are from searching the larger UniProt reference fungi protein database (~130 MB). Rows separate results by the type of alignment used to capture overlapping PSMs: "Aligned Spectra" indicates pairing by spectral alignment and "Aligned Seqs." indicates pairing by PSM sequence similarity.

		Small Database			Large Database		
		MS-GFDB	StarGF	% Increase	MS-GFDB	StarGF	% Increase
Aligned Spectra	Spectra	13305	18249	**37.2 %**	8799	13743	**56.2 %**
	Peptides	9653	12368	**28.1 %**	6439	9367	**45.5 %**
Aligned Seqs.	Spectra	32777	44621	**36.1 %**	20521	33973	**65.6 %**
	Peptides	26422	34116	**29.1 %**	16525	26689	**61.5 %**

Table 2. Spectrum and peptide-level identification rate of all (paired and unpaired) PSMs at 1% FDR. The "Small Database" column indicates results using the UniProt reference yeast protein database (~4 MB) while results in the "Large Database" column are from searching the larger UniProt reference fungi protein database (~130 MB). (top) Identification rates of all three search tools; numbers in bold indicate the increased percentage of IDs retained by StarGF compared to MS-GFDB. (bottom) Percent of PSMs and peptides lost by each search tool at 1% FDR as they moved from the small to large search space.

	Small Database			Large Database		
	X!Tandem	MS-GFDB	StarGF (% inc.)	X!Tandem	MS-GFDB	StarGF (% inc.)
Spectra	28923	40926	50310 **(22.9 %)**	13847	27128	40269 **(48.4 %)**
Peptides	23957	34165	39077 **(14.4 %)**	11483	22782	32891 **(44.4 %)**

	% lost from larger search space		
	X!Tandem	MS-GFDB	StarGF
Spectra	52.1 %	33.7 %	20.0 %
Peptides	52.1 %	33.3 %	15.8 %

Figure 4 illustrates the overlap between peptides identified by MS-GFDB against the small database and peptides identified by StarGF. The majority (74%) of peptides identified by StarGF against the small database were also identified by MS-GFDB. The remaining peptides that MS-GFDB did not identify were predominantly found in PSM pairs (96%), and thus assigned higher significance by StarGF. Of the peptides identified by StarGF against the large database, nearly all were "rescued" from sets of peptides identified against the small database by either MS-GFDB or StarGF.

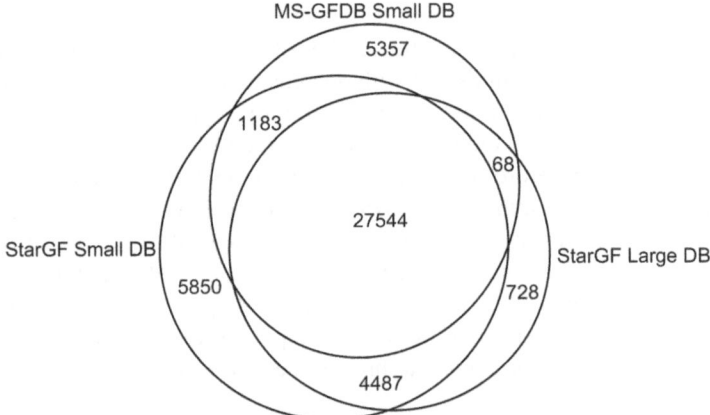

Fig. 4. Overlap of unique peptides identified at 1% peptide-level FDR. The top circle denotes peptides identified by MS-GFDB against the small database while the left and right circles denote peptides identified by StarGF against the small and large databases, respectively. Peptides that only differed by I/L or K/Q ambiguities were counted as the same. Figure is not drawn to exact scale.

4   Discussion

While MS-GF [7] demonstrated how de novo sequencing techniques could be used to greatly improve the state of the art in peptide identification by rigorously computing the score distribution of *all* peptides against every spectrum, it still misses as many as 38% (= ((26689 - 16525)) / 26689) of identifiable (unmodified) peptides when searching large databases by ignoring the significance of overlapping PSMs (see Table 1). By now extending this principle using a multi-spectrum approach to compute the probability distribution of PSM scores for all peptides against every pair of overlapping spectra, StarGF is able to assign higher significance p-values to true PSMs while only marginally increasing the significance of false PSMs. Thus, where traditional database search loses sensitivity in searching larger databases, we now show that it is possible to regain nearly all peptides that are lost by MS-GFDB when searching a database 32 times the size. Although StarGF performs best when paired with MS/MS protocols that maximize acquisition of spectra from partially-overlapping peptides, our results indicate that significant gains in identification rate can still be made by utilizing commonly observed pairs of spectra from the same peptide, particularly pairs of spectra with different precursor charge states.

Although StarGF significantly outperforms a state-of-the-art database search tool (MS-GFDB [4]) in identifying tandem mass spectra at an empirically validated FDR of 1% (confirmed here using matches to non-yeast peptides in the large fungi database), it would be useful to thoroughly assess the limitations of the Target/Decoy Approach when estimating FDR for searches against small databases, as previously done for MS-GFDB searches [27]. In some cases, the enforcement of overlapping PSMs may sometimes result in so few decoy PSMs that it becomes difficult to accurately estimate FDR [28]. A similar situation can also occur in searches with highly accurate parent masses since the number of high-scoring decoy peptides with a given parent mass becomes miniscule with decreasing parent mass tolerance.

While the generating function described here only supports unmodified peptides, it can be extended to analyze modified peptides by considering modified amino acid mass edges (as shown before [4]). Further improvements are foreseeable with additional support for high-resolution MS/MS peak masses and incorporation of alternative fragmentation modes (e.g. HCD, ETD) to improve of the quality of PRM spectra, especially if from highly charged precursors [29]. Given that MS-GFDB supports multiple fragmentation modes and that we utilize PepNovo⁺ to transform MS/MS spectra to PRM spectra, it is possible for this approach to support any fragmentation mode since PepNovo⁺ can be trained to process new types of spectra [12].

Acknowledgement. This work was partially supported by the National Institutes of Health Grant 8 P41 GM103485-05 from the National Institute of General Medical Sciences.

References

1. Eng, J.K., McCormack, A.L., Yates, J.R.: An approach to correlate tandem mass spectral data of peptides with amino acid sequences in a protein database. J. Am. Soc. Mass Spectrom. 5, 976–989 (1994)
2. Perkins, D.N., Pappin, D.J., Creasy, D.M., Cottrell, J.S.: Probability-based protein identification by searching sequence databases using mass spectrometry data. Electrophoresis 20, 3551–3567 (1999)
3. Agilent Technologies, `http://spectrummill.mit.edu/`
4. Kim, S., Mischerikow, N., Bandeira, N., Navarro, J.D., Wich, L., Mohammed, S., Heck, A.J.R., Pevzner, P.A.: The generating function of CID, ETD, and CID/ETD pairs of tandem mass spectra: applications to database search. Mol. Cell. Proteomics. 9, 2840–2852 (2010)
5. Nesvizhskii, A.I.: A survey of computational methods and error rate estimation procedures for peptide and protein identification in shotgun proteomics. J. Proteomics 73, 2092–2123 (2010)
6. Elias, J.E., Gygi, S.P.: Target-decoy search strategy for increased confidence in large-scale protein identifications by mass spectrometry. Nat. Methods 4, 207–214 (2007)
7. Kim, S., Gupta, N., Pevzner, P.A.: Spectral probabilities and generating functions of tandem mass spectra: a strike against decoy databases. J. Proteome Res. 7, 3354–3363 (2008)
8. Chourey, K., Nissen, S., Vishnivetskaya, T., Shah, M., Pfiffner, S., Hettich, R.L., Loffler, F.E.: Environmental proteomics reveals early microbial community responses to biostimulation at a uranium- and nitrate-contaminated site. Proteomics 13, 2921–2930 (2013)
9. Castellana, N.E., Payne, S.H., Shen, Z., Stanke, M., Bafna, V., Briggs, S.P.: Discovery and revision of Arabidopsis genes by proteogenomics. Proc. Natl. Acad. Sci. U. S. A. 105, 21034–21038 (2008)
10. Jagtap, P., McGowan, T., Bandhakavi, S., Tu, Z.J., Seymour, S., Griffin, T.J., Rudney, J.D.: Deep metaproteomic analysis of human salivary supernatant. Proteomics 12, 992–1001 (2012)

11. Guthals, A., Clauser, K.R., Bandeira, N.: Shotgun protein sequencing with meta-contig assembly. Mol. Cell. Proteomics 10, 1084–1096 (2012)
12. Guthals, A., Clauser, K.R., Frank, A.M., Bandeira, N.: Sequencing-Grade De novo Analysis of MS/MS Triplets (CID/HCD/ETD) From Overlapping Peptides. J. Proteome Res. 12, 2846–2857 (2013)
13. Bandeira, N., Tang, H., Bafna, V., Pevzner, P.A.: Shotgun protein sequencing by tandem mass spectra assembly. Anal. Chem. 76, 7221–7233 (2004)
14. Guthals, A., Watrous, J.D., Dorrestein, P.C., Bandeira, N.: The spectral networks paradigm in high throughput mass spectrometry. Mol. Biosyst. 8, 2535–2544 (2012)
15. Bandeira, N., Clauser, K.R., Pevzner, P.A.: Shotgun protein sequencing: assembly of peptide tandem mass spectra from mixtures of modified proteins. Mol. Cell. Proteomics 6, 1123–1134 (2007)
16. Edelmann, M.J.: Strong Cation Exchange Chromatography in Analysis of Posttranslational Modifications: Innovations and Perspectives (2011)
17. Swaney, D.L., Wenger, C.D., Coon, J.J.: Value of using multiple proteases for large-scale mass spectrometry-based proteomics. J. Proteome Res. 9, 1323–1329 (2010)
18. Bandeira, N., Tsur, D., Frank, A., Pevzner, P.A.: Protein identification by spectral networks analysis. Proc. Natl. Acad. Sci. U. S. A. 104, 6140–6145 (2007)
19. Pevzner, P.A., Dancík, V., Tang, C.L.: Mutation-tolerant protein identification by mass spectrometry. J. Comput. Biol. 7, 777–787 (2000)
20. Dancík, V., Addona, T.A., Clauser, K.R., Vath, J.E., Pevzner, P.A.: De novo peptide sequencing via tandem mass spectrometry. J. Comput. Biol. 6, 327–342 (1999)
21. Frank, A.M., Savitski, M.M., Nielsen, M.L., Zubarev, R.A., Pevzner, P.A.: De novo peptide sequencing and identification with precision mass spectrometry. J. Proteome Res. 6, 114–123 (2007)
22. Kessner, D., Chambers, M., Burke, R., Agus, D., Mallick, P.: ProteoWizard: open source software for rapid proteomics tools development. Bioinformatics 24, 2534–2536 (2008)
23. Frank, A.M., Bandeira, N., Shen, Z., Tanner, S., Briggs, S.P., Smith, R.D., Pevzner, P.A.: Clustering millions of tandem mass spectra. J. Proteome Res. 7, 113–122 (2008)
24. Bairoch, A., Apweiler, R., Wu, C.H., Barker, W.C., Boeckmann, B., Ferro, S., Gasteiger, E., Huang, H., Lopez, R., Magrane, M., Martin, M.J., Natale, D.A., O'Donovan, C., Redaschi, N., Yeh, L.-S.L.: The Universal Protein Resource (UniProt). Nucleic Acids Res. 35, 190–195 (2008)
25. Craig, R., Beavis, R.C.: TANDEM: matching proteins with tandem mass spectra. Bioinformatics 20, 1466–1467 (2004)
26. Tabb, D.L., MacCoss, M.J., Wu, C.C., Anderson, S.D., Yates, J.R.: Similarity among tandem mass spectra from proteomic experiments: detection, significance, and utility. Anal. Chem. 75, 2470–2477 (2003)
27. Jeong, K., Kim, S., Bandeira, N.: False discovery rates in spectral identification. BMC Bioinformatics 13(suppl. 1), S2 (2012)
28. Gupta, N., Bandeira, N., Keich, U., Pevzner, P.A.: Target-Decoy Approach and False Discovery Rate: When Things Go Wrong. J. Am. Soc. Mass Spectrom 22, 1111–1120 (2011)
29. Guthals, A., Bandeira, N.: Peptide identification by tandem mass spectrometry with alternate fragmentation modes. Mol. Cell. Proteomics 11, 550–557 (2012)

Decoding Coalescent Hidden Markov Models in Linear Time

Kelley Harris[1], Sara Sheehan[2], John A. Kamm[3], and Yun S. Song[2,3,4]

[1] Department of Mathematics, University of California, Berkeley
[2] Computer Science Division, University of California, Berkeley
[3] Department of Statistics, University of California, Berkeley
[4] Department of Integrative Biology, University of California, Berkeley
kharris@math.berkeley.edu, {ssheehan,yss}@eecs.berkeley.edu,
jkamm@stat.berkeley.edu

Abstract. In many areas of computational biology, hidden Markov models (HMMs) have been used to model local genomic features. In particular, coalescent HMMs have been used to infer ancient population sizes, migration rates, divergence times, and other parameters such as mutation and recombination rates. As more loci, sequences, and hidden states are added to the model, however, the runtime of coalescent HMMs can quickly become prohibitive. Here we present a new algorithm for reducing the runtime of coalescent HMMs from quadratic in the number of hidden time states to linear, without making any additional approximations. Our algorithm can be incorporated into various coalescent HMMs, including the popular method PSMC for inferring variable effective population sizes. Here we implement this algorithm to speed up our demographic inference method diCal, which is equivalent to PSMC when applied to a sample of two haplotypes. We demonstrate that the linear-time method can reconstruct a population size change history more accurately than the quadratic-time method, given similar computation resources. We also apply the method to data from the 1000 Genomes project, inferring a high-resolution history of size changes in the European population.

Keywords: Demographic inference, effective population size, coalescent with recombination, expectation-maximization, augmented hidden Markov model, human migration out of Africa.

1 Introduction

The hidden Markov model (HMM) is a natural and powerful device for learning functional and evolutionary attributes of DNA sequence data. Given an emitted sequence of base pairs or amino acids, the HMM is well-suited to locating hidden features of interest such as genes and promotor regions [2,5]. HMMs can also be used to infer hidden attributes of a collection of related DNA sequences. In this case, emitted states are a tuple of A's, C's, G's and T's, and the diversity of emitted states in a particular region can be used to infer the local evolutionary history of the sequences. When two sequences are identical throughout a long

R. Sharan (Ed.): RECOMB 2014, LNBI 8394, pp. 100–114, 2014.
© Springer International Publishing Switzerland 2014

genetic region, they most likely inherited that region identical by descent from a recent common ancestor. Conversely, high genetic divergence indicates that the sequences diverged from a very ancient common ancestor [1,15].

In recent years, coalescent HMMs such as the Pairwise Sequentially Markov Coalescent (PSMC) [15] have been used to infer the sequence of times to most recent common ancestor (TMRCAs) along a pair of homologous DNA sequences. Two other coalescent HMMs (CoalHMM [4,12,16] and diCal [24,25]) also tackle the problem of inferring genealogical information in samples of more than two haplotypes. These methods are all derived from the coalescent with recombination, a stochastic process that encapsulates the history of a collection of DNA sequences as an ancestral recombination graph (ARG) [13,29]. The hidden state associated with each genetic locus is a tree with time-weighted edges, and neighboring trees in the sequence are highly correlated with each other. Sequential changes in tree structure reflect the process of genetic recombination that slowly breaks up ancestral haplotypes over time.

The methods mentioned above all infer approximate ARGs for the purpose of demographic inference, either detecting historical changes in effective population size or estimating times of divergence and admixture between different populations or species. PSMC and CoalHMM have been used to infer ancestral population sizes in a variety of non-model organisms for which only a single genome is available [6,17,19,20,28,30], as well as for the Neanderthal and Denisovan archaic hominid genomes [18]. Despite this progress, the demographic inference problem is far from solved, even for extremely well-studied species like *Homo sapiens* and *Drosophila melanogaster* [7,9,15,23,27]. Estimates of the population divergence time between European and African humans range from 50 to 120 thousand years ago (kya), while estimates of the speciation time between polar bears and brown bears range from 50 kya to 4 million years ago [3,10,19]. One reason that different demographic methods often infer conflicting histories is that they make different trade-offs between the mathematical precision of the model and scalability to larger input datasets. This is even true within the class of coalescent HMMs, which are much more similar to each other than to methods that infer demography from summary statistics [8,11,21] or Markov chain Monte Carlo [7].

Exact inference of the posterior distribution of ARGs given data is a very challenging problem, the major reason being that the space of hidden states is infinite, parameterized by continuous coalescence times. In practice, when a coalescent HMM is implemented, time needs to be discretized and confined to a finite range of values. It is a difficult problem to choose an optimal time discretization that balances the information content of a dataset, the complexity of the analysis, and the desire to infer particular periods of history at high resolution. Recent demographic history is often of particular interest, but large sample sizes are needed to distinguish between the population sizes at time points that are very close together or very close to the present.

In a coalescent HMM under a given demographic model, optimal demographic parameters can be inferred using an expectation-maximization (EM) algorithm. The speed of this EM algorithm is a function of at least three variables: the length

L of the genomic region being analyzed, the number n of sampled haplotypes, and the number d of states for discretized time. In most cases, the complexity is linear in L, but the complexity in n can be enormous because the number of distinct n-leaved tree topologies grows super-exponentially with n. PSMC and CoalHMM avoid this problem by restricting n to be very small, analyzing no more than four haplotypes at a time. diCal admits larger values of n by using a *trunk genealogy* approximation (see [22,24,25] for details) which is derived from the diffusion process dual to the coalescent process, sacrificing information about the exact structure of local genealogies in order to analyze large samples which are informative about the recent past.

To date, all published coalescent HMMs have had quadratic complexity in d. This presents a significant limitation given that small values of d lead to biased parameter estimates [16] and limit the power of the method to resolve complex demographic histories. PSMC is typically run with a discretization of size $d = 64$, but diCal and CoalHMM analyses of larger datasets are restricted to coarser discretizations by the cost of increasing the sample size. In this paper, we exploit the natural symmetries of the coalescent process to derive an alternate EM algorithm with linear complexity in d. The speedup requires no approximations to the usual forward-backward probabilities; we perform an exact computation of the likelihood in $O(d)$ time rather than $O(d^2)$ time using an augmented HMM. We implement the algorithms presented in this paper to speed up our published method diCal, which is equivalent to PSMC when the sample size is two, yielding results of the same quality as earlier work in a fraction of the runtime. We have included the speedup in the most recent version of our program diCal; source code can be downloaded at http://sourceforge.net/projects/dical/.

2   Linear-Time Computation of the Forward and Backward Probabilities

We consider a coalescent HMM $\mathcal{M}$ with hidden states $S_1, \ldots, S_L$ and observations $x = x_1, \ldots, x_L$. For PSMC, S_ℓ is the discretized time interval in which two homologous chromosomes coalesce at locus ℓ, while x_ℓ is an indicator for heterozygosity. The method diCal is based on the conditional sampling distribution (CSD) which describes the probability of observing a newly sampled haplotype x given a collection $\mathcal{H}$ of n already observed haplotypes. In diCal, the hidden state at locus ℓ is $S_\ell = (H_\ell, T_\ell)$, where $H_\ell \in \mathcal{H}$ denotes the haplotype in the "trunk genealogy" (see [22]) with which x coalesces at locus ℓ and $T_\ell \in \{1, \ldots, d\}$ denotes the discretized time interval of coalescence; the observation $x_\ell \in \mathcal{A}$ is the allele of haplotype x at locus ℓ. For $n = |\mathcal{H}| = 1$, diCal is equivalent to PSMC. In what follows, we present our algorithm in the context of diCal, but we note that the same underlying idea can be applied to other coalescent HMMs.

2.1   A Linear-Time Forward Algorithm

We use $f(x_{1:\ell}, (h, j))$ to denote the joint forward probability of observing the partial emitted sequence $x_{1:\ell} := x_1, \ldots, x_\ell$ and the hidden state $S_\ell = (h, j)$ at

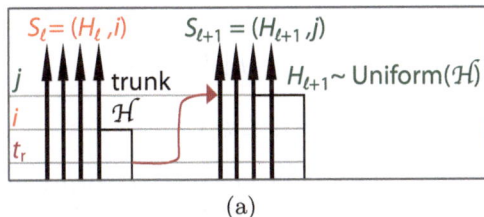 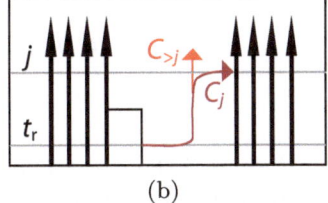

(a)   (b)

Fig. 1. (a). Here, we illustrate a transition from hidden state $S_\ell = (h_n, i)$ to hidden state $S_{\ell+1} = (h_k, j)$ that proceeds via recombination at time t_r. The probability of this transition does not depend on the identity of the haplotype h_k.(b). As a recombined lineage floats through time interval j, it can either coalesce with the trunk (event C_j) or keep floating (event $C_{>j}$) and eventually coalesce with the trunk in a more ancient time interval.

locus ℓ. The probability of transitioning from state (h', k) at locus ℓ to state (h, j) at locus $\ell + 1$ is denoted by $\phi(h, j \mid h', k)$, the stationary probability of state (h, i) is denoted $\zeta(h, i)$, and the emission probability of the observed allele $x_\ell = a$ given coalescence at $T_\ell = j$ onto haplotype h with allele $h_\ell = b$ at locus ℓ is denoted by $\xi(a \mid b, j)$. When ℓ is obvious from the context, we sometimes use $\xi(a \mid s) := \xi(a \mid h_\ell, j)$ for $s = (h, j)$. Explicit expressions for $\zeta(h, i)$, $\phi(h, j \mid h', k)$, and $\xi(a \mid b, j)$ in the context of our program diCal are given in [24].

The forward probabilities are computed using the recursion

$$f(x_{1:\ell+1}, (h, j)) = \xi(x_{\ell+1} \mid h_{\ell+1}, j) \cdot \sum_{k=1}^{d} \sum_{h' \in \mathcal{H}} f(x_{1:\ell}, (h', k)) \cdot \phi(h, j \mid h', k), \quad (1)$$

which contains nd terms. Since there are also nd possibilities for $S_{\ell+1} = (h, j)$, it should naively take $O(n^2 d^2 L)$ time to compute the entire forward dynamic programming (DP) table $\{f(x_{1:\ell}, S_\ell)\}_{\ell=1}^{L}$. The key to achieving a speed-up is to factor (1) in a way that reflects the structure of the coalescent, exploiting the fact that many transitions between different hidden states have identical probabilities.

After a sampled lineage recombines at time t_r between loci ℓ and $\ell + 1$, it will "float" backward in time from the recombination breakpoint until eventually coalescing with a trunk lineage chosen uniformly at random (Figure 1a). This implies that $\phi(h, j \mid h', k) = \phi(h, j \mid h'', k)$ whenever $h' \neq h$ and $h'' \neq h$, and exploiting this symmetry allows the forward table to be computed in $O(nd^2 L)$ time. This speed-up was already implemented in the algorithm described in Paul et al. [22].

Another symmetry of the transition matrix, not exploited previously, can be found by decomposing the transition from locus ℓ to locus $\ell + 1$ as a sequence of component events. In particular, let R_i be the event that a recombination occurs during time interval i, and let $\overline{R}$ be the event that no recombination occurs between ℓ and $\ell + 1$. Then we have that

$$\phi((h,j) \mid (h',k)) = \frac{1}{n} \sum_{i=1}^{\min(j,k)} (\mathbb{P}(R_i, T_{\ell+1} = j \mid T_\ell = k)$$

$$+ \mathbb{1}_{\{(h,j)=(h',k')\}} \mathbb{P}(\overline{R} \mid T_\ell = k)), \qquad (2)$$

where $\mathbb{1}_E = 1$ if the event E is true or 0 otherwise. The factor $1/n$ corresponds to the probability that the sampled lineage coalesces with haplotype $h \in \mathcal{H}$ in the trunk genealogy.

If a recombination occurs in time interval i, the sampled lineage will start to "float" freely back in time until it either coalesces in i or floats into the next time interval $i+1$ (Figure 1b). Specifically, we let $C_{>i}$ denote the event where the sampled lineage recombines at or before i and floats into $i+1$, and C_i denote the event where the recombined lineage coalesces back in interval i. Noting that $\mathbb{P}(R_i, C_i \mid T_\ell = i')$ and $\mathbb{P}(R_i, C_{>i} \mid T_\ell = i')$ are independent of i' whenever $i' > i$, and that coalescence happens as a Markov process backwards in time, we obtain

$$\mathbb{P}(R_i, T_{\ell+1} = j \mid T_\ell = k) = \mathbb{1}_{i=j=k} \cdot \mathbb{P}(R_i, C_i \mid T_\ell = i)$$

$$+ \mathbb{1}_{i=j<k} \cdot \mathbb{P}(R_i, C_i \mid T_\ell > i)$$

$$+ \mathbb{1}_{i=k<j} \cdot \mathbb{P}(R_i, C_{>i} \mid T_\ell = i) \cdot \prod_{k=i}^{j-1} \mathbb{P}(C_{>k+1} \mid C_{>k})$$

$$+ \mathbb{1}_{i<\min(j,k)} \cdot \mathbb{P}(R_i, C_{>i} \mid T_\ell > i) \cdot \prod_{k=i}^{j-1} \mathbb{P}(C_{>k+1} \mid C_{>k}). \qquad (3)$$

Explicit formulas (specific to the method diCal) for the above terms are provided in the supporting information available at http://www.eecs.berkeley.edu/~yss/publications.html.

By combining (2) with (3) and then collecting terms in (1), we can remove the sum over $T_\ell = k$ when computing $f(x_{1:\ell+1}, S_{\ell+1})$. In particular, we define additional forward probabilities

$$f(x_{1:\ell}, T_\ell = k) := \mathbb{P}(x_{1:\ell}, T_\ell = k) = \sum_{h' \in \mathcal{H}} f(x_{1:\ell}, S_\ell = (h', k)), \qquad (4)$$

$$f(x_{1:\ell}, T_\ell > k) := \mathbb{P}(x_{1:\ell}, T_\ell > k) = \sum_{k'=k+1}^{d} \sum_{h' \in \mathcal{H}} f(x_{1:\ell}, S_\ell = (h', k')), \qquad (5)$$

$$f(x_{1:\ell}, R_{\leq j}, C_{>j}) := \sum_{i=1}^{j} \mathbb{P}(x_{1:\ell}, R_i, C_{>i}, \ldots, C_{>j}) \qquad (6)$$

$$= \sum_{i=1}^{j} \left\{ \left[\prod_{i'=i+1}^{j} \mathbb{P}(C_{>i'} \mid C_{>i'-1}) \right] \right.$$

$$\times \left[f(x_{1:\ell}, T_\ell = i)\mathbb{P}(R_i, C_{>i} \mid T_\ell = i) + f(x_{1:\ell}, T_\ell > i)\,\mathbb{P}(R_i, C_{>i} \mid T_\ell > i) \right] \right\}.$$

Then, (1) can be written as

$$f(x_{1:\ell+1}, (h, j)) = \xi(x_{\ell+1} \mid h_{\ell+1}, j) \cdot \left[\frac{1}{n} f\left(x_{1:\ell}, R_{\leq j-1}, C_{>j-1}\right) \mathbb{P}(C_j \mid C_{>j-1}) \right.$$

$$+ \frac{1}{n} f\left(x_{1:\ell}, T_\ell > j\right) \mathbb{P}(R_j, C_j \mid T_\ell > j)$$

$$+ \frac{1}{n} f\left(x_{1:\ell}, T_\ell = j\right) \mathbb{P}(R_j, C_j \mid T_\ell = j)$$

$$\left. + f(x_{1:\ell}, (h, j)) \mathbb{P}(\overline{R} \mid T_\ell = j) \right]. \tag{7}$$

This can be seen by noting that the first three terms in the sum correspond to the terms for $i < j$, $i = j < k$, and $i = j = k$, respectively when putting together (1) and (2). Alternatively, (7) follows from directly considering the probabilistic interpretation of the terms $f(x_{1:\ell}, *)$ as given by (4), (5), and (6).

The required values of $f\left(x_{1:\ell}, R_{\leq i}, C_{>i}\right)$ and $f\left(x_{1:\ell}, T_\ell > i\right)$ can be computed recursively using

$$f\left(x_{1:\ell}, T_\ell > i\right) = f\left(x_{1:\ell}, T_\ell > i+1\right) + f\left(x_{1:\ell}, T_\ell = i+1\right), \tag{8}$$

$$f\left(x_{1:\ell}, R_{\leq i}, C_{>i}\right) = f\left(x_{1:\ell}, R_{\leq i-1}, C_{>i-1}\right) \mathbb{P}(C_{>i} \mid C_{>i-1})$$

$$+ f\left(x_{1:\ell}, T_\ell = i\right) \mathbb{P}(R_i, C_{>i} \mid T_\ell = i)$$

$$+ f\left(x_{1:\ell}, T_\ell > i\right) \mathbb{P}(R_i, C_{>i} \mid T_\ell > i), \tag{9}$$

with the base cases

$$f\left(x_{1:\ell}, T_\ell > d\right) = 0,$$

$$f\left(x_{1:\ell}, R_{\leq 1}, C_{>1}\right) = f\left(x_{1:\ell}, T_\ell > 1\right) \mathbb{P}(R_1, C_{>1} \mid T_\ell > 1)$$

$$+ f\left(x_{1:\ell}, T_\ell = 1\right) \mathbb{P}(R_1, C_{>1} \mid T_\ell = 1).$$

Hence, using the recursions (7), (8), and (9), it is possible to compute the entire forward DP table $\{f(x_{1:\ell}, S_\ell)\}_{\ell=1}^L$ exactly in $O(ndL)$ time.

2.2 A Linear-Time Backward Algorithm

The backward DP table $\{b(x_{\ell+1:L} \mid S_\ell)\}$ can be also computed in $O(ndL)$ time. Given the linear-time forward algorithm discussed in the previous section, the easiest way to compute the backward DP table is as follows: Let $x^{(r)} = x_1^{(r)}, x_2^{(r)}, \ldots, x_L^{(r)} = x_L, x_{L-1}, \ldots, x_1$ denote the reversed x and let $S_\ell^{(r)}$ denote the hidden states for the HMM generating $x^{(r)}$. Then, since the coalescent is reversible along the sequence,

$$b(x_{\ell+1:L}^{(r)} \mid s) = \frac{\mathbb{P}(x_{\ell+1:L}^{(r)}, S_\ell = s)}{\zeta(s)} = \frac{\mathbb{P}(x_{\ell:L}^{(r)}, S_\ell = s)}{\xi(x_\ell^{(r)} \mid s)\zeta(s)} = \frac{f(x_{1:L-\ell+1}^{(r)}, S_{L-\ell+1}^{(r)} = s)}{\xi(x_\ell^{(r)} \mid s)\zeta(s)}.$$

3   Linear-Time EM via an Augmented HMM

The primary application of PSMC and diCal is parameter estimation, specifically the estimation of demographic parameters such as changing population sizes. This is done through a maximum likelihood framework with the expectation maximization (EM) algorithm. In this section, we describe how to speed up the EM algorithm to work in linear time.

3.1   The Standard EM Algorithm with $O(d^2)$ Time Complexity

Let Θ denote the parameters we wish to estimate, and $\hat{\Theta}$ denote the maximum likelihood estimate:

$$\hat{\Theta} = \arg\max_{\Theta'} \mathcal{L}(\Theta') = \arg\max_{\Theta'} \mathbb{P}_{\Theta'}(x_{1:L}).$$

To find $\hat{\Theta}$, we pick some initial value $\Theta^{(0)}$, and then iteratively solve for $\Theta^{(t)}$ according to

$$\Theta^{(t)} = \arg\max_{\Theta'} \mathbb{E}_{S_{1:L};\Theta^{(t-1)}}[\log \mathbb{P}_{\Theta'}(x_{1:L}, S_{1:L}) \mid x_{1:L}],$$

where $S_{1:L} := S_1, \ldots, S_L$. The sequence $\Theta^{(0)}, \Theta^{(1)}, \ldots$ is then guaranteed to converge to a local maximum of the surface $\mathcal{L}(\Theta)$.

Since $(x_{1:L}, S_{1:L})$ forms an HMM, the joint likelihood $\mathbb{P}(x_{1:L}, S_{1:L})$ can be written as

$$\mathbb{P}_{\Theta'}(x_{1:L}, S_{1:L}) = \zeta_{\Theta'}(S_1) \left[\prod_{\ell=1}^{L} \xi_{\Theta'}(x_\ell \mid S_\ell) \right] \left[\prod_{\ell=2}^{L} \phi_{\Theta'}(S_\ell \mid S_{\ell-1}) \right].$$

Letting $\mathbb{E}[\#\ell : E \mid x_{1:L}]$ denote the posterior expected number of loci where event E occurs, and $\pi(x) := \mathbb{P}(x) = \sum_s f(x_{1:L}, s)$ denote the total probability of observing x, we then have

$$\mathbb{E}_{S_{1:L};\Theta}\left[\log \mathbb{P}_{\Theta'}(x_{1:L}, S_{1:L}) \middle| x_{1:L}\right]$$
$$= \sum_s (\log \zeta_{\Theta'}(s)) \, \mathbb{P}_\Theta(S_1 = s | x_{1:L})$$

$$+ \sum_{(h,i)} \sum_{a,b \in \mathcal{A}} (\log \xi_{\Theta'}(a|b,i)) \, \mathbb{E}_\Theta[\#\ell : \{S_\ell = (h,i), h_\ell = b, x_\ell = a\} | x_{1:L}]$$

$$+ \sum_{s,s'} (\log \phi_{\Theta'}(s' \mid s)) \, \mathbb{E}_\Theta[\#\ell : \{S_{\ell-1} = s, S_\ell = s'\} | x_{1:L}]$$

$$= \frac{1}{\pi_\Theta(x)} \left[\sum_s (\log \zeta_{\Theta'}(s)) \, f_\Theta(x_1, s) b_\Theta(x_{2:L}|s) \right.$$

$$\left. + \sum_{(h,i)} \sum_{a,b \in \mathcal{A}} (\log \xi_{\Theta'}(a|b,i)) \sum_{\substack{\ell : x_\ell = a \\ h_\ell = b}} f_\Theta(x_{1:\ell}, (h,i)) b_\Theta(x_{\ell+1:L}|h,i) \right.$$

$$+ \sum_{s,s'} (\log \phi_{\Theta'}(s' \mid s)) \left(\sum_{\ell=1}^{L-1} f_{\Theta}(x_{1:\ell}, s) \phi_{\Theta}(s' \mid s) \xi_{\Theta}(x_{\ell+1} \mid s') b_{\Theta}(x_{\ell+2:L} \mid s') \right) \Bigg].$$
(10)

Note that we have to compute the term $\sum_{\ell} f_{\Theta}(x_{1:\ell}, s) \phi_{\Theta}(s' \mid s) \xi_{\Theta}(x_{\ell+1} \mid s')$ $b_{\Theta}(x_{\ell+2:L} \mid s')$ for every pair of states s, s', which makes computing the EM objective function quadratic in the number d of discretization time intervals, despite the fact that we computed the forward and backward tables in linear time.

3.2 A Linear-Time EM Algorithm

By augmenting our HMM to condition on whether recombination occurred between loci ℓ and $\ell+1$, the EM algorithm can be sped up to be linear in d. We now describe this augmented HMM. Let $\mathcal{M}$ denote our original HMM, with states $S_{1:L}$ and observations $x_{1:L}$. Between loci ℓ and $\ell + 1$, define

$$\mathcal{R}_{l,l+1} = \begin{cases} \overline{R}, & \text{if no recombination,} \\ R_i, & \text{if recombination occurred at time } i. \end{cases}$$

Now let $S_1^* = S_1$, and $S_\ell^* = (\mathcal{R}_{\ell-1,\ell}, S_\ell)$ for $\ell > 1$. We let $\mathcal{M}^*$ be the HMM with hidden variables $S_{1:L}^* = S_1^*, \ldots, S_L^*$, observations $x_{1:L}$, transition probabilities $\mathbb{P}(S_\ell^* \mid S_{\ell-1}^*) = \mathbb{P}(S_\ell^* \mid S_{\ell-1})$, and emission probabilities $\mathbb{P}(x_\ell \mid S_\ell^*) = \mathbb{P}(x_\ell \mid S_\ell)$. Note that the probability of observing the data is the same under $\mathcal{M}$ and $\mathcal{M}^*$, i.e.,

$$\mathcal{L}(\Theta) = \mathbb{P}_{\Theta}(x_{1:L} \mid \mathcal{M}) = \mathbb{P}_{\Theta}(x_{1:L} \mid \mathcal{M}^*),$$

and so we may find a local maximum of $\mathcal{L}(\Theta)$ by applying the EM algorithm to the augmented HMM $\mathcal{M}^*$, instead of to the original HMM $\mathcal{M}$.

To compute the EM objective function for $\mathcal{M}^*$, we start by noting that the joint likelihood is

$$\mathbb{P}(x_{1:L}, S_{1:L}^*) = \zeta(S_1) \left[\prod_{\ell=1}^{L} \xi(x_\ell \mid S_\ell) \right] \left[\prod_{\ell:\mathcal{R}_{\ell,\ell+1}=\overline{R}} \mathbb{P}(\overline{R} \mid T_\ell) \right]$$
(11)

$$\times \left[\prod_{i=1}^{d} \prod_{\ell:\mathcal{R}_{\ell,\ell+1}=R_i} \mathbb{P}(R_i, T_{\ell+1} \mid T_\ell) \right] \left(\frac{1}{n} \right)^{\#\ell:\mathcal{R}_{\ell,\ell+1}\neq\overline{R}},$$

where we decomposed the joint likelihood into the initial probability, the emission probabilities, the transitions without recombination, and the transitions with recombination. We note that the initial probability can be decomposed as

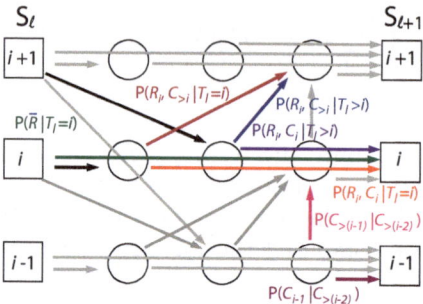

Fig. 2. This diagram illustrates the flow of transition probabilities through the augmented HMM. Lineages may transition between different coalescence times at loci ℓ and $\ell+1$ by recombining and passing through the floating states represented by circles. Each interval contains three distinct floating states to capture the the dependence of recombination and coalescence probabilities on whether any of these events occur during the same time interval.

$$\zeta(S_1 = (h, j)) = \frac{1}{n} \left[\prod_{i=1}^{j-1} \mathbb{P}(C_{>i} \mid C_{>i-1}) \right] \mathbb{P}(C_j \mid C_{>j-1}), \tag{12}$$

and from (3), we decompose the product of transition recombination probabilities as

$$\prod_{i=1}^{d} \prod_{\ell:\mathcal{R}_{\ell,\ell+1}=R_i} \mathbb{P}(R_i, T_{\ell+1} \mid T_\ell) = \prod_{i=1}^{d} \left\{ \left[\prod_{\substack{\ell:\mathcal{R}_{\ell,\ell+1}=R_i \\ T_\ell=T_{\ell+1}=i}} \mathbb{P}(R_i, C_i \mid T_\ell = i) \right] \right.$$

$$\times \left[\prod_{\substack{\ell:\mathcal{R}_{\ell,\ell+1}=R_i \\ T_\ell>T_{\ell+1}=i}} \mathbb{P}(R_i, C_i \mid T_\ell > i) \right] \left[\prod_{\substack{\ell:\mathcal{R}_{\ell,\ell+1}=R_i \\ T_{\ell+1}>T_\ell=i}} \mathbb{P}(R_i, C_{>i} \mid T_\ell = i) \right]$$

$$\times \left[\prod_{\substack{\ell:\mathcal{R}_{\ell,\ell+1}=R_i \\ T_\ell,T_{\ell+1}>i}} \mathbb{P}(R_i, C_{>i} \mid T_\ell > i) \right] \left[\prod_{\substack{\ell:T_{\ell+1}>i \\ \mathcal{R}_{\ell,\ell+1}\in R_{<i}}} \mathbb{P}(C_{>i} \mid C_{>i-1}) \right]$$

$$\times \left[\prod_{\substack{\ell:T_{\ell+1}=i \\ \mathcal{R}_{\ell,\ell+1}\in R_{<i}}} \mathbb{P}(C_i \mid C_{>i-1}) \right] \right\}, \tag{13}$$

where $R_{<i} := \cup_{j<i} R_j$. Figure 2 shows a graphical representation for the transitions of $\mathcal{M}^*$.

By plugging (12) and (13) into (11), then taking the posterior expected logarithm of (11), we obtain the EM objective function for $\mathcal{M}^*$:

$$\mathbb{E}_{S^*_{1:L};\Theta} \left[\log \mathbb{P}_{\Theta'}(x_{1:L}, S^*_{1:L}) \mid x_{1:L} \right] = -L \log n + \sum_{i=1}^{d} q_i(\Theta, \Theta'), \tag{14}$$

where

$$
q_i(\Theta, \Theta') := \sum_{a,b \in \mathcal{A}} \left[\frac{\log \xi_{\Theta'}(a \mid b, i)}{\pi_\Theta(x)} \sum_{\ell : x_\ell = a} \sum_{h : h_\ell = b} f_\Theta(x_{1:\ell}, (h, i)) b_\Theta(x_{\ell+1:L} \mid (h, i)) \right]
$$
$$
+ \left(\log \mathbb{P}_{\Theta'}(\overline{R} \mid T = i) + \log n \right) \mathbb{E}_\Theta \left[\#\ell : \{\mathcal{R}_{\ell,\ell+1} = \overline{R}, T_\ell = i\} \mid x_{1:L} \right]
$$
$$
+ \left(\log \mathbb{P}_{\Theta'}(R_i, C_i \mid T = i) \right) \mathbb{E}_\Theta \left[\#\ell : \{\mathcal{R}_{\ell,\ell+1} = R_i, T_\ell = T_{\ell+1} = i\} \mid x_{1:L} \right]
$$
$$
+ \left(\log \mathbb{P}_{\Theta'}(R_i, C_i \mid T > i) \right) \mathbb{E}_\Theta \left[\#\ell : \{\mathcal{R}_{\ell,\ell+1} = R_i, T_\ell > T_{\ell+1} = i\} \mid x_{1:L} \right]
$$
$$
+ \left(\log \mathbb{P}_{\Theta'}(R_i, C_{>i} \mid T = i) \right) \mathbb{E}_\Theta \left[\#\ell : \{\mathcal{R}_{\ell,\ell+1} = R_i, T_{\ell+1} > T_\ell = i\} \mid x_{1:L} \right]
$$
$$
+ \left(\log \mathbb{P}_{\Theta'}(R_i, C_{>i} \mid T > i) \right) \mathbb{E}_\Theta \left[\#\ell : \{\mathcal{R}_{\ell,\ell+1} = R_i, T_\ell > i, T_{\ell+1} > i\} \mid x_{1:L} \right]
$$
$$
+ \left(\log \mathbb{P}_{\Theta'}(C_{>i} \mid C_{>i-1}) \right) \mathbb{E}_\Theta \left[\#\ell : \{\mathcal{R}_{\ell,\ell+1} \in R_{<i}, T_{\ell+1} > i\} \mid x_{1:L} \right]
$$
$$
+ \left(\log \mathbb{P}_{\Theta'}(C_i \mid C_{>i-1}) \right) \mathbb{E}_\Theta \left[\#\ell : \{\mathcal{R}_{\ell,\ell+1} \in R_{<i}, T_{\ell+1} = i\} \mid x_{1:L} \right]
$$
$$
+ \mathbb{P}_\Theta(T_1 > i \mid x_{1:L}) + \mathbb{P}_\Theta(T_1 = i \mid x_{1:L}). \tag{15}
$$

The computation time for each of the posterior expectations $\mathbb{E}_\Theta[\#\ell : * \mid x_{1:L}]$ and $\mathbb{P}_\Theta(T_1 \mid x_{1:L})$ does not depend on d; full expressions are listed in the supporting information (http://www.eecs.berkeley.edu/~yss/publications.html). Hence, the number of operations needed to evaluate (14) is linear in d.

We note another attractive property of (14). By decomposing the EM objective function into a sum of terms $q_i(\Theta, \Theta')$, we obtain a natural strategy for searching through the parameter space. In particular, one can attempt to find the $\arg\max_{\Theta'}$ of (14) by optimizing the $q_i(\Theta, \Theta')$ one at a time in i. In fact, for the problem of estimating changing population sizes, $q_i(\Theta, \Theta')$ depends on Θ' almost entirely through a single parameter (the population size λ'_i in interval i), and we pursue a strategy of iteratively solving for λ'_i while holding the other coordinates of Θ' fixed, thus reducing a multivariate optimization problem into a sequence of univariate optimization problems.

Although both the linear and quadratic EM procedures are guaranteed to converge to local maxima of $\mathcal{L}(\Theta)$, they may have different rates of convergence, and may converge to different local maxima. The search paths of the two EM algorithms may differ for two reasons: first, the intermediate objective functions (10) and (14) are not equal, and secondly, as discussed above, we use different search strategies to find the optima of (10) and (14). We have no proven guarantee that either search should perform better than the other, but our observations indicate that the linear-time EM algorithm typically converges to a value of Θ with a equal or higher value of $\mathcal{L}(\Theta)$ than the quadratic-time algorithm, in a fraction of the time (see Figure 5 for an example).

4   Results

To confirm the decrease in runtime, we ran the linear-time diCal method on simulated data with $L = 2$ Mb of loci and 2 haplotypes (in which case diCal is equivalent to PSMC), using $d = 2, 4, 8, 16, 32, 48, 64, 80, 96, 112, 128$ discretization intervals. To simulate the data, we used ms [14] with a population-scaled recombination rate $\rho = 0.0005$ to generate an ARG, and then added mutations using a population-scaled mutation rate of $\theta = 0.0029$ and a finite-sites mutation matrix described in Sheehan *et al.* [24]. Figure 3(a) shows the time required to compute the table of forward probabilities. We also measured the time required for one EM iteration and then extrapolated to 20 iterations to approximate the time required to estimate an effective population size history (Figure 3(b)). In both figures, the linear runtime of our new algorithm is apparent and significantly improves our ability to increase the number of discretization intervals.

To assess the gain in accuracy of population size estimates that is afforded by more discretization intervals, we ran both the linear- and quadratic-time methods on simulated data with 10 haplotypes and $L = 2$ Mb. The conditional sampling distribution was used in a leave-one-out composite likelihood approach [24] in this experiment. To run each method for roughly the same amount of time (≈ 40 hours), we used $d = 9$ for the quadratic method and $d = 21$ for the linear method. For both methods, we ran the EM for 20 iterations and inferred $d/3$ size change parameters. As measured by the PSMC error function, which integrates the absolute value of the difference between the true size function and the estimated size function [15], larger values of d permit the inference of more accurate histories.

We also ran our method on 10 CEU haplotypes (Utah residents of European descent) sequenced during Phase I of the the 1000 Genomes Project [26] (Figure 4(b)). We can see that for the quadratic method with $d = 9$, we are unable

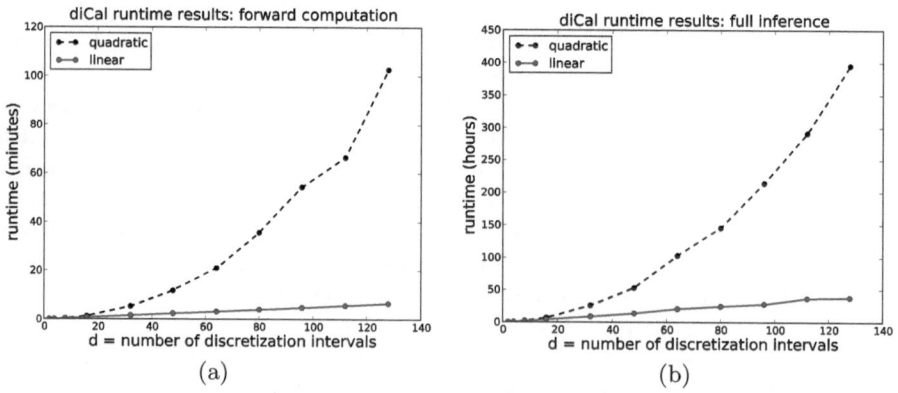

Fig. 3. Runtime results on simulated data with $L = 2$ Mb and 2 haplotypes, for varying number d of discretization intervals. (a) Runtime results (in minutes) for the forward computation. (b) Runtime results (in hours) for the entire EM inference algorithm (20 iterations) extrapolated from the time for one iteration.

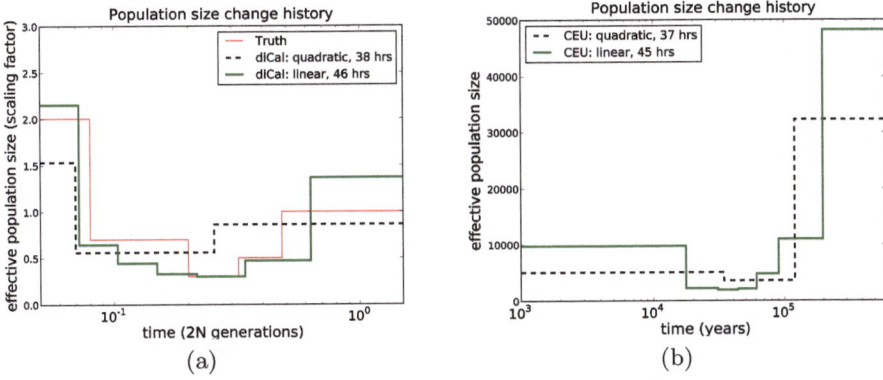

Fig. 4. Effective population size change history results. The speedup from the linear method allows us to use a finer discretization ($d = 21$) than the quadratic method ($d = 9$) for about the same amount of runtime. (a) Results on simulated data with $L = 2$ Mb and 10 haplotypes. Using the quadratic method with $d = 9$, the error was 0.148. Using the linear method with $d = 21$, the error dropped to 0.079. (b) Results on 10 European haplotypes over a 2 Mb region of chromosome 1. The out-of-Africa bottleneck is very apparent with $d = 21$, but is not as well characterized for $d = 9$.

to fully characterize the out-of-Africa bottleneck. In the same amount of computational time, we can run the linear method with $d = 21$ and easily capture this feature. The disagreement in the ancient past between the two methods is most likely due to diCal's lack of power in the ancient past when there are not many coalescence events. Using a leave-one-out approach with 10 haplotypes, the coalescence events in the ancient past tend to be few and unevenly spaced, resulting in a less confident inference.

The runtime of the full EM algorithm depends on the convergence of the M-step, which can be variable. Occasionally we observed convergence issues for the quadratic method, which requires a multivariate optimization routine. For the linear method, we used the univariate Brent optimization routine from Apache Math Commons (`http://commons.apache.org/proper/commons-math/`), which converges quickly and to a large extent avoids local maxima.

To examine the convergence of the two EM algorithms, we ran the linear and quadratic methods on the simulated data with 10 haplotypes and the same number of intervals $d = 16$. We examine the likelihoods in Figure 5(a). The linear method reaches parameter estimates of higher likelihood, although it is unclear whether the two methods have found different local maxima, or whether the quadratic method is approaching the same maximum more slowly. Figure 5(b) shows the inferred population sizes for each method, which although similar, are not identical.

We have also looked at the amount of memory required for each method, and although the difference is small, the linear method does require more memory to store the augmented forward and backward tables. A more thorough investigation of memory requirements will be important as the size of the data continues to increase.

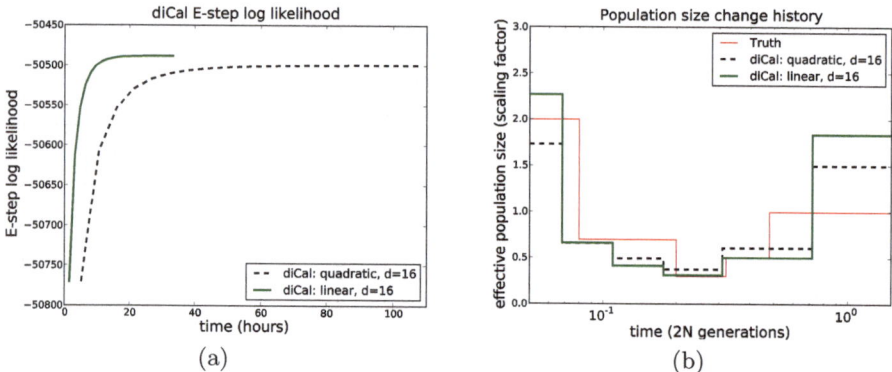

Fig. 5. Results on simulated data, using the same discretization for the linear and quadratic methods. Each method was run for 20 iterations. (a) The log likelihood of the EM algorithm, plotted against time, for both the linear and quadratic methods. (b) Population size change history results for the linear and quadratic methods, run with the same discretization using $d = 16$ and estimating 6 parameters.

5 Discussion

The improvement to diCal described in this paper will enable users to analyze larger datasets and infer more detailed demographic histories. This is especially important given that large datasets are needed to distinguish between histories with subtle or recent differences. By using samples of 10 haplotypes rather than 2, diCal v1.0 [24] was able to distinguish between histories that diverged from each other less than 0.1 coalescent time units ago, in which period PSMC tends to exhibit runaway behavior and hence cannot produce reliable population size estimates. The faster algorithm described here can handle samples of 30 haplotypes with equivalent computing resources. Our results indicate that this improves the method's ability to resolve rapid, recent demographic shifts.

In organisms where multiple sequenced genomes are not available, the resources freed up by $O(d)$ HMM decoding could be used to avoid grouping sites into 100-locus bins. This binning technique is commonly used to improve the scalability of PSMC, but has the potential to downwardly bias coalescence time estimates in regions that contain more than one SNP per 100 bp.

In general, it is a difficult problem to choose the time discretization that can best achieve the goals of a particular data analysis, achieving high resolution during biologically interesting time periods without overfitting the available data. Sometimes it will be more fruitful to increase the sample size n or sequence length L than to refine the time discretization; an important avenue for future work will be tuning L, n, and d to improve inference in humans and other organisms.

Another avenue for future work will be to develop augmented HMMs for coalescent models with population structure. Structure and speciation have been incorporated into several versions of CoalHMM and diCal, and the strategy presented in this paper could be used to speed these up, though a more elaborate network of hidden states will be required. We are hopeful that our new technique will help coalescent HMMs keep pace with the number and diversity of genomes being sequenced and tease apart the demographic patterns that differentiated them.

Acknowledgments. We are grateful to Matthias Steinrücken and other members of the Song group for helpful discussions. This research was supported in part by NSF Graduate Research Fellowships to K.H. and S.S., and by an NIH grant R01-GM094402 and a Packard Fellowship for Science and Engineering to Y.S.S.

References

1. Browning, B.L., Browning, S.R.: A fast, powerful method for detecting identity by descent. Am. J. Hum. Genet. 88, 173–182 (2011)
2. Burge, C., Karlin, S.: Prediction of complete gene structures in human genomic DNA. J. Mol. Biol. 268, 78–94 (1997)
3. Cahill, J.A., Green, R.E., Fulton, T.L., et al.: Genomic evidence for island population conversion resolves conflicting theories of polar bear evolution. PLoS Genetics 9, e1003345 (2013)
4. Dutheil, J.Y., Ganapathy, G., Hobolth, A., et al.: Ancestral population genomics: the coalescent hidden Markov model approach. Genetics 183, 259–274 (2009)
5. Ernst, J., Kellis, M.: ChromHMM: automating chromatin-state discovery and characterization. Nature Methods 9, 215–216 (2012)
6. Groenen, M.A., Archibald, A.L., Uenishi, H., et al.: Analyses of pig genomes provide insight into porcine demography and evolution. Nature 491(7424), 393–398 (2012)
7. Gronau, I., Hubisz, M.J., Gulko, B., et al.: Bayesian inference of ancient human demographic history from individual genome sequences. Nature Genetics 43, 1031–1034 (2011)
8. Gutenkunst, R.N., Hernandez, R.D., Williamson, S.H., Bustamante, C.D.: Inferring the joint demographic history of multiple populations from multidimensional SNP frequency data. PLoS Genetics 5, e1000695 (2009)
9. Haddrill, P.R., Thornton, K.R., Charlesworth, B., Andolfatto, P.: Multilocus patterns of nucleotide variability and the demographic selection history of *Drosophila melanogaster* populations. Genome Res. 15, 790–799 (2005)
10. Hailer, F., Kutschera, V.E., Hallstrom, B.M., et al.: Nuclear genomic sequences reveal that polar bears are an old and distinct bear lineage. Science 336, 344–347 (2012)
11. Harris, K., Nielsen, R.: Inferring demographic history from a spectrum of shared haplotype lengths. PLoS Genetics 9, e1003521 (2013)
12. Hobolth, A., Christensen, O.F., Mailund, T., Schierup, M.H.: Genomic relationships and speciation times of human, chimpanzee, and gorilla inferred from a coalescent hidden Markov model. PLoS Genetics 3, 294–304 (2007)
13. Hudson, R.R.: Properties of the neutral allele model with intergenic recombination. Theor. Popul. Biol. 23, 183–201 (1983)
14. Hudson, R.R.: Generating samples under a Wright–Fisher neutral model of genetic variation. Bioinformatics 18(2), 337–338 (2002)

15. Li, H., Durbin, R.: Inference of human population history from individual whole-genome sequences. Nature 10, 1–5 (2011)
16. Mailund, T., Dutheil, J.Y., Hobolth, A., et al.: Estimating divergence time and ancestral effective population size of Bornean and Sumatran orangutan subspecies using a coalescent hidden Markov model. PLoS Genetics 7, e1001319 (2011)
17. Mailund, T., Halager, A.E., Westergaard, M., et al.: A new isolation with migration model along complete genomes infers very different divergence processes among closely related great ape species. PLoS Genetics 8(12), e1003125 (2012)
18. Meyer, M., Kircher, M., Gansauge, M.T., et al.: A high-coverage genome sequence from an archaic Denisovan individual. Science 338, 222–226 (2012)
19. Miller, W., Schuster, S.C., Welch, A.J.: Polar and brown bear genomes reveal ancient admixture and demographic footprints of plast climate change. Proc. Natl. Acad. Sci. USA 109, 2382–2390 (2012)
20. Orlando, L., Ginolhac, A., Zhang, G., et al.: Recalibrating *Equus* evolution using the genome sequence of an early Middle Pleistocene horse. Nature 499, 74–78 (2013)
21. Palamara, P.F., Lencz, T., Darvasi, A., Pe'er, I.: Length distributions of identity by descent reveal fine-scale demographic history. Am. J. Hum. Genet. 91, 809–822 (2012)
22. Paul, J.S., Steinrücken, M., Song, Y.S.: An accurate sequentially Markov conditional sampling distribution for the coalescent with recombination. Genetics 187, 1115–1128 (2011)
23. Pritchard, J.: Whole-genome sequencing data offer insights into human demography. Nature Genetics 43, 923–925 (2011)
24. Sheehan, S., Harris, K., Song, Y.S.: Estimating variable effective population sizes from multiple genomes: A sequentially Markov conditional sampling distribution approach. Genetics 194, 647–662 (2013)
25. Steinrücken, M., Paul, J.S., Song, Y.S.: A sequentially Markov conditional sampling distribution for structured populations with migration and recombination. Theor. Popul. Biol. 87, 51–61 (2013)
26. The 1000 Genomes Project Consortium: A map of human genome variation from population-scale sequencing. Nature 467, 1061–1073 (2010)
27. Thornton, K., Andolfatto, P.: Approximate Bayesian inference reveals evidence for a recent, severe bottleneck in a Netherlands population of *Drosophila melanogaster*. Genetics 172, 1607–1619 (2006)
28. Wan, Q.H., Pan, S.K., Hu, L., et al.: Genome analysis and signature discovery for diving and sensory properties of the endangered chinese alligator. Cell Res. 23(9), 1091–1105 (2013)
29. Wiuf, C., Hein, J.: Recombination as a point process along sequences. Theor. Popul. Biol. 55, 248–259 (1999)
30. Zhao, S., Zheng, P., Dong, S., et al.: Whole-genome sequencing of giant pandas provides insights into demographic history and local adaptation. Nature Genetics 45, 67–71 (2013)

AptaCluster – A Method to Cluster HT-SELEX Aptamer Pools and Lessons from Its Application

Jan Hoinka[1,*], Alexey Berezhnoy[2,*], Zuben E. Sauna[3], Eli Gilboa[2,**], and Teresa M. Przytycka[1,**]

[1] National Center of Biotechnology Information, National Library of Medicine, NIH, Bethesda MD 20894, USA
przytyck@ncbi.nlm.nih.gov

[2] Department of Microbiology & Immunology, University of Miami Miller School of Medicine, Miami, Florida 33101, USA
EGilboa@med.miami.edu

[3] Laboratory of Hemostasis, Division of Hematology, Center for Biologics Evaluation and Research, Food and Drug Administration, Bethesda, Maryland, USA

Abstract. Systematic Evolution of Ligands by EXponential Enrichment (SELEX) is a well established experimental procedure to identify aptamers - synthetic single-stranded (ribo)nucleic molecules that bind to a given molecular target. Recently, new sequencing technologies have revolutionized the SELEX protocol by allowing for deep sequencing of the selection pools after each cycle. The emergence of High Throughput SELEX (HT-SELEX) has opened the field to new computational opportunities and challenges that are yet to be addressed. To aid the analysis of the results of HT-SELEX and to advance the understanding of the selection process itself, we developed AptaCluster. This algorithm allows for an efficient clustering of whole HT-SELEX aptamer pools; a task that could not be accomplished with traditional clustering algorithms due to the enormous size of such datasets. We performed HT-SELEX with Interleukin 10 receptor alpha chain (IL-10RA) as the target molecule and used AptaCluster to analyze the resulting sequences. AptaCluster allowed for the first survey of the relationships between sequences in different selection rounds and revealed previously not appreciated properties of the SELEX protocol. As the first tool of this kind, AptaCluster enables novel ways to analyze and to optimize the HT-SELEX procedure. Our AptaCluster algorithm is available as a very fast multiprocessor implementation upon request.

1 Introduction

Aptamers are short, ($\sim$20 to $\sim$100 nucleotides) synthetic, single-stranded (ribo)-nucleic molecules that can be generated to bind specifically to molecular targets. These binding targets can vary from small organic molecules [1], through proteins and protein complexes [2], to viruses [3], and cells [4]. Aptamers have high

---

* First authors.
** Corresponding authors.

R. Sharan (Ed.): RECOMB 2014, LNBI 8394, pp. 115–128, 2014.
© Springer International Publishing Switzerland 2014

structural stability over a wide range of pH and temperatures making them ideal reagents for a broad spectrum of in-vitro, ex-vivo, and in-vivo applications [5]. A pegylated aptamer that inhibits binding of Vascular Endothelial Growth Factor (VEGF) to the VEGF receptor (Macugen ®) is approved for the treatment of age-related macular degeneration [6]. Aptamers can also be used to monitor small changes in the conformation of proteins, a property that can be utilized for detecting the effect changes in the manufacturing process or during the development of generic versions of protein-therapeutics [7].

Aptamers are experimentally identified through a procedure known as Systematic Evolution of Ligands by EXponential Enrichment (SELEX) [8]. The traditional SELEX procedure iterates over five basic steps which together define one selection cycle: incubation, binding, partitioning and washing, target-bound elution, and amplification (Fig. 1). The process starts with a single-stranded (ribo)nucleic acid sequence library of, typically, 10^{15} random sequences of fixed length flanked by constant primer sites to aid amplification. Each random sequence permits the molecule to fold into a unique 3D shape or conformation. At the start of each cycle, such a RNA/ssDNA pool is incubated with a target of interest. Due to the large number of unique sequences in the library, the probability of at least some aptamer molecules to bind the target with specificity and affinity is quite high. At the end of each cycle, low affinity binders are removed from the solution whereas bound aptamer molecules are eluted and amplified, forming the input for the next round. Eventually, only molecules that bind the target with high affinity remain. The aptamer molecules thus selected for high affinity and specificity are then individually evaluated experimentally and optimized for specific properties, such as size or stability, depending on the intended application. The experimental optimization is often assisted by computational analysis. Such analysis includes finding minimum free energy secondary structures and the identification of sequence motifs common to the final pool of aptamers. Recently, Hoinka et al. developed AptaMotif, a computational method for the identification of sequence-structure motifs in SELEX-derived aptamers [9].

New sequencing technologies have revolutionized the SELEX protocol by allowing deep/next-generation sequencing of entire aptamer pools ([10], Fig. 1). This extension, the so-called HT-SELEX, holds the promise for greatly accelerating aptamer discoveries and expanding their applications. For example, in the special case where the target molecule is a transcription factor, a variant of HT-SELEX designed for double-stranded DNA aptamers has been successfully used to uncover transcription factor binding motifs [11–13].

Traditionally, the SELEX process has been treated as a black box and only a handful of binders elucidated in the last cycle were sequenced. In contrast, sequencing of earlier pools using HT-SELEX provides the opportunity to uncover potential binders that might otherwise have been lost in later steps of the selection process. More importantly, by analyzing the relative changes of consecutive selection rounds of properties such as sequence diversity and mutation rates, the method provides an unprecedented opportunity to gain deeper insights into the selection process *per se*. Thus, HT-SELEX coupled with computational

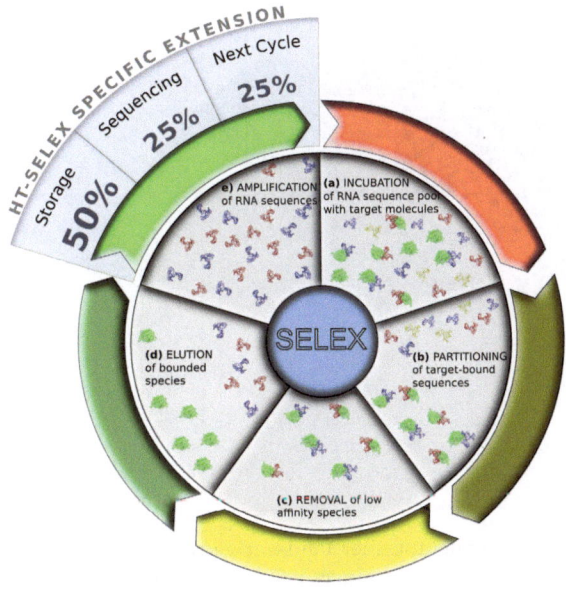

Fig. 1. The SELEX procedure iterates over five basic steps incubation, binding, partitioning and washing, target-bound elution, and amplification. Traditionally, only the binders elucidated in the last cycle were sampled and examined. The HT-SELEX includes sequencing of the final and intermediate selection pools.

assessment of the relation between sequences has the potential to trace the dynamics of the selection process and the rational selection of aptamers with desired properties making the SELEX process more rapid as well as more efficient.

Despite the success of HT-SELEX for drug design, efficient computational tools that exploit and encompass data from all sequenced rounds, therefore elucidating the selection process from the initial pool to the final cycles, have yet to be developed. Computational processing of HT-SELEX data is currently largely based on simple counting of aptamer species in the final round of selection, frequently discarding low-frequency species from the analysis, and choosing the sequences that occur in high counts for further investigation [10]. In addition, a small number of most frequent sequences from the final selection round might be used as seeds for similarity searches. The underlying postulate of these methods is that the best predictor of binding affinity is the frequency at which a particular aptamer occurs in a pool. While these approaches might be suitable for candidate identification, they lack the ability of providing insight into the mechanisms governing the selection process itself. Note that as the selection progresses, low affinity binders (Fig. 2 high z-coordinate) are eliminated from the pool leaving aptamers that sample local minima of binding energy. It is therefore expected that clustering of aptamers in consecutive cycles should provide valuable information about the selection process and should allow for the delineation of the entire aptamer landscape probed by the SELEX protocol. Hence, our primary

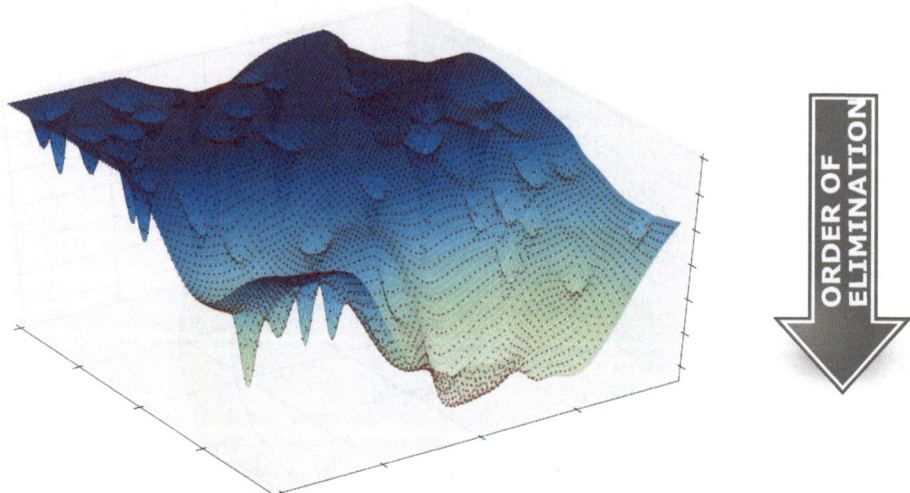

Fig. 2. A visualization of the aptamer landscape probed by the SELEX protocol. The surface represents all possible aptamers of fixed length and the red dots represent aptamers used in the initial pool. The distance on the surface is a conceptual projection of sequence similarity. Multiple local minima correspond to groups of aptamers that bind to the different areas of the targets surface or to the same region but are related by structure rather than sequence similarity.

objective is to cluster aptamers in all rounds of selection according to their sequence similarity. This task however could not be accomplished with previous clustering algorithms due to the enormous size (2-50 Million sequences per cycle) of the data set generated by high throughput sequencing, especially for early rounds of selection which feature a high degree of unique sequences ($\geq 90\%$). To address this challenge, we developed a novel approach, AptaCluster, capable of efficiently clustering entire aptamer pools.

Several sequence similarity measures are commonly found in clustering methods, of which the Hamming and Levenshtein (edit) distances are most prominent. However, full-scale clustering approaches are computationally untrackable for HT-SELEX data. Therefore we use the randomized dimensionality reduction technique, known as locality-sensitive hashing (LSH) [14], to implicitly approximate an upper bound to the edit distance for each sequence pair without the need of exhaustive pairwise comparison. In the subsequent step, we eventually compute precise sequence distances based on k-mer counting between pairs of aptamers below this bound, while the remaining distances are not relevant and might be arbitrarily assumed to be infinity.

We applied AptaCluster to analyze the results of the HT-SELEX experiment that we preformed using Interleukin 10 receptor alpha chain (IL-10RA) as the target molecule. IL-10 is considered to be a master regulator of immunity to infection and is an important therapeutic molecular target [15]. We preformed 5 cycles of HT-SELEX with a 40nt variable region, sequencing the samples of pools 2-5. AptaCluster has enabled us to analyze the results of HT-SELEX, revealed interesting properties of the selection landscape, and allowed for a better understanding of the HT-SELEX experiment. AptaCluster scales very well with data size. While the sequenced pools in our IL-10RA HT-SELEX experiment varies between 2 and 4.5 Million aptamers, we have applied AptaCluster to much larger pools of more than 20 Million sequences in the context of whole-cell HT-SELEX (data not shown) without loss of noticeable performance.

2   The AptaCluster Algorithm

Our approach is centered around a randomized dimensionality reduction technique, known as locality-sensitive hashing (LSH) [14]. First, a compressed representation of the data set is constructed by reducing the pool to non-redundant species and their corresponding frequency counts. We then apply a user-defined number of randomized locality-sensitive hash functions to the data set in order to distinguish sequence pairs that are potentially similar from those that are, with very high probability, not similar. Each function operates by selecting a small number of nucleotide positions from each aptamer and treats the substring, resulting from the concatenation of these bases, as input for the hashing procedure. Hence, aptamers with highly similar primary structure are likely to fall into the same group whereas dissimilar sequences rarely produce identical hash values. In the third step, the actual clustering step, we compute precise sequence distances between aptamers of identical hash value, while the distances between the aptamers never encountered in the same group are set to infinity. To accelerate the clustering, AptaCluster relies on a similarity measure based on k-mer counting. Thus the algorithm preforms three main steps outlined below. Relevant implementation details and the parameters used throughout this study can be found in the Methods section.

Dataset Compression. Data compression is achieved by using a hash map in which the keys correspond to the species in the pool and the values correspond to their respective frequency counts which can be done in $\mathcal{O}(N)$ time. In the following, let $s = (s_i)_{i=1}^{l}$ be an aptamer of of length l defined by the sequence of nucleotides s_i over the alphabet $\Omega = \{A, C, G, T\}$ where the index i corresponds to the ith position of the aptamer. Furthermore, we define $S = \{s^j \in P \parallel s^j \neq s^k \; \forall j, k \in [1, \ldots, |S|] \wedge \sum_{j=1}^{|S|} m(s^j) = N\}$, where $m(s^j)$ corresponds to the frequency of s^i, as the keys of the hash map, i.e. the set of unique aptamers for pool P.

Filtering Using Locality Sensitive Hashing. LSH is based on the idea that data points that are close in high dimension, after applying a probabilistic dimensionality reduction and using the reduced representation as the input to a hash function, are likely to obtain the same hash value and hence fall into the same bucket [16].

AptaCluster exploits this property by treating each sequence $s^j \in S$ as an l-dimensional vector and reducing this vector into d dimensions ($d < l$). This is done by generating a set I_d of d randomly sampled indices $i \in [1, \ldots, l]$ and, for each sequence s^j, only selecting those nucleotides s_i for which $i \in I_d$ as input for the hashing procedure. Hence, the more similar the primary structure of a set of aptamers, the higher the probability that they will produce the same mapping. Similarly, the choice of d controls the minimal degree of similarity between the members of each partition since these are guaranteed to differ in at most $l - d$ positions. In other words, our approach implicitly computes an upper bound to the edit distance. We iteratively improve this upper bound by repeating this procedure a user defined number of times, each time using a different hash function. With sufficient number of of iterations, if two sequences never fall into the same bucket they are assumed to be dissimilar with very high probability. The iterative computation of the upper bound is performed as follows. Let $d_{lsh}^k(s^1, s^2)$ be the upper bound computed after the kth iteration and let $L^k(s)$ be the value of the kth hash function for sequence s. We assume that, by default, we have for all pairs $d_{lsh}^0(s^1, s^2) = \infty$. Then

$$d_{lsh}^k(s^1, s^2) = \begin{cases} l - d & L^k(s^1) = L^k(s^2) \\ d_{lsh}^{k-1}(s^1, s^2) & L^k(s^1) \neq L^k(s^2) \end{cases} \tag{1}$$

Clearly, only the assignment in the first line needs to be executed. To define $L^k(s)$, for each iteration k we randomly select a mapping h from a family of functions

$$F = \{h : \mathbb{N}^l \to \mathbb{N}^d \, \| \, h(I) = I_d\} \tag{2}$$

where $I = (1, \ldots, l)$ represents the nucleotide positions of an aptamer of size l, and apply the function

$$L = \{\Omega^l \to \Omega^d \, \| \, L(s) = (s_i) \, \forall \, i \in I_d\} \tag{3}$$

to each aptamer s, creating a sub-string $\hat{s}$ comprised of the concatenation of the nucleotides at the positions defined in I_d. Finally, traditional hashing is performed on the set $\hat{S} = \{\hat{s}^i\}, i = 1, \ldots, |S|$. $I_d = (i_0, \ldots, i_d)$ can be efficiently computed as follows: Let $i_0 \in [1, l]$ be a randomly selected index of I and define $x \in [2, l-1]$ as a random number co-prime to l. Then, the remaining positions can be generated with

$$i_j = (i_{j-1} + x) \mod l, \quad j = 1, \ldots, d-1 \tag{4}$$

and

$$I_d = (i_j)_{j=0}^{d-1}, \quad i_j < i_{j+1} \, \forall \, j \tag{5}$$

corresponds to the sequence of indices after sorting these in ascending order. Using this scheme guarantees that each index in I is selected exactly once and avoids scenarios in which only adjacent positions of the sequence are chosen.

Cluster Extraction. Based on the assumption that high-frequency of a sequence in a selection pool is related to its selective advantage due to its binding affinity, we build the clusters iteratively around these high frequency aptamers. We repeatedly choose the highest frequency sequence s not assigned to any cluster, making it a seed of the new cluster. We then we employ a k-mer based distance function [17] to compute the distance of the selected seeds to all other sequences for which the upper bound estimated with LSH was finite and include it in the cluster if d_{kmer} is smaller than a user defined cutoff. In particular,

$$d_{kmer}(s^x, s^y) = \sum_{i=1}^{4k} \left| \frac{X_i}{|s^x| - k + 1} - \frac{Y_i}{|s^y| - k + 1} \right|^2 \tag{6}$$

where X_i and Y_i denotes the number of times the i-th k-mer occurs in sequence s^x and s^y respectively and $|s^i|$ corresponds to the length of the aptamer. Since we compare only sequences that are in the same bucket in at least one iteration, this approach allows us to extract clusters in $\mathcal{O}(N * m * k)$ where m denotes to the maximum number of seed sequences in a bucket which is bounded by the size of the largest bucket generated during LSH.

3   Results of Application to HT-SELEX Experiment for IL-10RA

We performed 5 rounds of HT-SELEX experiment with Interleukin 10 receptor alpha chain (IL-10RA) as the target molecule. Here, we summarize the insights obtained using AptaCluster.

3.1   Validating Clustering Results

The main advantage of AptaCluster is that to cluster an aptamer pool it does not need to compute the distances between all pairs of sequences but instead uses locality-sensitive hashing to filter out pairs that do not need to be compared. However, the filtering step is heuristic and its outcome might depend on the number of LHS iterations and properties of the dataset. Therefore we started by confirming that the filtering step produces correct results, i.e. that sequences filtered out as not potentially similar are indeed remote from the seed sequences in terms of exact distance. Since the dataset size prohibits an exhaustive computation of all distances, we used 400 aptamers (the 20 most frequent species from the top 20 clusters) and computed their edit distances to all other aptamers. We then computed the distribution of the distances to the members of the same cluster to the distances to the rest of the aptamers. The former group sampled the sequences whose distances to the reference sequences has been computed and found to be below the clustering threshold. The latter group sampled two types of sequences: the sequences whose distance to the reference sequence has been computed but found to be above the threshold and the sequences filtered out without computing the distance based on our locality sensitive hashing function.

The results for all selection cycles are summarized in Fig. 3 for a set of default and relaxed parameters (see Parameters section). The results demonstrate that no sequence that was filtered out using locality sensitive hashing is close to the seed sequences of the clusters. In addition, it also demonstrates that SELEX derived aptamer clusters are well separated. Indeed, relaxing the locality-sensitive hashing based filtering and increasing clustering threshold did not change the clustering results appreciatively (Fig. 3 (b)).

3.2   Distribution of Aptamers within Clusters

Next, we examined the distribution of aptamers within the clusters. Interestingly, we found that the distribution of these frequencies was very skewed (Fig. 4). Except for a handful of highly abundant aptamers, most of the species in a cluster had low frequencies. Such extreme differences in frequencies is consistent with a situation in which most of the cluster diversity can be attributed to mutations caused by Polymerase errors. To test this hypothesis, we investigated whether aptamers with a maximal count of 5 from the top 20 clusters in cycle 5 were also present in the sequenced portion of the selection pool from cycle 2. Indeed, the vast majority of these sequences (99% of singletons, 97% for frequency 5) where absent in this pool (Supplementary Table 1). Note that the sequences introduced by Polymerase errors can be subsequently selected and amplified providing an important source of cluster's diversity. However, due to the late introduction, their frequency count might not correctly reflect their binding affinity.

3.3   Frequency Counts versus Binding Affinity

It is often assumed that an aptamer sequence's frequency in the pool later cycles provides a good predictor of its binding affinity. Indeed this would be a reasonable expectation under the assumption that the selection process is free of any artifacts, all aptamers are present in the initial pool with the same frequency, and there was no stochastic variability during the above mentioned partitioning. However the realization that a large fraction of sequences in the final pool might have been absent from the initial pool but introduced in a later stage made us to reexamine this assumption. We measured disassociation constant K_d for 30 Aptamers including the most frequent ones. We found that cycle-to-cycle enrichment of aptamer frequencies, i.e. their relative increase in multiplicity, from cycle 4 to 5 is a better predictor of binding than the frequency in the final pool (data not shown). Specifically, taking 125 K_d as a reasonable threshold between binders and non-binders, sorting by cycle-to-cycle enrichment separates binders form non-binders while sorting by frequency leaves these two groups randomly mixed.

In addition to the emergence of new sequences, another source of dissonance between aptamer frequency and its binding potential could also be the differences in their frequencies in the initial pools due to the stochastic nature of partitioning the pool into groups to be used for storage/sequencing/next cycle. Looking at

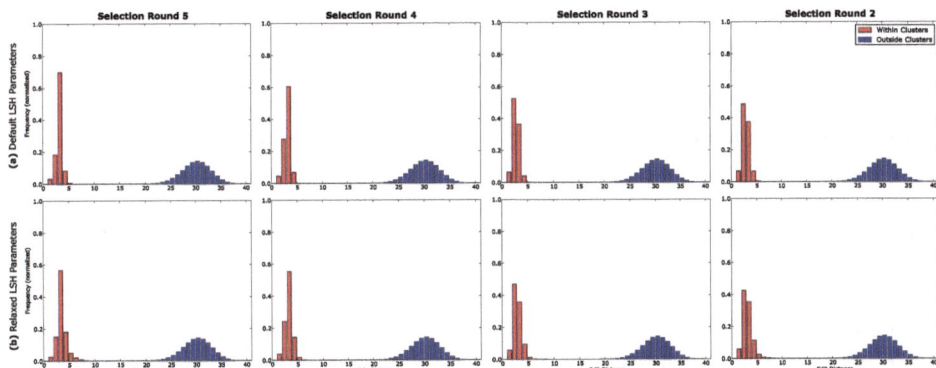

Fig. 3. Distribution of the edit distances between aptamers belonging to a cluster (red) and distances between cluster members and all non-cluster sequences (blue) for selection rounds 2 to 5. Within each of the top 20 clusters, the 20 most frequent aptamers where compared against all other cluster members as well as the remaining aptamers of the pool. (a) Distributions using the defaults parameters of AptaCluster as described in the Parameter section is shown in the top panel. (b) Relaxed parameters as depicted in the bottom panel in which only 40% of the randomized region was sampled during LSH.

cycle-to-cycle sequence enrichment instead of counts permits a resolution of this problem. However, other artifacts exist that can affect aptamer frequencies as well. In particular, we also tested the K_d values for non IL-10RA specific binding using binding to IgG as proxy for such non-specificity (data not shown). We found for example that cluster with ID 3 has high frequency in cycle 5 but it is not IL-10RA specific.

4   Conclusions and Discussion

Given the great promise of the HT-SELEX approach and rapidly diminishing costs of next generation sequencing, the usage of this method is likely to increase rapidly. Therefore it is imperative that researchers are able to analyze and correctly interpret HT-SELEX results. We have developed a new approach, AptaCluster, that allows for clustering based on primary structure of pools of aptamers sequenced using Hi-Seq technology.

Until now, a typical HT-SELEX analysis was reduced to counting the frequency of each aptamer and using such counts as a predictor of binding affinity. However our results indicate that such counting is actually not as good of a predictor as it has been anticipated. Instead, a predictor that utilizes the dynamics of the cycle-to-cycle enrichment holds greater promise.

Our results of applying AptaCluster to the outcome of the IL-10RA HT-SELEX experiment revealed important properties of the resulting clusters. We found the clusters to be well separated, and typically dominated by one or a few individuals. Relaxing the parameters to allow for larger intra-cluster distances did not change the results significantly. Consequently, sequence profiles of individual clusters were dominated by one or a few of the most abundant sequences. We have also implemented a procedure that enables the tracing of the clusters over consecutive selection cycles and, consistently with the observation above, we found that the clusters' sequence profile did not change much during consecutive selection steps.

The distribution of frequency counts within clusters suggests that cluster diversity is, in a large part, a result of Polymerase errors. The emergence of such Polymerase mutants creates an interesting opportunity to sample around local minima. This is strengthened by the observation that the number of mutations correlates with the frequency of the cluster seeds: the more frequent the seed, the more frequent the mutants. How to design the dynamics of the selection process to optimally utilize these emerging mutants is an open question. One possibility is to replace the typical selection procedure where selection pressure increases in each cycle by an approach that alternates between stronger and weaker selection.

AptaCluster provides a valuable tool which will help us and others to analyze and to optimize the HT-SELEX procedure. It has enabled us to analyze the results of HT-SELEX for IL-10 and allowed for a better understanding of the HT-SELEX experiment. We expect that the properties of the clusters obtained with AptaCluster will vary depending on the experimental details of HT-SELEX protocol in use, the length of the variable region, error rate of Polymerase, and properties of the target. Independently of this expected variability, AptaCluster can be used as the first step towards understanding the aptamer binding landscape, and for the identification of a broad spectrum of potential binders. We point out that AptaCluster is not intended to elucidate complex, indel-containing motifs but rather to operate on sequences of equal length. It is designed to serve as a pre-processing step for approaches to uncover sequence-structure motifs such as the planned extension of our AptaMotif algorithm to high throughput sequencing data [9].

5   Materials and Methods

5.1   Dataset Description

We applied 4 rounds of selection and cDNA generated from round 5 bound fractions as well as RNA recovered from bound fractions at rounds 2, 3 and 4 was amplified and sequenced using Illuminas HiSeq 2500 device with 100-cycle paired-end sequencing protocol (see HT-SELEX Experiment section for the experimental protocol). Aptamers were extracted by aligning the transcribed, inverted sequence of the reverse run to the corresponding forward lane and only retaining those sequences with less then 5 mismatches between the actual primers/tags and the identified primer region. Furthermore we restricted the number of allowed mismatches between the sequences of the forward and reverse lane in the randomized region to four. Mismatches in the randomized region were corrected by choosing the nucleotide with higher Illumina quality score. For the entire experiment, a total of 12895554 sequences where retrieved of which 4621438 species belonged to round 5, 1923823 to round 4, 2181720 to round 3, and 4168573 to round 2. Out of these respectively 617220, 1021668, 1902904, and 3857210 were unique.

5.2 Implementation Details

AptaCluster is currently available as a multi-threaded implementation in C++ using the OpenMP and Boost libraries for its parallel programming operations and hashing procedures, respectively [18, 19]. It features a complete, highly modular pipeline from data input and parsing, over cluster extraction, to result visualization and database storage. We implemented threaded parsers for a number of file formats, including FASTA, FASTQ, and RAW sequence files, both for paired-end and single-end sequencing data as well as automatic multiplexing procedures for separating the individual SELEX rounds when sequenced together. Depending on the number of available CPUs, clustering and distance calculations are performed in parallel for each pool. Cluster families and their evolution from cycle to cycle are currently visualized in HTML format. Finally, the algorithms behavior can be controlled using a configuration file allowing for the assignment of most parameters used for parsing and clustering, among others. We have empirically determined a set of default values, of which the most relevant are discussed below.

5.3 Parameters

For the experiments described in this paper, we performed a total of $r = 10$ iterations of LSH sampling 60% of the randomized region (i.e. $l = 24$). The parameter $d = 4$ is set in terms of the maximal number of point mutations any pair of sequences should have and is converted into the k-mer distance cutoff by sampling a user defined number of aptamers from the pool (10000 by default), artificially mutating that sequence up to d times, and averaging over all d_{kmer} between these mutants and the wild-type. Furthermore we set $k = 3$ for the computation of d_{kmer} which has shown to give reasonable results for aptamer-sized sequences.

5.4 HT-SELEX Experiment

Selection Details. A DNA template for the selection library was ordered from IDT (Coralville, IA). 1 nM of each N_{40} template (5-TCTCGATCTCAGCGAGTCGTCG $-N_{40}$-CCCATCCCTCTTCCTCTCTCCC-3) and 5 primer (5-GGGGGAATTCTAATACGACTC ACTATAGGGAGAGAGGAAGAGGGATGGG-3) were annealed together, extended with Taq polymerase (Life Science), and transcribed in vitro using Durascribe (in-vitro transcription) IVT kit (Illumina). The random R0 RNA was purified by denaturing PAGE and, after preclearing with human IgG-coated (Sigma) beads (GE Healthcare), used for in-vitro selection. 1 nM of R0 RNA was used in a first round of selection to coincubate with 0.3 nM of bead-bound human IL-10RA-Fc fusion protein (Novus Biologicals) in 100 mM NaCl selection buffer. After washes, a recovered bound RNA fraction was reverse transcribed using the cloned AMV RT kit (Life Science). cDNA was amplified by either emulsion or open PCR using Platinum Taq PCR kit (Life Science) as described below. The DNA template was used to IVT RNA for the next round. During subsequent rounds, amount of protein was reduced 25% each time, while concentration of NaCl was gradually increased to 150 mM.

Emulsion PCR. cDNA was amplified using Platinum Taq PCR kit with addition of 10% PCRx enhancer solution and following primers: 5-GGGGGAATTCTAAT ACGACTCACTATAGGGAGAGAGGAAGAGGGATGGG-3 and 5-TCTCGATCTCAGCGAGTCGTCG-3. After preparing the master mix PCR reaction solution, it was separated to 100 μL aliquots and each aliquot was mixed with 600 μL ice-cold oil fraction assembled from components supplied with emulsion PCR kit (EURx) according to manufacturers instructions. Water and oil mixture was emulsified by 5 vortexing at +4C and amplified in standard PCR machine for 25 cycles. Control open PCR reaction was carried with aqueous phase only for 16 cycles.

Preparing Libraries for HTS. After 4 rounds of selection, 3 nM of RNA was prepared for round 5. The RNA was pre-cleared using IgG-coated beads and separated into three identical aliquots. Each aliquot was incubated with either human IL10RA protein, murine IL10RA protein or human IgG. After standard washes, bound RNA fraction was extracted from beads and reverse transcribed as described previously. A cDNA generated from round 5 bound fractions, as well as RNA recovered from bound fractions at rounds 2, 3 and 4, was amplified by emulsion PCR with two sets of primers as described previously [2]. Amplified DNA was purified by 2% agarose gel electrophoresis and sequenced using Illuminas HiSeq 2500 device with 100-cycle paired-end sequencing protocol.

Acknowledgments. We thank Rolf Backofen and Fabrizio Costa, Freiburg University for a stimulating discussion. This work was supported in part by the Intramural Research Program of the NIH, National Library of Medicine (JH, TMP), in part funds from the Laboratory of Hemostasis and the Center for Biologics Evaluation and Research, Food and Drug Administration's Modernization of Science program (ZES), and in part by bequest from the Dodson estate and the Sylvester Comprehensive Cancer Center, Medical School, University of Miami (AB and EG). The funders had no role in study design, data collection and analysis, decision to publish, or preparation of the manuscript. The findings and conclusions in this article have not been formally disseminated by the Food and Drug Administration and should not be construed to represent any Agency determination or policy.

References

1. Kim, Y.S., Gu, M.B.: Advances in aptamer screening and small molecule aptasensors. Advances in Biochemical Engineering/Biotechnology (July 2013) PMID: 23851587

2. Berezhnoy, A., Stewart, C.A., Mcnamara, J.O.N., Thiel, W., Giangrande, P., Trinchieri, G., Gilboa, E.: Isolation and optimization of murine il-10 receptor blocking oligonucleotide aptamers using high-throughput sequencing. Molecular Therapy: The Journal of the American Society of Gene Therapy 20(6), 1242–1250 (2012) PMID: 22434135

3. Binning, J.M., Wang, T., Luthra, P., Shabman, R.S., Borek, D.M., Liu, G., Xu, W., Leung, D.W., Basler, C.F., Amarasinghe, G.K.: Development of rna aptamers targeting ebola virus vp35. Biochemistry (September 2013) PMID: 24067086

4. Shi, H., Cui, W., He, X., Guo, Q., Wang, K., Ye, X., Tang, J.: Whole cell-selex aptamers for highly specific fluorescence molecular imaging of carcinomas in vivo. PloS One 8(8), e70476 (2013) PMID: 23950940

5. Cerchia, L., Hamm, J., Libri, D., Tavitian, B., de Franciscis, V.: Nucleic acid aptamers in cancer medicine. FEBS Letters 528(1-3), 12–16 (2002) PMID: 12297271

6. Macugen: Fda approves new drug treatment for age-related macular degeneration

7. Zichel, R., Chearwae, W., Pandey, G.S., Golding, B., Sauna, Z.E.: Aptamers as a sensitive tool to detect subtle modifications in therapeutic proteins. PloS One 7(2), e31948 (2012) PMID: 22384109

8. Ellington, A.D., Szostak, J.W.: In vitro selection of rna molecules that bind specific ligands. Nature 346(6287), 818–822 (1990) PMID: 1697402

9. Hoinka, J., Zotenko, E., Friedman, A., Sauna, Z.E., Przytycka, T.M.: Identification of sequence-structure rna binding motifs for selex-derived aptamers. Bioinformatics 28(12), i215–i223 (2012) PMID: 22689764

10. Kupakuwana, G.V., Crill, J.E.N., McPike, M.P., Borer, P.N.: Acyclic identification of aptamers for human alpha-thrombin using over-represented libraries and deep sequencing. PloS One 6(5), e19395 (2011) PMID: 21625587

11. Zhao, Y., Granas, D., Stormo, G.D.: Inferring binding energies from selected binding sites. PLoS Computational Biology 5(12), e1000590 (1999) PMID: 19997485

12. Ogawa, N., Biggin, M.D.: High-throughput selex determination of dna sequences bound by transcription factors in vitro. Methods in Molecular Biology 786, 51–63 (2012) PMID: 21938619

13. Jolma, A., Kivioja, T., Toivonen, J., Cheng, L., Wei, G., Enge, M., Taipale, M., Vaquerizas, J.M., Yan, J., Sillanp, M.J., et al.: Multiplexed massively parallel selex for characterization of human transcription factor binding specificities. Genome Research 20(6), 861–873 (2010) PMID: 20378718

14. Gionis, A., Indyk, P., Motwani, R.: VLDB 1999. In: Similarity Search in High Dimensions via Hashing, pp. 518–529. Morgan Kaufmann Publishers Inc. (1999)

15. Couper, K.N., Blount, D.G., Riley, E.M.: Il-10: the master regulator of immunity to infection. Journal of Immunology 180(9), 5771–5777 (2008) PMID: 18424693

16. Andoni, A., Indyk, P.: Near-optimal hashing algorithms for approximate nearest neighbor in high dimensions. Commun. ACM 51(1), 117–122 (2008)

17. Yang, K., Zhang, L.: Performance comparison between k-tuple distance and four model-based distances in phylogenetic tree reconstruction. Nucleic Acids Research 36(5), e33 (2008) PMID: 18296485

18. OpenMP Architecture Review Board: OpenMP application program interface version 3.0 (May 2008)

19. Schling, B.: The Boost C++ Libraries. XML Press (2011)

AptaCluster – A Method to Cluster HT-SELEX Aptamer Pools and Lessons from Its Application Supplementary Materials

6   Supplementary Figures and Tables

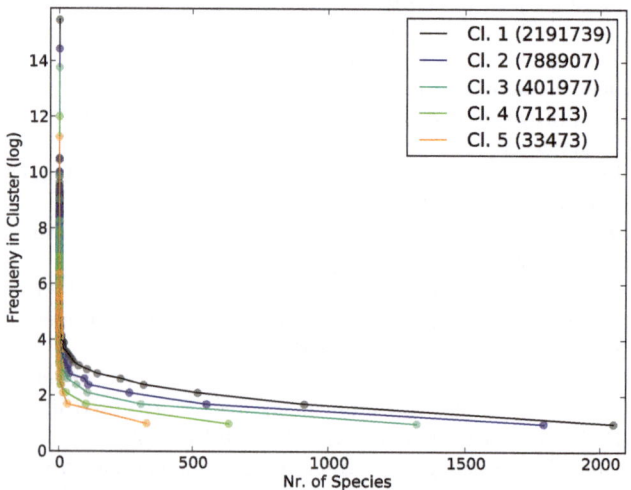

Fig. 4. The frequency distribution of the members of the 5 largest clusters. The cluster sizes are given in the brackets.

Table 1. Number of species with counts 1 to 5 present in the top 20 clusters of selection round 5 compared to the frequency of their occurrence in selection round 2. The overwhelming majority of the sequences are not present in the latter.

	Nr. of aptamers with frequency				
	1	2	3	4	5
Top 20, cycle 5	8529	2202	1074	614	465
Found in cycle 2	61	36	27	18	16

Learning Sequence Determinants of Protein: Protein Interaction Specificity with Sparse Graphical Models

Hetunandan Kamisetty[1,*,**], Bornika Ghosh[2], Christopher James Langmead[3], and Chris Bailey-Kellogg[2,*]

[1] Department of Biochemistry, University of Washington
hetu@uw.edu
[2] Department of Computer Science, Dartmouth
cbk@cs.dartmouth.edu
[3] School of Computer Science, Carnegie Mellon University
cjl@cs.cmu.edu

Abstract. In studying the strength and specificity of interaction between members of two protein families, key questions center on *which* pairs of possible partners actually interact, *how well* they interact, and *why* they interact while others do not. The advent of large-scale experimental studies of interactions between members of a target family and a diverse set of possible interaction partners offers the opportunity to address these questions. We develop here a method, DGSPI (Data-driven Graphical models of Specificity in Protein:protein Interactions), for learning and using graphical models that explicitly represent the amino acid basis for interaction specificity (*why*) and extend earlier classification-oriented approaches (*which*) to predict the ΔG of binding (*how well*). We demonstrate the effectiveness of our approach in analyzing and predicting interactions between a set of 82 PDZ recognition modules, against a panel of 217 possible peptide partners, based on data from MacBeath and colleagues. Our predicted ΔG values are highly predictive of the experimentally measured ones, reaching correlation coefficients of 0.69 in 10-fold cross-validation and 0.63 in leave-one-PDZ-out cross-validation. Furthermore, the model serves as a compact representation of amino acid constraints underlying the interactions, enabling protein-level ΔG predictions to be naturally understood in terms of residue-level constraints. Finally, the model, DGSPI readily enables the design of new interacting partners, and we demonstrate that designed ligands are novel and diverse.

Keywords: protein:protein interaction, specificity, ΔG prediction, graphical model, PDZ.

1 Introduction

The molecular machinery of the cell is driven largely by protein:protein interactions. Traditional high-throughput technologies [5] provide evidence for the

---

* Now at Facebook Inc.
** Corresponding author.

R. Sharan (Ed.): RECOMB 2014, LNBI 8394, pp. 129–143, 2014.
© Springer International Publishing Switzerland 2014

existence of interactions that existing computational systems biology techniques utilize to build global networks of interacting proteins. However, finer-grained methods are necessary in order to better understand, predict, and control these interactions. Fortunately, appropriate experimental methodologies are rapidly developing, e.g., using protein microarrays to isolate numerous pairs of possible partners, and fluorescence polarization to assess their interaction strength (Fig. 1, left). Several large-scale studies have been conducted using such techniques for particular families of interacting proteins, including PDZ domains and their peptide ligands [4,40], and human basic-region leucine zippers (bZIPs) and their coiled-coil partners [6,8]. In lieu of large-scale studies, the aggregation of a large number of smaller-scale experiments can also yield extensive amounts of detailed binding data, e.g., for major histocompability complex (MHC) and ligands [29,27,41,2,43], and serine proteases and inhibitors [21,18].

As one particular example, consider the specific recognition between PDZ domains and their peptide ligands. PDZs are small peptide recognition modules that bind specific C-terminal peptides of other proteins (Fig. 1, right), in order to mediate protein:protein interactions (e.g., in signaling networks). Early studies of PDZ:peptide recognition developed consensus motifs to capture the common amino acids comprising the ligands of different PDZ "classes" (e.g., class I = S/T-X-Φ vs. class II = Φ-X-Φ, where Φ is a hydrophobic residue). More recent studies yielded more refined statistical *binary* interaction predictors (interact or not?), based on analysis of amino acid pairs (across the PDZ:peptide interface) in curated datasets of experimentally identified PDZ:ligand partners [3,38]. MacBeath and colleagues then made the leap to large-scale quantitative data, determining the ΔG of binding for 829 PDZ:peptide pairs from 96 PDZs (from mouse, fly, and worm) against a panel of 259 possible peptide partners [36]. They used this data to develop a binary interaction predictor, based on the constituent PDZ:peptide amino acid pairs like the predictors mentioned above, but taking advantage of the quantitative and negative data [4]. More recently, Bader and co-workers used the MacBeath data to train a type of support vector regression model for predicting ΔG of binding for PDZ:peptide pairs [33].

Motivated by the exciting growth in quantitative studies of protein:protein interactions, we have developed a data-driven, sequence-based model that directly and compactly reveals and represents the amino acid interactions underlying experimentally measured ΔG values of binding (henceforth just ΔG) and enables efficient, accurate, robust, and transparent prediction of ΔGs for new pairs of possible partners (Fig. 1). We employ a graph-structured model (which we refer to simply as a graphical model) that explicitly models amino acid interactions and provides a probabilistic interpretation for them. Sequence-based graphical models of protein families have been used to capture amino acid interactions in order to predict protein structure [26,10,12] and function [37,1] and design new proteins [39,15]. We build here on our sequence-based models of interacting protein families for binary prediction of interaction [38], significantly extending that approach to incorporate quantitative data and thereby predict ΔG.

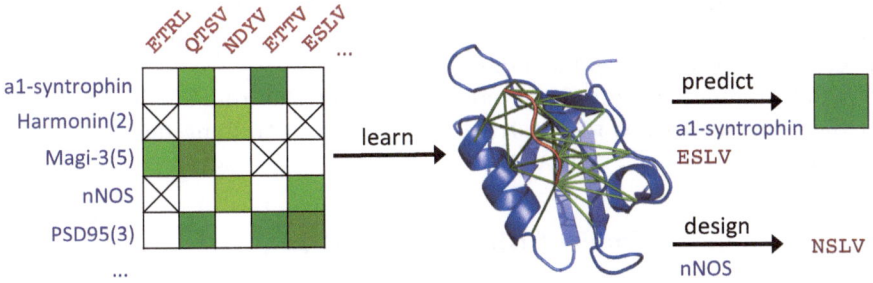

Fig. 1. DGSPI: Data-driven Graphical models of Specificity in Protein:protein Interactions. A graphic model of PDZ:peptide interactions encapsulates the amino acid constraints conferring the strength and specificity of the interactions in an input dataset. **(left)** The dataset has ΔG values (shades of green) or "non-interacting" indications ('X's) for some PDZ (blue) peptide (red) pairs. **(middle)** We learn a graphical model with bipartite nodes for some residues in the PDZ (blue) and peptide (red), with edges (green) encapsulating and providing a probabilistic interpretation for amino acid constraints. **(right)** We use the model to predict novel interactions as well as to design novel peptide partners for PDZs.

We call our approach DGSPI, for Data-driven Graphical models of Specificity in Protein:protein Interactions. Using the PDZ data from MacBeath and co-workers, we demonstrate that DGSPI is highly predictive of ΔG, obtaining predicted-experimental correlation coefficients of up to 0.69 in a ten-fold cross-validation and 0.63 in leave-one-PDZ-out cross-validation. This performance is essentially equivalent to that obtained by Bader and colleagues, but importantly, our approach provides a readily interpretable model of the amino acid contributions underlying specific interactions. Furthermore, since our graphical models can be used in designing new interacting partners (again, interpretable in terms of the amino acid contributions), and we show that there is a diversity of novel peptides that are predicted to bind well against any given PDZ and thus provide worthwhile hypotheses for experimental testing.

2   Methods

DGSPI takes as input (Fig. 1, left) two sets of protein sequences; for simplicity but without implications about function, we refer to one set as the "receptor" and the other as the "ligand"; e.g., the PDZ protein recognition modules as receptors and corresponding peptides as ligands. In addition to the sequences, there are experimental binding measurements for some of the pairs (one from each set); the measurement is either a ΔG value or an indication of "non-interacting" within the sensitivity of the experiment. Our goal is to be able to predict the ΔG of interaction for a previously untested receptor:ligand pair and to design new ligands partners for a specified receptor (Fig. 1, right). To do this, we seek a method that admits explanation of predictions in terms of the underlying amino acid-level interactions conferring specificity of interaction. Thus we employ a graph-structured, or graphical, model (Fig. 1, middle) with nodes for the receptor and ligand residues, and bipartite edges capturing the amino acid constraints between receptor and ligand residues.

We first summarize a graphical model to predict ΔG from a pair of sequences, and then the algorithms to construct a model from training data of sequence pairs with observed ΔG.

2.1   A Graphical Model of Binding Free Energy

We assume the receptors have been multiply aligned to p informative (non-gappy) columns, and the ligands likewise to q residues. Let $\mathbf{X} = \{X_1, X_2, \ldots, X_p\}$ be a set of p random variables representing the receptor amino acid composition, with X_i a discrete random variable for the amino acid type at position i. Each X_i takes values in $\mathcal{A} = \{$ala, arg, $\ldots$, val, -$\}$, corresponding to the 20 amino acid types and an additional '-' for a gap in the multiple sequence alignment. Similarly, let $\mathbf{Y} = \{Y_1, Y_2, \ldots, Y_q\}$ be a set of q random variables representing the ligand composition.

Given a receptor sequence $\mathbf{x} = \{x_1, x_2, \ldots, x_p\}$ (i.e., amino acid values for the random variables in $\mathbf{X}$), along with a ligand sequence $\mathbf{y} = \{y_1, y_2, \ldots, y_q\}$, we want to predict the strength of a possible interaction between the two proteins, $\Delta G_{pred}(\mathbf{x}, \mathbf{y})$. Our goal here is to develop a robust predictive model that is interpretable in terms of the amino acid interactions driving specific protein:protein recognition. Therefore we model ΔG_{pred} with a bipartite graphical model, with nodes for $\mathbf{X}$ and $\mathbf{Y}$ representing the amino acids and edges $\mathcal{E} \subset \mathbf{X} \times \mathbf{Y}$ representing their dependencies. Nodes x_i, y_j have associated $|\mathcal{A}| \times 1$ vectors $V_i[a], V_j[b]$ to capture position specific environment effects to binding. Edge (i, j), between nodes x_i and y_j, has an associated $|\mathcal{A}| \times |\mathcal{A}|$ matrix $W_{i,j}[a, b]$ of weights for $a, b \in \mathcal{A}$, holding the position-specific contributions to binding for each possible pair of amino acids, intended to capture electrostatics, van der Waals, hydrogen bonding, and other such interactions, which depend on the composition of the amino acids involved. We point out that these physical justifications for the parameters of our model are descriptive rather than prescriptive: they guide the intuition for learning and interpreting the model, but we make no assumptions on sources of these interactions beyond what we can learn from data.

In summary, then, given two protein sequences $\mathbf{x}$ and $\mathbf{y}$, we predict their binding free energy as:

$$\Delta G_{pred}(\mathbf{x}, \mathbf{y}) = \sum_{i=1}^{p} V_i[x_i] + \sum_{j=1}^{q} V_j[y_j] + \sum_{(i,j) \in \mathcal{E}} W_{i,j}[x_i, y_j] \qquad (1)$$

2.2   Training Objective

Our data-driven approach to modeling protein:protein interaction specificity uses experimental data to learn the parameters V and W that define the model in Eq. 1. We assume that the experimental data is partitioned into interactions $\mathcal{I}_+$, with ΔG values, and non-interactions $\mathcal{I}_-$, where the binding was weaker than ΔG_{max}, a maximum experimentally detectable ΔG value. Thus each member of $\mathcal{I}_+$ is of the form $(\mathbf{x}, \mathbf{y}, \Delta G)$, giving a pair of sequences and the measured ΔG value, while each member of $\mathcal{I}_-$ is simply an $(\mathbf{x}, \mathbf{y})$ pair.

We take as our primary objective minimizing the squared error between the observed and predicted ΔG values for members of $\mathcal{I}_+$. For the non-interactions in $\mathcal{I}_-$, we incorporate a penalty for an incorrect prediction, i.e., for ΔG_{pred} better than ΔG_{max}. In particular, we use a one-sided squared penalty for non-interactions predicted as interactions. Compared with the hinge-loss commonly used in SVMs, this tends to penalize small differences to a lesser extent, which is a desirable property in cases such as ours where the focus is on the regression error and not the misclassification cost. The one-sided square error has no points of discontinuity, making optimization easier as well.

Thus our objective function for a specific set of parameters V, W is:

$$L(V, W) = \sum_{(\mathbf{x}, \mathbf{y}, \Delta G) \in \mathcal{I}_+} (\Delta G_{pred}(\mathbf{x}, \mathbf{y}; V, W) - \Delta G)^2$$

$$+ \sum_{\substack{(\mathbf{x}, \mathbf{y}) \in \mathcal{I}_- \ s.t. \\ \Delta G_{pred}(\mathbf{x}, \mathbf{y}; V, W) < \Delta G_{max}}} \gamma_- \cdot (\Delta G_{pred}(\mathbf{x}, \mathbf{y}; V, W) - \Delta G_{max})^2 \quad (2)$$

where we emphasize the dependence of ΔG_{pred} in Eq. 1 on V and W by including them as parameters. The parameter γ_- sets the relative weighting between the contributions from interactions and non-interactions.

2.3 Block-Sparse Regularization

A suitable model can be learned from the data by minimizing the objective function in Eq. 2. However, directly optimizing this function is likely to lead in overfitting as there are usually far more parameters in the model than there are data points available with which to fit them. To circumvent this problem, we instead optimize a regularized objective function. Regularization is usually described as a penalty to the objective function; an alternate but equivalent view of the regularization is that it is a Bayesian prior on the models that biases the learning method towards models consistent with the prior. Protein:protein interactions can be reasonably expected to display structural sparsity —due to spatial restrictions, only a few of all possible bipartite interactions between the partners are likely to be important in biochemical interactions. Motivated by this prior belief, we employ block-L1 regularization, a form of regularization that penalizes the number of non-zero edges (or "blocks"), so that each edge (i, j) is penalized unless all parameters within the edge, $W_{i,j}$, are zero. This promotes a sparser structure promoting interpretability; furthermore, by reducing the number of non-zero parameters in the model, it helps avoid overfitting. For our model, the block-L1 regularization term is:

$$R_{1,2}(V, W) = \frac{1}{\sqrt{|\mathcal{A}|}} \left(\sum_{i=1}^{p} \|V_i\| + \sum_{j=1}^{q} \|V_j\| \right) + \sum_{(i,j) \in \mathcal{E}} \|W_{i,j}\| \quad (3)$$

where $\| \cdot \|$ refers to the vector two-norm of the corresponding set of parameters and the fraction $\frac{1}{\sqrt{|\mathcal{A}|}}$ (number of amino acids plus gap) is a correction factor to account for the different degrees of freedom in V and W.

Our learning objective is then:

$$\arg\min_{V,W} \left(L(V,W) + \lambda_{1,2} \cdot R_{1,2}(V,W) \right) \tag{4}$$

where $\lambda_{1,2}$ sets the relative weight between the learning objective and the regularization term.

2.4   Learning Algorithms

Schmidt *et al.* [32] developed a Limited-Memory Projected Quasi-Newton (PQN) approach suitable for squared error objectives. We customize their method for our graphical models of protein-protein interactions. A constrained optimization to incorporate the block-L1 regularization is performed by a projected gradient method that iterates between unconstrained gradient descent updates to the parameter values, and constrained projections of the parameter values onto the constrained space.

While this approach can be used to learn the structure and parameters of the model (i.e., which vertices and edges, along with their weights), in practice, the resulting procedure can result in biased weights for the non-zero parameters, despite identifying the correct structure [1,23]. To avoid this, after learning the structure of the model, we re-learn the non-zero parameters with L2 regularization. That is, in a second stage, we restrict the optimization to the vertices and edges contributing in the first stage, but reoptimize their weights using a modified version of Eq. 4, replacing $R_{1,2}$ with: $R_2(V,W) = \sum_{i=1}^{p} \|V_i\|^2 + \sum_{j=1}^{q} \|V_j\|^2 + \sum_{(i,j)\in\mathcal{E}} \|W_{i,j}\|^2$, weighted by a corresponding λ_2. Since $R_2(V,W)$ penalizes the square of the vector 2-norms, each element of each parameter vector is penalized independent of any group membership; the regularization is thus independent of the degrees of freedom in the corresponding groups. This two-stage approach finds sparse models with small edge weights, regularizing a pseudo-likelihood objective similarly to the approach of [1]. We find in practice that this approach yields models that are both interpretable and predictive of ΔG.

3   Results

Our goal is to make quantitative predictions of the ΔG of PDZ:peptide interactions, interpretable in terms of underlying amino acid constraints. This is in contrast to the approach of MacBeath and co-workers [4], who studied the ability of a computational method to classify interaction vs. non-interaction. (A graphical model approach to do that has been previously described [38]; we have found that classifying based on predicted ΔG is not as robust.) It is also in contrast to the Support Vector Regression approach of Bader and co-workers [33], in that while our method achieves comparable predictive accuracy, it has the added benefit of being able to automatically identify the amino acid-level interactions with the greatest impact, and directly characterize their contributions. These interactions not only allow us to characterize the sequence determinants of binding affinity and specificity, but also allow us to design new interacting partners based on the derived "rules" of good interactions.

We apply DGSPI to the extensive PDZ dataset collected by MacBeath and colleagues [36,4]. To enable appropriate comparison of results, we use the processed version of the dataset provided by Bader and colleagues [33]. The dataset includes 82 mouse PDZs and 217 peptides, with a reported 560 interactions and 1167 non-interactions. We obtained a structure- and sequence-based multiple sequence alignment of 225 columns where the peptides were represented by 5 C-terminal residues. We then removed highly conserved and highly gap-ful columns, reducing the alignment to 114 PDZ positions and 5 peptide positions.

3.1 ΔG Prediction

10-fold cross-validation. To test the ability of our model to predict the affinity of PDZ-peptide interaction, we first performed a ten-fold cross-validation (i.e. we learned the model with 90% of the data and tested it on the left-out 10%, doing this with each 10% left out). This represents the scenario in which data are available for some interactions, and we want to make predictions for others.

Our learning approach has three main parameters: γ_-, a parameter trading off the relative importance of positive and negative interactions in the objective function; $\lambda_{1,2}$, the strength of the block-L1 regularization used to determine the non-zero parameters of the model and λ_2, the regularization weight used to estimate the values of the non-zero parameters. γ_- and λ_2 were set to 0.05 and 1 respectively based on our initial experiments using one train-test split. The small value of γ_- reflects the relative abundance of non-interactions in our dataset and our emphasis on modeling interactions comprehensively since they are biologically more interesting. For each training split, we varied $\lambda_{1,2}$ generating multiple models spanning the spectrum from models with no interactions to models where nearly all possible interactions were allowed.

Fig. 2 summarizes trends over values of $\lambda_{1,2}$. The top-left panel characterizes the increase in number of interactions with decreasing $\lambda_{1,2}$, and top-right panel the corresponding increase in the average Pearson correlation coefficient, from 0.49 when there are no edges in the model and all contributions are due to the V_i, V_j, to 0.66 when ~ 60 interactions are included, to a maximum of 0.69 when ~ 300 interactions are included. The relatively large increase in model accuracy when the number of edges increases from 0 to 60 suggests that these edges make important contributions to binding affinity and specificity. In contrast, the relatively small increase in accuracy of the model as the number of edges increases beyond 60 to being a completely connected model suggests that the edges introduced later have relatively low importance. The bottom panel shows the average strength of each edge, calculated as the norm of $W_{i,j}$, as a function of $\lambda_{1,2}$. Each line represents a separate unique interacting pair of residues; interactions that

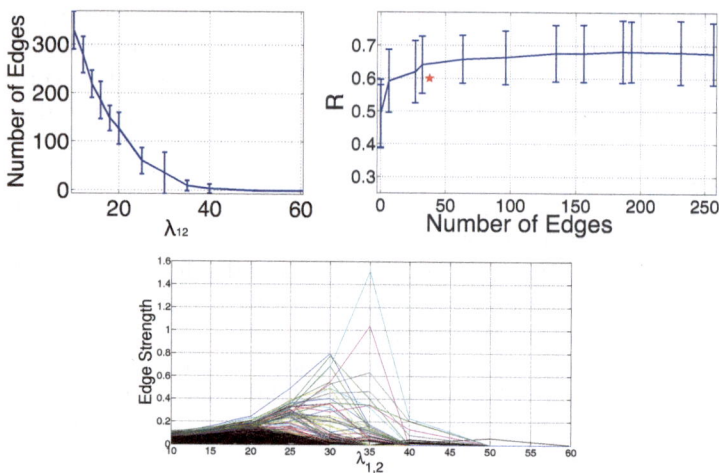

Fig. 2. Trends with varying regularization weight (parameter $\lambda_{1,2}$), with higher values yielding sparser models. **(top-left)** Number of edges in models learned with varying $\lambda_{1,2}$. **(top-right)** Regularization path, showing each edge's strength as a function of $\lambda_{1,2}$. The red star indicates a model with edges fixed according to contacts in a crystal structure. **(bottom)** Test correlation coefficient as a function of the average number of edges in trained model.

have high weight at $\lambda = 25$ are highlighted in color, while the remaining interactions are shown in black. When the model is sparse (high $\lambda_{1,2}$), there are few, strong, interactions; as the density of the model increases (low $\lambda_{1,2}$), most interactions have non-zero strength but are very weak.

Fig. 3 shows the prediction results for one 10-fold repetition at $\lambda_{1,2} = 20$. The overall correlation coefficient across the dataset was 0.67 while the root mean square error between experiment and prediction was 0.62. Most errors were equally distributed around zero, and actually within typical experimental error. However, there were a few clear outliers where the model under-predicted binding energies.

Contact-based model structure. When an experimentally determined 3D structure of the protein-protein interface is available, an alternate approach to determining the structure (edges) of the graphical model could be to restrict the non-zero interactions to the pairs of residues close to each other in the 3D structure. The parameters of this model with fixed structure can then be readily learned with L2 regularization, as before. Macbeath and colleagues identified 38 contacts between 16 PDZ residues and 5 peptide [4]. We repeated our 10 fold cross-validation experiments, using these 21 positions and 38 contacting residues as the set of vertices and edges in the model (instead of identifying them using the block-L1 penalty), and estimated their parameters with L2 regularization. The average correlation coefficient of the contact-based models is 0.60, which, while good, is lower than the 0.66 correlation obtained by models with about

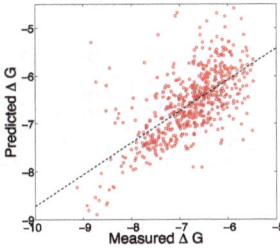 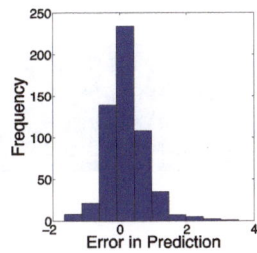

Fig. 3. Example prediction results, combined across 10 splits in one repetition at $\lambda_{1,2} = 20$. **(left)** Scatterplots of experimental vs. predicted ΔG. Pearson Correlation Coefficient across entire test-split was 0.67. **(right)** Histogram of prediction errors.

60 interactions. Could the difference in accuracy be due to the difference in the number of interactions? The top-right panel in Fig. 2 highlights the accuracy of this model (shown as a red star), compared to the correlation coefficients obtained by varying $\lambda_{1,2}$. We see that the models with learned structure can achieve accuracy similar to the contact-structure model but using *fewer* interactions; alternativeely, a model with learned structure and a comparable number of interactions to that of the contact structure achieves higher correlation. Thus our data-driven approach to learning model structure can identify important interactions beyond those that might be inferred by inspection of the 3D structure.

Leave-one PDZ out. To test the scenario where the model is applied to make predictions for a new PDZ, we performed "leave-one-PDZ-out" cross-validation following the approach of Bader and colleagues [33]. We held out data for each of the 23 PDZ domains with at least 10 interactions, training the model on the remaining data and testing on the held-out domain. Since the effect of $\lambda_{1,2}$ on the sparsity of the model depends on the number of sequences in the training set, instead of choosing the same value of as selected by ten-fold cross validation, we performed a grid search on $\lambda_{1,2}$ and used the value that gives a model of similar sparsity as the cross-validated models. This process allows us to parameterize the model by the number of edges as opposed to the less natural $\lambda_{1,2}$. Using this procedure we obtained an average correlation coefficient of 0.61 across the 23 PDZs that had at least 10 interactions. Again, allowing for denser models by changing the regularization weight slightly improved the average correlation coefficient to 0.63 which is comparable to the 0.65 obtained by Bader and colleagues by Support Vector Regression [33]; when restricting to contact edges, we obtain 0.54 (about the same as the 0.56 of Bader and colleagues).

3.2   Model Analysis

A key feature of DGSPI is that a model can be easily "opened up" to characterize the amino acid determinants of binding. To illustrate, we characterized the models trained at $\lambda_{1,2} = 25$ across the 10 folds, computing the average

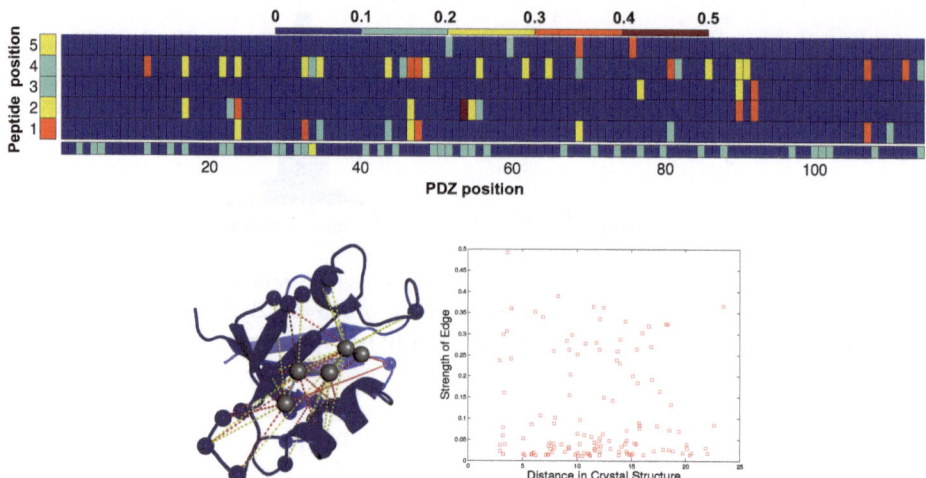

Fig. 4. Model analysis. **(top)** Average strength of the vector 2-norms for the PDZ positions (i.e., V_i), peptide positions (i.e., V_j), and potentially interacting pairs (i.e., $W_{i,j}$) in the model trained at $\lambda_{1,2} = 25$. **(bottom-left)** Strong interactions highlighted in top panel, displayed on the NMR structure of the alpha syntropin PDZ (pdb id: 2PDZ). Color scheme same as above. **(bottom-right)** Average edge strength across 10 training splits plotted against distance in the 3D structure.

strength of the vector 2-norms for the protein positions (i.e., V_i), peptide positions (i.e., V_j), and potentially interacting pairs (i.e., $W_{i,j}$). Fig. 4-top shows these values: the strengths of the vertex terms appear along the axes (x-axis for PDZ positions and y-axis for peptide positions), while the strengths of the PDZ:peptide edge terms appear in the heat map. As might be expected for inter-action affinities, the position-based terms are relatively weaker, with most being less than 0.2. In contrast, more than 40 interaction terms have norms larger than this value with a large fraction of them between position 4 of the peptide and the protein. Fig. 4-bottom-left overlays these strong interactions on the structure of the murine al-syntrophin PDZ (colored blue to light pink according to position) complexed with the peptide KESLV (colored in red). Fig. 4-bottom-right plots the edge strength (y-axis) against the distance of the corresponding residue pairs (x-axis) Interestingly, while most of the strong edges tend to be between posi-tions less than 15 Å apart in the crystal, there are a few edges that are at a longer-range that appear consistently.

Despite the fact that no 3D structure information was used in learning the model, our method identifies several contacting residues as important for de-termining interaction specificity. This suggests that our data-driven approach might be capturing physically important interactions. To test this hypothesis further, we determined the average weight assigned to each possible amino acid pair for the top three interacting residue pairs across the models for the 10 train-ing folds at $\lambda_{1,2} = 25$. Fig. 5 shows these weights with strong negative energies

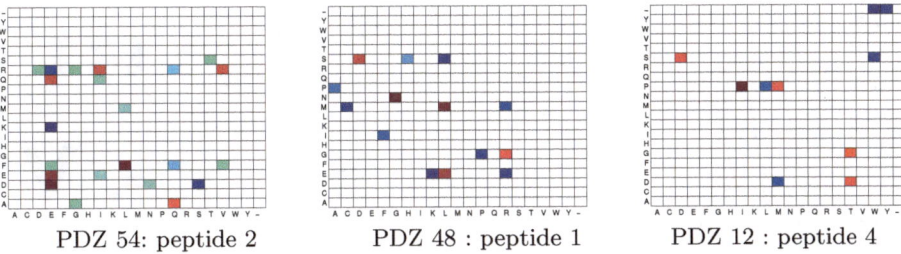

PDZ 54: peptide 2   PDZ 48 : peptide 1   PDZ 12 : peptide 4

Fig. 5. Average weights for amino acid pairs for the top three interacting residue pairs

(i.e., favoring binding) in shades of blue and strong positive energies in shades of red. Darker shades correspond to stronger effects in both cases. The strongest interacting residue pair (PDZ position 54 : peptide position 2) strongly favors interactions between oppositely charged Arginine/Lysine in the PDZ and Gluta-mate in the peptide, while strongly penalizing Aspartate/Glutamate : Glutamate between pairs of negatively charged residues, suggesting a strong electrostatic effect between these positions. Similar effects are seen in the other two inter-actions with Glutamate:Lysine favored between 48:1 and Aspartate:Threonine penalized between 12:4. Our method can thus provide structural information as well as insights into the biochemical determinants of binding affinity.

In summary, our results suggest that a large fraction of the binding affinity is due to interactions between a relatively small set of positions, not all spatially adjacent to the binding pocket. A larger set of weak interactions might have an additional small effect on binding; these might effect particular sub-families of PDZs or might reflect allosteric affects related to alternate conformational states of the protein previously described in this family [20].

3.3   From Sequence Determinants to Sequence Design

The accuracy and simplicity of our model allows us to rapidly evaluate the binding affinity of any PDZ-peptide pair. We demonstrate the utility of this approach by "designing" optimal binders for a given PDZ sequence. Using a model learned from the entire training set with $\lambda_{1,2}=25$, we searched all five residue peptides and determined the top 10 peptides by their predicted ΔG for each PDZ sequence. Fig. 6-left shows the density of *predicted* binding energies of these PDZ:peptide pairs in blue and that of natural PDZ-peptide pairs bind-ing energies in our training dataset in red. The predicted binding affinities of designed sequences are considerably lower than those of the natural sequences. .

While the designed sequences include the natural substrates (at close to their predicted affinities, as discussed above), they also include a diverse array of alternatives. Fig. 6-right shows sequence logos of the top 10 designed peptides for three different PDZs. Even among these sets of top predicted binders, we see interesting diversity among the peptides, suggesting novel designs potentially worthy of experimental evaluation.

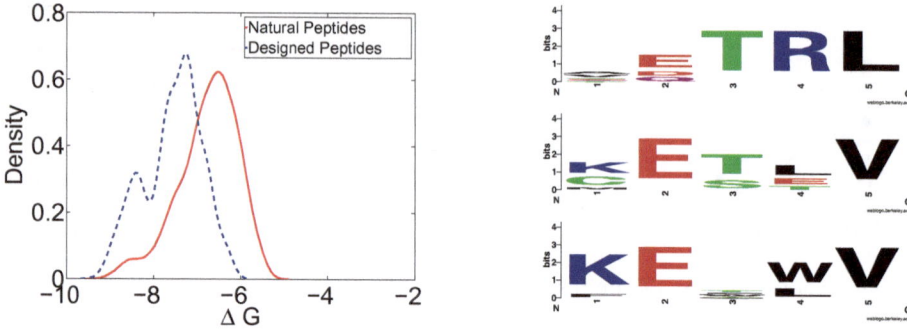

Fig. 6. (left) Density of predicted PDZ-peptide ΔG for designed peptides (blue) and experimental ΔG for natural PDZ-peptide pairs (red). **(right)** Sequence logos for the top 10 peptide designs for SHANK1, CHAPSYN, and PSD95 (top, middle and bottom).

4   Discussion and Conclusion

We have developed a graphical model which is highly predictive of the ΔG of binding in protein:protein interactions, while providing an interpretable and designable basis for its predictions. The notion of modularity is fundamental to the idea of a graphical model. Hence these models form a powerful and natural tool to solve problems involving complex probability distributions over many random variables, like the ones here. Due to the natural equivalence between the graph structure of a model and the structure of spatial interactions in proteins, graphical models have seen considerable use in modeling various aspects of proteins: in recognizing structural motifs [19,24,25], in protein structure alignments [42], and in modeling dynamics [30]. A growing body of work using graphical models to capture correlated mutations in protein families has also seen substantial success in predicting residue-residue contacts in the protein structure [10,12,22,26,28], highlighting the power of these models.

While basing the modeling of ΔG on sequence and data is fundamentally different from structure-based predictors, which employ physics-based models and analysis of side-chain (and potentially backbone) conformations to assess interactions (e.g., [9,16,35]), we note that structure-based undirected graphical models have been used to predict ΔG [13,14]. The integration of the structure-based approach and the sequence+data-based approach provides an interesting future direction. Our preliminary work on such integration for individual proteins [11] provides evidence that the two viewpoints can be complementary and enable better prediction than either alone.

The method we developed here could be applied to any pair of interacting protein families with a similar extent of quantitative binding data. Due to their size and easy availability, PDZ domains form a "model system" for studying protein-protein interactions, [4,17,34,40]. They are involved in formation of protein complexes which are involved in cellular signal transduction and neural circuitry [34] and so make an interesting test case from the point of view of protein-engineering [7] and drug design [31].

We demonstrated that our models can be used to design novel peptides that interact strongly with a given PDZ domain. This approach could be extended using sampling or other inferential techniques to design a desired interaction, rather than only the peptide, and to scale up to larger sets of involved residues.

Acknowledgements. This work is supported in part by US NSF grant IIS-0905193 (CJL and CBK) and US NIH P41 GM103712 (CJL).

References

1. Balakrishnan, S., Kamisetty, H., Carbonell, J., Lee, S., Langmead, C.: Learning generative models for protein fold families. Proteins: Structure, Function, and Bioinformatics 79(4), 1061–1078 (2011)
2. Bordner, A., Mittelmann, H.: MultiRTA: A simple yet accurate method for predicting peptide binding affinities for multiple class II MHC allotypes. BMC Bioinformatics 11, 482 (2010)
3. Brannetti, B., Via, A., Cestra, G., Cesareni, G., Citterich, M.H.: SH3-SPOT: an algorithm to predict preferred ligands to different members of the SH3 gene family. Journal of Molecular Biology 298(2), 313–328 (2000)
4. Chen, J., Chang, B., Allen, J., Stiffler, M., MacBeath, G.: Predicting PDZ domain-peptide interactions from primary sequences. Nat. Biotechnol. 26(9), 1041–1045 (2008)
5. Fields, S., Song, O.: A novel genetic system to detect protein-protein interactions. Nature 340(6230), 245–246 (1989)
6. Fong, J., Keating, A., Singh, M.: Predicting specificity in bZIP coiled-coil protein interactions. Genome Biology 5(2), R11 (2004)
7. Fuh, G., Pisabarro, M., Li, Y., Quan, C., Lasky, L., Sidhu, S.: Analysis of PDZ domain-ligand interactions using carboxyl-terminal phage display. J. Biol. Chem. 275(28), 21486–21491 (2000)
8. Grigoryan, G., Reinke, A., Keating, A.: Design of protein-interaction specificity gives selective bZIP-binding peptides. Nature 458(7240), 859–864 (2009)
9. Guerois, R., Nielsen, J.E., Serrano, L.: Predicting changes in the stability of proteins and protein complexes: a study of more than 1000 mutations. Journal of Molecular Biology 320, 369–387 (2002)
10. Jones, D.T., Buchan, D.W., Cozzetto, D., Pontil, M.: Psicov: precise structural contact prediction using sparse inverse covariance estimation on large multiple sequence alignments. Bioinformatics 28(2), 184–190 (2012)
11. Kamisetty, H., Ghosh, B., Bailey-Kellogg, C., Langmead, C.: Modeling and Inference of Sequence-Structure Specificity. In: Proc. of the 8th International Conference on Computational Systems Bioinformatics (CSB), pp. 91–101 (2009)
12. Kamisetty, H., Ovchinnikov, S., Baker, D.: Assessing the utility of coevolution-based residue–residue contact predictions in a sequence-and structure-rich era. Proceedings of the National Academy of Sciences 110(39), 15674–15679 (2013)
13. Kamisetty, H., Ramanathan, A., Bailey-Kellogg, C., Langmead, C.: Accounting for conformational entropy in predicting binding free energies of protein-protein interactions. Proteins 79(2), 444–462 (2011)

14. Kamisetty, H., Xing, E., Langmead, C.: Free Energy Estimates of All-atom Protein Structures Using Generalized Belief Propagation. J. Comp. Bio. 15(7), 755–766 (2008)
15. Kamisetty, H., Xing, E., Langmead, C.: Approximating Correlated Equilibria using Relaxations on the Marginal Polytope. In: Proc. of the 28th International Conference on Machine Learning (ICML), pp. 1153–1160 (2011)
16. Kortemme, T., Baker, D.: A simple physical model for binding energy hot spots in protein-protein complexes. Proceedings of the National Academy of Sciences 99(22), 14116–14121 (2002)
17. Kurakin, A., Swistowski, A., Wu, S., Bredesen, D.: The pdz domain as a complex adaptive system. PLoS One 2(9) 2(9), e953 (2007)
18. Li, J., Yi, Z.P., Laskowski, M., Laskowski Jr., M., Bailey-Kellogg, C.: Analysis of sequence-reactivity space for protein-protein interactions. Proteins: Structure, Function, and Bioinformatics 58(3), 661–671 (2005)
19. Liu, Y., Carbonell, J., Gopalakrishnan, V., Weigele, P.: Conditional graphical models for protein structural motif recognition. Journal of Computational Biology 16(5), 639–657 (2009)
20. Lockless, S.W., Ranganathan, R.: Evolutionarily conserved pathways of energetic connectivity in protein families. Science 286(5438), 295–299 (1999)
21. Lu, S., Lu, W., Qasim, M., Anderson, S., Apostol, I., Ardelt, W., Bigler, T., Chiang, Y., Cook, J., James, M., Kato, I., Kelly, C., Kohr, W., Komiyama, T., Lin, T., Ogawa, M., Otlewski, J., Park, S., Qasim, S., Ranjbar, M., Tashiro, M., Warne, N., Whatley, H., Wieczorek, A., Wieczorek, M., Wilusz, T., Wynn, R., Zhang, W., Laskowski Jr., M.: Predicting the reactivity of proteins from their sequence alone: Kazal family of protein inhibitors of serine proteinases. Proceedings of the National Academy of Sciences 98(4), 1410–1415 (2001)
22. Marks, D.S., Colwell, L.J., Sheridan, R., Hopf, T.A., Pagnani, A., Zecchina, R., Sander, C.: Protein 3d structure computed from evolutionary sequence variation. PLoS One 6(12), e28766 (2011)
23. Meinshausen, N., Bühlmann, P.: High-dimensional graphs and variable selection with the lasso. The Annals of Statistics 34(3), 1436–1462 (2006)
24. Menke, M., Berger, B., Cowen, L.: Markov random fields reveal an n-terminal double beta-propeller motif as part of a bacterial hybrid two-component sensor system. PNAS 107(9), 4069–4074 (2010)
25. Moitra, S., Tirupula, K., Klein-Seetharaman, J., Langmead, C.: A minimal ligand binding pocket within a network of correlated mutations identified by multiple sequence and structural analysis of G protein coupled receptors. BMC Biophysics 5(13) (2012), doi:10.1186/2046-1682-5-13
26. Morcos, F., Pagnani, A., Lunt, B., Bertolino, A., Marks, D.S., Sander, C., Zecchina, R., Onuchic, J.N., Hwa, T., Weigt, M.: Direct-coupling analysis of residue coevolution captures native contacts across many protein families. Proceedings of the National Academy of Sciences 108(49), E1293–E1301 (2011)
27. Nielsen, M., Lundegaard, C., Blicher, T., Lamberth, K., Harndahl, M., Justesen, S., Roder, G., Peters, B., Sette, A., Lund, O., Buus, S.: NetMHCpan, a method for quantitative predictions of peptide binding to any HLA-A and -B locus protein of known sequence. PLoS One 2, e796 (2007)
28. Nugent, T., Jones, D.T.: Accurate de novo structure prediction of large transmembrane protein domains using fragment-assembly and correlated mutation analysis. Proceedings of the National Academy of Sciences 109(24), E1540–E1547 (2012)

29. Peters, B., Sidney, J., Bourne, P., Bui, H.H., Buus, S., Doh, G., Fleri, W., Kronenberg, M., Kubo, R., Lund, O., Nemazee, D., Ponomarenko, J.V., Sathiamurthy, M., Schoenberger, S., Stewart, S., Surko, P., Way, S., Wilson, S., Sette, A.: The immune epitope database and analysis resource: from vision to blueprint. PLoS Biol. 3, e91 (2005)
30. Razavian, N., Kamisetty, H., Langmead, C.: Learning generative models of molecular dynamics. BMC Genomics 13(suppl. 1) (2012), doi:10.1186/1471-2164-13-S1-S5
31. Saro, D., Li, T., Rupasinghe, C., Paredes, A., Caspers, N., Spaller, M.: A thermodynamic ligand binding study of the third pdz domain (pdz3) from the mammalian neuronal protein psd-95. Biochemistry 46(21), 6340–6352 (2007)
32. Schmidt, M., van der Berg, E., Friedlander, M.P., Murphy, K.: Optimizing costly functions with simple constraints:a limited-memory projected quasi-newton algorithm. AISTATS 5, 456–463 (2009)
33. Shao, X., Tan, C., Voss, C., Li, S., Deng, N., Bader, G.: A regression framework incorporating quantitative and negative interaction data improves quantitative prediction of PDZ domain-peptide interaction from primary sequence. Bioinformatics 27(3), 383–390 (2010)
34. Sheng, M., Sala, C.: Pdz domains and the organization of supramolecular complexes. Annu. Rev. Neurosci. 24, 1–29 (2001)
35. Smith, C., Kortemme, T.: Structure-based prediction of the peptide sequence space recognized by natural and synthetic pdz domains. Journal of Molecular Biology 402(2), 460–474 (2010)
36. Stiffler, M., Chen, J., Grantcharova, V., Lei, Y., Fuchs, D., Allen, J., Zaslavskaia, L., MacBeath, G.: Pdz domain binding selectivity is optimized across the mouse proteome. Science 317(5836), 364–369 (2007)
37. Thomas, J., Ramakrishnan, N., Bailey-Kellogg, C.: Graphical models of residue coupling in protein families. IEEE/ACM Transactions on Computational Biology and Bioinformatics 5(2), 183–197 (2008)
38. Thomas, J., Ramakrishnan, N., Bailey-Kellogg, C.: Graphical models of protein-protein interaction specificity from correlated mutations and interaction data. Proteins: Structure, Function, and Bioinformatics 76(4), 911–929 (2009)
39. Thomas, J., Ramakrishnan, N., Bailey-Kellogg, C.: Protein design by sampling an undirected graphical model of residue constraints. IEEE/ACM Transactions on Computational Biology and Bioinformatics 6(3), 506–516 (2009)
40. Tonikian, R., Zhang, Y., Sazinsky, S., Currell, B., Yeh, J., Reva, B., Held, H., Appleton, B., Evangelista, M., Wu, Y., Xin, X., Chan, A., Seshagiri, S., Lasky, L., Sander, C., Boone, C., Bader, G., Sidhu, S.: A specificity map for the PDZ domain family. Plos Biology 6(9), e239 (2008)
41. Wang, P., Sidney, J., Dow, C., Mothe, B., Sette, A., Peters, B.: A systematic assessment of MHC class II peptide binding predictions and evaluation of a consensus approach. PLoS Comp. Biol. 4, e1000048 (2008)
42. Xu, J., Jiao, F., Berger, B.: A parameterized algorithm for protein structure alignment. In: Apostolico, A., Guerra, C., Istrail, S., Pevzner, P.A., Waterman, M. (eds.) RECOMB 2006. LNCS (LNBI), vol. 3909, pp. 488–499. Springer, Heidelberg (2006)
43. Zhang, L., Udaka, K., Mamitsuka, H., Zhu, S.: Toward more accurate pan-specific MHC-peptide binding prediction: a review of current methods and tools. Brief. Bioinform. 13, 350–364 (2012)

On Sufficient Statistics
of Least-Squares Superposition of Vector Sets

Arun S. Konagurthu[1], Parthan Kasarapu[1], Lloyd Allison[1], James H. Collier[1],
and Arthur M. Lesk[2]

[1] Faculty of Information Technology, Monash University, Clayton VIC 3800 Australia
[2] The Huck Institute of Genomics, Proteomics and Bioinformatics and the
Department of Biochemistry and Molecular Biology, Pennsylvania State University,
University Park PA 16802 USA
arun.konagurthu@monash.edu

Abstract. Superposition by orthogonal transformation of vector sets by
minimizing the least-squares error is a fundamental task in many areas
of science, notably in structural molecular biology. Its widespread use
for structural analyses is facilitated by exact solutions of this problem,
computable in linear time. However, in several of these analyses it is common to invoke this superposition routine a very large number of times,
often operating (through addition or deletion) on previously superposed
vector sets. This paper derives a set of *sufficient statistics* for the least-
squares orthogonal transformation problem. These sufficient statistics
are additive. This property allows for the superposition parameters (rotation, translation, and root mean square deviation) to be computable as
constant time updates from the statistics of partial solutions. We demonstrate that this results in a massive speed up in the computational effort,
when compared to the method that recomputes superpositions *ab initio*.
Among others, protein structural alignment algorithms stand to benefit
from our results.

1 Introduction

Optimal superposition through orthogonal transformation of vector sets forms
the linchpin of macromolecular structure comparison [1, 2]. This task is ubiq-
uitously used to analyse globular three-dimensional structures of proteins [3].
Orthogonal transformation involves finding the best rigid-body rotation and
translation of two vector sets that are in one-to-one correspondence so that
they can be superimposed. This superposition immediately provides a qualitative (through visual inspection) as well as a quantitative measure of shape
similarity.

An almost universally used criterion to define the *best* superposition of vector sets is the one that minimizes the *sum of square errors* over the entire
search space of possible rotations and translations. This results in a quantitative measure, *root mean square deviation* (or r.m.s.d.) after best superposition.
This measure is central in assessing the quality of superposition with attractive
metrical properties.

R. Sharan (Ed.): RECOMB 2014, LNBI 8394, pp. 144–159, 2014.
© Springer International Publishing Switzerland 2014

Superpositions pervade protein structural analyses because they provide essential information about comparisons of conformations of structures and substructures; it is remarkable and comes in handy that optimal superposition of aligned sets of points can be computed exactly and efficiently [3]. Given the importance of this routine, several approaches have been proposed to address this problem over the years [4–14]. However, among the most-widely used approach to solve this problem is the method of Kabsch [5] that solves this problem using Lagrange multipliers that constrain the search to *pure rotations* (and avoid improper ones).

An equivalent, but a more elegant, approach to solving the same problem was proposed by Kearsley [11] using the mathematical object called *quaternions* [15]. Quaternions are generalizations of complex numbers with direct applications to transformations in three dimensional space. Specifically, the space group corresponding to unit quaternions is equivalent to the group of all possible pure rotations in three dimensions (3D) defined about an arbitrary origin. That is, any 3D pure rotation by an angle θ about some normalized axis $\hat{n}$ passing through the origin can be represented using a unit quaternion as follows: $\left[\cos\left(\frac{\theta}{2}\right), \hat{n}\sin\left(\frac{\theta}{2}\right)\right]$. Among the key advantages of using Kearsley's quaternion method to solve the least-squares superposition problem are: (1) the problem can be solved analytically in quaternion parameters, and (2) the method avoids problems with singularities (and rotoinversions) that can result from using Kabsch's approach, where these oddities are handled explicitly after the solution is found [11, 13]. In general, the least-squares superposition involves a computational effort that asymptotically grows *linearly* with the number of corresponding points being superimposed.

Many methods that facilitate analyses involving protein structures employ least-squares superpositions. Among the primary example of this is when computing the residue-residue correspondences betweeen two or more protein structures – *the structural alignment problem*. Many popular methods build an alignment between structures using orthogonal superpositions of fragments [9, 16–23]. The general strategy involves finding aligned (contiguous) fragment pairs that are often maximally extended, one residue-residue correspondence at a time starting from some minimum fragment size, until the fragment pairs superposes within some specified threshold of r.m.s.d. This results in a library of well-fitting fragment pairs, construction effort of which grows as a cubic in the length of the structures being aligned ($O(n^2)$ number of superpositions, each taking $O(n)$ superposition effort, where n is the number of residues in the structures being aligned). Further, by computing the joint superpositions of these well-fitting maximal fragment pairs, a structural alignment is *assembled* by collecting fragment pairs that superpose consistently. This involves repeated concatenation and superposition calls using the fragment pairs in the library. Such superpositions are currently recomputed from scratch (even though the previous superpositions provide a wealth of information about the joint superposition, as we shall demonstrate in the forthcoming sections). It can be seen that the number of joint superpositions grows (at least) quadratically in the size of the fragment library, with each joint superposition taking a linear effort in the size of the concatenated vector sets.

Although the optimal solution of the least-squares superposition problem can be computed extremely efficiently, the algorithmic complexity term hides a sizeable constant factor. This imposes a significant computational demand when performing a large number of superpositions, as required for computing pairwise structural alignments. The amount of time spent in superposing fragments quickly becomes computationally impractical when aligning multiple protein structures simultaneously, where the multiple structural alignment is commonly built using all-vs-all pairwise structural alignments, each of which makes a very large number of calls to the superposition routine.

Contribution of This Work: In this paper we explore the theoretical underpinning of the orthogonal superposition problem and derive a set of statistics that are sufficient to compute the r.m.s.d of best superposition, and its corresponding rotation and translation parameters). We demonstrate that these *sufficient statistics* [24] are additive. Thus these statistics can be used to compute new superpositions as *constant time* updates using the statistics of the partial solutions. Using such an approach results in a drastic speed up in comparison with the approach that recomputes the new superposition from scratch.

Organization of This Paper: Section 2 gives the basic background of the orthogonal superposition problem using the widely-used least-squares criterion. Section 3 introduces the statistical aspects of sufficient statistics, and derives the full set of sufficient statistics for the optimal orthogonal superposition problem. Section 4.1 provides the mechanics of performing constant-time updates to superpositions building on the sufficient statistics of previous (partial) superpositions. Section 5 describes an approach to speed up the diagonalization step used in the Kearsley approach. Finally, the paper ends with an experimental evaluation of computing optimal superpositions using sufficient statistics.

2   Orthogonal Superposition

Formally let $\mathcal{U} = \{u_1, \cdots, u_n\}$ and $\mathcal{V} = \{v_1, \cdots, v_n\}$ denote two vector sets with one-to-one correspondence. In this paper we consider vectors in three dimensions. Let the (x, y, z) components of each u_i be represented here as $(u_i(x), u_i(y), u_i(z))$. (Similar representation holds for v_i or any other vector.)

The rigid-body least-squares superposition problem is a constrained optimization problem that involves finding the best rotation (matrix) $\mathbf{R}$ and translation (vector) t with the optimality criterion defined as:

$$\mathcal{E} = \min |\mathbf{R}\mathcal{U} + t - \mathcal{V}|^2 = \min \sum_{i=1}^{n} |\mathbf{R}u_i + t - v_i|^2 = \min \sum_{i=1}^{n} \langle \mathbf{R}u_i + t - v_i, \mathbf{R}u_i + t - v_i \rangle$$

where $\langle \cdot, \cdot \rangle$ denotes the inner product between the stated terms, $\mathbf{R}$ is a 3×3 pure rotation matrix, and t is a translation vector.

Under this least-squares criterion, the translation with respect to the optimal superposition is independent of rotation. This can be easily seen by differentiating $\mathcal{E}$ with respect to t and evaluating it at its extremum:

$$\frac{\partial \mathcal{E}}{\partial t} = \frac{\partial}{\partial t} \sum_{i=1}^{n} \langle \mathbf{R}u_i + t - v_i, \mathbf{R}u_i + t - v_i \rangle = \sum_{i=1}^{n} 2 \frac{\partial(\mathbf{R}u_i + t - v_i)}{\partial t}(\mathbf{R}u_i + t - v_i) = 0$$

$$\implies \sum_{i=1}^{n} \mathbf{R}u_i + t - v_i = 0$$

$$\implies t = \frac{\sum_{i=1}^{n} v_i}{n} - \mathbf{R}\frac{\sum_{i=1}^{n} u_i}{n} = \mathbf{Centroid}(\mathcal{V}) - \mathbf{R} \, \mathbf{Centroid}(\mathcal{U})$$

It follows that moving each of the vector sets to an origin at its centroid, about which the rotation is defined, gives us a modified (but equivalent) objective which is independent of the translation t:

$$\mathcal{E} = \min \sum_{i=1}^{n} |\mathbf{R}u_i' - v_i'|^2$$

where, $u_i' = u_i - \dfrac{\sum_{i=1}^{n} u_i}{n}$ and $v_i' = v_i - \dfrac{\sum_{i=1}^{n} v_i}{n}$.

Kearsley [11] proposed an elegant method that removes the non-linear aspect to this least-squares problem and transforms it to an eigenvalue problem of the form $\mathbf{Q}q = \lambda q$, where $\mathbf{Q}$ is a 4×4 square symmetric matrix

$$\begin{pmatrix} \sum(x_m^2 + y_m^2 + z_m^2) & \sum(y_p z_m - y_m z_p) & \sum(x_m z_p - x_p z_m) & \sum(x_p y_m - x_m y_p) \\ \sum(y_p z_m - y_m z_p) & \sum(x_m^2 + y_p^2 + z_p^2) & \sum(x_m y_m - x_p y_p) & \sum(x_m z_m - x_p z_p) \\ \sum(x_m z_p - x_p z_m) & \sum(x_m y_m - x_p y_p) & \sum(x_p^2 + y_m^2 + z_p^2) & \sum(y_m z_m - y_p z_p) \\ \sum(x_p y_m - x_m y_p) & \sum(x_m z_m - x_p z_p) & \sum(y_m z_m - y_p z_p) & \sum(x_p^2 + y_p^2 + z_m^2) \end{pmatrix}, (1)$$

$$q = (q_1, q_2, q_3, q_4)^T = \left(\cos\left(\frac{\theta}{2}\right), \hat{n}(x)\sin\left(\frac{\theta}{2}\right), \hat{n}(y)\sin\left(\frac{\theta}{2}\right), \hat{n}(z)\sin\left(\frac{\theta}{2}\right) \right)^T$$

are the (unknown or to be solved) quaternion components associated with some rotation θ about a normalized axis $\hat{n}$, and λ is an (unknown) eigenvalue. In Equation 1, we use the notation x_m to denote the component-wise difference $v_i'(x) - u_i'(x)$ (and similarly y_m and z_m) and x_p to denote the component-wise sum $v_i'(x) + u_i'(x)$ (similarly y_p and z_p). From this point onwards, we use the term *quaternion matrix* to indicate the 4×4 square symmetric matrix in Equation 1 and denote it as $\mathbf{Q}$.

Diagonalizing this matrix yields four eigenvalues and (corresponding) eigenvectors. The eigenvector corresponding to the smallest eigenvalue, $\lambda_{\min}$, corresponds to the rotation producing the least-squares error, and the r.m.s.d is computed as $\sqrt{\dfrac{\lambda_{\min}}{n}}$.

Time Complexity. The computational effort that takes to solve the rigid-body superposition problem using Kearsley's quaternion approach (or equivalently Kabsch's approach) grows linearly with the number of vectors being superimposed. In Kearsley's approach this is dominated by the computation of the $\mathbf{Q}$ where each of 10 distinct terms in the matrix requires $O(n)$ effort. The diagonalization of $\mathbf{Q}$ is independent of n and shows a rapid convergence with numerical methods such as Jacobi's diagonalization algorithm [25].

3   Sufficient Statistics

We note that this rigid-body superposition problem is a geometric instance of the general regression problem using total least-squares, where a regression line is determined that minimizes the sum of the squared errors of the observed data with respect to it.

It is widely known that solution of the regression problem produces error terms that are normally distributed as $\mathcal{N}(0, \sigma)$ where the mean μ is 0 and σ is the standard deviation which is minimized by the problem. In fact, the least squares estimator of σ is also its maximum likelihood estimator.

More formally, consider the standard normal distribution of some random variable x:

$$\mathcal{N}(x|\mu, \sigma) = \frac{1}{\sqrt{2\pi}\sigma} \exp\left[-\frac{(x-\mu)^2}{2\sigma^2}\right]$$

This normal density can be reparameterized into a general form denoting the family of exponential distributions:

$$f(x|\boldsymbol{\eta}) = h(x)g(\boldsymbol{\eta})\exp(\boldsymbol{\eta}^T \boldsymbol{U}(x))$$

where $h(x) = \dfrac{1}{\sqrt{\pi}}$, $g(\eta_2) = \sqrt{-\eta_2}\exp\left(\dfrac{\eta_1^2}{4\eta_2}\right)$, $\boldsymbol{\eta}^T = (\dfrac{\mu}{\sigma^2}, -\dfrac{1}{2\sigma^2})$, $\boldsymbol{U}^T(x) = (x, x^2)$.

This transformation can be used to show certain important properties that allows efficient computation of maximum likelihood estimators of μ and σ.

Considering a sample set of observations that are normally distributed $\mathbf{X} = \{x_1, x_2, \cdots, x_n\}$. The likelihood for these samples is given by:

$$f(\mathbf{X}|\boldsymbol{\eta}) = \left(\prod_{i=1}^{n} h(x_i)\right)(g(\boldsymbol{\eta}))^n \exp(\boldsymbol{\eta}^T \sum_{i=1}^{n} u(x_i))$$

Taking natural logarithms on both sides gives us the log likelihood:

$$\log(f(\mathbf{X}|\boldsymbol{\eta})) = \kappa + n\log(g(\boldsymbol{\eta})) + \boldsymbol{\eta}^T \sum_{i=1}^{n} \boldsymbol{U}(x_i)$$

where $\kappa = \sum_{i=1}^{n} \log(h(x_i))$ is a term independent of $\boldsymbol{\eta}$.

To find the maximum likelihood estimators $\hat{\eta}$, take the gradient with respect to η and set to 0. This results in:

$$n\nabla_{\hat{\eta}}\left[\log\left(g(\hat{\eta})\right)\right] + \sum_{i=1}^{n} U(x_i) = 0$$

$$\implies -\nabla_{\hat{\eta}}\left[\log\left(g(\hat{\eta})\right)\right] = \frac{1}{n}\sum_{i=1}^{n} U(x_i)$$

$$= \frac{-1}{g(\hat{\eta})}\nabla_{\hat{\eta}}g(\hat{\eta}) = \frac{1}{n}\sum_{i=1}^{n} U(x_i)$$

Notice that maximum likelihood estimate $\hat{\eta}$ depends on the statistic $\sum_{i=1}^{n} U(x_i)$ rather than the individual data. This suggests that to obtain the maximum likelihood estimate we do not need the data explicitly as it can be derived from that statistic. This sufficiency to derive the maximum likelihood estimator without explicit consideration of data makes $\sum_{i=1}^{n} U(x_i)$ a *sufficient statistic* for the exponential family of functions. For normal distribution, we saw earlier that $U(x_i) = (x_i, x_i^2)$ gives the sufficient statistics of $\sum_{i=1}^{n} x_i$ and $\sum_{i=1}^{n} x_i^2$ [24].

Sufficient Statistics for Orthogonal Superposition

We note that each error term, $\varepsilon_i = \mathbf{R}u_i' - v_i'$, is assumed to be normally distributed: *i.e.*, $\varepsilon_i \sim \mathcal{N}(\mu = 0, \sigma)$. We now derive the sufficient statistics for σ of ε_is, which is equivalent to the r.m.s.d. after least-squares superposition. The likelihood of the observed normally distributed errors after superposition, $\mathbf{E} = \{\varepsilon_1, \ldots, \varepsilon_n\}$, can be written as:

$$f(\varepsilon_1, \ldots, \varepsilon_n | \sigma) = \prod_{i=1}^{n}(2\pi\sigma^2)^{-\frac{1}{2}}\exp\left(-\frac{1}{2\sigma^2}\|\mathbf{R}u_i' - v_i'\|^2\right)$$

$$= (2\pi\sigma^2)^{-\frac{n}{2}}\exp\left(-\frac{1}{2\sigma^2}\sum_{i=1}^{n}\|\mathbf{R}u_i' - v_i'\|^2\right) \tag{2}$$

Let's examine the decomposition of

$$\varepsilon_i^2 = \|\mathbf{R}u_i' - v_i'\|^2 = \|u_i'\|^2 + \|v_i'\|^2 - 2v_i'^{\mathrm{T}}\mathbf{R}u_i' \tag{3}$$

From Equation 1, the matrix $\mathbf{Q}$ is made up of terms of the form

$$A_m = v_i'(A) - u_i'(A) \text{ and } A_p = v_i'(A) + u_i'(A)$$

where each A and B take the values $\{x, y, z\}$ denoting vector components. Rewriting, we get

$$v_i'(A) = \frac{A_p + A_m}{2} \text{ and } u_i'(A) = \frac{A_p - A_m}{2}$$

The first two terms on the right hand side of Equation 3 can be expanded as follows:

$$\|{\boldsymbol u_i}'\|^2 + \|{\boldsymbol v_i}'\|^2 = (u_i'(x)^2 + u_i'(y)^2 + u_i'(z)^2) + (v_i'(x)^2 + v_i'(y)^2 + v_i'(z)^2)$$

$$= \frac{1}{2}(x_m^2 + x_p^2 + y_m^2 + y_p^2 + z_m^2 + z_p^2)$$

$$= \frac{1}{2} \sum_{A \in \{x,y,z\}} A_m^2 + \frac{1}{2} \sum_{A \in \{x,y,z\}} A_p^2 \tag{4}$$

The last term on the right hand side of Equation 3 can be expanded as ${\boldsymbol v_i}'^{\mathrm{T}} \mathbf{R} {\boldsymbol u_i}' = {\boldsymbol v_i}'^{\mathrm{T}} [\mathbf{r_1}\ \mathbf{r_2}\ \mathbf{r_3}] {\boldsymbol u_i}'$ where $\mathbf{r_1}, \mathbf{r_2}, \mathbf{r_3}$ are column vectors of the 3×3 rotation matrix $\mathbf{R}$. Therefore,

$${\boldsymbol v_i}'^{\mathrm{T}} \mathbf{R} {\boldsymbol u_i}' = ({\boldsymbol v_i}'.\mathbf{r_1}) u_i'(x) + ({\boldsymbol v_i}'.\mathbf{r_2}) u_i'(y) + ({\boldsymbol v_i}'.\mathbf{r_3}) u_i'(z) \tag{5}$$

Take the first term on the right hand side of Equation 5. This can be expanded as:

$$({\boldsymbol v_i}'.\mathbf{r_1}) u_i'(x) = r_{11} v_i'(x) u_i'(x) + r_{12} v_i'(y) u_i'(x) + r_{13} v_i'(z) u_i'(x)$$

$$= \frac{r_{11}}{4}(x_p + x_m)(x_p - x_m) + \frac{r_{12}}{4}(y_p + y_m)(x_p - x_m) + \frac{r_{13}}{4}(z_p + z_m)(x_p - x_m)$$

$$= \frac{r_{11}}{4}(x_p^2 - x_m^2) + \frac{r_{12}}{4}(y_p x_p - y_p x_m + y_m x_p - y_m x_m)$$

$$+ \frac{r_{13}}{4}(z_p x_p - z_p x_m + z_m x_p - z_m x_m)$$

where r_{11}, r_{12}, r_{13} are the terms in the $\mathbf{r_1}$ column vector in $\mathbf{R}$. More generally,

$$({\boldsymbol v_i}'.\mathbf{r_1}) u_i'(x) = c_1 A_p^2 + c_2 A_m^2 + c_3 A_p B_p + c_4 A_m B_m + c_5 A_m B_p \tag{6}$$

where c_k are constants in terms of components of $\mathbf{r_1}$.

Similarly, $({\boldsymbol v_i}'.\mathbf{r_2}) u_i'(y)$ and $({\boldsymbol v_i}'.\mathbf{r_3}) u_i'(z)$ can be expanded as above and will have the same form as (6) but with different constants. Therefore, combining Equations 4-5, the equation 3 can be written as

$$\varepsilon_i^2 = \zeta_1 \sum_A A_p^2 + \zeta_2 \sum_A A_m^2 + \zeta_3 \sum_{\forall A \neq B} A_p B_p + \zeta_4 \sum_{\forall A \neq B} A_m B_m + \zeta_5 \sum_{\forall A \neq B} A_m B_p$$

where ζ_k are constants. Hence, the likelihood function can be written as

$$f(\varepsilon_1, \ldots, \varepsilon_n | \sigma) = (2\pi\sigma^2)^{-\frac{n}{2}} \exp\left(-\frac{1}{2\sigma^2} \mathbf{U}\right) \tag{7}$$

where

$$\mathbf{U} = \sum_{i=1}^{n} \left(\zeta_1 \sum_A A_p^2 + \zeta_2 \sum_A A_m^2 + \zeta_3 \sum_{\forall A \neq B} A_p B_p + \zeta_4 \sum_{\forall A \neq B} A_m B_m + \zeta_5 \sum_{\forall A \neq B} A_m B_p \right)$$

and $A, B \in \{x, y, z\}$

Using Equation 7, the negative log-likelihood is given as:

$$\mathcal{L}(\varepsilon_1,\ldots,\varepsilon_n|\sigma) = \frac{n}{2}\log(2\pi) + n\log\sigma + \frac{1}{2\sigma^2}\mathbf{U} \tag{8}$$

The maximum likelihood estimate $\hat{\sigma}$ can be determined by minimising Equation 8 and evaluating the corresponding σ, *i.e.*

$$\frac{\partial\mathcal{L}}{\partial\sigma} = 0 \implies \hat{\sigma}^2 = \frac{\mathbf{U}}{n} \tag{9}$$

$\mathbf{U}$ involve statistics that do not take into account the data explicitly, and are sufficient to estimate σ (or r.m.s.d). Therefore the set of *sufficient statistics* for the least-squares superposition problem can be defined as:

$$\mathbf{\Psi} = \left\{\sum_{i=1}^{n}A_m, \quad \sum_{i=1}^{n}A_p, \quad \sum_{i=1}^{n}A_mB_m, \quad \sum_{i=1}^{n}A_mB_p, \quad \sum_{i=1}^{n}A_pB_p\right\} \tag{10}$$

where A and B take the values $\{x,y,z\}$, $A_m = v_i'(A) - u_i'(A)$ is the component-wise difference (similarly B_m), and $A_p = v_i'(A) + u_i'(A)$ is the component-wise sum (similarly B_p). Altogether, the set $\mathbf{\Psi}$ consists of 24 distinct statistics.

In addition, using the same notation, the statistics required to compute the centroid are of the form $\sum_{i=1}^{n}u_i'(A)$ and $\sum_{i=1}^{n}v_i'(A)$, and these are equivalent to $\sum_{\forall A}A_m$ and $\sum_{\forall A}A_p$.

4   Updating Sufficient Statistics

4.1   Addition Operation on Vector Sets Using Sufficient Statistics

Consider two pairs of corresponding vector sets: $\mathcal{Q} \leftrightarrow \mathcal{R}$ containing n_1 correspondences and $\mathcal{S} \leftrightarrow \mathcal{T}$ containing n_2 correspondences. Let $\mathcal{U}$ be defined as a combination of vectors $\mathcal{Q}$ and $\mathcal{S}$) and similarly $\mathcal{V}$ as a combination of $\mathcal{R}$ and $\mathcal{T}$. Let $\mathbf{\Psi}_1$ denote the sufficient statistics of superposing the first pair and $\mathbf{\Psi}_2$ denote the same for the second pair. Define these as:

$$\mathbf{\Psi}_1 = \left\{\sum_{i=1}^{n_1}C_m, \sum_{i=1}^{n_1}C_p, \sum_{i=1}^{n_1}C_mD_m, \sum_{i=1}^{n_1}C_mD_p, \sum_{i=1}^{n_1}C_pD_p\right\} \tag{11}$$

$$\mathbf{\Psi}_2 = \left\{\sum_{i=1}^{n_2}E_m, \sum_{i=1}^{n_2}E_p, \sum_{i=1}^{n_2}E_mF_m, \sum_{i=1}^{n_2}E_mF_p, \sum_{i=1}^{n_2}E_pF_p\right\} \tag{12}$$

Where C, D, E and F are all either $\{x,y,z\}$ denoting the components of the corresponding vectors in the vector sets under consideration. Consistent with the previous notation (see Equation 10), C_p and C_m (similarly D_p and D_m)

are the component-wise sums and differences between corresponding vectors in $\mathcal{Q}$ and $\mathcal{R}$. The same definitions hold for E_m (and E_p) and F_m (and F_p), with respect to corresponding vectors in $\mathcal{S}$ and $\mathcal{T}$.

We want to use Ψ_1 and Ψ_2 to compute a new set of sufficient statistics Ψ (defined in Equation 10) for the superposition of vector sets $\mathcal{U} = \mathcal{Q} + \mathcal{S}$ with $\mathcal{V} = \mathcal{R} + \mathcal{T}$. Below we derive the construction of the new sufficient statistics.

The statistics involved in computing the new centroids of the sets $\mathcal{U}$ and $\mathcal{V}$,

$$\sum_{i=1}^{n=n_1+n_2} u(A) \text{ and } \sum_{i=1}^{n=n_1+n_2} v(A), \text{ can be trivially updatated using the statistics}$$

$$\sum_{i=1}^{n_1} q(C), \sum_{i=1}^{n_1} r(D), \sum_{i=1}^{n_2} s(E), \text{ and } \sum_{i=1}^{n_2} t(F).$$

To compute the remaining statistics in Ψ, define vectors:

$$\begin{aligned} \alpha_1 &= \mathbf{Centroid}(\mathcal{U}) - \mathbf{Centroid}(\mathcal{Q}) & \beta_1 &= \mathbf{Centroid}(\mathcal{V}) - \mathbf{Centroid}(\mathcal{R}) \\ \alpha_2 &= \mathbf{Centroid}(\mathcal{U}) - \mathbf{Centroid}(\mathcal{S}) & \beta_2 &= \mathbf{Centroid}(\mathcal{V}) - \mathbf{Centroid}(\mathcal{T}). \end{aligned}$$

These vectors define the corrections that are required to be made to the previous centroids to recover the updated ones.

Lemma 1. $\displaystyle\sum_{i=1}^{n=n_1+n_2} A_m = \left[\sum_{i=1}^{n_1} C_m + n_1 \Delta_m^C\right] + \left[\sum_{i=1}^{n_2} E_m + n_2 \Delta_m^E\right]$, where $\Delta_m^C = \beta_1(C) - \alpha_1(C)$ and $\Delta_m^E = \beta_2(E) - \alpha_2(E)$ and $A = C = E \in \{x, y, z\}$

Proof

$$\begin{aligned} \sum_{i=1}^{n=n_1+n_2} A_m &= \left[\sum_{i=1}^{n_1} [(r'(C) + \beta_1(C)) - (q'(C) + \alpha_1(C))]\right] \\ &+ \left[\sum_{i=1}^{n_2} [(t'(E) + \beta_2(E)) - (s'(E) + \alpha_2(E))]\right] \\ &= \left[\sum_{i=1}^{n_1} (r'(C) - q'(C)) + (\beta_1(C) - \alpha_1(C))\right] \\ &+ \left[\sum_{i=1}^{n_2} (t'(C) - s'(C)) + (\beta_2(E) - \alpha_2(E))\right] \\ &= \left[\sum_{i=1}^{n_1} C_m + \sum_{i=1}^{n_1} \Delta_m^C\right] + \left[\sum_{i=1}^{n_2} E_m + \sum_{i=1}^{n_2} \Delta_m^E\right] \\ &= \left[\sum_{i=1}^{n_1} C_m + n_1 \Delta_m^C\right] + \left[\sum_{i=1}^{n_2} E_m + n_2 \Delta_m^E\right] \end{aligned}$$

Corollary 1. $\displaystyle\sum_{i=1}^{n} A_p = \left[\sum_{i=1}^{n_1} C_p + n_1 \Delta_p^C\right] + \left[\sum_{i=1}^{n_2} E_p + n_2 \Delta_p^E\right]$

Lemma 2

$$\sum_{i=1}^{n=n_1+n_2} A_m B_m = \left[\sum_{i=1}^{n_1} C_m D_m + \Delta_m^C \sum_{i=1}^{n_1} D_m + \Delta_m^D \sum_{i=1}^{n_1} C_m + n_1 \Delta_m^C \Delta_m^D\right]$$

$$+ \left[\sum_{i=1}^{n_2} E_m F_m + \Delta_m^E \sum_{i=1}^{n_2} F_m + \Delta_m^F \sum_{i=1}^{n_2} E_m + n_2 \Delta_m^E \Delta_m^F\right]$$

where $\Delta_m^C = \beta_1(C) - \alpha_1(C)$, $\Delta_m^D = \beta_1(D) - \alpha_1(D)$, $\Delta_m^E = \beta_2(E) - \alpha_2(E)$, *and* $\Delta_m^F = \beta_2(F) - \alpha_2(F)$ $A = C = E \in \{x, y, z\}$ *and* $B = D = F \in \{x, y, z\}$

Proof

$$\text{Updated}(\sum_{i=1}^{n_1} C_m D_m) = \sum_{i=1}^{n_1} [(r'(C) + \beta_1(C)) - (q'(C) + \alpha_1(C))]$$

$$[(r'(D) + \beta_1(D)) - (q'(B) + \alpha_1(D))]$$

$$= \sum_{i=1}^{n_1} [(r'(C)r'(D) - q'(C)q'(D) - r'(C)q'(D) + q'(C)r'(D)]$$

$$+ \sum_{i=1}^{n_1} [(\beta_1(C)r'(D) - \alpha_1(C)r'(D) - \beta_1(C)q'(D) + \alpha_1(C)q'(D)]$$

$$+ \sum_{i=1}^{n_1} [(r'(C)\beta_1(D) - r'(C)\alpha_1(D) - q'(C)\beta_1(D) + q'(C)\alpha_1(D)]$$

$$+ \sum_{i=1}^{n_1} [(\beta_1(C)\beta_1(D) - \alpha_1(C)\beta_1(D) - \beta_1(C)\alpha_1(D) + \alpha_1(C)\alpha_1(D)]$$

$$= \sum_{i=1}^{n_1} C_m D_m + \sum_{i=1}^{n_1} \Delta_m^C D_m + \sum_{i=1}^{n_1} \Delta_m^D C_m + \sum_{i=1}^{n_1} \Delta_m^C \Delta_m^D$$

$$= \sum_{i=1}^{n_1} C_m D_m + \Delta_m^C \sum_{i=1}^{n_1} D_m + \Delta_m^D \sum_{i=1}^{n_1} C_m + n_1 \Delta_m^C \Delta_m^D$$

Similarly, we can show that:

$$\text{Updated}(\sum_{i=1}^{n_2} E_m F_m) = \sum_{i=1}^{n_2} E_m F_m + \Delta_m^E \sum_{i=1}^{n_2} F_m + \Delta_m^F \sum_{i=1}^{n_2} E_m + n_2 \Delta_m^E \Delta_m^F$$

Adding the two updated statistics, the lemma follows.

Corollary 2

$$\sum_{i=1}^{n=n_1+n_2} A_m^2 = \sum_{i=1}^{n=n_1+n_2} A_m A_m = \left[\sum_{i=1}^{n_1} C_m C_m + 2\Delta_m^C \sum_{i=1}^{n_1} C_m + n_1 \left(\Delta_m^C\right)^2\right]$$
$$+ \left[\sum_{i=1}^{n_2} E_m E_m + 2\Delta_m^E \sum_{i=1}^{n_2} E_m + n_2 \left(\Delta_m^E\right)^2\right]$$

Corollary 3

$$\sum_{i=1}^{n=n_1+n_2} A_p B_p = \left[\sum_{i=1}^{n_1} C_p D_p + \Delta_p^C \sum_{i=1}^{n_1} D_p + \Delta_p^D \sum_{i=1}^{n_1} C_p + n_1 \Delta_p^C \Delta_p^D\right]$$
$$+ \left[\sum_{i=1}^{n_2} E_p F_p + \Delta_p^E \sum_{i=1}^{n_2} F_p + \Delta_p^F \sum_{i=1}^{n_2} E_p + n_2 \Delta_p^E \Delta_p^F\right]$$

Corollary 4

$$\sum_{i=1}^{n=n_1+n_2} A_p^2 = \sum_{i=1}^{n=n_1+n_2} A_p A_p = \left[\sum_{i=1}^{n_1} C_p C_p + 2\Delta_p^C \sum_{i=1}^{n_1} C_p + n_1 \left(\Delta_p^C\right)^2\right]$$
$$+ \left[\sum_{i=1}^{n_2} E_p E_p + 2\Delta_p^E \sum_{i=1}^{n_2} E_p + n_2 \left(\Delta_p^E\right)^2\right]$$

Lemma 3

$$\sum_{i=1}^{n=n_1+n_2} A_m B_p = \left[\sum_{i=1}^{n_1} C_m D_p + \Delta_m^C \sum_{i=1}^{n_1} D_p + \Delta_p^D \sum_{i=1}^{n_1} C_m + n_1 \Delta_m^C \Delta_p^D\right]$$
$$+ \left[\sum_{i=1}^{n_2} E_m F_p + \Delta_m^E \sum_{i=1}^{n_2} F_p + \Delta_p^F \sum_{i=1}^{n_2} E_m + n_2 \Delta_m^E \Delta_p^F\right]$$

where $\Delta_m^C = \beta_1(C) - \alpha_1(C)$, $\Delta_m^D = \beta_1(D) - \alpha_1(D)$, $\Delta_m^E = \beta_2(E) - \alpha_2(E)$, and $\Delta_m^F = \beta_2(F) - \alpha_2(F)$ $A = C = E \in \{x, y, z\}$ and $B = D = F \in \{x, y, z\}$

Proof

$$\text{Updated}(\sum_{i=1}^{n_1} C_m D_p) = \sum_{i=1}^{n_1} [(r'(C) + \beta_1(C)) - (q'(C) + \alpha_1(C))]$$
$$[(r'(D) + \beta_1(D)) + (q'(B) + \alpha_1(D))]$$

$$= \sum_{i=1}^{n_1} [(r'(C)r'(D) - q'(C)q'(D) + r'(C)q'(D) - q'(C)r'(D)]$$

$$+ \sum_{i=1}^{n_1} [(\beta_1(C)r'(D) - \alpha_1(C)r'(D) + \beta_1(C)q'(D) - \alpha_1(C)q'(D)]$$

$$+ \sum_{i=1}^{n_1} [(r'(C)\beta_1(D) - r'(C)\alpha_1(D) + q'(C)\beta_1(D) - q'(C)\alpha_1(D)]$$

$$+ \sum_{i=1}^{n_1} [(\beta_1(C)\beta_1(D) - \alpha_1(C)\beta_1(D) + \beta_1(C)\alpha_1(D) - \alpha_1(C)\alpha_1(D)]$$

$$= \sum_{i=1}^{n_1} C_m D_p + \sum_{i=1}^{n_1} \Delta_m^C D_p + \sum_{i=1}^{n_1} \Delta_m^D C_p + \sum_{i=1}^{n_1} \Delta_m^C \Delta_p^D$$

$$= \sum_{i=1}^{n_1} C_m D_p + \Delta_p^C \sum_{i=1}^{n_1} D_m + \Delta_m^D \sum_{i=1}^{n_1} C_p + n_1 \Delta_m^C \Delta_p^D$$

Similarly, we can show that:

$$\text{Updated}(\sum_{i=1}^{n_2} E_m F_p) = \sum_{i=1}^{n_2} E_m F_p + \Delta_m^E \sum_{i=1}^{n_2} F_p + \Delta_p^F \sum_{i=1}^{n_2} E_m + n_2 \Delta_m^E \Delta_p^F$$

Adding the two updated statistics, the lemma follows.

4.2   Deletion Operation of Vector Sets Using Sufficient Statistics

Let us consider the case where we want to find a superposition under a deletion operation. That is, let $\mathcal{Q} \leftrightarrow \mathcal{R}$ and $\mathcal{S} \leftrightarrow \mathcal{T}$ denote two pairs of vector sets that are in correspondence. Let $\mathcal{S} \subset \mathcal{Q}$ and $\mathcal{T} \subset \mathcal{R}$. Under this assumption, let us define $\mathcal{U} = \mathcal{Q} - \mathcal{S}$ and $\mathcal{V} = \mathcal{R} - \mathcal{T}$.

Using the same notations as in the previous section, it is straightforward to see that the sufficient statistics $\mathbf{\Psi}$ of the superposition of $\mathcal{U}$ with $\mathcal{V}$ can be derived from the sufficient statistics $\mathbf{\Psi_1}$ (of $\mathcal{Q} \leftrightarrow \mathcal{R}$) and $\mathbf{\Psi_2}$ (of $\mathcal{S} \leftrightarrow \mathcal{T}$). The update rules defining the deletion operation are similar to the ones described above, so we leave these rules to the reader as an exercise.

5   Computing the r.m.s.d. from Updated Sufficient Statistics

It is easy to see that Kearsley's 4×4 quaternion matrix $\mathbf{Q}$ given in Equation 1 can be constructed using the updated sufficient statistics $\mathbf{\Psi}$ derived from $\mathbf{\Psi_1}$ and $\mathbf{\Psi_2}$. The matrix $\mathbf{Q}$ contains 10 distinct elements (given that $\mathcal{Q}$ is square symmetric) which can be computed in constant time.

In practice, $\mathbf{Q}$ is diagonalized using the Jacobi's iterative rotation approach, which with each rotation annihilates an off-diagonal element. This approach has a fast convergence, and requiring no additional optimization. However, in many cases the updated superposition shows only a marginal change from the previous one. For example, if we were to extend a current superposition by one pair of residues, the resultant new transformation will often, in practice, be very close to the previously computed one. This allows the diagonalisation to build on the previous solution.

Let $\mathbf{Q}$ denote the Kearsley's 4×4 matrix corresponding to the superposition of corresponding vector sets $\mathcal{U}$ and $\mathcal{V}$. From eigen decomposition theorem, we get $\mathbf{Q} = S \Lambda S^{-1}$, where S is the matrix of eigenvectors and Λ is the diagonal matrix of eigenvalues. Also note that $\mathbf{Q}$ is positive semidefinite matrix with the property $\mathbf{Q}^T \mathbf{Q} = \mathbf{Q} \mathbf{Q}^T$. This implies that all the eigenvectors are orthogonal to each other. This further simplifies the decomposition to $\mathbf{Q} = S \Lambda S^T$. Also, since S is an orthogonal matrix, $\mathbf{Q} = S \Lambda S^T \implies \Lambda = S^T \mathbf{Q} S$.

Now, assume that the corresponding vector sets are augmented from $\mathcal{U}$ and $\mathcal{V}$ to $\mathcal{U}'$ and $\mathcal{V}'$, resulting in an updated Kearsley's matrix $\mathbf{Q}'$. We want to diagonalize this matrix into $S' \Lambda' S'^T$. Instead of starting the Jacobi's iterative process from scratch, we use the previously computed eigenvectors (before the vector sets were augmented), S, and compute $\tilde{\Lambda}$ as $S^T \mathbf{Q}' S$. Notice that if the augmentation does not include drastic changes, then $\tilde{\Lambda}$ is nearly diagonal (that is, $\tilde{\Lambda} \approx \Lambda'$), thus requiring very few iterations to fully diagonalize $\tilde{\Lambda}$. This provides a further optimization to the diagonalization step under update operations on vector sets.

6   Experiments

C++ programs were developed to compare the performance gain using sufficient statistics, when compared with the approach which recomputes the superposition *ab initio*.

8992 ASTRAL SCOP [26, 27] domains were as the source structures from which superposable fragments are randomly sampled. The general procedure of sampling is as follows. From the list of source structures, uniformly randomly choose a particular structure. Within this structure choose 2 random fragments of lengths l_1 and l_2, where the length is between 10 and 40 residues. These chosen fragments form the sets $\mathcal{Q}$ and $\mathcal{S}$. Yet another structure is again randomly chosen, and two fragments are sampled from it such that their lengths are strictly l_1 and l_2 respectively. These form the sets $\mathcal{R}$ and $\mathcal{T}$.

Assuming one-to-one correspondence between $\mathcal{Q} \leftrightarrow \mathcal{R}$ we compute the sufficient statistics $\mathbf{\Psi}_1$ of their orthogonal superposition. Similarly the sufficient statistics $\mathbf{\Psi}_2$ is computed for the orthogonal superposition between $\mathcal{R} \leftrightarrow \mathcal{T}$. Define $\mathcal{U} = \mathcal{Q} + \mathcal{S}$ and $\mathcal{V} = \mathcal{R} + \mathcal{T}$.

Iterating this process over 100 million such random samples, we compute:

1. The time it takes to superpose $\mathcal{U} \leftrightarrow \mathcal{V}$ and compute r.m.s.d from scratch.
2. The time it takes to superpose the same and compute r.m.s.d using the sufficient statistics $\boldsymbol{\Psi}_1$ and $\boldsymbol{\Psi}_2$
3. The difference between the two r.m.s.d values. (This is performed to ascertain the numerical stability involved in computing the r.m.s.d. values from sufficient statistics.)

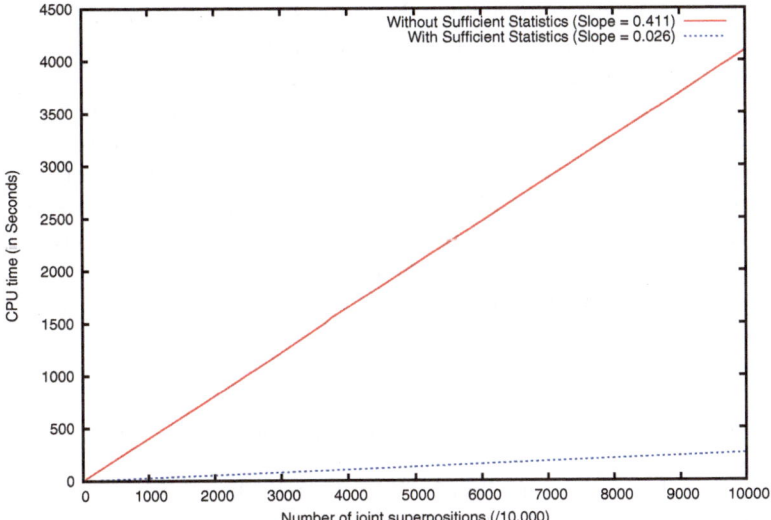

Fig. 1. The CPU times (in seconds) performing joint superpositions from scratch (Red line) compared against the same using sufficient statistics (Blue line) over 100 million random fragment data sets derived from ASTRAL SCOP domains. The X-axis reports the number of joint superpositions divided by 10,000.

Figure 1 compares the run times for the data set discussed above. Without sufficient statistics the run times takes 1.15 hours to conduct 100 million joint superpositions, while the same task is be achieved in 261 seconds ($\approx$4 minutes) using sufficient statistics. This shows a drastic improvement in the run time.

These empirical runtime results demonstrate what we have shown in Section 4.1, that the updates using sufficient statistics can be performed in constant time. If $|J|$ is the number of joint superpositions and n is the (average) number of points being superposed, then the first method grows as $O(n|J|)$. Since $n \ll |J|$ we see a linear trend (with a steeper gradient accounting for the multiplier n in the complexity term). In comparison, the results with sufficient statistics grow simply as $O(|J|)$ with a small gradient, made possible due to constant time computation of r.m.s.d values (using sufficient statistics) in each iterations.

To assess the numerical stability of our approach, we computed the r.m.s.d. values using the two approaches. The mean and standard deviation of the *difference* between the two r.m.s.d values were then computed. Both the mean and the standard deviation are zero Å up to double precision. This demonstrates the numerical stability of computing r.m.s.d. using sufficient statistics.

7 Conclusion

Optimal superpositions of vector sets provide the foundation to determine similarities and differences between spatial objects, especially for macromolecular structures. We derived a set of sufficient statistics for the orthogonal superposition problem minimizing the sum of squares error. These statistics provide a highly efficient method to operate (via addition and deletion of vectors) on the existing superpositions. Our results demonstrate a drastic improvement in the computational effort required to compute r.m.s.d. using sufficient statistics. These results are relevant to many analyses involving structural data. These include the plethora of algorithms to construct pairwise and multiple protein structural alignments by assembling fragment pairs. Source code (written in C++) to undertake superpositions of vector sets using sufficient statistics can be downloaded from http://www.csse.monash.edu.au/~karun/suffStatSuperpose.html

References

1. Lesk, A.M.: Introduction to protein architecture: the structural biology of proteins. Oxford University Press (2001)
2. Eidhammer, I., Jonassen, I., Taylor, W.R.: Protein Bioinformatics: An algorithmic approach to sequence and structure analysis. J. Wiley & Sons (2004)
3. Lesk, A.M.: The unreasonable effectiveness of mathematics in molecular biology. The Mathematical Intelligencer 22(2), 28–37 (2000)
4. Kabsch, W.: A solution for the best rotation to relate two sets of vectors. Acta Crystallographica Section A: Crystal Physics, Diffraction, Theoretical and General Crystallography 32(5), 922–923 (1976)
5. Kabsch, W.: A discussion of the solution for the best rotation to relate two sets of vectors. Acta Crystallographica Section A: Crystal Physics, Diffraction, Theoretical and General Crystallography 34(5), 827–828 (1978)
6. McLachlan, A.D.: Rapid comparison of protein structures. Acta Crystallographica Section A: Crystal Physics, Diffraction, Theoretical and General Crystallography 38(6), 871–873 (1982)
7. KenKnight, C.: Comparison of methods of matching protein structures. Acta Crystallographica Section A: Foundations of Crystallography 40(6), 708–712 (1984)
8. Mackay, A.L.: Quaternion transformation of molecular orientation. Acta Crystallographica Section A: Foundations of Crystallography 40(2), 165–166 (1984)
9. Lesk, A.: A toolkit for computational molecular biology. II. on the optimal superposition of two sets of coordinates. Acta Crystallographica Section A: Foundations of Crystallography 42(2), 110–113 (1986)
10. Diamond, R.: A note on the rotational superposition problem. Acta Crystallographica Section A: Foundations of Crystallography 44(2), 211–216 (1988)
11. Kearsley, S.K.: On the orthogonal transformation used for structural comparisons. Acta Crystallographica Section A: Foundations of Crystallography 45(2), 208–210 (1989)

12. Cohen, G.: ALIGN: a program to superimpose protein coordinates, accounting for insertions and deletions. Journal of Applied Crystallography 30(6), 1160–1161 (1997)
13. Coutsias, E.A., Seok, C., Dill, K.A.: Using quaternions to calculate RMSD. Journal of Computational Chemistry 25(15), 1849–1857 (2004)
14. Koehl, P.: Protein structure similarities. Current opinion in Structural Biology 11(3), 348–353 (2001)
15. Hamilton, W.R., Hamilton, W.E.: Elements of quaternions. Longmans, Green, & Company (1866)
16. Shindyalov, I.N., Bourne, P.E.: Protein structure alignment by incremental combinatorial extension (ce) of the optimal path. Protein Engineering 11(9), 739–747 (1998)
17. Ye, Y., Godzik, A.: Flexible structure alignment by chaining aligned fragment pairs allowing twists. Bioinformatics 19(suppl. 2), ii246–ii255 (2003)
18. Shatsky, M., Nussinov, R., Wolfson, H.J.: A method for simultaneous alignment of multiple protein structures. Proteins: Structure, Function, and Bioinformatics 56(1), 143–156 (2004)
19. Konagurthu, A.S., Whisstock, J.C., Stuckey, P.J., Lesk, A.M.: MUSTANG: a multiple structural alignment algorithm. Proteins: Structure, Function, and Bioinformatics 64(3), 559–574 (2006)
20. Shatsky, M., Nussinov, R., Wolfson, H.J.: Flexible protein alignment and hinge detection. Proteins: Structure, Function, and Bioinformatics 48(2), 242–256 (2002)
21. Vriend, G., Sander, C.: Detection of common three-dimensional substructures in proteins. Proteins: Structure, Function, and Bioinformatics 11(1), 52–58 (1991)
22. Lackner, P., Koppensteiner, W.A., Sippl, M.J., Domingues, F.S.: Prosup: a refined tool for protein structure alignment. Protein Engineering 13(11), 745–752 (2000)
23. Kolodny, R., Koehl, P., Levitt, M.: Comprehensive evaluation of protein structure alignment methods: scoring by geometric measures. Journal of Molecular Biology 346(4), 1173–1188 (2005)
24. Hogg, R.V., Craig, A.: Introduction to mathematical statistics. Prentice Hall (1994)
25. Jacobi, C.G.J.: Über ein leichtes Verfahren, die in der Theorie der Säkularstörungen vorkommenden Gleichungen numerisch aufzulösen. Journal für die Reine und Angewandte Mathematik 30(1846), 51–95 (1846)
26. Murzin, A.G., Brenner, S.E., Hubbard, T., Chothia, C.: SCOP: a structural classification of proteins database for the investigation of sequences and structures. Journal of Molecular Biology 247(4), 536–540 (1995)
27. Chandonia, J.M., Hon, G., Walker, N.S., Conte, L.L., Koehl, P., Levitt, M., Brenner, S.E.: The ASTRAL compendium in 2004. Nucleic Acids Research 32(suppl. 1), D189–D192 (2004)

IDBA-MTP: A Hybrid MetaTranscriptomic Assembler Based on Protein Information

Henry C.M. Leung, S.M. Yiu, and Francis Y.L. Chin

Department of Computer Science, The University of Hong Kong, Pokfulam Road, Hong Kong
{cmleung2,smyiu,chin}@cs.hku.hk

Abstract. Metatranscriptomic analysis provides information on how a microbial community reacts to environmental changes. Using next-generation sequencing (NGS) technology, biologists can study microbe community by sampling short reads from a mixture of mRNAs (metatranscriptomic data). As most microbial genome sequences are unknown, it would seem that de novo assembly of the mRNAs is needed. However, NGS reads are short and mRNAs share many similar regions and differ tremendously in abundance levels, making de novo assembly challenging. The existing assembler, IDBA-MT, designed specifically for the assembly of metatranscriptomic data only performs well on high-expressed mRNAs.

This paper introduces IDBA-MTP, which adopts a novel approach to metatranscriptomic assembly that makes use of the fact that there is a database of millions of known protein sequences associated with mRNAs. How to effectively use the protein information is non-trivial given the size of the database and given that different mRNAs might lead to proteins with similar functions (because different amino acids might have similar characteristics). IDBA-MTP employs a similarity measure between mRNAs and protein sequences, dynamic programming techniques and seed-and-extend heuristics to tackle the problem effectively and efficiently. Experimental results show that IDBA-MTP outperforms existing assemblers by reconstructing 14% more mRNAs. **Availability:** www.cs.hku.hk/~alse/hkubrg/.

Keywords: metatranscriptomic reads, assembling, next-generation sequencing, protein sequence alignment.

1   Introduction

The traditional approach for studying microorganisms is to isolate and cultivate each single microorganism and then study its behavior, such as gene expression levels, under different environments. As different microbes usually live together to form a microbial community, isolating a single microbe is usually impossible [4] and, even possible, changes the microbe's living behavior in a microbial community. Metatranscriptomic studies in the past have been based on microarrays or cDNA clone libraries [2,23,29]. The microarray-based approach [17] requires knowledge of target mRNA sequences, which limits its usefulness in relation to novel mRNAs.

R. Sharan (Ed.): RECOMB 2014, LNBI 8394, pp. 160–172, 2014.
© Springer International Publishing Switzerland 2014

cDNA clone libraries, on the other hand, can be applied to novel mRNAs, but the approach is labor-intensive and the estimations of expression levels of mRNAs are inaccurate.

High-throughput next-generation sequencing (NGS) technology [3,7,22,18] introduces a new and better approach for studying metatranscriptomic data. By sequencing reads from mRNA sequences of a sample, scientists can reconstruct novel mRNA sequences by assembling reads and can estimate the expression levels of each mRNA by the number of reads aligned to the mRNA sequence. Currently, there are two main NGS technologies for metatranscriptomic data: pyrosequencing technology and synthesis technology. Pyrosequencing technology [6,8,24,30] produces long reads (of length about 400 bp) with relatively higher cost (over 40 times higher for the same throughput). Since the read length is long, no or limited assembly is required. Pyrosequencing technology has achieved promising results for soil samples [30] and marine samples [6,8]. Synthesis technology, on the other hand, produces relatively short reads (of length varying from 75 bp to 150 bp) at much lower cost. Since the length of reads produced by synthesis technology is much shorter than the length of the mRNA sequence (about 1000 bp), the reads need to be assembled into longer sequences (contigs) before analysis.

Compared with assembling genomic, transcriptomic or metagenomic data, assembling metatranscriptomic data is much more difficult because of the following reasons.

1. *Repeat patterns across different mRNAs.* Repeat patterns usually introduce ambiguity during assembly and are a common problem in all types of assembling. However, the problem is more serious in metatranscriptomes than in other data. Many genes exist in multiple species with similar functions and the resultant proteins share common protein domains [9]. As a result, in the metatranscriptomic data, many different mRNAs have similar patterns. According to analysis of genBank [1], based on known gene information, 24.53% of bacteria genes contain at least one repeat pattern of length longer than 100 bp (note that, in this analysis, different versions of the same genes from the same bacteria were ignored and only the repeat patterns in genes from different bacteria were considered). In these circumstances, assemblers, not specially designed for metatranscriptome data, produce either short contigs or chimeric contigs that merge mRNA sequences from more than one gene [15]. This is consistent with our experiments (see Table 2): these assemblers can either only recover 31% of mRNAs with average contig length of 194bp and 4.14% error rate (Oasis), or recover more mRNAs (59.29%) with longer average contig length (395bp) but the error rate is increased to 10.73% (IDBA-UD).

2. *Extreme differences in abundances.* For the DNA genome assembly problem of a single species, this is not a problem because there is one abundance only. On the other hand, in transcriptomic data and metagenomic data, since the abundances of different mRNAs and the number of genomes vary (can be 100 times and 1,000 times different, respectively [25]) because of different expression levels and abundances of species, erroneous reads cannot be identified easily by sampling

rates. In metatranscriptomic data, this problem becomes more serious. Since both the abundances of species and the expression levels of mRNAs from the same species may vary, the abundances of different mRNAs can vary much more significantly (over 100,000 times). Thus low-expressed mRNA sequences are very difficult to reconstruct as correct reads from these sequences and erroneous reads are very difficult to distinguish. As Table 1 shows for our experiments on low-expressed mRNAs, the performance of existing assemblers suffers.

Thus, existing assemblers for genomic, transcriptomic and metagenomic data do not perform well on metatranscriptomic data especially for the low-expressed transcriptomes [15]. To our best knowledge, IDBA-MT [15] is the only assembler designed for metatranscriptomic data. IDBA-MT aims at solving the repeat pattern problem. By applying information from paired-end reads, IDBA-MT can resolve some of the chimeric contigs (See Table 2, IDBA-MT can recover more mRNAs while decreasing the error rate from about 10% to 5% when compared to IDBA-UD). However, this approach can only work for high-expressed transcriptomes with high sequencing depths as it relies on paired-end data and fails when there are insufficient sampling reads from the mRNAs (i.e., low-expressed mRNAs).

Similar to genome assembly, besides de novo assembly, one can apply the reference-based approach. Existing work tries to reconstruct mRNAs by aligning metatranscriptomic reads to known genomes or gene DNA sequences. However, this approach has had only limited success [32] as the genomes of most microbes are still unknown [4] and the microbe gene sequences mutate frequently.

Our Observations on the Reference-Based Approach: Although the aforementioned reference-based approach has limited success, about 60% to 70% of the proteins in bacteria have similar sequences as some known proteins [5, 30], thus known reference protein sequences could help in the assembling of novel mRNAs. There are two difficulties to resolve in order to make use of the protein sequences. First, we need to consider amino acid instead of nucleotides. Even if we consider amino acid, it is not trivial due to the following. For proteins with similar functionalities, even though their structures are similar and their sequences share some conserved regions, the amino acid sequences corresponding to these conserved regions might not be exactly the same. Second, to consider amino acid, the information contained in a single read becomes much less (3 nucleotides converted to 1 amino acid). Since one read only corresponds to about 25 amino acids (aa), it is difficult to have a confident alignment [32]. Another approach is to align contigs, instead of reads, to proteins. However, as the performance of existing assemblers is not good, the resultant contigs are short or incorrect and not many confident alignments can be obtained.

Our Contributions: To overcome the first problem of amino acid similarity, we found that even though the amino acid sequences may not be exactly the same, it is known that some amino acids, though different, have similar chemical properties and functionality [11]. Consequently, the mRNA can be reconstructed using the approach of first decoding the reads into peptide sequences and then aligning these peptide sequences to protein sequences based on the similarity of amino acids (e.g. Blosom 62). Thus, we incorporate the similarity of amino acids into our alignment algorithm. To solve the problem of short reads and low-expressed mRNAs, we make use of the paths of the de Bruijn graph with a small k.

Our proposed assembler, IDBA-MTP, reconstructs mRNA sequences from metatranscriptomic reads, especially for low-expressed mRNAs, using the information of known microbial protein sequences to guide the construction of contigs as follows. IDBA-MTP first constructs a de Bruijn graph from the input reads using a relatively small k ($k = 21$ bp) to compensate for the missing long k-mers in low-expressed mRNAs. Since k is small, the de Bruijn graph, though connected, has many branches representing repeat regions in the mRNA sequences (due to problem 1 and 2) and with each mRNA represented by one of its paths. In order to determine whether a path represents an mRNA sequence or not, IDBA-MTP will decode the path into a peptide sequence and then align it with known protein sequences. Those paths, which can be aligned to known protein sequences, should be potential candidates for mRNA sequences depending on their similarity and sequencing depths. However, since the number of paths is huge (many paths will not represent any mRNA sequences) and the alignment with the protein sequences is not straightforward because of the similar chemical properties of amino acids, a dynamic programming approach with a seed-and-extend (with the seed derived from the known protein sequences) heuristics is employed to reduce the complexity of the problem.

Since the candidate mRNA sequences are constructed by aligning known protein sequences, mRNA sequences for novel proteins cannot be reconstructed using this approach.   An intuitive idea is to run IDBA-MT for novel mRNAs, then combine the results of IDBA-MT and the output from our reference-based approach. However, some mRNAs sequences may be reconstructed by both approaches, which results in redundant or similar contigs. To prevent having redundant contigs, IDBA-MTP will treat those mRNAs sequences reconstructed by alignment of known reference proteins as long input reads for IDBA-MT, i.e. the output of the first approach will be the input of the second approach. Experiments on simulated data show that even though 48% regions of the mRNAs can be aligned to known reference proteins, existing assemblers can only reconstruct contigs representing at most 62.9% of these regions. IDBA-MTP can reconstruct contigs covering 77.6% of these regions and some novel mRNAs using protein reference sequences. As a result, IDBA-MTP can reconstruct 14% more mRNAs (in term of the total length of mRNAs) than existing assemblers.

The paper is organized as follows. The IDBA-MTP algorithm is described in Section 2. Experimental results for IDBA-MTP and other existing assemblers on both simulated and real metatranscriptomic data are presented in Section 3. Conclusions are drawn on the performance of IDBA-MTP in Section 4.

2   Methodology

Given a set of reads sampled from a set of mRNA sequences (with nucleotides A, C, G and U), we can construct a de Bruijn graph where each vertex v represents a length-k substring (k-mer) of the reads and where an edge connects vertex u to vertex v if and only if the corresponding k-mers for vertex u and vertex v overlap at $k-1$ positions and appear in a read. An mRNA sequence can be represented by a path of k-mers in the de Bruijn graph. Since there are many paths in the de Bruijn graph and most of them do not represent any mRNA, a correct mRNA sequence R can be reconstructed from the de Bruijn graph if some known protein sequence P can be aligned to the path. If the alignment similarity between R and P is high, R will likely be an mRNA sequence in the sample.

A protein or peptide sequence is represented by a sequence of amino acids (of which there are 20 kinds). Given a length-$3m$ mRNA sequence R, we can decode it into a length-m sequence $D(R)$ of amino acids by converting each non-overlapping coden (length-3 substring) in R into an amino acid character. Given a protein sequence P and an mRNA sequence R, P and $D(R)$ can be aligned by inserting space characters in P and $D(R)$ to form P' and $D(R)'$ of equal length respectively, and the similarity score based on this alignment is defined as follows:

$$score_a(P', D(R)') = \sum \delta(P'[i], D(R)'[i]) + p_{\text{open}} \cdot \text{number of gaps} \qquad (1)$$

where $P'[i]$ and $D(R)'[i]$ are the i-th amino acid in P' and $D(R)'$ respectively, $\delta(x,y)$ is the similarity score between amino acids x and y (which depends on their chemical properties and roles in the protein's functionality), p_{open} is the gap penalty and a gap is defined as consecutive space characters in P' or $D(R)'$ (the gap penalty can be refined to take the gap size into consideration). Note that the similarity score $\delta(x,y)$ can be negative and is $-\infty$ whenever a stopping coden in $D(R)'$ is compared to space or any amino acid in P'. The optimal global similarity score between P and $D(R)$ is defined as the highest similarity score of all alignments of P and $D(R)$.

$$score_g(P, D(R)) = max_{\text{all alignment } P' \text{ and } D(R)'}\{score_a(P', D(R)')\} \qquad (2)$$

Since the decoded protein from an mRNA usually does not exist in the protein database but some part of the decoded protein sequence might match with some regions of some proteins in the database because of their functional similarity, instead of aligning the whole sequence of P and $D(R)$, the optimal local alignment between all substrings of P and $D(R)$ is considered in IDBA-MTP and this information, in terms of contigs, will be needed for mRNA assembly later (see Section 2.3). The optimal local similarity score is defined as:

$$score_l(P, D(R)) = max_{\text{all substrings } p_s \text{ and } d_s \text{ of P and } D(R)}\{score_g(p_s, d_s)\} \qquad (3)$$

The **Protein-Graph Alignment (PGA) Problem** can be defined as follows: given a de Bruijn graph G and a protein P, find a path in G (representing a substring in an mRNA sequence R) such that $score_l(P, D(R))$ is maximized.

2.1   Dynamic Programming

The PGA problem can be solved by dynamic programming based on the principle of optimality. Consider an optimal global alignment Opt of the substring d_s (represented by a path $Q(d_s)$) of the decoded protein $D(R)$ for an mRNA sequence R with the substring p_s of protein sequence P. The same alignment Opt for any subpath of $Q(d_s)$ and the corresponding substring of p_s should also be optimal.

Let $S(v, i)$ define the maximum global similarity score between a substring of P ending at $P[i]$ and all decoded sequences $D(R)$ for path R in the de Bruijn graph G ending at vertex v. Similarly, we define $S_M(v, i)$, $S_P(v, i)$ and $S_R(v, i)$ to be the maximum global similarity score with the following restrictions respectively: (1) $P[i]$ is aligned with the last amino acid of the corresponding protein sequence decoded from the path ending at vertex v, (2) $P[i]$ is aligned with the space character and (3) the last amino acid of the corresponding protein sequence decoded from the path ending at vertex v is aligned with the space character. The value of $S(v, i)$ is the maximum of 0 (alignment of two null substring), $S_M(v, i)$, $S_P(v, i)$ and $S_R(v, i)$. The value of $S_M(v, i)$, $S_P(v, i)$ and $S_R(v, i)$ can be calculated by considering the alignment of the last coden, any length-3 path $s \rightarrow v$ with $D(s \rightarrow v)$ represent the decoded amino acid of path $s \rightarrow v$, and the subproblem of alignment ending as vertex s.

$S(v, i)$, $S_M(v, i)$, $S_P(v, i)$ and $S_R(v, i)$ can be calculated as follows:

$$S(v, i) = \begin{cases} 0 & \text{no path ending at } v \text{ can be decoded to an amino acid} \\ \max\{0, S_M(v, i), S_P(v, i), S_R(v, i)\} & \text{otherwise} \end{cases}$$

$$S_M(v, i) = \begin{cases} -\infty & \text{no path ending at } v \text{ can be decoded to an amino acid} \\ S(s, i-1) + \delta\big(P[i], D(s \rightarrow v)\big) & \text{otherwise} \end{cases}$$

$$S_P(v, i) = \begin{cases} -\infty & i = 0 \\ \max\{S_P(v, i-1), S_M(v, i-1) + p_{\text{open}}\} + \delta(P[i], space) & \text{otherwise} \end{cases}$$

$$S_R(v, i)$$
$$= \begin{cases} -\infty & \text{no path ending at } v \text{ can be decoded to an amino acid} \\ \max\{S_P(s, i), S_M(s, i) + p_{\text{open}}\} + \delta(space, D(s \rightarrow v)) & \text{otherwise} \end{cases}$$

If $D(s \rightarrow v)$ represents the stopping coden, $\delta(D(s \rightarrow v), x) = -\infty$. $\max_{v,i}\{S(v, i)\}$ represents the optimal local similarity score and the corresponding aligned mRNA sequence can be obtained by backtracking. Note that care should be taken for the starting vertex of the path. Since the starting vertex of a path in de Bruijn graph represents the length-k prefix of an mRNA and each subsequent vertex represents an extra nucleotide of the mRNA, we modify zero in-degree vertices in the de Bruijn graph implicitly such that each vertex only represents one single nucleotide (the last nucleotide of the k-mer) of an mRNA. Note that since the protein sequence P is fixed, the dynamic programming is correct even there is loop in the de Bruijn graph.

Since there are at most $4^3 = 64$ length-3 paths $s{\rightarrow}v$ to a vertex v, each entry $S(v, i)$, $S_M(v, i)$, $S_P(v, i)$ and $S_R(v, i)$ can be computed in constant time by preprocessing. The time complexity for aligning a length-$|P|$ protein P is $O(n|P|)$ and for a set of protein sequences with total length m is $O(nm)$, where n is the total number of vertices in the de Bruijn graph.

2.2   Seed-and-Extend Heuristic

Although the dynamic programming approach can solve the PGA problem in $O(nm)$ time, n and m are usually large for real biological data (in the order of millions and thousand millions respectively) and the running time for the above dynamic programming approach is too long for practical use. In order to speed up the running time, IDBA-MTP applies on seed-and-extend heuristic to speed up the process. Assume that the optimal local alignment of an mRNA and a protein has at least one aligned region with t consecutive matches of amino acids (with similarity score larger than a predefined threshold), the PGA problem can be solved by a seed-and-extend heuristic. Given a simple path (a path with all intermediate vertices have exactly one incoming and one outgoing edge) or a k-mer in the de Bruijn graph representing a length-t peptide (sequence of amino acids), the reference protein sequences containing this peptide can be obtained in constant time after $O(m)$ preprocessing, where m is the total length of the reference proteins. By considering these positions as the starting alignment positions (seeds) and extending the alignment in both forward and backward directions using dynamic programming, a small subset of paths containing the seed as a subpath will be considered and the running time can be greatly reduced in practice.

2.3   Preventing Redundant mRNAs

As some reference proteins could have similar sequences, these similar proteins might align to overlapping paths in the de Bruijn graph and similar mRNA sequences may be obtained. Among these similar mRNA sequences, it is likely that only one of them is correct while the others are only artifacts caused by sequencing errors or misalignment. However, duplicate genes and genes with similar functions in different species may also introduce similar mRNA sequences. IDBA-MTP applies two techniques to remove artifacts. The first approach is to prevent aligning multiple proteins with seeds on the same simple path in the de Bruijn graph. Simple paths in the de Bruijn graph are sorted in decreasing order of lengths and are considered one by one. Once a protein is aligned to a path R (with the maximum alignment score among all proteins) in the de Bruijn graph, all substrings in R are removed from the seed table and will not be considered as starting positions for alignment. Note that these simple paths could still be considered when extending the alignment of other proteins using dynamic programming. Although the first approach can determine some redundant contigs represent the same mRNAs, sequence error could introduce error paths in the de Bruijn graph result as alignment of similar proteins to overlapped but similar paths in the de Bruijn graph. In our experiment, there can be 50 similar paths represented by the correct and erroneous paths corresponding to the same mRNA. Thus, we should not output the aligned mRNAs directly. The second approach was considering these mRNAs as long reads and treating them as input to IDBA-MT for de novo assembly. By using these extra long reads, paired-end reads and sequencing depths information, IDBA-MT avoids assembling redundant mRNAs and can reconstruct novel mRNAs with no similar reference proteins.

3   Experiments

We compared the performances of Oases [26], Trinity [10], IDBA-UD [21], IDBA-MT[15] and IDBA-MTP on a real dataset from mouse gut [32] and two simulated datasets generated from known bacteria gene sequences obtained from genBank [1]. Oases and Trinity were designed for assembling transcriptomic data, IDBA-UD for assembling metagenomic data, and IDBA-MT for assembling metatranscriptomic data. All bacteria gene sequences with known sources in the genBank were downloaded. To prevent selecting mRNAs from the same species (either from the same or different strains), duplicated sequences were removed and only one version was kept. Note that similar mRNAs obtained from different bacteria would be kept. Similar to [15], mRNAs sharing at least half of the sequences with other mRNAs were selected for generating a difficult dataset (mRNAs which do not share common sequence regions with others would be isolated in the de Bruijn graph and can be assembled easily). The resultant 658 mRNA sequences were used to generate the simulated data. Although the number of mRNA sequences selected is small compared with the real experiments, this small subset of mRNAs sequences with long repeats represents the most difficult part of assembling metatranscriptomic data. The reference bacteria protein sequences for IDBA-MTP was downloaded from NCBI database and we used the Blosum-62 scoring matrix, open gap penalty $= -10 - (-1) = -9$ and gap extend penalty $= -1$ for calculating the similarity scores of protein sequences. In all experiments on simulated data, all the corresponding protein sequences of the 658 mRNA sequences were removed from the reference protein sequences for testing the performance of IDBA-MTP.

For each simulated dataset, we randomly picked length-75 bp paired-end reads from the RNA sequence with 1% sequencing error according to the predefined abundances. The mean insert distance of each paired-end read was 200 bp with a standard deviation of 10 bp. Two sets of simulated data were generated: (1) Low abundance - 100 mRNAs were sampled with 3x sequencing depth for evaluating the performances of the assemblers for mRNAs with low expression levels. (2) Mixture abundance - 658 mRNAs were sampled from 1000x to 3x sequencing depth with the number of mRNAs following the power law (number of mRNAs with a certain abundance is directly proportional to the negative of abundance ratio) for evaluating the performances of the assemblers for mRNAs with different expression levels.

All assemblers were tested on simulated data using default parameters. Each contig produced by the assemblers was aligned to the 658 mRNAs in the samples using Blat [13]. A contig was considered correct if and only if at least 95% of the contig region could be aligned to the mRNA sequence with 95% similarity. Some short, even correct, contigs which could not align confidently to the 658 mRNAs were considered incorrect. Regions of mRNAs aligned by correct contigs were considered covered and the coverage of an assembler was calculated as the ratio of regions in the mRNAs covered by the contigs produced by the assembler. Although Oases, Trinity, IDBA-UD could produce scaffolds using paired-end reads, the scaffolds performed worse than the contigs in all simulated data because these assemblers connected contigs wrongly and produced long but incorrect scaffolds. Thus, we compared the performances of the assemblers based on the resultant contigs and the experimental results are shown in Table 1 and 2.

Table 1. Experimental Result on simulated data with low abundance ratios

Softare	Coverage	Max. Len.	Avg. Len.	# of wrong contig (len.)	# of correct contig (len.)	Error Rate
Oases	25.99%	524 bp	172 bp	9 (1,063 bp)	149 (25,690 bp)	3.97%
Trinity	9.85%	497 bp	287 bp	17 (7,362 bp)	34 (9,837 bp)	42.80%
IDBA-UD	48.26%	783 bp	342 bp	8 (4,425 bp)	83 (28,480 bp)	13.45%
IDBA-MT	52.68%	900 bp	279 bp	8 (3,194 bp)	136 (37,993 bp)	7.75%
IDBA-MTP	66.00%	916 bp	273 bp	5 (1,057 bp)	156 (42,771 bp)	2.40%

Table 2. Experimental Result on simulated data with mixed abundance ratios

Softare	Coverage			Max. Len.	Avg. Len.	# of wrong contig (len.)	# of correct contig (len.)	Error Rate
	total	≤5x	> 5x					
Oases	31.00%	22.46%	8.45%	676 bp	194 bp	63 (8,471 bp)	1009 (196,162 bp)	4.14%
Trinity	15.10%	11.28%	3.80%	1,270 bp	319 bp	106 (75,713 bp)	310 (99,603 bp)	43.18%
IDBA-UD	59.29%	42.74%	16.38%	1,430 bp	395 bp	43 (28,837 bp)	606 (239,887 bp)	10.73%
IDBA-MT	64.07%	46.53%	17.37%	1,511 bp	310 bp	37 (18,023 bp)	1005 (312,500 bp)	5.45%
IDBA-MTP	69.62%	51.29%	18.33%	1,615 bp	368 bp	41 (23,461 bp)	1127 (415,813 bp)	5.34 %

3.1   Low Abundance mRNAs

When the abundances of mRNAs were low, Oases and Trinity did not perform well in assembly because of the low sequencing depths and the similarity of mRNAs. Oases tended to produce confident but shorter contigs. As a result, it had a low error rate (3.97%) but the lengths of contigs were short (average length = 172 bp) and the coverage was not high (25.99%). Since Trinity was designed for assembling transcriptomic data for eukaryotic mRNAs and was not suitable for assembling prokaryotic mRNAs, the error rate of Trinity was high (42.80%) and the coverage was low (9.85%). IDBA-UD, which was designed for assembling metagenomic data, performed better than Oases and Trinity because it applied various technologies, e.g. multiple k-mers, local assembling and local coverage of contigs for assembling reads sampled from low abundance genomes (mRNAs in this case). However, since the mRNAs had many similar sequences, IDBA-UD could not determine these chimeric contigs and the error rate was high (13.45%) but the coverage was acceptable (48.26%). IDBA-UD has such high error rate because it merged two or more mRNA sequences incorrectly to produce chimeric contigs. IDBA-MT, which was designed for assembling metatranscriptomic data, outperformed IDBA-UD because it used paired-end reads information to resolve chimeric contigs. It achieved a relatively high coverage (52.68%) with low error rate (7.75%). With the information from known protein sequences, IDBA-MTP further improved the coverage to 66.00% and had the lowest error rate (2.40%).

Table 3. Experimental Result on real mouse gut data

Softare	Maximum Length	Average Length	Contigs number	Total Length	# of contig aligned to known proteins (length)
Oases	693 bp	127 bp	99,611	12,655,199 bp	489 (84,044 bp)
Trinity	15,857 bp	500 bp	19,721	9,862,469 bp	7,188 (2,994,588 bp)
IDBA-UD	10,741 bp	490 bp	18,951	9,287,101 bp	9,510 (4,178,162 bp)
IDBA-MT	8,863 bp	490 bp	18,972	9,301,484 bp	9,515 (4,181,949 bp)
IDBA-MTP	9,070 bp	477 bp	20,062	9,581,626 bp	10,429 (4,712,857 bp)

3.2   mRNAs with Different Abundances

For the simulated data with mixed abundances, the overall performance of the assemblers improved because of the mRNAs with high abundances. We have also analysed the coverage of low-abundance mRNAs (76% mRNA with sequencing depth $\leq 5x$) and high-abundance mRNAs (24% mRNA with sequencing depth $> 5x$). As expected, the high-abundance mRNAs had better overall results than the low-abundance mRNAs. Again, Oases produced short but confident contigs, achieved higher coverage (31.00%) than Trinity and had the lowest error rate (4.14%). Trinity, which assembled many long and wrong contigs, had the lowest coverage (15.10%) and the highest error rate (43.18%). IDBA-UD had higher coverage (59.29%) and moderate error rate (10.73%). By resolving some chimeric contigs, IDBA-MT had slightly higher coverage (64.07%) and lower error rate (5.45%) than IDBA-UD. IDBA-MTP had the highest coverage (69.62%) and a low error rate (5.34%). Considering the performance of mRNAs with different abundances, IDBA-MTP could reconstruct 5% and 1% more mRNAs with low and high abundances respectively than the best existing assembler IDBA-MT. By using protein reference information, the performance of IDBA-MTP improved not only for the low-abundance mRNAs, but also for the high-abundance mRNAs.

3.3   Real Metatranscriptomic Data

Xiong et al. [32] isolated mRNAs from the lumen of the cecum and colon of 4 mice at 12 weeks old, colonized with an Altered Schaedler flora (ASF) containing eight known species without reference genomes. A total of 3.3 million paired-end reads were generated using Illumina sequencing technology. The read length was about 75 bp and the insert distance was about 300 bp. Similar to [15], we merged the reads sampled from the 4 mice into a single dataset as the number of reads in each sample was small. The reads were inputted to existing assemblers for comparison. Since there were no reference genomes, we evaluated the accuracy of output contigs by aligning them to known protein sequences using Blastx with default parameters. A contig was considered "correct" if at least 90% of the contig sequence could be aligned to a single protein sequence. We used number of aligned contigs instead of number of aligned proteins to evaluating the result because a contigs can be aligned to hundred of similar proteins and it is difficult to evaluate the softwares based on the number of discovered proteins. Noted that IDBA-UD, IDBA-MT and IDBA-MTP consider each k-mer in the de Bruijn should belong to at most one contigs, they should not output redundant contigs represents the same protein or protein regions.

Similar to simulated data, Oases produced very short contigs. Since it was difficult to obtain confident alignment for short contigs, only 489 (out of 99,611) contigs produced by Oases could be aligned to known protein sequences. Trinity produced longer contigs than other assemblers. However, over half of them (7,188 out of 19,721 can be aligned) could not be aligned to known protein sequences although the contigs were long enough for confident alignment. The performances of IDBA-UD and IDBA-MT were similar with half of the contigs aligned to known protein sequences. IDBA-MTP produced a thousand more contigs than IDBA-UD and IDBA-MT. Since the extra contigs constructed mainly due to using protein reference sequences, most of these extra contigs could be aligned to known protein sequences.

4   Conclusions

Existing assemblers do not perform well on metatranscriptomic data, especially on low-expressed mRNAs. In this paper, we have proposed IDBA-MTP to assemble mRNAs, making use of information from the database of millions of known protein sequences. In particular, dynamic programming technique with a seed-and-extend heuristics was introduced to reconstruct mRNA sequences from paths in the de Bruijn graph with maximum similarity scores when aligned with the known protein sequences. Experimental results on both simulated and real biological data showed that IDBA-MTP outperformed existing assemblers on metatranscriptomic data.

However, when applying IDBA-MTP on metatranscriptomic data, there is an issue of running time when compared with existing assemblers. Since the reference proteins databae is big and highly redundant, i.e. many proteins with very similar sequences exist, IDBA-MTP takes one or two days for aligning reference proteins to de Bruijn graph even using the seed-and-extend heuristic. This is much longer that existing assemblers which takes one or two hours to assemble the reads. Although it may not be a problem at current state because it takes weeks to generate a metatranscriptomic dataset, further research should be performed to increase the speed of IDBA-MTP by preprocessing the reference proteins or parallel processing.

The techinque of assembly based on known protein sequence information is applicable not only on metatranscriptomic data. It can also improve the performance on transcriptomic data of single species. We plan to study the usage of protein reference sequence information on transcriptomic assembly of single species.

Acknowledgement. This work was supported by Hong Kong GRF HKU 7111/12E, HKU 719709E and 719611E, Shenzhen basic research project (NO. JCYJ20120618143038947) and NSFC(11171086).

References

1. Benson, D., Karsch-Mizrachi, I., Lipman, D., Ostell, J., Rapp, B., Wheeler, D.: GenBank. Nucleic Acids Research 28(1), 15–18 (2000)
2. Booijink, C., Boekhorst, J., Zoetendal, E., Smidt, H., Kleerebezem, M., de Vos, W.: Metatranscriptome Analysis of the Human Fecal Microbiota Reveals Subject-Specific Expression Profiles, with Genes Encoding Proteins Involved in Carbohydrate Metabolism Being Dominantly Expressed. Appl. Environ. Microbiol. 76(16), 5533–5540 (2010)
3. ten Bosch, J., Grody, W.: Keeping up with the next generation: massively parallel sequencing in clinical diagnostics. J. Mol. Diagn. 10, 484–492 (2008)
4. Eisen, J.: Environmental shotgun sequencing: its potential and challenges for studying the hidden world of microbes. PLoS Biology 5(3), e82 (2007)
5. Finn, R., Tate, J., Mistry, J., et al.: The Pfam Protein Families Database. Nucleic Acids Research 28(1), 263–266 (2000)
6. Frias-Lopez, J., Shi, Y., Tyson, G., et al.: Microbial community gene expression in ocean surface waters. Proc. Natl. Acad. Sci. 105, 3805–3810 (2008)
7. Fullwood, M., Wei, C., Liu, E., Ruan, Y.: Next-generation DNA sequencing of paired-end tags (PET) for transcriptome and genome analyses. Genome Res. 19, 521–532 (2009)
8. Gilbert, J., Field, D., Huang, Y., et al.: Detection of large numbers of novel sequences in the metatranscriptomes of complex marine microbial communities. PLoS One 3, e3042 (2008)
9. Glazer, A., Kechris, K.: Conserved Amino Acid Sequence Features in the α Subunits of MoFe, VFe, and FeFe Nitrogenases. PLoS One 4(7), e6136 (2009)
10. Grabherr, M., Haas, B., Yassour, M., et al.: Full-length transcriptome assembly from RNA-seq data without a reference genome. Nat. Biotechnol. 29(7), 644–652 (2011)
11. Henikoff, S., Henikoff, J.: Amino Acid Substitution Matrices from Protein Blocks. PNAS 89(22), 10915–10919 (1992)
12. Huang, X., Wang, J., Aluru, S., Yang, S., Hillier, L.: PCAP: AWhole-Genome Assembly Program. Genome Research 13, 2164–2170 (2003)
13. Kent, J.: BLAT–the BLAST-like alignment tool. Genome Research 12(4), 656–664
14. Leininger, S., Urich, T., Schloter, M., et al.: Archaea predominate among ammonia-oxidizing prokaryotes in soils. Nature 442, 806–809 (2006)
15. Leung, H., Yiu, S., Parkinson, J., Chin, F.: IDBA-MT: de novo assembler for metatranscriptomic data generated from next-generation sequencing technology. Journal of Computational Biology 20(7), 540–550 (2013)
16. Khachatryan, Z., Ktsoyan, Z., Manukyan, G., Kelly, D., Ghazaryan, K., Aminov, R.: Predominant role of host genetics in controlling the composition of gut microbiota. PLoS One 3(8), e3064 (2008)
17. Parro, V., Moreno-Paz, M., Gonzalez-Toril, E.: Analysis of environmental transcriptomes by DNA microarrays. Env. Microbiol. 9, 453–464 (2007)
18. Morozova, O., Marra, M.: Applications of next-generation sequencing technologies in functional genomics. Genomics 92, 255–264 (2008)
19. Mullikin, J., Ning, Z.: The Phusion Assembler. Genome Research 13, 81–90 (2003)

20. Peng, Y., Leung, H., Yiu, S., Chin, F.: Meta-IDBA: a de Novo assembler for metagenomic data. Bioinformatics 27(13), i94–i101 (2011)
21. Peng, Y., Leung, H., Yiu, S., Chin, F.: IDBA-UD: a de novo assembler for single-cell and metagenomic sequencing data with highly uneven depth. Bioinformatics 28(11), 1420–1428 (2012)
22. Pettersson, E., Lundeberg, J., Ahmadian, A.: Generations of sequencing technologies. Genomics 93, 105–111 (2009)
23. Poretsky, R., Bano, N., Buchan, A., et al.: Analysis of microbial gene transcripts in environmental samples. Appl. Environ. Microbiol. 71, 4121–4126 (2005)
24. Poretsky, R., Sun, S., Mou, X., Moran, M.: Transporter genes expressed by coastal bacterioplankton in response to dissolved organic carbon. Environ. Microbiol. 12, 616–627 (2010)
25. Qin, J., Li, R., Raes, J., et al.: A human gut microbial gene catalogue established by metagenomic sequencing. Nature 464(7285), 59–65 (2010)
26. Schulz, M., Zerbino, D., Vingron, M., Birney, E.: Oases: Robust de novo RNA-seq assembly across the dynamic range of expression levels. Bioinformatics 28(8), 1086–1092 (2012)
27. Simpson, J., Durbin, R.: Efficient construction of an assembly string graph using the FM-index. Bioinformatics 26(12), i367–i373 (2010)
28. Simpson, J., Wong, K., Jackman, S., Schein, J., Jones, S., Birol, I.: Assembly By Short Sequences - a de novo, parallel, paired-end sequence assembler. Genome Res. 19(6), 1117–1123 (2009)
29. Tartar, A., Wheeler, M., Zhou, X., Coy, M., Boucias, D., Scharf, M.: Parallel metatranscriptome analyses of host and symbiont gene expression in the gut of the termite Reticulitermes flavipes. Biotechnology for Biofuels 2, 25 (2009)
30. Tatusov, R., Koonin, E., Lipman, D.: A Genomic Perspective on Protein Families. Science 278(5338), 631–637 (1997)
31. Urich, T., Lanzen, A., Qi, J., Huson, D., Schleper, C., Schuster, S.: Simultaneous Assessment of Soil Microbial Community Structure and Function through Analysis of the Meta-Transcriptome. PLoS One 3(6), e2527 (2008)
32. Xiong, X., Frank, D., Robertson, C., et al.: Generation and Analysis of a Mouse Intestinal Metatranscriptome through Illumina Based RNA-Sequencing. PLoS One 7(4), e36009 (2012)
33. Zerbino, D., Birney, E.: Velvet: Algorithms for de novo short read assembly using de Bruijn graphs. Genome Research 18(5), 821–829 (2008)

MRFalign: Protein Homology Detection
through Alignment of Markov Random Fields

Jianzhu Ma, Sheng Wang, Zhiyong Wang, and Jinbo Xu[*]

Toyota Technological Institute at Chicago, 6045 Kenwood Ave, Chicago, Illinois, USA
{majianzhu,wangsheng,j3xu}@ttic.edu, jinboxu@gmail.com

Sequence-based protein homology detection has been extensively studied, but it still remains very challenging for remote homologs with divergent sequences. So far the most sensitive method for homology detection is based upon comparison of protein sequence profiles, which are usually derived from multiple sequence alignment (MSA) of sequence homologs in a protein family and represented as a position-specific scoring matrix (PSSM) or an HMM (Hidden Markov Model). HMM is more sensitive than PSSM because the former contains position-specific gap information and also takes into account correlation among sequentially adjacent residues. The main issue with HMM lies in that it makes use of only position-specific amino acid mutation patterns and very short-range residue correlation, but not long-range residue interaction. However, remote homologs may have very divergent sequences and are only similar at the level of (long-range) residue interaction pattern, which is not encoded in current popular PSSM or HMM models.

To significantly advance homology detection, this paper presents a Markov Random Fields (MRFs) modeling of an MSA. MRFs can model long-range residue interactions and thus, encodes information for the global 3D structure of a protein family. In particular, MRF is a graphical model encoding a probability distribution over the MSA by a graph and a set of preset statistical functions. A node in the MRF corresponds to one column in the MSA and one edge specifies correlation between two columns. Each node is associated with a function describing position-specific amino acid mutation pattern. Similarly, each edge is associated with a function describing correlated mutation statistics between two columns. With MRF representation, alignment of two proteins or protein families becomes that of two MRFs.

To score the similarity of two MRFs, we use both node and edge alignment potentials, which measure the node (i.e., residue) similarity and edge (i.e., interaction pattern) similarity, respectively. To derive the node alignment potential, we use a set of 1400 protein pairs as the training data, which covers 458 SCOP folds. The reference alignment for a protein pair is generated by a structure alignment tool DeepAlign [1]. The edge alignment potential is derived from a software package EPAD [2], which takes as input PSSM and residue interaction strength and outputs the inter-residue distance probability distribution. The interaction strength of two residues can be calculated by different ways. In current implementation we calculate the mutual information matrix (MI) and its power series (MI^2, MI^3,..., MI^{11}) as interaction strength. The MI power series are much more informative than MI alone, as shown in [3].

---

[*] Corresponding author.

R. Sharan (Ed.): RECOMB 2014, LNBI 8394, pp. 173–174, 2014.
© Springer International Publishing Switzerland 2014

It is computationally challenging to optimize the MRFalign scoring function due to the edge alignment potential. We formulate this problem as an integer programming problem and then develop an ADMM (Alternative Direction Method of Multipliers) algorithm to solve it efficiently to a suboptimal solution. ADMM divides the MRF alignment problem into two tractable sub-problems and then solve them iteratively.

Alignment Accuracy. We use reference-dependent alignment recall to measure the alignment accuracy, which is the fraction of alignable residues that are correctly aligned. To reduce bias, we use three structure alignment tools TM-align, Matt, and DeepAlign to generate reference alignments. Two datasets from [4] are used to test the alignment accuracy. As shown in Table 1, MRFalign greatly outperforms HMMER and HHalign. MRFalign also has better alignment precision (not shown).

Table 1. Reference-dependent alignment recall on Set3.6K and Set2.6K. Three tools are used to generate reference alignments.

	Set3.6K			Set2.6K		
	TMalign	Matt	DeepAlign	TMalign	Matt	DeepAlign
HMMER	22.9%	24.1%	25.5%	36.5%	38.6%	40.4%
HHalign	36.3%	37.0%	38.4%	62.5%	63.2%	64.0%
MRFalign	**47.4%**	**47.5%**	**49.2%**	**72.8%**	**73.5%**	**74.2%**

Homology Detection Performance. We employ two benchmarks SCOP20 and SCOP40 to test homology detection rate at the superfamily and fold levels, respectively, as shown in Table 2.

Table 2. Homology detection performance at the superfamily and fold levels

	Superfamily level				Fold level			
	SCOP20		SCOP40		SCOP20		SCOP40	
	Top1	Top5	Top1	Top5	Top1	Top5	Top1	Top5
hmmscan	35.2%	36.5%	40.2%	41.7%	5.2%	6.1%	6.2%	6.9%
FFAS	48.6%	54.4%	52.1%	56.3%	13.1%	18.7%	10.4%	14.5%
HHsearch	51.6%	57.3%	55.8%	60.8%	16.3%	24.7%	17.6%	25.3%
HHblits	51.9%	56.3%	56.0%	59.8%	17.4%	25.2%	19.1%	26.0%
MRFalign	**58.2%**	**61.7%**	**59.3%**	**63.6%**	**27.2%**	**36.8%**	**28.3%**	**37.9%**

Availability. The MRFalign server will be available at http://raptorx.uchicago.edu/.

References

1. Wang, S., Ma, J., Peng, J., Xu, J.: Protein structure alignment beyond spatial proximity. Scientific Reports 3 (2013)
2. Zhao, F., Xu, J.: A Position-Specific Distance-Dependent Statistical Potential for Protein Structure and Functional Study. Structure 20(6), 1118–1126 (2012)
3. Wang, Z., Xu, J.: Predicting protein contact map using evolutionary and physical constraints by integer programming. Bioinformatics 29(13), i266–i273 (2013)
4. Ma, J., Wang, S., Zhao, F., Xu, J.: Protein threading using context-specific alignment potential. Bioinformatics 29(13), i257–i265 (2013)

An Integrated Model of Multiple-Condition ChIP-Seq Data Reveals Predeterminants of Cdx2 Binding

Shaun Mahony[1,*,**], Matthew D. Edwards[2,**], Esteban O. Mazzoni[3], Richard I. Sherwood[4], Akshay Kakumanu[1], Carolyn A. Morrison[5], Hynek Wichterle[5], and David K. Gifford[2,*]

[1] The Pennsylvania State University, University Park, PA 16802, USA
[2] Massachusetts Institute of Technology, Cambridge, MA 02139, USA
[3] New York University, New York, NY 10003, USA
[4] Harvard Medical School, Boston, MA 02115, USA
[5] Columbia University Medical Center, New York, NY 10032, USA
mahony@psu.edu, dkg@mit.edu

Introduction: Profiling the genomic binding activity of regulatory proteins in multiple cell types is important for understanding cellular function, as a single regulator can bind to distinct sets of genomic targets depending on the cellular context in which it is expressed. Characterizing the determinants of such differential binding specificity is key to understanding how a single regulator can play multiple roles during development and other dynamic cellular processes. For example, pre-existing chromatin context such as chromatin accessibility or the binding of other regulators may determine the binding of some developmental transcription factors (TFs) [1,2], while other pioneer TFs may find their binding targets independently of the established chromatin state [3]. In order to reliably characterize such condition-specific regulatory activity, we require methods that can integrate analysis across multiple related experiments in a principled way.

We present MultiGPS, an integrated machine learning approach for the analysis of multiple related ChIP-seq experiments. MultiGPS is based on a generalized Expectation Maximization framework that enables the principled incorporation of external information in binding event discovery, which we use to share information across multiple experiments. MultiGPS performs binding event analysis across multiple conditions, sharing information across conditions to produce accurate joint binding estimates while simultaneously allowing for condition-specific binding events. MultiGPS leverages a flexible framework for incorporating prior information into binding event discovery, allowing models of joint binding and sequence dependence to be used. In analyzing multiple-condition ChIP-seq datasets, MultiGPS encourages consistency in the reported binding event locations across conditions and provides accurate estimation of ChIP enrichment levels at each event.

---

* Corresponding authors.
** Joint first authors.

R. Sharan (Ed.): RECOMB 2014, LNBI 8394, pp. 175–176, 2014.
© Springer International Publishing Switzerland 2014

Results: MultiGPS uses a joint multi-experiment model that considers the read data from all experiments to produce accurate location estimates of punctate binding events. The multi-experiment model is one aspect of a novel modeling approach that enables external sources of information to be included as priors in binding event identification. We demonstrate that our framework enables the simultaneous modeling of sparse condition-specific binding changes, sequence dependence, and replicate-specific noise sources. On simulated binding data, MultiGPS encourages consistent binding event locations across experiments and improves the quantification of condition-specific binding events when compared to single-condition peak callers. The use of the multi-condition prior along with a motif prior fully utilizes available information and allows almost all example binding events to be aligned to consistent (typically motif-associated) locations. On experimental data from 14 factors studied in 3 cell lines, MultiGPS improves the cross-replicate correlation of binding event quantification estimates and reduces the effects of inter-replicate noise when compared to peak quantifications from another peak caller and a window-based analysis.

As a case study of our framework's sensitive and accurate multi-condition analysis, we applied MultiGPS to the analysis of Cdx2 ChIP-seq data generated in three developmentally relevant cellular contexts. We found that condition-specific Cdx2 binding events are predicted by preexisting chromatin state. Surprisingly, condition-independent Cdx2 binding events that are bound in multiple contexts do not appear to be predetermined by accessibility or other chromatin signatures, and instead may be predicted on the basis of cognate motif occurrence. Our results suggest that Cdx2 can act as a pioneer factor at a subset of sites, while also being influenced by preexisting chromatin context at other sites. Therefore, our results have consequences for understanding where TFs will bind when introduced into an established regulatory state during development, or when induced artificially during cellular programming techniques.

Conclusions: MultiGPS provides a principled platform for the analysis of differential protein-DNA binding across multiple experimental conditions by preferring consistent binding locations across related experiments while also modeling condition-specific experimental parameters. MultiGPS is freely available from `http://mahonylab.org/software`.

References

1. Mahony, S., Mazzoni, E.O., McCuine, S., Young, R.A., Wichterle, H., et al.: Ligand-dependent dynamics of retinoic acid receptor binding during early neurogenesis. Genome. Biol. 12, R2 (2011), doi:10.1186/gb-2011-12-1-r2
2. John, S., Sabo, P.J., Thurman, R.E., Sung, M.-H., Biddie, S.C., et al.: Chromatin accessibility pre-determines glucocorticoid receptor binding patterns. Nat. Genet. 43, 264–268 (2011), doi:10.1038/ng.759
3. Zaret, K.S., Carroll, J.S.: Pioneer transcription factors: establishing competence for gene expression. Genes. Dev. 25, 2227–2241 (2011), doi:10.1101/gad.176826.111

PASTA: Ultra-Large Multiple Sequence Alignment

Siavash Mirarab, Nam Nguyen, and Tandy Warnow

University of Texas at Austin - Department of Computer Science
2317 Speedway, Stop D9500 Austin, TX 78712
{smirarab,namphuon,tandy}@cs.utexas.edu

Abstract. In this paper, we introduce a new and highly scalable algorithm, PASTA, for large-scale multiple sequence alignment estimation. PASTA uses a new technique to produce an alignment given a guide tree that enables it to be both highly scalable and very accurate. We present a study on biological and simulated data with up to 200,000 sequences, showing that PASTA produces highly accurate alignments, improving on the accuracy of the leading alignment methods on large datasets, and is able to analyze much larger datasets than the current methods. We also show that trees estimated on PASTA alignments are highly accurate – slightly better than SATé trees, but with substantial improvements relative to other methods. Finally, PASTA is very fast, highly parallelizable, and requires relatively little memory.

Keywords: Multiple sequence alignment, Ultra-large, SATé.

1 Introduction and Motivation

Multiple sequence alignment (MSA) is a basic step in many bioinformatics analyses, including predicting the structure and function of RNAs and proteins and estimating phylogenies. Yet only a handful of the many MSA methods are able to analyze large datasets with 10,000 or more sequences. Performance studies evaluating MSA methods on large datasets have shown that some MSA methods can produce highly accurate alignments, as measured by traditional alignment criteria (sum-of-pairs or column scores) for sufficiently slowly evolving sequence datasets (e.g., [1]). However, other studies, focusing on phylogeny estimation from nucleotide datasets, have found that only a handful of MSA methods can provide good enough alignments on nucleotide datasets with 10,000 or more sequences to produce highly accurate trees, even when trees are estimated using the best maximum likelihood heuristics [2–5]. However, these studies have relied upon benchmarks with at most 28,000 sequences, and so little is known about alignment accuracy and its impact on tree accuracy for larger datasets. Yet, phylogenetic analyses of sequence datasets containing more than 100,000 nucleotide sequences are being attempted by at least two groups that we are aware of: the iPTOL project [6] and the Thousand Transcriptome project (1KP)[7].

R. Sharan (Ed.): RECOMB 2014, LNBI 8394, pp. 177–191, 2014.
© Springer International Publishing Switzerland 2014

In this paper we present PASTA, "Practical Alignments using SATé and TrAnsitivity", a new method for ultra-large multiple sequence alignment of nucleotide sequence datasets. PASTA begins with an alignment and tree estimated using a very simple profile HMM-based technique and then re-aligns the sequences using the tree. If desired, a new tree can be estimated on the new alignment, and the algorithm can iterate.

The key to the accuracy and scalability of PASTA is the novel technique it uses for estimating an alignment on a guide tree. As in SATé [3], PASTA uses the centroid edge dataset decomposition technique and computes MAFFT -linsi [8] alignments on the subsets; however, PASTA and SATé merge these subset alignments into an alignment on the full dataset using very different techniques. While SATé uses Opal [9] (or Muscle [10, 11], if the dataset is too large) to hierarchically merge all the subset alignments into a single alignment, PASTA uses Opal only to merge pairs of adjacent subset alignments, producing overlapping alignments, then treats each resultant alignment as an equivalence relation and uses transitivity to merge these larger alignments. The result is a very fast re-alignment method that is highly parallelizable and easily scales to large datasets. Furthermore, this re-alignment step in PASTA is a negligible fraction of the PASTA analysis, whereas the re-alignment step in SATé is the majority of its running time on large datasets (44% of the running time for datasets with 10K sequences, and 78% of the time for datasets with 50K sequences). Thus, PASTA is dramatically faster than SATé on large datasets. Interestingly, PASTA produces more accurate alignments and trees than SATé. We demonstrate PASTA's speed and accuracy on a collection of datasets, including a 200K-sequence RNASim dataset [12], which we align in less than 24 hours using PASTA on a 12-core machine.

2   PASTA

We describe PASTA and present some theorems about its performance guarantees and running time; due to space limits, proofs are provided in the appendix.

PASTA uses an iterative divide-and-conquer strategy to align an input set S of sequences, and uses the following input parameters: a starting tree (default our HMM-based profile alignment technique, described below), subset size k (default 200), a subset alignment technique (default MAFFT -linsi), an alignment merger technique (default OPAL), and a stopping rule (default 3 iterations).

The first iteration begins with the starting tree, and subsequent iterations begin with the tree estimated in the previous iteration. Each iteration involves six steps, shown in Figure 1.

Starting Tree: PASTA can begin with any reasonable starting tree, but here we describe the simple technique, similar to that used in [14, 15], that we use in these experiments. We take a random subset X of 100 sequences from S and compute a SATé alignment A on the set; this is called the "backbone alignment". We then use HMMER [16, 17] to compute a Hidden Markov Model on A, and to align all sequences in $S-X$ one by one (and independently) to A, and hence build

---

Step 1 Decompose the input set S into subsets $S_1 \ldots S_m$ of size at most k.
Step 2 Compute a spanning tree T^* to connect the subsets $S_1 \ldots S_m$.
Step 3 Align each subset using the subset alignment technique.
Step 4 Merge the two alignments on endpoints of each edge in T^*.
Step 5 Use successive applications of transitive closure to merge the overlapping and compatible alignments obtained in Step 4.
Step 6 Compute a maximum likelihood (ML) tree on the full MSA using FastTree-2 [13].

Fig. 1. Algorithmic steps of PASTA for each iteration

an alignment of the full dataset. We then construct a ML tree on this alignment using FastTree-2 [13]. If this technique fails to produce an alignment on the full set of sequences (which can happen if HMMER considers some sequences unalignable), we randomly add the unaligned sequences into the tree obtained on the alignment obtained by HMMER.

Step 1: We use the centroid decomposition technique in SATé [3] on the current guide tree to divide the sequence set into disjoint sets, $S_1, \ldots, S_m$, each with at most k sequences. If the tree has at most k leaves, we return the set of sequences; otherwise, we find an edge in the tree that splits the set of leaves into roughly equal sizes, remove it from the tree, and then recurse on each subtree.

Step 2: We compute a spanning tree T^* on the subsets, $S_1, S_2, \ldots, S_m$, as follows. For every i, we compute the set of nodes v in the guide tree that are on a path between two leaves that both belong to S_i, and we label all these nodes by S_i; thus, if v is a leaf and belongs to S_i, we label v by S_i. Then, if some nodes are not yet labelled, we propagate labels from nodes to unlabelled neighbors (breaking ties by using the closest neighbor according to branch lengths in the guide tree) until all nodes are labelled. We then collapse edges that have the same label at the endpoints. The result is a spanning tree on $S_1, S_2, \ldots, S_m$.

Step 3: We compute MSAs on each S_i using the subset alignment method specified by the user. We refer to each such alignment as a "Type 1 sub-alignment".

Step 4: Every node in T^* is labelled by an alignment subset for which we have a Type 1 sub-alignment from Step 3. For every edge (v, w) in T^*, we use the specified alignment merger technique to merge the Type 1 sub-alignments at v and w; this produces a new set of alignments, each containing at most $2k$ sequences, which are called "Type 2 sub-alignments". We require that the merger technique used to compute Type 2 sub-alignments not change the alignments on the Type 1 sub-alignments; therefore, Type 2 sub-alignments induce the Type 1 sub-alignments computed in Step 2.

Step 5: We compute the transitivity merge through a sequence of pairwise transitivity mergers. To motivate this technique, we note that every MSA defines an equivalence relation on the letters within its sequences, whereby two letters are in the same equivalence class *if and only if* they are in the same column [18]. Hence given two alignments A and B that induce identical alignments on their shared sequences (called *overlapping compatible* alignments henceforth), we can define an equivalence relation on the union of the letters from their sequence subsets, as follows: a and b are in the same equivalence class for the merged alignment if and only if at least one of the following is true: (1) they are in the same equivalence class in A or B, or (2) there is some letter c such that a and c are in the same equivalence class in one alignment, and b and c are in the same equivalence class for the other alignment. This is the basis for the pairwise transitivity merger, but the spanning tree enables this technique to extend to a set of alignments through a sequence of pairwise transitivity mergers in a computationally efficient manner. We provide details for the transitivity merge in the appendix.

Step 6: If an additional iteration (or a tree on the alignment) is desired, we run FastTree-2 [13] to estimate a maximum likelihood tree on the MSA produced in the previous step. To speed up this step, we mask all sites that have more than 99.9% gaps in the alignment obtained in Step 5. Note that PASTA's alignment merging step is conservative in introducing new homologies (only what is necessary through transitivity is added) and thus PASTA tends to produce many gappy columns. Masking these highly gapped columns is harmless for the tree estimation step but has a dramatic effect on the running time.

Running Time Considerations. PASTA keeps alignments in the memory in a condensed format by representing each sequence by its unaligned sequence and column indices for each letter. So, for example, (`'ACCA'`,`[1,3,5,6]`) corresponds to `'A-C-CA'`. This format reduces the memory requirement of PASTA, as well as the running time for each transitivity merge.

As long as each pairwise transitivity merge is performed correctly, the final output multiple sequence alignment does not depend on the order in which edges of the spanning tree are processed, but the order can impact the running time. However, if we merge sub-alignments using the reverse order of the centroid edge deletions, then the running time can be bounded, as follows:

Theorem 1. *Given m Type 1 alignments and $m-1$ Type 2 alignments, the algorithm to compute the transitivity merge of these alignments uses $O(Km \log m + Lm)$ time, where K is the maximum length of any sequence (not counting gaps) in any Type 1 alignment, and L is the length of output alignment.*

The bound given in the theorem is achievable using an order of edge contractions that reverses the order of centroid edge deletions; however, an *arbitrary* order of edge contractions can result in a worst case $O(Km^2 + Lm)$ running time; see Supplementary Material [19] for discussion and proof.

Table 1. Empirical statistics of large datasets based on reference alignments. Number of sequences, number of sites, proportion of gap characters, maximum p-distance, and average p-distance are given for each dataset (the p-distance is the fraction of sites in which two sequences differ). For 10K RNASim datasets, the given values are averages over 10 replicates. For RNASim datasets with 100K and 200K sequences, pairwise distances could not be computed; however, since 100K and 200K datasets are random subsamples from the same alignment as 10K and 50K datasets, their alignment statistics are likely very close to those of 10K and 50K.

	Dataset	# Sequences	# Sites	Proportion gaps	Max p-dist	Avg p-dist
Gutell	16S.B.ALL	27,643	6,857	0.800	0.769	0.210
Gutell	16S.T	7,350	11,856	0.874	0.900	0.345
Gutell	16S.3	6,323	8,716	0.821	0.832	0.315
RNASim	10,000	10,000	8,637	0.820	0.616	0.410
RNASim	50,000	50,000	12,400	0.875	0.620	0.410
RNASim	100,000	100,000	14,316	0.891	≈ 0.62	≈ 0.410
RNASim	200,000	200,000	16,365	0.905	≈ 0.62	≈ 0.410

3   Experimental Setup

Datasets. To evaluate the accuracy on moderate size datasets, we use 1000-taxon simulated datasets from [2]. For evaluating performance on larger datasets, we used RNASim, a simulated RNA dataset with 1,000,000 sequences [12], subsampling it to create datasets with 10,000, 50,000, 100,000, and 200,000 sequences. For the 10K case we created 10 different replicates, but for other cases, due to running time requirements, we created only one replicate. Finally, we use three large 16S biological datasets obtained from the Gutell lab [20], and previously studied in [2]. These datasets include 16S.3 with 6323 sequences, 16S.T with 7350 sequences, and 16S.B.ALL with 27,643 sequences. The reference alignments for the biological datasets are based on secondary structure, and the reference trees are computed on these reference alignments using RAxML [21], with all edges having bootstrap support less than 75% contracted; using other thresholds produces similar results (see online supplementary materials [19]). The reference alignments and trees for the simulated datasets are the true alignment and true trees, which are known because they are the result of a simulation process. Table 1 shows more statistics about the reference alignments.

Methods. We compare PASTA to SATé, Muscle, MAFFT-Profile [22], ClustalW (quicktree algorithm) [23], and also to our approach for obtaining the starting alignment and tree. PASTA results are based on the default settings: three iterations, subset size set to 200, MAFFT-linsi used on subsets, and Opal used for merging alignments. The starting tree is obtained by the technique described in Section 2. MAFFT-Profile is a version of MAFFT that can add new sequences into an existing backbone alignment [22]; we provide MAFFT-Profile the same backbone alignment that we use for the starting tree of PASTA. We run SATé

with identical starting trees as PASTA, and we also run SATé for three iterations. Due to high computational costs of OPAL on large datasets, we use Muscle
for merging alignments inside SATé for datasets with 5,000 sequences or more,
and otherwise we use the default settings in SATé. Finally, we use FastTree-2
to compute ML trees on each alignment. See supplementary material [19] for
commands and version numbers.

Criteria. We measure the alignment accuracy, tree error, and running time.
Alignment accuracy is measured using FastSP [18] with two different metrics: the
SP-score (the percentage of homologies in the reference alignment recovered in
the estimated alignment) and the modeler score (the percentage of homologies in
the estimated alignment that are correct), averaged together to get one measure.
Note that SP-score is the complement of the SP-FN error rate, and the modeler-
score is the complement of the SP-FP error rate; thus our measure of alignment
accuracy is influenced equally by false positive and false negative homologies.
In addition, we also report the number of columns that are recovered entirely
correctly in the estimated alignment (TC score). The standard error metric for
tree estimation is the bipartition distance, also known as the Robinson-Foulds
(RF) rate; however, this metric is not appropriate when the reference tree is
not fully resolved, as is the case for the biological datasets. Therefore, we use
the False Negative (FN) rate, which is the percentage of true tree edges missing
in the estimated trees, to evaluate estimated trees. Note that the FN rate is
identical to the RF rate when both estimated and reference trees are binary.

Computational Platform. We ran all methods on the Lonestar Linux cluster at
TACC [24], and each run was given one node with 12 cores and 24 GB of memory.
Since running time on Lonestar is limited to 24 hours, we were only able to run
techniques that could finish in 24 hours (see below). However, PASTA and SATé
are iterative techniques, and we allowed them to perform as many iterations (but
no more than three) as they could complete within 24 hours. We report the wall
clock time in all cases.

4   Results

Ability to complete analyses. We report which methods completed analyses
within 24 hrs using 12 cores and 24 GB of memory. All methods completed
on all datasets with at most 10,000 sequences. On 16S.B.ALL, all methods except for ClustalW finished. However, ClustalW, Muscle, and SATé-2 failed to
complete on the RNASim datasets with 50,000 sequences or more, and MAFFT-
profile failed to complete on the RNASim dataset with 200K sequences. On 100k
RNAsim, PASTA finished 2 iterations in 24 hours, and on 200k, PASTA was able
to complete one iteration and was the only method that could run.

1000-Taxon datasets from SATé papers. We studied PASTA on the 1000-taxon
simulated datasets from [2, 3] and observed that PASTA trees matched the

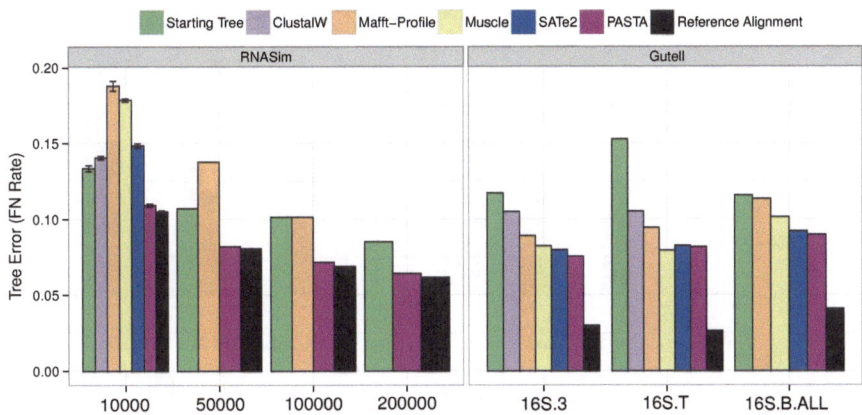

Fig. 2. Tree error rates on RNASim 10K-200K (left) and biological (right) datasets. We show missing branch rates for maximum likelihood trees estimated on the reference alignment as well as alignments computed using PASTA and other methods; results not shown indicate failure to complete within 24 hours using 12 cores on the datasets. Error bars on 10,000 RNASim dataset show standard error over 10 replicates.

accuracy of SATé trees and had improved alignment accuracy; PASTA was also more accurate than the other methods we tested (Opal, Muscle, and MAFFT); see supplementary materials for these results [19].

Tree Error on RNASim and biological datasets. Figure 2 reports results for tree error rates of ML trees on the reference alignments and on the different estimated alignments for the RNASim and biological datasets. On the RNASim data, PASTA returns the most accurate trees, coming very close to FastTree trees on the reference (true) alignment. The difference between PASTA and trees on the next most accurate alignment is very large. Note also that only PASTA and its starting tree complete within the time limit on the 100K and 200K sequence datasets. Furthermore, PASTA has very low error rates overall (e.g., only 6.4% tree error on the 200K dataset).

We also show results for the biological datasets using the 75%-support reference trees. FastTree-2 trees computed on the reference alignments had the best accuracy. PASTA, SATé, and Muscle came next, and the remaining methods had poorer accuracy.

Alignment Accuracy on the RNASim and biological datasets. Figure 3 compares methods with respect to two ways of evaluating accuracy - the total column score (TC) and the average of the SP-score and modeler score. On the RNASim data, PASTA returns by far the most accurate alignments of all methods tested according to TC, and its SP-scores are better than all other methods except the starting alignment. Furthermore, the PASTA alignment had high accuracy: on the 200K dataset, its sum-of-pairs accuracy was 88% and more than 800 columns

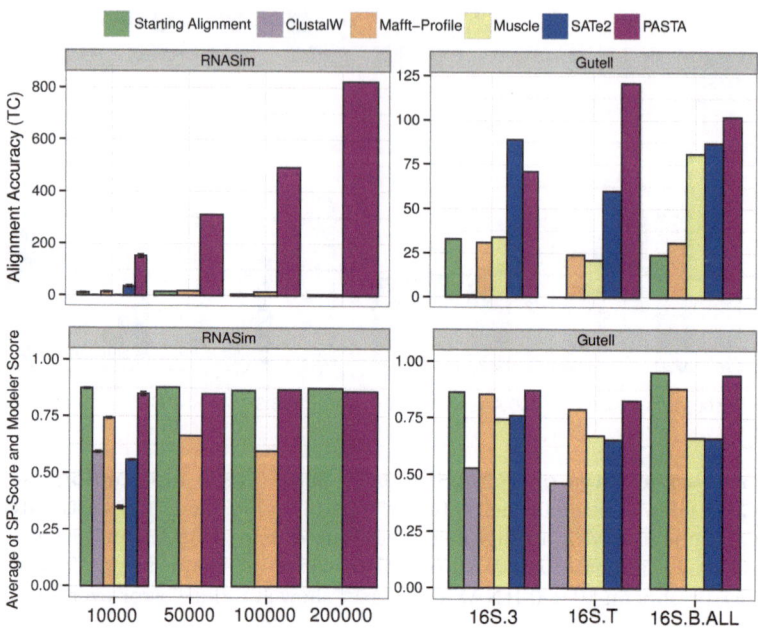

Fig. 3. Alignment accuracy on the RNASim 10K-200K (left) and biological (right) datasets. We show the number of correctly aligned sites (top) and the average of the SP-score and modeler score (bottom). The starting alignment was incomplete on the 16S.T dataset, and so no result is shown for the starting alignment on that dataset.

were recovered entirely correctly. Another interesting trend is that as the number of sequences increases, the alignment accuracy decreased for MAFFT-profile but not for PASTA.

The biological datasets are smaller (see Table 1) and so are not as challenging. On the 16S.T dataset, the starting alignment did not return an alignment with all the sequences on the 16S.T dataset because HMMER considered one of the sequences unalignable. However, the starting alignment technique had good SP-scores for the other two datasets. Of the remaining methods, PASTA has the best sum-of-pairs scores (bottom panel), and MAFFT-profile has only slightly poorer scores; the other methods are substantially poorer. With respect to TC scores, on 16S.B.ALL and 16S.T, PASTA is in first place and SATé is in second place, but they swap positions on 16S.3. TC scores for the other methods are clearly less accurate, though Muscle does fairly well on the 16S.B.ALL dataset.

Comparison to SATé on 50,000 taxon dataset. SATé could not finish even one iteration on the RNASim with 50,000 sequences running for 24 hours and given 12 CPUs on TACC. However, we were able to run two iterations of SATé on a separate machine with no running time limits (12 Quad-Core AMD Opteron(tm) processors, 256GB of RAM memory). Given 12 CPUs, each iteration of SATé takes roughly 70 hours, compared to 5 hours for PASTA, and as shown next, the

Table 2. Two iterations of PASTA compared to SATé on 50,000 RNASim given more than 24 hours of running time (outside TACC)

	Alignment Accuracy			Tree Error	Running Time
	SP-score	Modeler score	TC	FN	(hours)
PASTA-2iter	**80.2%**	**81.8%**	**311**	**8.2%**	10
SATé-2iter	20.5%	55.9%	30	12.6%	137

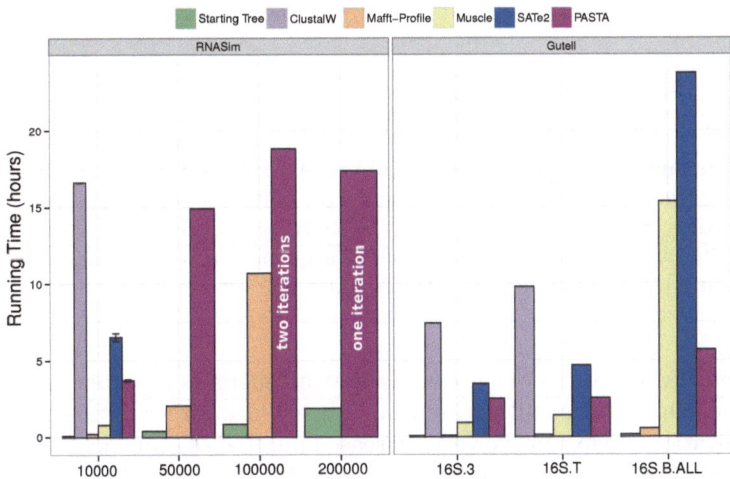

Fig. 4. Alignment running time (hours). Note that PASTA is run for three iterations everywhere, except on 100,000-sequence RNASim dataset where it is run for two iterations, and on the 200,000-sequence RNASim dataset where it is run for one iteration.

majority of SATé running time is spent in the merge step. However, the resulting SATé alignment is much less accurate, and produces trees that are substantially less accurate than PASTA (see Table 2).

Running Time. Figure 4 compares the running time (in hours) of different alignment methods. PASTA is faster than SATé, and MAFFT-Profile is faster than PASTA on the smallest datasets. However, the running time of MAFFT-Profile grows faster than PASTA so that at 200,000 sequences it is not able to finish in 24 hours, while PASTA can. Muscle is faster than PASTA on datasets with 10,000 sequences or less, but is slower on 16S.B.ALL, the only dataset above 10,000 sequence where it can actually run. Our approach for producing the starting tree is the fastest method on all datasets, and ClustalW is always the slowest. However, note that neither Clustalw or Muscle is parallelized and so these methods cannot take advantage of the multiple cores.

Figure 5a presents the running time comparison to SATé. Note that merging subset alignments is the majority of the time used by SATé to analyze the 50K RNASim dataset, but a very small fraction of the time used by PASTA.

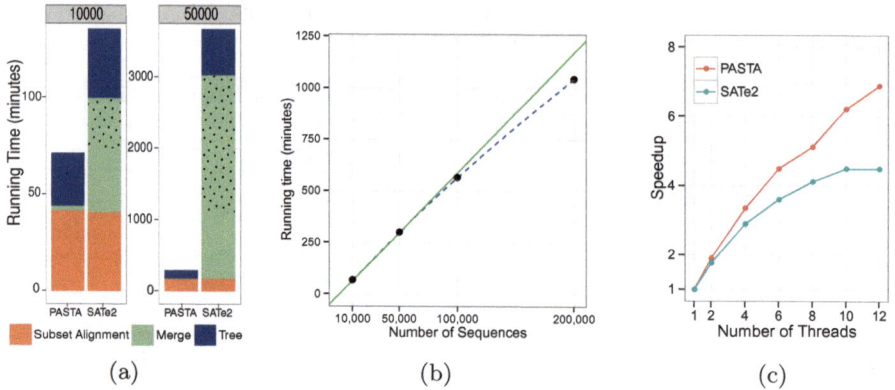

Fig. 5. Running time comparison of PASTA and SATé. (a) Running time profiling on one iteration for RNASim datasets with 10K and 50K sequences (the dotted region indicates the last pairwise merge). (b) Running time for one iteration of PASTA with 12 CPUs as a function of the number of sequences (the solid line is fitted to the first two points). (c) Scalability for PASTA and SATé with increased number of CPUs.

The reason SATé uses so much time is that all mergers are done hierarchically using either Opal (for small datasets) or Muscle (on larger datasets), and both are computationally expensive with increased number of sequences. For example, the last pairwise merge within SATé, shown by the dotted area in Figure 5a, is entirely serial and takes up a large chunk of the total time. PASTA solves this problem by using transitivity for all but the initial pairwise mergers, and therefore scales well with increased dataset size, as shown in Figure 5b (the sub-linear scaling is due to a better use of parallelism with increased number of sequences). Finally, Figure 5c shows that PASTA is highly parallelizable, and has a much better speed-up with increasing number of threads than SATé does. While PASTA has a much improved parallelization, it does not quite scale up linearly, because FastTree-2 does not scale up well with increased thread count.

Divide-and-Conquer strategy: impact of guide tree. We also investigated the impact of the use of the guide tree for computing the subset decomposition, and hence defining the Type 1 sub-alignments. We compared results obtained using three different decompositions: the decomposition computed by PASTA on the HMM-based starting tree, the decomposition computed by PASTA on the true (model) tree, and a random decomposition into subsets of size 200, all on the RNASim 10k dataset. PASTA alignments and trees had roughly the same accuracy when the guide tree was either the true tree or the HMM-based starting tree (Table 3). However, when based on a random decomposition, tree error increased dramatically from 10.5% to 52.3%, and alignment scores also dropped substantially. Thus, the guide-tree based dataset decomposition used by PASTA provides substantial improvements over random decompositions, and the default technique for getting the starting tree works quite well.

Table 3. Effect of subset decomposition in PASTA algorithm, based on one iteration of PASTA on one replicate of the 10k RNASim dataset

	Alignment Accuracy			Tree Accuracy
	SP-score	Modeler score	TC	FN
Random	78.4%	81.4%	2	52.3%
Phylogeny-based (estimated tree)	**86.3%**	**87.3%**	**138**	**10.5%**
Phylogeny-based (true tree)	85.5%	86.7%	133	**10.5%**

5   Discussion and Future Work

One of the intriguing observations in this study is that alignment accuracy measures are not always predictive of tree accuracy. For example, on the Gutell datasets, MAFFT-profile produced less accurate trees than Muscle, yet had better sum-of-pairs alignment accuracy scores. Similarly, the PASTA starting alignment is typically among the best in terms of alignment accuracy but far from the best in terms of tree error. Most likely this is because not all pairwise homologies are equally important for phylogeny estimation, and alignment accuracy measures treat pairwise homologies identically. Failing to recover some homologies may not have much impact on tree estimation, while other homologies may be essential for phylogenetic accuracy. Furthermore, alignment methods that aim to recover the conserved regions may be able to have high alignment accuracy scores but fail to produce good trees - because conserved regions may not be as useful for phylogeny estimation as regions that change. Thus, the sites and even specific homologies that are most informative of the phylogenetic branching process may not be the homologies that many alignment methods are trained to recover. More generally, then, this disconnect suggests a real challenge in using alignment metrics to predict the utility of an alignment, especially if the purpose of the alignment is phylogeny estimation.

We have shown results for the current default version of PASTA; however, we also explored variants where we changed some algorithmic parameters (see supplementary material [19]). We found PASTA to be robust to the choice of the starting tree. Interestingly, while varying the alignment subset size (between 50 and 200) had only a small impact on accuracy, PASTA run with smaller alignment subsets is much faster, raising the possibility that comparable accuracy at reduced running time might be achievable through smaller alignment subsets.

Finally, we note that PASTA, like SATé, is a method that "boosts" the performance (accuracy and/or scalability) of the base method used to align subsets. The good performance using MAFFT as the base method suggests the possibility that PASTA could be used to extend computationally intensive statistical methods, such as BAli-Phy [25], to large datasets, while maintaining their accuracy. Our future work will explore this possibility.

6   Conclusions

PASTA is a new method for nucleotide sequence alignment and tree estimation that is designed for speed, scalability, and accuracy, especially for large datasets.

PASTA is based on SATé, but its design allows it to provide improved accuracy while using only a fraction of the time on large datasets. The key algorithmic contribution is the new technique for aligning sequences on a given guide tree. This algorithmic design addresses computational limitations in SATé and other methods, however it also provides improved accuracy on large datasets because it uses transitivity to extend highly accurate overlapping alignments rather than trying to directly infer homologies between distantly related sets of sequences.

PASTA is fast and also scales well with the number of processors, so that datasets with even 200,000 sequences can be analyzed in less than a day with a small number of processors. Thus, highly accurate alignment and phylogeny estimation is possible, even on hundreds of thousands of sequences, without supercomputers.

PASTA software is implemented by extending the SATé code, and is publicly available at `https://github.com/smirarab/pasta`. Datasets are available at `http://www.cs.utexas.edu/users/phylo/software/pasta/`.

Acknowledgments. This research was supported in part by NSF grant DBI 0733029 to TW, by an International Predoctoral Fellowship to SM from the HHMI, and by a subgrant from the University of Alberta to TW, made possible through a donation from Musea Ventures, which is held by Professor Gane Ka-Shu Wong. The authors wish to thank the anonymous referees for their helpful comments. This research was supported in part by the iPlant Collaborative, NSF award number DBI-1265383.

A   Appendix: Computing the Transitivity Merge

We compute the transitivity merge through a sequence of pairwise transitivity mergers. Recall that every node v in the spanning tree T^* computed in Step 2 is labelled by an alignment subset (i.e., a subset of the input sequence dataset on which we have a Type 1 sub-alignment). In addition, during Step 4, we computed Type 2 sub-alignments for every pair of Type 1 sub-alignments whose alignment subsets are adjacent in the spanning tree T^*. We now define a set $S(v)$ for every vertex v and $Label(e)$ for every edge e, as follows. For node v in T^*, we define the set $S(v) = \{X_v\}$ where X_v is the alignment subset associated to the node v, and for edge $e = (v, v')$, we set $Label(e) = (X_v, X_{v'})$. Note that $S(v)$ is a set containing one element - the alignment subset associated to v - and that $Label(e)$ is a pair of alignment subsets. Furthermore, we have computed Type 2 sub-alignments for each $X \cup Y$ where $Label(e) = (X, Y)$.

We will use T^* to guide a sequence of pairwise transitivity mergers, resulting finally in an MSA for the full set of sequences. As we do so, we will modify T^* through a sequence of edge contractions, until there is only one vertex left. The contraction of an edge $e = (v, w)$ will create a new vertex x with a new label

$S(x) = S(v) \cup S(w)$, but will not modify the labels at the edges. Therefore, at every point in the process, each edge will be labelled by a pair of alignment subsets for which we have a Type 2 sub-alignment, and each vertex will be labelled by a set of alignment subsets. Some edge contractions will require that we compute a transitivity merge of two overlapping compatible alignments. The new sub-alignments that result from transitivity mergers are called "Type 3 sub-alignments", and these Type 3 sub-alignments are defined by transitivity applied to some subset of the Type 2 sub-alignments.

For an edge $e = (v, w)$ and $Label(e) = (S_i, S_j)$, we have a Type 2 sub-alignment A^{ij} on $S_i \cup S_j$, and $S(v) \cap S(w) = \emptyset$. If $S(v)$ and $S(w)$ are singletons, then collapse the edge, and label the new vertex by the union of the labels at the endpoints. Otherwise, at least one endpoint of e is labelled by a set containing two or more alignment subsets, and the alignments A^v and A^{ij} are overlapping compatible alignments. Therefore, the three alignments A^{ij}, A^v, and A^w are all compatible, and so we can use transitivity (i.e., treating each alignment as an equivalence relation) to define the "transitivity merge" of these three alignments. To compute this transitivity merge, we first merge A^v and A^{ij}, and then we merge the resulting alignment with A^w (each step involves merging two overlapping compatible alignments). The result of each merger of these three MSAs creates a Type 3 sub-alignment on $S(v) \cup S(w)$. We contract the edge (v, w) to create the new node x, and we set $S(x) = S(v) \cup S(w)$.

Transitivity merge of two alignments. To compute the transitivity merge of two overlapping compatible alignments A and B, given two columns (one in A and the other in B) that share a common letter (i.e. the ith character of the jth sequence) we simply merge the two columns into one column.

Theorem 2. *Given m Type 1 alignments and $m-1$ Type 2 alignments, the algorithm to compute the transitivity merge of these alignments uses $O(Km \log m + mL)$ time, where K is the maximum length of any sequence (not counting gaps) in any Type 1 alignment and L is length of output alignment.*

Proof (Sketch): We begin with the following observation, which we provide without a proof due to space limitations (but see our supplementary material [19]). *Lemma: Let X, Y, and Z be disjoint sequence datasets, and alignments A and A' be alignments on $X \cup Z$ and $Y \cup Z$, respectively, that induce identical alignments on Z. Let K be the length of the longest sequence in X, Y, and Z, and L be the total number of sites in A and A'. Then we can merge alignments A and A' using transitivity in $O(L + (|X| + |Y| + |Z|)K)$.*

Let our dataset consist of N sequences, with each sequence of length at most K, and for the sake of simplicity, assume that our decomposition produces m subsets, all with equal sizes (note that centroid decomposition produces balanced subsets, so this assumption is justified). As described before, in Step 5, we chose an edge $e = (v, w)$ from the spanning tree, contract that edge, and perform two transitivity merges: one between $S(v)$ and $Label(e)$, and another between the result of the first merger and $S(w)$.

Based on the Lemma above, the first transitivity merge will have a running time of $O(K(|S(v)| + 2) + L)$, and the second merge will have a cost of $O(K(|S(v)| + |S(w)|) + L)$, and thus the cost of each edge contraction is $O(K(2 * |S(v)| + |S(w)|) + L)$. Now, imagine the case where the spanning tree is a path. If we start merging from one end to the other end, we get the total running time of $O(K(3 + 4 + \ldots + m)) = O(Km^2)$; however, we can improve on that. The important observation is that the spanning tree should be traversed such that transitivity mergers are between alignments with balanced number of sequences on each side.

The order in which edges are processed in PASTA is obtained by a recursive approach. Given the spanning tree, we divide it into two halves on the centroid edge, and thus obtain two roughly equal size subtrees. We process each half recursively using the same strategy, and thus get two single leaves at the endpoints of the centroid edge. Each leaf would represent the merger of all alignments in each half, and by construction they would have roughly equal size. We then contract the centroid edge, merge the two sides, and obtain the full alignment. If each half has roughly x sequences, the cost of the final edge contraction is $O(K(2x + x) + L) = O(3Kx + L)$ (as shown before). If $f(x)$ denotes the cost of applying our transitivity merger on a spanning tree with x nodes, we have

$$f(2x) = 2f(x) + 3kx + L$$

which has a $O(x \log(x) + xL)$ solution. Therefore, our particular order of traversing the spanning tree results in a total cost of $O(Km \log(m) + mL)$.

References

1. Sievers, F., Dineen, D., Wilm, A., Higgins, D.G.: Making automated multiple alignments of very large numbers of protein sequences. Bioinformatics 29(8), 989–995 (2013)
2. Liu, K., Raghavan, S., Nelesen, S., Linder, C.R., Warnow, T.: Rapid and accurate large-scale coestimation of sequence alignments and phylogenetic trees. Science 324(5934), 1561–1564 (2009)
3. Liu, K., Warnow, T., Holder, M., Nelesen, S., Yu, J., Stamatakis, A., Linder, C.: SATé-II: Very fast and accurate simultaneous estimation of multiple sequence alignments and phylogenetic trees. Syst. Biol. 61(1), 90–106 (2011)
4. Nelesen, S., Liu, K., Wang, L.S., Linder, C., Warnow, T.: DACTAL: divide-and-conquer trees (almost) without alignments. Bioinformatics 28(12), i274–i282 (2012)
5. Liu, K., Linder, C., Warnow, T.: Multiple sequence alignment: a major challenge to large-scale phylogenetics. PLoS Currents: Tree of Life (2010)
6. iPlant Collaborative: iPTOL, Assembling the Tree of Life for the Plant Sciences (2013), https://pods.iplantcollaborative.org/wiki/display/iptol/Home
7. Wong, G.K.S.: The Thousand Transcriptome (1KP) Project (2013), http://www.onekp.com/project.html
8. Katoh, K., Kuma, K., Toh, H., Miyata, T.: MAFFT version 5: improvement in accuracy of multiple sequence alignment. Nucl. Acids. Res. 33(2), 511–518 (2005)

9. Wheeler, T., Kececioglu, J.: Multiple alignment by aligning alignments. In: Proceedings of the 15th ISCB Conference on Intelligent Systems for Molecular Biology, pp. 559–568 (2007)
10. Edgar, R.C.: MUSCLE: a multiple sequence alignment method with reduced time and space complexity. BMC Bioinformatics 5(113), 113 (2004)
11. Edgar, R.C.: MUSCLE: a multiple sequence alignment with high accuracy and high throughput. Nucleic Acids Res. 32(5), 1792–1797 (2004)
12. Guo, S., Wang, L.S., Kim, J.: Large-scale simulating of RNA macroevolution by an energy-dependent fitness model. arXiv:0912.2326 (2009)
13. Price, M., Dehal, P., Arkin, A.: FastTree-2 approximately maximum-likelihood trees for large alignments. PLoS One 5(3), e9490 (2010)
14. Matsen, F., Kodner, R., Armbrust, E.: pplacer: linear time maximum-likelihood and Bayesian phylogenetic placement of sequences onto a fixed reference tree. BMC Bioinformatics 11, 538 (2010)
15. Mirarab, S., Nguyen, N., Warnow, T.: SEPP: SATé-enabled phylogenetic placement. In: Pacific Symposium on Biocomputing, pp. 247–258 (2012)
16. Eddy, S.: A new generation of homology search tools based on probabilistic inference. Genome Inform. 23, 205–211 (2009)
17. Finn, R., Clements, J., Eddy, S.: HMMER web server: interactive sequence similarity searching. Nucleic Acids Research 39, W29–W37 (2011)
18. Mirarab, S., Warnow, T.: FastSP: Linear-time calculation of alignment accuracy. Bioinformatics 27(23), 3250–3258 (2011)
19. Mirarab, S., Nguyen, N., Warnow, T.: Supplementary Online Material, PASTA: ultra-large multiple sequence alignment. figshare (2014), http://dx.doi.org/10.6084/m9.figshare.899770 (retrieved January 13, 2014)
20. Cannone, J., Subramanian, S., Schnare, M., Collett, J., D'Souza, L., Du, Y., Feng, B., Lin, N., Madabusi, L., Muller, K., Pande, N., Shang, Z., Yu, N., Gutell, R.: The Comparative RNA Web (CRW) Site: An Online Database of Comparative Sequence and Structure Information for Ribosomal, Intron and Other RNAs. BioMed. Central Bioinformatics 3(15) (2002)
21. Stamatakis, A.: RAxML-VI-HPC: Maximum likelihood-based phylogenetic analyses with thousands of taxa and mixed models. Bioinf. 22, 2688–2690 (2006)
22. Katoh, K., Frith, M.C.: Adding unaligned sequences into an existing alignment using MAFFT and LAST. Bioinformatics 28(23), 3144–3146 (2012)
23. Larkin, M.A., Blackshields, G., Brown, N.P., Chenna, R., McGettigan, P.A., McWilliam, H., Valentin, F., Wallace, I.M., Wilm, A., Lopez, R., Thompson, J.D., Gibson, T.J., Higgins, D.G.: Clustal W and Clustal X version 2.0. Bioinformatics 23(21), 2947–2948 (2007)
24. Boisseau, J., Stanzione, D.: TACC: Texas Advanced Computing Center (2013), http://www.tacc.utexas.edu
25. Suchard, M.A., Redelings, B.D.: BAli-Phy: simultaneous Bayesian inference of alignment and phylogeny. Bioinformatics 22, 2047–2048 (2006)

Fast Flux Module Detection
Using Matroid Theory

Arne C. Müller[1,2,3], Frank J. Bruggeman[8], Brett G. Olivier[4,5,7],
and Leen Stougie[4,6]

[1] Department of Mathematics and Computer Science, Freie Universität Berlin,
Arnimallee 6, 14195 Berlin, Germany
arne.mueller@fu-berlin.de
[2] International Max Planck Research School for Computational Biology and
Scientific Computing (IMPRS-CBSC), Max Planck Institute for Molecular Genetics,
Ihnestrasse 63-73, 14195 Berlin, Germany
[3] Berlin Mathematical School (BMS), Berlin, Germany
[4] Centre for Mathematics and Computer Science (CWI),
Science Park 123, 1098 XG Amsterdam, The Netherlands
[5] Molecular Cell Physiology, VU University,
De Boelelaan 1087, 1081 HV, Amsterdam, The Netherlands
[6] Operations Research, VU University,
De Boelelaan 1085, 1081 HV, Amsterdam, The Netherlands
[7] Netherlands Institute for Systems Biology, Amsterdam, The Netherlands
[8] Systems Bioinformatics, VU University,
De Boelelaan 1087, 1081 HV, Amsterdam, The Netherlands

Abstract. *Flux balance analysis* (FBA) is one of the most often applied
methods on genome-scale metabolic networks. Although FBA uniquely
determines the optimal yield, the pathway that achieves this is usually
not unique. The analysis of the optimal-yield flux space has been an open
challenge. *Flux variability analysis* is only capturing some properties of
the flux space, while *elementary mode analysis* is intractable due to the
enormous number of elementary modes. However, it has been found by
Kelk et al. 2012, that the space of optimal-yield fluxes decomposes into
flux modules. These decompositions allow a much easier but still comprehensive analysis of the optimal-yield flux space.

Using the mathematical definition of module introduced by Müller
and Bockmayr 2013, we discovered that flux modularity is rather a local than a global property which opened connections to matroid theory.
Specifically, we show that our modules correspond one-to-one to so-called
separators of an appropriate matroid. Employing efficient algorithms developed in matroid theory we are now able to compute the decomposition into modules in a few seconds for genome-scale networks. Using
that every module can be represented by one reaction that represents
its function, in this paper, we also present a method that uses this decomposition to visualize the interplay of modules. We expect the new
method to replace flux variability analysis in the pipelines for metabolic
networks.

Keywords: metabolic networks, FBA, flux modules, matroid theory.

R. Sharan (Ed.): RECOMB 2014, LNBI 8394, pp. 192–206, 2014.
© Springer International Publishing Switzerland 2014

1   Introduction

The metabolic capabilities and behaviors of biological cells are often modeled using metabolic networks. A metabolic network is constituted of a set of chemical compounds and a set of reactions describing the possible transformations of compounds. In the last years it became possible to reconstruct such networks on the genome-scale. This means that on one hand nearly all the reactions that can happen in a biological cell are included. On the other hand such networks consist of thousands of reactions.

Constraint based methods have proven to be very successful in the analysis of metabolic networks [19,20]. In constraint based methods no detailed information on reaction kinetics is needed. Often, the knowledge of reaction stoichiometries is sufficient. Each reaction is described by a column of the stoichiometric matrix S, which has an entry for each chemical compound; s_{ij} the i,j-th entry of S is the number of molecules of compound i consumed ($s_{ij} < 0$), produced ($s_{ij} > 0$), or not involved ($s_{ij} = 0$) in reaction j. Typically, the network is assumed to be in equilibrium, i.e. every metabolite is produced at the same rate as it is consumed. This gives rise to linear constraints on the set of feasible flux-vectors v (pathways) through the network, formally written as

$$Sv = 0$$

Together with bounds ℓ, u on reaction rates (thermodynamic information in form of irreversibilities, limitation of nutrient uptake rates, etc.) we obtain a polyhedron of feasible flux-vectors:

$$P = \{v : Sv = 0, \ell \le v \le u\} \tag{1}$$

Among the most prominent analysis methods is *flux balance analysis* (FBA) [28,17,21]. It is, for example, used to compute the optimal biomass yield that can be achieved by a cell under some growth medium [5]. It also computes an optimal flux distribution.

Such an optimum is easily found by solving a linear programming problem of the following type:

$$\max\{cv : Sv = 0, \ell \le v \le u\}$$

However, in general such optimal flows are not unique [13]. If this is ignored, it can lead to wrong predictions of by-product flux rates [11].

Kelk et al. [9] showed that many reactions have fixed flux rate in all optimal solutions. The remaining variability is due to variability of the fluxes on a number of relatively small subnetworks, which we call *flux modules*. Each such a module has in each optimum specific fixed input and output compounds. Müller and Bockmayr [15] used this property to formalize the notion of flux-module in a mathematically rigorous way. This allowed them to show that every optimal yield *elementary flux mode* (EFM) [23] is a concatenation of reactions with fixed flux and an elementary mode of each of the flux modules.

While the method by Kelk et al. [9] required the enumeration of exponentially many vertices of a flux polyhedron (which are related to the optimal yield EFMs),

Müller and Bockmayr [15] showed a way to find the modules without needing to compute all extreme solutions. Their method however relied on many runs of *flux variability analysis* (FVA) [3,13]. In FVA for each reaction r of the network the following two linear optimization problems are solved to obtain the minimal and maximal possible flux rate:

$$\max / \min\{v_r : Sv = 0, \ell \le v \le u\}$$

Although faster than EFM enumeration, the method is very sensitive to numerical instabilities and analyses of genome-scale networks could still take several hours.

The most important result in this paper is an extremely simple method allowing to compute the flux-modules in a few seconds for genome-scale metabolic networks. The method, described in Section 2, is based on the observation that the modules correspond to the separators of the linear matroid defined by the columns of the stoichiometric matrix that belong to reactions with variable optimal flux. We will explain all these technical concepts in Section 1.1. The efficiency of our method is demonstrated in Section 3 by application to several genome-scale metabolic networks.

Flux modularity highly depends on the growth conditions. In particular, interesting flux modules can usually only be found in the optimal flux space. Hence, it is of high importance to understand how the decomposition of modules changes under different growth conditions and objective functions. Since with our new method, module computation has become so fast, we can simply compute and compare modules under many different growth conditions and compare the results. Essential for this is a visualization method that shows the interplay of modules in the context of the whole network. In Sec. 2.2 we present a method that automatically generates such a visualization using a clever compression based on flux modules. Results of that method applied to a set of genome-scale metabolic networks can be found in Sec. 3.2.

1.1 Definitions and Preliminaries

We use $\mathcal{M}$ to denote the set of metabolites, $\mathcal{R}$ to denote the set of reactions. We abuse the notation for sets also for their size. $S \in \mathbb{R}^{\mathcal{M} \times \mathcal{R}}$ denotes the stoichiometric matrix. For $b \in \mathbb{R}^{\mathcal{M}}$, we analyze flux spaces $P \subseteq \{v \in \mathbb{R}^{\mathcal{R}} : Sv = b\}$. We observe that $b = 0$ leads to the standard steady-state assumption. Here, we also allow $b \ne 0$ to simplify notation in the context of modules. Furthermore, the space of optimal-yield fluxes is again a polyhedron and can be written in this form, too [15]. In this paper we will show that we can reduce the analysis of P to the analysis of linear vector spaces, i.e. flux spaces defined by the kernel of S: $\ker(S) := \{v \in \mathbb{R}^{\mathcal{R}} : Sv = 0\}$. We use subscripts to index flux through reactions, i.e., v_r denotes flux through reaction r. The support of flux-vector v is denoted by $\mathrm{supp}(v) := \{r \in \mathcal{R} : v_r \ne 0\}$. We will also be interested in the flux through a subset of reactions $A \subseteq \mathcal{R}$. Hence, we write v_A to denote the components of v corresponding to the reactions in A and we use S_A to denote the stoichiometric matrix that only contains the columns corresponding to the reactions in A. We define the projection $\mathrm{pr}_A(P) := \{v_A : v \in P\}$.

Definition 1. *[Flux Module, [15]] $A \subseteq \mathcal{R}$ is a P-module if there exists a $d \in \mathbb{R}^{\mathcal{M}}$ s.t. $S_A v_A = d$ for all $v \in P$. We call d the interface flux of the module.*

In contrast to the definition in [15], we also allow $A = \emptyset$ to be a module, which together with $\mathcal{R}$ we call the trivial modules. We present here some useful properties of modules proven in [15]. They may also help the reader to get some intuition on the concept of module.

Proposition 1. *Properties of Modules.*

(i) *If disjoint sets A and B are P-modules then $A \cup B$ is a P-module;*
(ii) *If A and B are P-modules and $B \subset A$ then $A \setminus B$ is a P-module.*

The rest of this section is devoted to an introduction to the relevant concepts from Matroid Theory [18], which is a generalization of graph theory and linear algebra. A matroid is defined by a universe of elements and subsets of them that have some independence structure.

Definition 2. *Given a universe $\mathcal{U}$ and a family $\mathcal{A}$ of independent subsets of U. Then $\{\mathcal{U}, \mathcal{A}\}$ is a matroid if it satisfies the following conditions.*

- $\emptyset \in \mathcal{A}$;
- *If $A \in \mathcal{A}$ and $A' \subset A$, then $A' \in \mathcal{A}$;*
- *If $A, A' \in \mathcal{A}$ and A' contains more elements than A, then there exists an element $e \in A' \setminus A$, such that $A \cup \{e\} \in \mathcal{A}$.*

As a very relevant example, a set of vectors in $\mathbb{R}^{\mathcal{R}}$, together with their linearly independent subsets form a matroid; a so-called *linear matroid*. Another example, less relevant here, is a graph with subsets of its edges that form forests as independent sets; a so-called *graphic matroid*. Matroid theory has already been used in the past to describe metabolic networks [16,2]. Indeed, many concepts from metabolic networks also exist in matroid theory. For example, flux modes in metabolic networks correspond to cycles in matroid theory; i.e., dependent sets of a matroid. Elementary flux modes correspond to circuits; i.e., minimal dependent sets. The only difference is, that in matroid theory we only talk about the support. I.e., $A \subseteq \mathcal{R}$ is a cycle if and only if there exists a flux mode v with $A = \mathrm{supp}(v)$. Similarly, a circuit $C \subseteq \mathcal{R}$ is a cycle with minimal support.

However, matroid theory is a wide field with additional concepts that have not yet been studied for metabolic networks. We show some of these concepts, that originate from linear algebra, directly in terms of the linear matroid represented by the columns of the stoichiometrix matrix S, which is the matroid that we use to describe metabolic networks. In our notation $\mathcal{R}$ is the set of columns of S.

- $A \subseteq \mathcal{R}$ is called a basis if it is a maximal independent set.
- $r(A)$ is the rank of S_A, i.e. the size of the largest independent subset of A.
- The dual matroid is the matroid represented by a kernel matrix of S.

Also graph theory introduces some further useful concepts into matroid theory. Typical examples are the circuits and cycles. Another property is connectivity. A graph is 2-connected if for any two edges, there exists a circuit that contains both. If a graph is not 2-connected, then the union of connected components is called a separator; i.e., any of the two sides of a partition of a graph into two parts A, B that are not 2-connected to each other. We can understand this as that there exists no flow circulation between A and B. It follows that the interface flux of A and B has to be zero. In matroid theory separators are defined using the matroid rank function:

Definition 3 (Separator). *A set A is defined to be a separator if and only if*

$$r(A) + r(\mathcal{R} \setminus A) = r(\mathcal{R}).$$

In Sec. 2.1 we show how the flux modules of a metabolic network correspond one-to-one to the separators of the corresponding matroid. We then use matroid theory to derive a very fast and simple algorithm for finding modules. It is based on a result by Krogdahl [12]. The runtime results on a set of genome-scale metabolic networks are presented in Sec. 3.1.

2   Methods

2.1   Finding Modules Efficiently

We first show that it is sufficient to analyze modularity as a local property of one point in the inside of the flux space, implying that we can ignore reaction reversibilities and simply analyze a subvector-space (Thm. 1). This allows to describe modularity in terms of matroid separators (Thm. 2), which we then exploit in designing an efficient algorithm to compute modules.

To make the first step, consider a point x *inside* the flux space and a neighbourhood of it (Fig. 1). This neighbourhood captures all the characteristics needed to analyse modularity of the whole flux space. We only have to deal with the term "inside". Since $P \subseteq \{v \in \mathbb{R}^{\mathcal{R}} : Sv = b\}$, it follows that P is of lower dimension in $\mathbb{R}^{\mathcal{R}}$. Hence, we will only consider the interior relative to $\ker(S)$. However, if we have reactions with fixed flux rate, P will also have lower dimension than

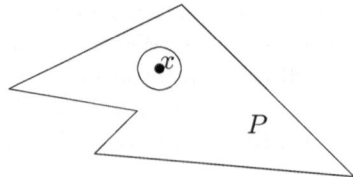

Fig. 1. Viewed from a point x inside the flux space, the flux space looks like a linear vector space and the bounds are not important

$\ker(S)$. Therefore, we will restrict to reactions with variable flux rate, which we define by:

$$V := \{r \in \mathcal{R} : v_r^{\max} \neq v_r^{\min}\}, \text{where} \tag{2}$$
$$v_r^{\max} := \sup\{v_r : v \in P\}$$
$$v_r^{\min} := \inf\{v_r : v \in P\}$$

This restriction does not destroy the module property:

Observation 1. *It holds for all $A \subseteq V$ that A is P-module $\Leftrightarrow A$ is $\mathrm{pr}_V(P)$-module.*

To guarantee that we can find a x inside the flux space after we restricted to reactions with variable flux rate, we require that P is convex. Theoretically, there are weaker conditions that are also sufficient.

Theorem 1. *If $P \subseteq \{v \in \mathbb{R}^{\mathcal{R}} : Sv = b\}$, is convex, it holds for all $A \subseteq \mathcal{R}$*

$$A \text{ is } P\text{-module} \Leftrightarrow A \cap V \text{ is } \ker(S_V)\text{-module.}$$

The proof can be found on the supplementary website.

By Thm. 1 we can restrict our attention to the analysis of linear vector spaces. Hence, in the following we will only analyse polyhedra of the form $P = \ker(S)$. We will relate modules of $\ker(S)$ to separators of the matroid defined by the columns of S. Remember the explanation of a separator in a graph in terms of the non-existence of a flow circulation in Section 1.1 and observe, that every module in $\ker(S)$ also has interface flux 0 since $0 \in \ker(S)$.

Formally, we obtain the following theorem, the proof of which is deferred to the supplementary website `https://sourceforge.net/projects/fluxmodules/`.

Theorem 2. *$A \subseteq \mathcal{R}$ is a $\ker(S)$-module if and only if A is a separator in the matroid represented by S.*

The characterization of modules as separators of matroids allows to compute the flux-modules of a metabolic network efficiently. Since separators and modules are closed under disjoint union, it suffices to describe the set of *minimal nontrivial separators* (modules).

Definition 4 (Minimal Module). *A P-module $\emptyset \neq A \subseteq \mathcal{R}$ is called minimal if there exists no P-module $B \neq \emptyset$ with $B \subset A$.*

To understand the algorithm for finding the modules, we observe that the minimal non-trivial separators are the connected components of the matroid. In the contex of graph-theory these are called 2-connected components (Note the inconsistency of the terminology between matroid and graph theory. The connected components in graph-theory are something different.) Formulated in matroid-terminology we recall the following graph-theoretic characterization of 2-connected component: For any 2 elements (columns of S in the linear matroid, edges in the graph) in the same connected component there exists a minimal dependent set (circuit) that contains them both. For pairs of elements of different connected components this is not true. It turns out that this also holds for matroids in general (Proposition 4.3.4 in [18]).

Theoretically, we could now build a graph $G = (V, E)$, where V is the set of reactions defined in (2) and there is an edge between two reactions (columns of S_V) if and only if there exists a circuit that contains both. The connected components (in the graph-theoretic sense) of G will be the minimal separators. Since the number of circuits explodes exponentially, it is not efficient to enumerate all circuits in order to compute the connected components of the graph G. Indeed, this is also not necessary and it suffices to look at a special set of circuits, so called *fundamental circuits* [27].

A set of fundamental circuits is obtained as follows: We start by finding a basis X of the matroid; i.e., a maximal independent set, which we compute by Alg. 1. Notice that, starting from the empty set, the algorithm grows X by adding elements only if this keeps X independent. Since we try to add all elements to X, it follows that at the end of the algorithm, X will be a basis of the linear matroid represented by S_V.

Let $Y := V \backslash X$. Clearly, for every $r \in Y$, adding r to X will create a cycle $C^r \subseteq X \cup \{r\}$. It is easy to see that C^r is actually a circuit, which is called fundamental circuit. In Alg. 1 the fundamental circuits are constructed simultaneously with constructing X. This gives us a so-called *partial representation*.

We now build, by Alg. 2, the graph $G' = (V, E')$, where two reactions are connected by an edge if there exists a fundamental circuit that contains both. Krogdahl and Cunningham showed that the connected components of G', found by Alg. 2, are precisely the minimal separators of the matroid [4,12].

To each circuit C there exists a flux vector v that is unique up to scaling with $C = \text{supp}(v)$, $Sv = 0$. If we enter for every circuit in B the corresponding flux values from v, we obtain a null-space matrix of S. Hence, this approach can be understood as computing a block-diagonalization of the null-space matrix. Approaches like this in the context of stoichiometric matrices have already been studied in [24]. However, [24] does not use matroid theory and it is unclear whether their method will always compute the finest block-diagonalization.

Here we recapitulate all the steps for finding the modules of the optimal flux space of a metabolic network.

1. Determine the optimal value by LP;
2. Set the objective function equal to the optimum value and add it as a constraint;
3. For each reaction r maximize and minimize the flux through r in the optimal flux space;
4. Determine the set V of reactions for which the maximum and the minimum are not equal;
5. Select the set of columns S_V corresponding to V of the stoichiometric matrix S and neglect the non-negativity constraints; i.e., irreversibilities, directions of the reactions;

Algorithm 1. Computes a basis X and its set of fundamental circuits of a matroid represented by S

```
function ComputePartialRepresentation(S)
C = ∅
X = ∅
for r ∈ V do
    check feasibility of S_X v = -S_r
    if feasible then
        C := supp(v) ∪ {r}
        C := C ∪ {C}
    else
        X := X ∪ {r}
    end if
end for
return C
```

Algorithm 2. Computes the modules of $\{v : S_V v = 0\}$

```
function ComputeModules()
C = ComputePartialRepresentation(S_V)
Build Graph G = {V, E} with (x, y) ∈ E iff there exists C ∈ C with x, y ∈ C.
A = find connected components of G (e.g. using depth-first search).
return A
```

6. Apply Alg. 2 to compute the minimal modules $\mathcal{A}$ of $\{v \in \mathbb{R}^V : S_V v = 0\}$.
7. $\mathcal{A}$ is the set of minimal modules that contain reactions in V. The reactions with fixed flux are all minimal modules by themselves.

We notice that steps 3 (and therefore 4) of the algorithm can be parallelized in a trivial way, reducing the computation times even further.

2.2 Visualization

We develop a visualization tool to help us understand how the decomposition of modules changes under different growth conditions and objective functions. By the definition of module, the reactions inside a module have together a fixed function (the interface flux). Hence, we can represent the module by a single reaction with a fixed flux in the genome-scale network. The stoichiometry of the representing reaction is precisely the interface flux of the module.

This way we can create a compressed network that contains all the reactions with fixed flux rates and artificial reactions that represent the modules. This compressed network has the following advantages:

- The number of reactions carrying flux is compressed (a module with many reactions, is represented by a single reaction).
- All the reactions in the compressed network have a fixed flux rate.

Unfortunately, the number of fixed reactions is still very large. This prevents automatic visualization of the network and the role of the modules containing variable reactions is obfuscated. However, reactions that have a fixed flux rate can also be grouped together into modules by Prop. 1.

Theoretically, we could group all reactions with a fixed flux rate into 1 module. This would result in a compressed metabolic network consisting of $k+1$ reactions, where k is the number of minimal modules containing reactions with variable flux rates. In particular, the module containing all fixed reactions will likely also contain the biomass- and nutrient-uptake reactions. If we want to understand the role of the modules for biomass production or nutrient uptake, this is not very useful. Moreover, modules of variable reactions may disconnect reactions with fixed flux rates from each other. Such disconnected reactions are important for the mediation between modules and should also be displayed separately. Hence, we decided to build a compressed network as follows:

1. Given: A collection Mod of interesting modules (selected by the user). Mod has to cover all reactions with variable flux rates. Typically Mod contains all minimal modules of variable reactions, a module containing the biomass reaction and modules containing the nutrient uptake reactions.
2. We compute the set $\mathcal{R}_{\text{Mod}} := \{r \in \mathcal{R} : r \in M \ \exists \ M \in \text{Mod}\}$ of reactions in interesting modules.
3. We compute the set $\mathcal{R}_B := \{r \in \mathcal{R} \setminus \mathcal{R}_{\text{Mod}} : v_r = 0 \ \forall v \in P\}$ of blocked reactions.
4. We compute the set $\mathcal{M}_{\text{Mod}} := \{m \in \mathcal{M} : \exists \ r \in \mathcal{R}_{\text{Mod}} \text{ such that } m \in \text{supp}(S_r)\}$ of metabolites involved in the interesting modules.
5. We consider the metabolic network, where $\mathcal{R}_{\text{Mod}}, \mathcal{R}_B$ and $\mathcal{M}_{\text{Mod}}$ are removed. It is represented by the stoichiometric matrix $S' := S_{\mathcal{M} \setminus \mathcal{M}_{\text{Mod}}, \mathcal{R} \setminus (\mathcal{R}_{\text{Mod}} \cup \mathcal{R}_B)}$.
6. We compute the connected components Mod_F of S'. We do so by defining the incidence matrix of a bipartite graph, the nodes of which on one side of the bipartition correspond to the rows of S', and the ones on the other side tot the columns of S', and there is an edge between row-node i and column-node j if and only if $S'_{ij} \neq 0$. The column-nodes represent the reactions in $\mathcal{R} \setminus (\mathcal{R}_{\text{Mod}} \cup B)$, and the corresponding reactions of the connected components of this bipartite graph, whence Mod_F, forms a partition of $\mathcal{R} \setminus (\mathcal{R}_{\text{Mod}} \cup B)$. Clearly, every $A \in \text{Mod}_F$ is a module, since Mod_F only contains fixed reactions.
7. We represent each module in Mod, Mod_F by a single reaction with the corresponding interface flux. Let $\mathcal{M}_0$ be the set of metabolites that have a net interface flux of 0 in all these modules. We suppress $\mathcal{M}_0$, since they would just show up as isolated metabolites. We obtain a metabolic network with metabolites $\mathcal{M}' := \mathcal{M} \setminus \mathcal{M}_0$ and reactions $\mathcal{R}' := \text{Mod} \cup \text{Mod}_F$.
8. We remove reactions disconnected from the network that contain the target reaction, e.g. because of modules that form thermodynamically infeasible cycles or otherwise have no role in the metabolism.

In practice, this results in medium-scale networks that can automatically be visualized with graph-drawing software like GraphViz [8].

3   Results

3.1   Runtime of Module Finding

With the new method we can compute all flux modules for the optimal flux space of genome scale networks in about the same time as is needed for conventional flux variability analysis. In Table 1 we see that the new method using matroid theory outperforms the previous methods in orders of magnitude. We used the metaopt toolbox [14] to solve the flux variability subproblems. Unfortunately, we did not have access to all the runtime data of [9] which is why some of the data is missing and the reported runtimes may be only from some of steps in the pipeline. The computations for the matroid approach were obtained by computations on a 4-core desktop computer.

Table 1. Comparison of runtimes for computing modules in the optimal flux space of genome scale networks

Network	using [9]	using [15]	using matroids
E. coli iAF1260	133495sec	755sec	6.4sec
E. coli iJR904	1906sec	162sec	1.9sec
E. coli iJO1366			8.4sec
H. pylori iIT341		55.5sec	0.8sec
H. sapiens recon. 1			153.3sec
H. sapiens recon. 2			1131sec
M. barkeri iAF692	1088sec	941sec	1.4sec
M. tuberculosis iNJ661	9317sec	1623sec	4.3sec
S. aureus iSB619		127.8sec	1.2sec
S. cerevisiae iND750			3.0sec

In particular notice that large networks like *Human recon 2* can now also be analyzed. In addition, the new method is numerically much more stable. In the method introduced by [15] it often happens that error tolerances are chosen too small or too large, which causes that linear programs that should be feasible are detected as infeasible etc. This then usually caused the algorithm to abort and the tolerance sometimes needed to be adjusted according to the problem instance.

We experienced that the new matroid based method is much more robust in this respect. Our initial tolerances of 10^{-20} for the optimization step, 10^{-8} for the flux variability and 10^{-9} for the final module computation worked in all cases.

Note, that the other two methods are solving slightly different problems. In [15] we were actually looking for modules in the thermodynamically constrained flux space and in [9], rays and linealities are eliminated prior to module computation.

A comparison between the results of [15] and the new method on *E. coli*
iAF1260 revealed that 7 of the modules coincide, 2 modules from the new method
contain additional reactions (which have fixed flux under thermodynamic constraints). The remaining modules are computed by the new method, but not by
[15] since they again only contain reactions that have fixed flux by thermodynamic constraints (usually those modules are formed by a splitted pair of forward
and backward reactions). The differences seem to be small, but a detailed analysis will be subject to future work. Also, we want to point out that, for computing
modules, the method by Kelk et al. [9] has to enumerate all the extreme points of
the flux polyhedron of optimal fluxes (after some preprocessing), a much harder
task. As a result more information than modules is obtained. Hence, the previous
works still remain useful.

3.2   Visualization

We used the visualization method presented in Section 2.2 to create visualizations of the above mentioned genome scale networks. The results can be found
on the supplementary website. In Tab. 2, we compare the original size of the
networks with the size of the compressed networks that are used to visualize
the interplay of the flux modules with variable flux rates. Each reaction of the
compressed network is a flux module. Every minimal flux module containing reactions with variable flux rates is represented by exactly one reaction. Reactions
with fixed flux rate are grouped together. It is interesting to see that although
the networks have quite different sizes originally, the compressed sizes do not
vary very much.

Table 2. Size of the compressed networks

Network	No. Metabolites (original)	No. Reactions (original)	No. Metabolites (compressed)	No. Reactions (compressed)
E. coli iAF1260	1668	2382	46	25
E. coli iJR904	761	1075	42	17
E. coli iJO1366	1805	2583	49	27
H. pylori iIT341	485	554	32	20
M. barkeri iAF692	628	690	35	13
M. tuberculosis iNJ661	826	1025	58	26
S. aureus iSB619	655	743	39	22
S. cerevisiae iND750	1061	1266	57	24

Visualizations of some of the example networks and their modules, using the
tool `dot` [7] from the `GraphViz` toolbox, can be found on the supplementary
website `https://sourceforge.net/projects/fluxmodules/`. The MATLAB
scripts for module detection can be found there as well.

4   Discussion

4.1   Enumeration of Optimal-Yield Pathways

We showed that flux modules [9,15] of genome-scale metabolic networks can efficiently be computed using matroids. We confirmed the previous results that the optimal flux space of most genome-scale metabolic networks decomposes into modules. If we want to compute the set of all optimal yield elementary modes, we theoretically can do this by simply computing the optimal yield elementary modes for each module. Then, we can use the decomposition theorem of [15] and obtain all optimal yield elementary modes of the whole network. There is only a small numerical barrier to be climbed to do this in practice: The EFM enumeration problem for each module appears to be numerically very unstable. Hence it is likely that EFMs are missed if not everything is computed using precise rational arithmetic.

We noted that the previous methods [9,15] were computing flux modules on slightly different flux spaces (in [9] rays and linealities were removed, in [15] we worked on the thermodynamically feasible flux space). These differences seem to be small but could be of significant biological importance. For example, it could be that due to thermodynamic constraints a reaction is blocked and hence, we can refine the modules. In a follow up work we will (mathematically and empirically) analyse the impact of these differences.

The full flux space is usually not decomposable into modules. In a follow up paper we will generalize the notion of module. This will allow us to find interesting modules also for the full flux space. Furthermore, this will have the potential to derive similar decomposition theorems as in [15] that then will work on the full flux space as well. We think this will be a major step towards EFM enumeration of genome-scale networks.

4.2   Modularity under Different Growth-Conditions

It has been observed that the decomposition into modules depends on the growth condition [9,15]. If we want to understand how the optimal flux space changes if the growth condition is modified, we have to recompute the decomposition into modules. Previously, this was a tedious task. Now it is very simple and fast and it can be done even for very small changes.

We presented a visualization method that shows the interplay of the modules and how they contribute to optimal biomass production. We think that this visualization will be very helpful to detect when a change in a growth condition significantly changes the structure of the optimal flux space.

For the visualization we use the definition of module to lump reactions together. This way we compute a compressed metabolic network that shows the optimal flux distribution with only a small number of reactions. These networks were small enough to be visualized using automated graph drawing tools. Currently, we have only little control on how these networks are drawn, causing the visualization to seem to be very sensitive to changes. In particular it would be interesting if we could get more robust drawing results for small changes in the network.

4.3 Applications Outside Metabolic Networks

Note that the definition of module of a polyhedron is context free. Every polyhedron can be written in the form $P = \{v \in \mathbb{R}^n : Sv = b, \ell \leq v \leq u\}$, where $S \in \mathbb{R}^{m \times n}, b \in \mathbb{R}^m, \ell \in \{R \cup \{-\infty\}\}^n, u \in \{\mathbb{R} \cup \{\infty\}\}^n$ by addition of slack variables. The enumeration of extreme points and extreme rays of such polyhedra is a classical problem in polyhedral combinatorics and computational geometry. In fact, it is a major open problem if this can be done in total polynomial time, i.e., polynomial in input and output. Khachiyan et al. [10] showed that this is impossible if only the extreme points (so without the rays) of an unbounded polyhedron are to be enumerated, unless P=NP.

Still, methods have been proposed that do this enumeration. The most prominent ones are variations of the so-called Double Description Method and a method introduced by Fukuda [1,6] (see also the work of Terzer and Stelling [26,25] in the context of metabolic networks). Such methods are much faster on smaller polyhedra, and therefore it makes sense to subject any polyhedron to be enumerated first to our method to see if smaller modules can be found and then apply a favorite enumeration method on the smaller parts.

4.4 Conclusion

In this paper we presented a new method that allows us to compute flux modules very efficiently. This allows us to compute flux modules of many metabolic networks under a large set of different conditions to compare flux modules with existing classical metabolic subsystems like Glycolysis.

Compared to classical metabolic subsystems that, at worst, are arbitrary functional groupings of metabolic reactions/species, flux modules are mathematically well defined. They are structural features only depending on a defined set of conditions (inputs, optimality). This qualifies them as a performance and quality metric for genome-scale metabolic networks. Furthermore, it allows us to investigate the modularity, and simplify genome metabolic networks without the risk of a bias from conventional biological interpretation.

Acknowledgments. We thank Timo Maarleveld for providing us with runtime data of the method by [9].

Authors Contributions

AM and LS found the connection to matroid theory and took care of mathematical correctness. AM provided all detailed proofs. LS supervised and organised the collaboration process. AM, FB and BO developed the visualization method. AM, LS and FB worked on the manuscript. All authors read and approved the manuscript.

References

1. Avis, D., Fukuda, K.: A pivoting algorithm for convex hulls and vertex enumeration of arrangements and polyhedra. Discrete & Computational Geometry 8(1), 295–313 (1992)
2. Beard, D.A., Babson, E., Curtis, E., Qian, H.: Thermodynamic constraints for biochemical networks. Journal of Theoretical Biology 228, 327–333 (2004)
3. Burgard, A.P., Vaidyaraman, S., Maranas, C.D.: Minimal reaction sets for escherichia coli metabolism under different growth requirements and uptake environments. Biotechnology Progress 17, 791–797 (2001)
4. Cunningham, W.H.: A combinatorial decomposition theory. PhD thesis, University of Waterloo, Ontario, Canada (1973)
5. Feist, A.M., Palsson, B.Ø.: The biomass objective function. Current Opinion in Microbiology 13, 344–349 (2010)
6. Fukuda, K., Prodon, A.: Double description method revisited. In: Deza, M., Manoussakis, I., Euler, R. (eds.) CCS 1995. LNCS, vol. 1120, pp. 91–111. Springer, Heidelberg (1996)
7. Gansner, E.R., Koutsofios, E., North, S.C., Vo, K.-P.: A technique for drawing directed graphs. IEEE Transactions on Software Engineering 19(3), 214–230 (1993)
8. Gansner, E.R., North, S.C.: An open graph visualization system and its applications to software engineering. Software - Practice and Experience 30(11), 1203–1233 (2000)
9. Kelk, S.M., Olivier, B.G., Stougie, L., Bruggeman, F.J.: Optimal flux spaces of genome-scale stoichiometric models are determined by a few subnetworks. Scientific Reports 2, 580 (2012)
10. Khachiyan, L., Boros, E., Borys, K., Elbassioni, K., Gurvich, V.: Generating all vertices of a polyhedron is hard. Discrete Computational Geometry 39, 174–190 (2008)
11. Khannapho, C., Zhao, H., Bonde, B.L., Kierzek, A.M., Avignone-Rossa, C.A., Bushell, M.E.: Selection of objective function in genome scale flux balance analysis for process feed development in antibiotic production. Metabolic Engineering 10(5), 227–233 (2008)
12. Krogdahl, S.: The dependence graph for bases in matroids. Discrete Mathematics 19, 47–59 (1977)
13. Mahadevan, R., Schilling, C.H.: The effects of alternate optimal solutions in constraint-based genome-scale metabolic models. Metabolic Engineering 5, 264–276 (2003)
14. Müller, A.C., Bockmayr, A.: Fast thermodynamically constrainted flux variability analysis. Bioinformatics 29(7), 903–909 (2013)
15. Müller, A.C., Bockmayr, A.: Flux modules in metabolic networks. Journal of Mathematical Biology (2013) (in press, preprint), http://nbn-resolving.de/urn:nbn:de:0296-matheon-12084
16. Oliveira, J.S., Bailey, C.G., Jones-Oliveira, J.B., Dixon, D.A.: An algebraic-combinatorial model for the identification and mapping of biochemical pathways. Bulletin of Mathematical Biology 63, 1163–1196 (2001)
17. Orth, J.D., Thiele, I., Palsson, B.Ø.: What is flux balance analysis. Nature Biotechnology 28, 245–248 (2010)
18. Oxley, J.: Matroid Theory, 2nd edn. Oxford Graduate Texts in Mathematics. Oxford University Press, New York (2011)

19. Jason Papin, A., Stelling, J., Price, N.D., Klamt, S., Schuster, S., Palsson, B.Ø.: Palsson. Comparison of network-based pathway analysis methods. Trends in Biotechnology 22(8), 400–405 (2004)
20. Price, N.D., Reed, J.L., Palsson, B.Ø.: Genome-scale models of microbial cells: evaluating the consequences of constraints. Nature Reviews Microbiology 2, 886–897 (2004)
21. Santos, F., Boele, J., Teusink, B.: A practical guide to genome-scale metabolic models and their analysis. Methods in Enzymology 500, 509 (2011)
22. Schellenberger, J., Que, R., Fleming, R.M.T., Thiele, I., Orth, J.D., Feist, A.M., Zielinski, D.C., Bordbar, A., Lewis, N.E., Rahmanian, S., Kang, J., Hyduke, D.R., Palsson, B.Ø.: Quantitative prediction of cellular metabolism with constraint-based models: the COBRA toolbox v2.0. Nature Protocols 6(9), 1290–1307 (2011)
23. Schuster, S., Hilgetag, C.: On elementary flux modes in biochemical systems at steady state. J. Biol. Systems 2, 165–182 (1994)
24. Schuster, S., Schuster, R.: Detecting strictly detailed balanced subnetworks in open chemical reaction networks. Journal of Mathematical Chemistry 6, 17–40 (1991)
25. Terzer, M.: Large scale methods to enumerate extreme rays and elementary modes. PhD thesis, Diss., Eidgenössische Technische Hochschule ETH Zürich, Nr. 18538, 2009 (2009)
26. Terzer, M., Stelling, J.: Large-scale computation of elementary flux modes with bit pattern trees. Bioinformatics 24(19), 2229–2235 (2008)
27. Truemper, K.: Partial matroid representations. European Journal of Combinatorics (1984)
28. Varma, A., Palsson, B.Ø.: Metabolic flux balancing: Basic concepts, scientific and practical use. Nature Biotechnology 12, 994–998 (1994)

Supplementary Material

For the supplementary material we have created a sourceforge project `https://sourceforge.net/p/fluxmodules/wiki/Home/`. There you can find

- The proofs for Thm. 1 and Thm. 2.
- The MATLAB code for computing the modules (requires the `openCOBRA` toolbox [22])
- The MATLAB code for vizualising modules (requires the tool `dot` from `GraphViz` [7]).
- Computational results and visualizations for the example networks. For each network the results are grouped into one .zip archive.

Building a Pangenome Reference
for a Population

Ngan Nguyen[1], Glenn Hickey[1], Daniel R. Zerbino[2], Brian Raney[1], Dent Earl[1],
Joel Armstrong[1], David Haussler[1,3], and Benedict Paten[1]

[1] Center for Biomolecular Science and Engineering,
University of California Santa Cruz, CA, USA
[2] EMBL-EBI, Wellcome Trust Genome Campus, Cambridge CB10 1SD, UK
[3] Howard Hughes Medical Institute, University of California,
Santa Cruz, CA 95064, USA

Abstract. A reference genome is a high quality individual genome that is used as a coordinate system for the genomes of a population, or genomes of closely related subspecies. Given a set of genomes partitioned by homology into alignment blocks we formalise the problem of ordering and orienting the blocks such that the resulting ordering maximally agrees with the underlying genomes' ordering and orientation, creating a pangenome reference ordering. We show this problem is NP-hard, but also demonstrate, empirically and within simulations, the performance of heuristic algorithms based upon a cactus graph decomposition to find locally maximal solutions. We describe an extension of our Cactus software to create a pangenome reference for whole genome alignments, and demonstrate how it can be used to create novel genome browser visualizations using human variation data as a test.

1 Introduction

A reference genome is a genome assembly used to represent a species. Reference genomes are indispensable to contemporary research primarily because they act as a common coordinate system, so that genes, variations, and other functional annotations can be described in common terms [1–3].

In this paper we explore creating a pangenome reference from a set of genomes. We start from a set of genomes in a genome alignment [4], which partitions the genomes subsequences into homology sets that we term blocks. The problem is to find an ordering of the blocks that as closely as possible reflects the ordering of the underlying genome sequences. We call such an ordering a pangenome reference because it indexes every block, something that any individual genome within the population almost certainly does not. Though potentially useful for many of the things reference genomes are currently used for, our principal motivation is to produce better visualisations of variation and closely related species data within a genome browser [5], in which one reference genome is used as the coordinates to display data for a species and for which a pangenome display would allow a more complete view of the data.

R. Sharan (Ed.): RECOMB 2014, LNBI 8394, pp. 207–221, 2014.
© Springer International Publishing Switzerland 2014

Closely analogous to the problem of building a pangenome reference of aligned input genomes, a great deal of previous work has focused on methods for ancestral reconstruction. Most relevant to this work is the (rearrangement) median problem. The median problem is, informally, to find for a set of genomes and a given edit operation a median genome whose total pairwise edit distance from each of the other sequences is minimal [6]. Naively it might be assumed that good solutions to the median problem might have utility for finding an intra-species pangenome reference. However, in the median problem the edit operations are not necessarily restricted to maintain sequence colinearity while during evolution complex selective pressures often work to achieve exactly this [7]. For example, consider the three signed permutations: A, d, B, e, C and $A, -e, B, -d, C$ and $A, B, e, -d, C$. Assume that the capital letters, A, B and C represent very large subsequences of the genome and the lower case letters, d and e, represent short subsequences. In each of the sequences the large subsequences maintain their colinearity with respect to one another, and ignoring the short subsequences no edits appear to have occurred. However, incorporating the short subsequences the optimal median sequence under either the double-cut-and-join (DCJ) or reversal edit operations is $A, -e, -B, -d, C$ (the other sequences are each one operation away), despite the inversion of the large sequence B, which may be biologically implausible as a common ancestor. This tendency to lose colinearity has led to the study of 'perfect' rearrangement scenarios, in which common intervals of ordered subsequences present in the input are conserved [8]. However, current algorithms for finding perfect rearrangement scenarios require the common intervals to be pre-specified, do not allow copy number variation and require the common intervals to exist in all the inputs. This makes them inappropriate when there is no prior expert knowledge to define the intervals, or when representing large populations, where copy number variation is present and missing data and unusual variants break many intervals that would otherwise be common.

Related to our approach, methods to derive consensus orderings of sets of total and partial orders have been extensively considered, particularly in the domain of social choice [9, 10]. In general, the inputs to such problems are sequences or structures equivalent (in their most general form) to directed acyclic graphs (DAG), and the output is a consensus (partial) ordering. In such work, algorithms often work to minimise the consensus' (weighted) symmetric difference distance or Kemeny tau distance [10] (informally, the number of out of order (discordant) pairs). Recently, such consensus ordering procedures have been adapted to create consensus genetic maps from sets of individual subpopulation maps [11]. The problem formalised here has similarity to such approaches, with the important difference that it explicitly models the double stranded nature of DNA, allowing us to account for the cost of sequences being inverted with respect to one another. In what follows we will formalise the basic problem, prove its NP-hardness, describe a principled heuristic decomposition of the problem using cactus graphs [12], give heuristic algorithms for the problem's solution, demonstrate the algorithms performance using simulation and show a pangenome reference visualisation of variation data.

2 Results

2.1 The Pangenome Reference Problem

Genome Sequences. Let $S = \{\sigma_1, \sigma_2, \ldots, \sigma_k\}$ be the input DNA sequences, with lengths $(n_1, n_2, \ldots, n_k)$. For simplicity we assume here that the DNA sequences are linear, though extensions to allow additional circular sequences are straightforward. Due to the double stranded nature of DNA, we distinguish the $5'$ and $3'$ ends of each sequence element. We denote a tuple $(x \in \{1, 2, \ldots, k\}, i \in (1, 2, \ldots, n_x), a \in \{5', 3'\})$ as x_i^a, giving the coordinate of the a end of the ith element in σ_x. For any DNA sequence σ_x the ends are oriented consistently, so that for all $i > 1$ the $x_i^{5'}$ end is adjacent (contiguous) in the sequence to the $x_{i-1}^{3'}$ end and, for all $i < n_x$ the $x_i^{3'}$ end is adjacent in the sequence to the $x_{i+1}^{5'}$ end. We use signed notation to distinguish ends, hence $-x_i^{5'} = x_i^{3'}$ and $x_i^{5'} = -x_i^{3'}$. The set of all end coordinates is $\mathbf{S}$.

Alignment. The end coordinates in $\mathbf{S}$ are partitioned by their alignment relationships. To represent this we define the alignment relation $\sim \subset \mathbf{S}^2$. The alignment relation is an equivalence relation, i.e. one that is transitive, symmetric and reflexive. We denote the equivalence classes for $\sim$ as $\mathbf{S}/\sim$, and write $[x_i^a]$ to represent an equivalence class containing x_i^a. We also constrain the alignment relation to force the pairing of opposite ends. Firstly, we assume if $x_i^a \sim y_j^b$ then $-x_i^a \sim -y_j^b$, we call this *strand consistency*. Secondly, we assume if $x_i^a \sim y_j^b$ then neither $-x_i^a \sim y_j^b$ or $x_i^a \sim -y_j^b$, which we term *strand exclusivity*. Due to strand consistency, for all $[x_i^a]$ in $\mathbf{S}/\sim$ there exists $[-x_i^a] = \{-y_j^b : y_j^b \in [x_i^a]\}$, the reverse complement of $[x_i^a]$. Due to strand exclusivity, for all x_i^a, $[x_i^a] \neq [-x_i^a]$. Combining these two statements it follows that $|\mathbf{S}/\sim|$ is even. The set $[-x_i^a]$ can be equivalently denoted $-[x_i^a]$, so that the reverse complement of X in $\mathbf{S}/\sim$ is $-X$. We call each member of $\mathbf{S}/\sim$ a *side*, and each pair set of forward and reverse complement sides a *block*. Note that the alignment relation allows for copy number variation, i.e. arbitrary numbers of coordinates from sequences in the same genome can be present in a block.

Sequence Graphs. Let $G = (V, E)$ be a *(bidirected) sequence graph*. A bidirected graph is a graph in which each edge is given an independent orientation for each of its endpoints [13]. The vertices are the set of blocks, $V = \{\{X, -X\} : X \in \mathbf{S}/\sim\}$. The edges, $E = \{\{[x_i^{3'}], [x_{i+1}^{5'}]\} : \sigma_x \in S \wedge i \in (1, 2, \ldots, n_x - 1)\}$, encode the adjacencies (biologically the covalent bonds) between contiguous ends of sequence elements. Each edge is a pair set of sides rather than a pair set of vertices, therefore giving each endpoint its orientation, see Fig. 1(A). The cardinality and size of G are clearly at most linear in the size of $\mathbf{S}$.

A sequence of sides $(X_1, X_2, \ldots, X_n)$ is a *thread*. If the elements in $\{-X_1, X_2\}, \{-X_2, X_3\}, \ldots, \{-X_{n-1}, X_n\}$ are edges in the graph then the thread is a *thread path*. We use a sequence of sides, rather than vertices, because the sides orient the vertices, distinguishing forward and reverse complement orientations.

For example, for each sequence $\sigma^x \in S$, $[x_1^{5'}], [x_2^{5'}], \ldots, [x_{n_x}^{5'}]$ is a thread path in G, because for all $i \in 1, 2, \ldots, n_{x-1}$, $\{[x_i^{3'}], [x_{i+1}^{5'}]\}$ (equivalently $\{-[x_i^{5'}], [x_{i+1}^{5'}]\}$) is an edge in G.

A *transitive sequence graph*, $\hat{G} = (V, \hat{E} = \{\{[x_i^{3'}], [x_j^{5'}]\} : \sigma_x \in S \wedge i < j\})$, includes the sequence graph G as a subgraph but additionally includes edges defined by *transitive adjacencies*, that is pairs of ends connected by a thread path. The cardinality (vertex number) of $\hat{G}$ is the same as G, but the size (edge number) of $\hat{G}$ is worst-case quadratic in the size of **S**. A sequence graph encodes input sequences and an alignment, a transitive sequence graph models the complete set of ordering and orientation relationships between the blocks implied by the input sequences.

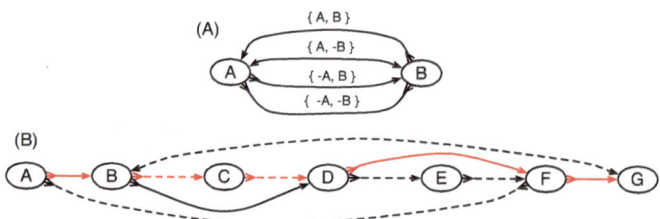

Fig. 1. (A) A bidirected graph representing the four ways two blocks can be connected. The arrowheads on the edges indicate their endpoints: the sides of the vertices. (B) An example pangenome reference on a sequence graph. There are two sequences, indicated by the colour of the edges. The red sequence, represented by the thread A, B, C, D, F, G and the black sequence, represented by the thread $A, -F, -E, -D, -B, G$. Neither includes all the blocks. A pangenome reference, indicated by the dotted edges, is $A, -F, -E, -D, -C, -B, G$. The dotted edges and the edges $\{-B, D\}$ and $\{-D, F\}$ are the edges consistent with the given pangenome reference.

Pangenome References. A *pangenome reference* F is a set of non-empty threads such that each block is visited exactly once, see Fig. 1(B). Intuitively, not all pangenome references are equally reasonable as a way of summarising S, because they will not all be equally "consistent" with the set of adjacencies, $\hat{E}$. An edge $\{X, Y\}$ is *consistent* with a pangenome reference F if and only if there exists a thread in F containing the subsequence $-X, \ldots, Y$, see Fig. 1(B). Given a weight function $z : \hat{E} \to \mathbb{R}_+$, which maps edges to positive real valued weights, the *pangenome reference problem* is to find a pangenome reference in $\mathbf{F} = \arg\max_F \sum_{e \in \hat{E}_F} z(e)$, where $\hat{E}_F$ is the subset of $\hat{E}$ consistent with F.

Exponential Weight Function. Although many possible weight functions exist, inspired by the nature of genetic linkage, we define $z(\{X, Y\}) = z'(X, Y) + z'(Y, X)$, where $z'(X, Y) = \sum_{\sigma_x \in S} \sum_{x_i^{3'} \in X} \sum_{x_j^{5'} \in Y} (1 - \theta)^{j-i} I_{\{i<j\}}$, in which $I_{\{i<j\}}$ is the indicator function that is 1 for pairs of i and j for which $i < j$ else 0, and the parameter θ is a real number in the interval $[0, 1)$. The θ

parameter intuitively represents the likelihood that an adjacency between two directly abutting sequence elements is broken or absent in any other randomly chosen sequence, and is defined analogously to its use in the LOD score [14] used in genetics. For $\theta > 0$, the score given to keeping elements in a sequence in the same order and orientation in the pangenome reference declines exponentially with distance separating them.

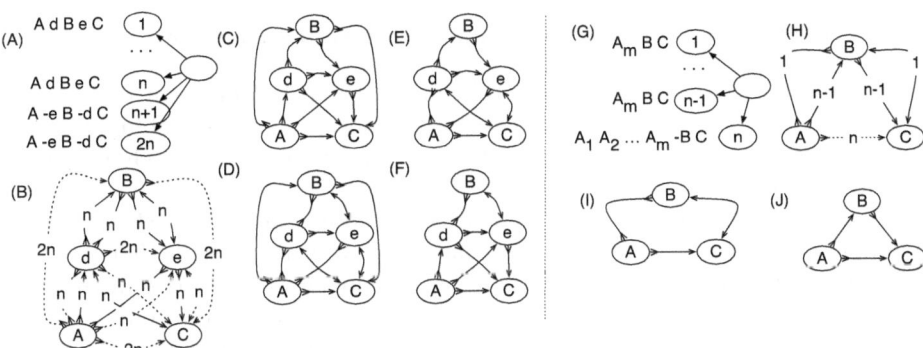

Fig. 2. Top: An illustration of why it is not always sufficient to consider only abutting adjacencies. (A) There are five blocks, A, B, C, d and e, reprising their roles from the example given in the introduction. The input contains n copies of the sequence A, d, B, e, C and n copies of the sequence $A, -e, B, -d, C$. (B) The bidirected graph representation of this problem, with the number of adjacencies supporting each edge labeled, the abutting adjacencies shown as solid lines and the non-abutting adjacencies shown as dotted lines. If we are only interested in solutions that start with A and end with C there are 4 maximal solutions, shown in (C,D,E,F). For θ less than 1 the (C) and (D) solutions are optimal, but as θ approaches 1 all four solutions become equally weighted, despite (E) and (F) having B in the reverse orientation. **Bottom:** An illustration of why θ should be greater than 0. (G) There are $m + 2$, blocks, the input contains $n - 1$ copies of the sequence A_m, B, C and 1 copy of the sequence $A_1, A_2, \ldots, A_m, -B, C$. (H) The bidirected graph representation of the problem, where the sequence of $A_1, A_2, \ldots, A_m$ blocks has been reduced to just a single vertex for convenience. The two maximal solutions are shown in (I,J), corresponding to the two distinct input sequences. If $m > n$ and θ is 0 then the solution with B in the reverse orientation (D) is optimal, despite this orientation being observed only once. By increasing θ the alternative solution with B in the forward orientation becomes optimal.

To make it clear that an intermediate value of θ is desirable we can look at what happens at extreme values of the parameter. As θ approaches 1 the weight function become dependent only on edges in the sequence graph. Fig. 2 demonstrates a limitation with considering only these edges, which is similar to that described for edit operations in the introduction. At $\theta = 0$ all transitive adjacencies are equally weighted, however this can lead to longer sequences having undue influence on the solution; Fig. 2 also gives an example of this limitation when weighting all adjacencies equally. One issue not dealt with by the definition of z are the evolutionary interdependencies between the input sequences. It is possible to adjust the weights given to adjacencies given a phylogenetic tree that relates the input sequences (or the genomes they derive from). However, where homologous recombination is present a weighting based upon a phylogenetic tree is insufficient and yet more complex strategies are needed.

2.2   NP-Hardness of the Pangenome Reference Problem

We show the pangenome reference problem is NP-hard, demonstrating that the pangenome reference problem can be projected onto the problem of finding maximum weight subgraphs of a bidirected graph that do not contain characteristic classes of simple cycle.

A M, N *bidirected simple cycle*, henceforth abbreviated to a M, N-cycle, is a simple cycle in a bidirected graph containing M vertices such that $M \geq N$, $M - N$ of the vertices have both their sides incident with an edge in the cycle (we call these *balanced vertices*) and the other N vertices have only one side incident with edges in the cycle (we call these *unbalanced vertices*). A M, N-cycle is odd if N is odd, else we call it even. We say a bidirected graph is *strongly acyclic* if it contains no $M, 0$-cycles or odd M, N-cycles. Let $\hat{\mathbf{G}}$ be the set of all strongly acyclic subgraphs of $\hat{G}$ of maximum weight. The following lemma shows the relationship between maximum weight strongly acyclic subgraphs and maximum weight pangenome references.

Lemma 1. *There exists a surjection* $f : \mathbf{F} \twoheadrightarrow \hat{\mathbf{G}}$, *such that for all F in* $\mathbf{F}$, $f(F) = (V, \hat{E}_F)$.

Proof. Let $F \in \mathbf{F}$, the threads in F orient all the vertices, partitioning the sides into two sets according to if they appear in a pangenome reference thread or not. By definition, the consistent edges and this bipartition of the sides form a bipartite graph. If there exists an odd M, N-cycle in $f(R)$, then it defines an odd cycle in this bipartite graph (a contradiction), hence $f(R)$ contains no odd M, N-cycles.

A pangenome reference induces a partial $<_F$ order on the vertices. If there exists a $M, 0$-cycle $\{\{X_1, -X_2\}, \{X_2, -X_3\}, \ldots, \{X_n, -X_1\}\} \in f(R)$, as these edges are consistent with F, this implies that both $\{X_1, -X_1\} <_F \{X_n, -X_n\}$ and $\{X_n, -X_n\} <_F \{X_1, -X_1\}$, but a partial order is asymmetric (a contradiction), therefore $f(R)$ contains no $M, 0 - cycles$.

As $f(F)$ is strongly acyclic, if it is not in $\hat{\mathbf{G}}$ then it must be possible to add an edge to $f(F)$ without creating a $M, 0$-cycle or odd M, N-cycle. Assume therefore that $f(F)$ is a subgraph of some $\hat{G}' \in \hat{\mathbf{G}}$. Let $\{X, Y\}$ be an edge in $\hat{G}'$ but not in $f(F)$. By definition, $\{X, Y\}$ has non-zero weight. Between $\{X, -X\}$ and $\{Y, -Y\}$ of the three other possible edges, $\{\{X, -Y\}, \{-X, Y\}, \{-X, -Y\}\}$, one must be in $\hat{E}_F$, else F is not a maximum weight solution to the pangenome reference problem, because in this case there must exist two threads in F, one that contains X

or $-X$ and one that contains Y or $-Y$, and these two threads can be concatenated together to create a new pangenome reference additionally consistent with
one of the four possible edges between $\{X, -X\}$ and $\{Y, -Y\}$. If $\{X, -Y\} \in \hat{E}_F$
then $\hat{G}'$ contains a $2, 1$-cycle $\{\{X, -Y\}, \{Y, X\}\}$, if $\{-X, -Y\}$ then $\hat{G}'$ contains
a $2, 0$-cycle $\{\{-X, -Y\}, \{Y, X\}\}$ and if $\{-X, Y\}$ then $\hat{G}'$ contains a $2, 1$-cycle
$\{\{-X, Y\}, \{Y, X\}\}$. In all cases therefore we derive a contradiction, therefore
$f(R) \in \hat{\mathbf{G}}$.

It remains to prove that for every member of $\hat{G}'$ in $\hat{\mathbf{G}}$ there exists F such
that $f(F) = \hat{G}'$. For $\hat{G}' = (\hat{V}', \hat{E}') \in \hat{\mathbf{G}}$ a *side bicolouring* is a labelling function
colour, such that each vertex and edge's sides are coloured such that one is *black*
and the other is *red*, i.e. it creates a bipartition of the sides of the graph.

To construct such a colouring for $\hat{G}'$ use a depth first search. In each connected
component of $\hat{G}'$ pick an unlabeled vertex and colour one of it sides red and the
other black. The depth first search then recurses from this vertex such that for
each edge of the form $\{X, Y\}$ if X is coloured red and Y is unlabelled then Y
is coloured black and $-Y$ is coloured red and vice versa if X is coloured black.
If during this recursion an edge is encountered such that both sides are already
labelled then the depth first search has traversed a M, N-cycle. Further, if the
sides of this edge are labelled with the same colour then the depth first search
has failed to produce a side bicolouring. Suppose we encounter such a cycle in
$\hat{G}'$, either there are two excess black sides or two excess red sides, as only the last
edge encountered does not have sides of distinct colours. Each balanced vertex
contributes a black and a red side while each unbalanced vertex contributes either
two black sides or two red sides, therefore $N \geq 1$. Furthermore, as there are only
two excess vertices of one colour N must be odd, implying $\hat{G}'$ is not strongly
acyclic, therefore there exists a side bicolouring of $\hat{G}'$. Given a side bicolouring
of $\hat{G}'$ let $\hat{G}'' = (\hat{V}'', \hat{E}'')$ be a digraph, such that $\hat{V}'' = \{X : \{X, -X\} \in \hat{V}' \wedge$
$colour(X) = \text{red}\}$ and $\hat{E}'' = \{(X, Y) : \{X, -Y\} \in \hat{E}' \wedge colour(X) = \text{red} \wedge$
$colour(-Y) = \text{black}\}$, where (a, b) is a directed edge from a to b. The graph $\hat{G}''$
is isomorphic to $\hat{G}'$, except that the arbitrary orientations of the sides within the
vertices have been reassigned so that there is only one type of edge in the graph
(Fig. 3). A directed cycle in $\hat{G}''$ would be a $M, 0$-cycle, but as $\hat{G}''$ is strongly
acyclic it must contain no directed cycles, therefore $\hat{G}''$ is a DAG. Any topological
sort $F = \{X_1, X_2, \dots, X_n\}$ of the vertices of $\hat{G}''$ is a pangenome reference for
which $f(F) = \hat{G}'$. □

Theorem 1. *The pangenome reference problem is NP-hard.*

Proof. The problem of finding a maximum weight strongly acyclic subgraph
of a bidirected graph is polynomial-time reducible to the pangenome reference
problem, because, by the previous lemma, the consistent subgraph of any solution to the pangenome reference problem is a maximum weight strongly acyclic
subgraph. It remains to prove that the problem of finding a maximum weight
strongly acyclic subgraph of a bidirected graph is NP-hard. We prove this by
reduction of the minimum feedback arc set problem [15], which is to find the
smallest set of edges in a directed graph that when removed result in a graph

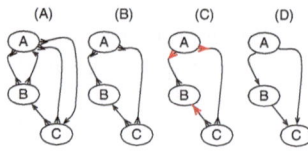

Fig. 3. (A) A bidirected graph with three vertices A, B and C. (B) A subgraph of (A) containing no $M, 0$-cycles or odd M, N-cycles. (C) A side bicolouring of (B). (D) A digraph for (C).

containing no directed cycles. Using the demonstration in the previous lemma, a digraph can be equivalently represented as a side bicoloured bidirected graph. An unbalanced vertex in an M, N-cycle is red if the endpoints of the edges incident with it in the cycle are colored red, else it is black. Suppose there exists an M, N-cycle in a side bicoloured bidirected graph with i balanced vertices, j unbalanced red vertices and k unbalanced black vertices. As in a side bicoloured bidirected graph each edge has one red endpoint and one black endpoint the total number of red and black endpoints is equal, therefore $i + 2j = i + 2k$, thus $k = j$ and therefore it is not possible to construct an odd M, N-cycle in a side bicoloured bidirected graph. As a directed cycle in a digraph corresponds to an $M, 0$-cycle in the equivalent side bicoloured bidirected graph, the minimum feedback arc set problem is thus polynomial-time reducible to the problem of finding a maximum weight strongly acyclic subgraph of a side bicolored bidirected graph (i.e. eliminating $M, 0$-cycles). □

An alternative, similarly simple proof of NP-hardness uses the elimination of odd M, N-cycles rather than the $M, 0$-cycles, reducing the maximum bipartite subgraph problem[16].

2.3   Algorithms for the Pangenome Reference Problem

We have established the pangenome reference problem is NP-hard, and now, given that knowledge, we describe a principled, and to our knowledge novel, heuristic to decompose the problem using cactus graphs, and briefly describe two straightforward algorithms to build and refine a pangenome reference.

Cactus Decomposition of the Pangenome Reference Problem. A cactus graph of the type introduced in [12] describes a sequence graph in a hierarchical form. For a sequence graph G, a pair of sides X and Y form a *chain interval* if there exists one or more thread paths of the form $-X, \ldots, Y$, but no thread paths of the form $-X, \ldots, -Y$ or $X, \ldots, Y$. Chain intervals represent intervals that are "fundamental", in the sense that all the simple threads for all the sequences in S follow the traversal rules defined above. It is reasonable therefore to search for reference sequences that preserve all such intervals.

The chain interval relation defines a partition of the vertices into a set of disjoint *chains*. A *chain* is a thread $(X_1, X_2, \ldots, X_n)$ such that all and only pairs of form $(-X_i, X_j)$ for which $j - i \geq 1$ define a chain interval; we call each chain interval of the form $(-X_i, X_{i+1})$ a *link*. Chains can be arranged hierarchically, because one *child* chain may be contained within the link of a *parent* chain. We call two chains *siblings*, if either they are both children of the same parent chain link, or both are not contained within any parent chain link (i.e. they are at the highest level of the hierarchy). For a thread $(X_1, X_2, \ldots, X_n)$ the two sides X_1 and $-X_n$ are *stubs*. A *net* is an induced subgraph of G defined by the set of stubs for a maximal set of sibling chains and (if they exist) the pair of sides that define the containing parent link, see Fig. 4(A). A graph in which the nodes are the nets and the edges are the oriented vertices of a sequence graph forms a cactus graph, see Fig. 4(B).

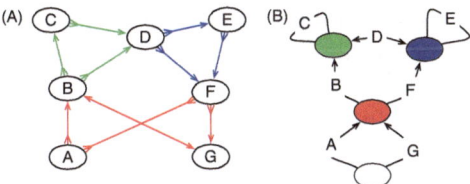

Fig. 4. (A) The bidirected graph from Fig. 1(B) rewritten to show the nets as colored side subgraphs. (B) The cactus graph representation of the blocks and nets in (A), with the white net containing the highest level chains. The edges represent the blocks, the vertices the nets. The arrow heads on the edges indicate endpoints that are links.

To construct a reference that respects all chain intervals we create a pangenome reference independently for each net, treating each pair of chain stubs as equivalent to blocks in the previous exposition. Additionally, the pangenome reference for each child net with a parent link must be composed of a single thread whose stubs are the sides of the parent link. This reduces the maximum size of the pangenome reference problem to that of the largest net in the sequence graph, which as the sequence graph for alignments of variation data is often relatively sparse, has (in our experience and in accordance to elementary random graph theory [17]) size only approximately logarithmically proportional to the number of vertices in the graph. It also facilitates parallel execution, because each net can be computed in parallel.

Greedy and Iterative Sampling Algorithms for the Pangenome Reference Problem. Given the decomposition, we build a pangenome reference for each subproblem using an initial greedy algorithm, before iterative refinement that employs simulated annealing.

In overview (see the source-code for more details), a pangenome reference F is composed, starting from the empty set, by greedily adding one member of V to

F at a time, each time picking the combination of insertion point and member of V that maximises consistency with elements already in the F. The algorithm is naively $|V|^3$ time (as each insertion is $|V|^2$ time), though by heuristically ignoring weights less than a specified threshold (the weight declines exponentially with sequence separation), and using a priority queue to decide which member of V to add next, it can be improved $|V|log(|V|)$ in practice.

Given an initial reference F the procedure progressively searches through a sequence of neighbouring permutations, where for a reference F a *neighbouring permutation* is created by removing an element from F and then inserting it either in the positive or negative orientation as a prefix, suffix or coordinate between elements in the reduced F, potentially including the elements original coordinate. The algorithm incorporates simulated annealing by using a monotonically decreasing temperature function to control the likelihood of choosing neighbouring, lower scoring permutations. As the temperature tends to zero the algorithm becomes greedy and we can search for a local minima, while as the temperature tends to positive infinity all permutations become equally probable and the search becomes a random walk. Each iteration of sampling, in which the repositioning of every block is considered once, is naively $|V|^2$ time, but is improved to $|V|log(|V|)$ in practice.

2.4   Simulation Experiments

To test the algorithms described we use a simple simulation of a rearrangement median problem. We start with a single linear chromosome, represented as a signed permutation of 250 elements, which we call the original median. We then simulate either 3, 5 or 10 leaves, treating each leaf with a set number of random edits. For convenience we simulate only translocations and inversions, which results in each leaf remaining a single contiguous chromosome, and apply an equal number of translocations and inversions. Note, for simplicity, we did not assess copy number changes (e.g. duplicative rearrangements), but doing so would be interesting.

We performed two sets of simulations, in the first we did not constrain the length of the subsequence of elements inverted or translocated. In such a scenario only a few edits are sufficient to radically reorder the genome and break many resulting ordering relationships. In the second scenario we constrained the lengths of inverted subsequences to 2 or 1, and constrained the length of translocated subsequences to just 1. In this scenario relatively large numbers of rearrangements are required to breakup the ordering of the original median.

To find solutions to the pangenome reference problem we use a combination of the algorithms described above, first using the greedy algorithm, then refining it with iterative sampling, performing 1000 iterations of improvement and setting $\theta = 0.1$ (values of theta between 0.5 and 0.001 made little difference to the result). We call this combination Ref. Alg. in the results that follow. To compare performance of our solutions we compare them to the original median, and to a median genome inferred using the AsMedian program [18] (using default parameters), which finds optimal solutions to the DCJ median problem with three leaves. We assess performance by looking at two metrics. Firstly, the DCJ distance, which gives the minimum number of edits needed to translate one genome into another by DCJ edits. Secondly, viewing the medians as two signed, partial order relations A and B on the blocks, the symmetric difference distance, defined as $\frac{|A \triangle B|}{|(A \cup B)|}$. This gives the proportion of order plus orientation relationships not common to the two medians.

The top panel of Fig. 5 shows the results of simulating unconstrained, arbitrary translocations and inversions. Unsurprisingly, Ref. Alg. constructs medians that are substantially farther from the leaves or the original median in terms of DCJ distance than the results of AsMedian (avg. 44% and 103% more overall than Ref. Alg. with 3 leaves, respectively, from the leaves and original median). Furthermore, in terms of symmetric difference distance, the AsMedian solutions are on average 52% closer to the original median (though not the leaves) than those constructed using Ref. Alg. with 3 leaves. This clearly demonstrates that using the Ref. Alg. for sequences whose ordering have been turned over by large rearrangements produces poor results, and that ancestral reconstruction algorithms can be used more effectively for moderate numbers of edits in this scenario, with the caveat that they may construct a multi-chromosomal ordering of the data.

The bottom panel of Fig. 5 shows the results of simulating short edits, demonstrating a striking converse to the unconstrained case. In terms of DCJ distance, the Ref. Alg. with 5 and 10 leaves is actually able to outperform the AsMedian program in terms of distance to the original median (Ref. Alg. with 5 leaves requires 20% on avg. fewer DCJ edits than AsMedian), while in terms of symmetric distance Ref. Alg. with 3 leaves is able to find solutions that are as close to the leaves as the original median and substantially closer to the original median than the AsMedian results (Ref. Alg. with 3 leaves is 31% closer on avg. than AsMedian in terms of symmetric difference distance to the original median). Furthermore, adding more leaves improves the results substantially (Ref. Alg. with 10 leaves is 52%, 44% closer, on average, to the original median in terms of avg. DCJ and symmetric difference than Ref. Alg. with 3 leaves). These results demonstrate that if edits have largely maintained the linear ordering of the sequences then, even when the sequences have been subject to substantial numbers of edits, Ref. Alg. is competitive with an ancestral reconstruction method in terms of DCJ, while ensuring that all elements appear in an ordering that is closer, in terms of ordering and orientation, than an optimal ancestral reconstruction method.

2.5 Creating a Pangenome Visualisation for the Major Histocompatibility Complex (MHC)

We recently introduced HAL format and associated APIs [19], which represent a multiple genome alignment as a series of pairwise alignments between ancestor and descendant genomes related by a tree. Using the algorithms and decomposition described here, we have adapted the Cactus alignment program to generate HAL format alignments, by additionally imputing a pangenome reference that

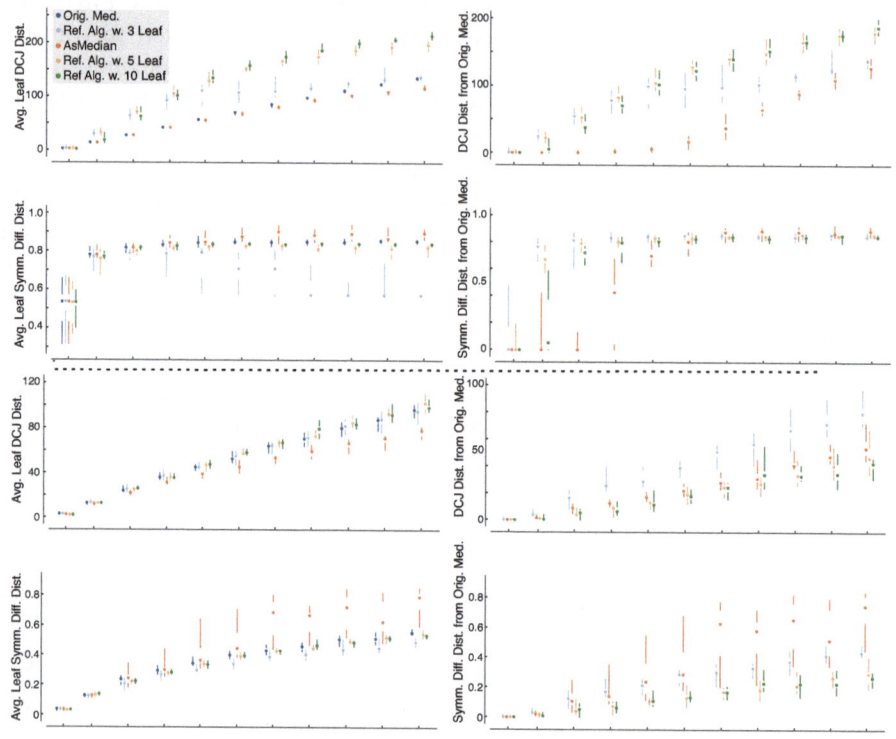

Fig. 5. Top: Simulation results using arbitrary inversion and translocation operations. Each plot shows the total number of operations (a mixture of 50% inversions and 50% translocations) vs. either the DCJ distance (top two plots) or symmetric difference distance (bottom two plots). The left plots give the average distance from the leaf genomes and the right plots give the distance from the original "true" median genome. Series shown include the original median genome (left plots only), the inferred median genome from the AsMedian program [18] using three leaves, and the inferred median genomes using our combined reference algorithms, using, separately, three, five and ten leaf genomes as input. Simulations used ten replicates for each fixed number of edits, points give median result, lines show max and min quartiles. **Bottom:** Simulation results using short inversion and translocation operations, laid out as in top panel.

serves as the common ancestral genome of the aligned genomes. To visualize HAL alignments we have developed *comparative assembly hubs* (to be described in full elsewhere), a UCSC browser framework that allows the loading of alignments as HAL files from user defined URLs, and their display using a new track that we call an *alignment snake*.

To demonstrate its performance with resequencing data, we created a pangenome reference for the MHC region, where there is a wealth of public data available. We used 16 human assemblies (see Supp. Data) plus the chimpanzee genome as an outgroup. To generate the Cactus alignment we used its default parameters and set $\theta = 10^{-4}$ (tested θ values between 10^{-2} and 10^{-6} produce similar results). The pangenome reference covers just over 5 megabases, and with respect to it the samples collectively contain tens of thousands of indels and hundreds of more complex rearrangements. The complete alignment and comparative assembly hub files are in the supplementary material, and can be used to load the demoed browsers. Fig. 6 shows example browser screenshots for the MHC. Each shows the alignment of a genome with respect to the pangenome reference (along the horizontal axis), which is converted into a consensus nucleotide sequence by creating a consensus sequence for each block and then concatenating these subsequences together in the order of the pangenome reference. Each screenshot shows a sequence of alignment snake tracks, arranged vertically, one for each of the aligned target genomes. Each alignment snake is a sequence of blue/red rectangles (depending on strand) connected together by lines. The rectangles represent subsequences of a target genome aligned to the reference, the lines represent the adjacencies between these aligned subsequences. The red tick marks within the rectangles represent single nucleotide variations (SNVs). The top panel of Fig. 6 shows indels; as no sequence apart from the pangenome reference contains all the blocks, only the pangenome reference browser can show the contents of all the segregating indel subsequences. The middle panel of the figure shows a segregating combination of an inversion and deletion. The chromosome reference sequence for the human reference genome (GRCh37) has the inversion, but the pangenome reference, being a comprehensive consensus, both includes the subsequences missing the chromosome reference sequence and orients the inversion according to the majority of the samples. In the bottom panel a tandem duplication is shown. All the human assemblies are either incomplete or have two copies of the tandemly duplicated subsequence, however, the pangenome reference, containing a single copy of the tandem duplication subsequence is able to cleanly display the event using the basic semantics of the alignment snake track.

3   Discussion and Conclusion

We have defined a problem useful for creating a pangenome reference between closely related genomes, proved it is NP-hard and described principled heuristics to find (approximate) solutions. We demonstrated in simulations the tradeoffs between optimising for conserved order relationships and minimising DCJ operations. Finally, we have demonstrated the method's utility in constructing visualizations of variation data in the UCSC browser, providing a view of the alignment not typically possible from any input genome.

We have demonstrated the relationship of the pangenome reference problem to a method for ancestral reconstruction, but the pangenome reference problem also has close similarities with sequence assembly problems, which have variants explicitly described on bidirected graphs [13] In particular, the scaffolding problem given paired reads involves arranging a set of "scaffold" sequences in a partial order to essentially maximise the numbers of consistently ordered, oriented and spaced paired reads. Apart from the additional constraint on spacing,

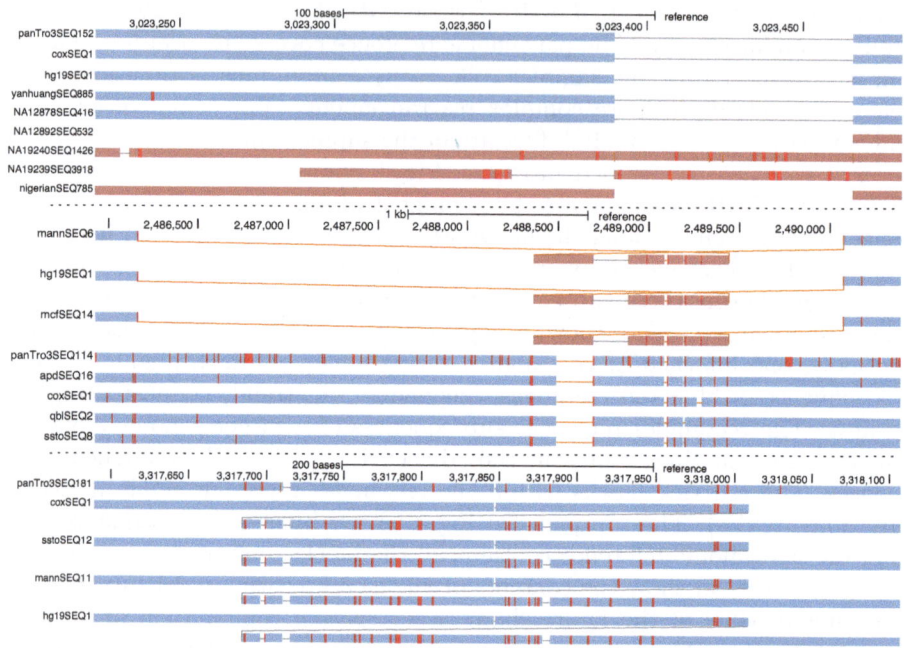

Fig. 6. UCSC pangenome reference browser screenshots. Top: Indels. Middle: A segregating inversion. Bottom: An apparently fixed tandem duplication. For reasons of space some samples are omitted from the screenshots. The human reference genome is labeled with the prefix hg19, the chimpanzee sequences are labelled with the prefix panTro and details of the other samples are in the supplementary material.

the scaffolding problem with paired reads can be defined equivalently to the pangenome reference problem.

Though our primary motivation in this paper is visualisation, pangenome references, being comprehensive and consensus orderings, are likely to prove useful for other purposes. Where reference genomes are currently used for computational convenience, for example in read compression, and are not integral for biological interpretation, a pangenome reference may present a useful alternative to current reference genomes. Additionally, given that (sequence) graphs do not have a implicit linear decomposition, having a pangenome coordinate system on such graphs could prove useful in processing multiple alignments.

The source code for this project is at: `https://github.com/benedictpaten/matchingAndOrdering/tree/development`.

Supplementary Material. The Cactus alignment source is in one convenient distribution at: `https://github.com/glennhickey/progressiveCactus`. It includes the source code for this project, as well as the HAL source code, which contains the hal2AssemblyHub.py script, for generating comparative assembly hubs that can be viewed using the UCSC browser. The MHC assemblies and the comparative assembly hub is at: `http://hgwdev.cse.ucsc.edu/ benedict/ MHCBrowserForRecomb`.

References

1. Coffey, A.J., Kokocinski, F., Calafato, M.S., Scott, C.E., Palta, P., Drury, E., Joyce, C.J., Leproust, E.M., Harrow, J., Hunt, S., Lehesjoki, A.E., Turner, D.J., Hubbard, T.J., Palotie, A.: The gencode exome: sequencing the complete human exome. Eur. J. Hum. Genet. 19(7), 827–831 (2011)
2. Myers, R.M., Stamatoyannopoulos, J., Snyder, M., Dunham, I., Hardison, R.C., Bernstein, B.E., Gingeras, T.R., Kent, W.J., Birney, E., Wold, B., Crawford, G.E.: A user's guide to the encyclopedia of dna elements (encode). PLoS Biol. 9(4), e1001046 (2011); ENCODE-Project-Consortium
3. 1000-Genomes-Project-Consortium: A map of human genome variation from population-scale sequencing. Nature 467(7319), 1061–1073 (October 2010)
4. Paten, B., Earl, D., Nguyen, N., Diekhans, M., Zerbino, D., Haussler, D.: Cactus: Algorithms for genome multiple sequence alignment. Genome Res. 21(9), 1512–1528 (2011)
5. Meyer, L.R., et al.: The UCSC Genome Browser database: extensions and updates 2013. Nucleic Acids Research, 64–69 (2013)
6. Tannier, E., Zheng, C., Sankoff, D.: Multichromosomal median and halving problems under different genomic distances. BMC Bioinformatics 10, 120 (2009)
7. Kirkpatrick, M.: How and why chromosome inversions evolve. PLoS Biol. 8(9) (January 2010)
8. Berard, S., Chateau, A., Chauve, C., Paul, C., Tannier, E.: Computation of perfect dcj rearrangement scenarios with linear and circular chromosomes. Journal of Computational Biology 16(10), 1287–1309 (2009)
9. Fagin, R., Kumar, R., Sivakumar, D.: Comparing Top k Lists. SIAMJ. Discrete Math. 17(1), 134–160 (2002)
10. Kendall, M.: A new measure of rank correlation. Biometrika 30(1/2), 81–93 (1938)
11. Bertrand, D., Blanchette, M., El-Mabrouk, N.: Genetic map refinement using a comparative genomic approach. J. Comput. Biol. 16(10), 1475–1486 (2009)
12. Paten, B., Diekhans, M., Earl, D., John, J.S., Ma, J., Suh, B., Haussler, D.: Cactus graphs for genome comparisons. J. Comput. Biol. 18(3), 469–481 (2011)
13. Medvedev, P., Brudno, M.: Maximum likelihood genome assembly. J. Comput. Biol. 16(8), 1101–1116 (2009)
14. Griffiths, A.J.F., Miller, J.H., Suzuki, D.T.: An introduction to genetic analysis (January 1999)
15. Karp, R.: Reducibility among combinatorial problems. Plenum (Complexity of Computer Computations), 85–103 (January 1972)
16. Newman, A.: Max-cut. Encyclopedia of Algorithms 1, 489–492 (2008)
17. Erdos, P., Rényi, A.: On the evolution of random graphs. Publications of the Mathematical Institute of the Hungarian Academy of Sciences 5, 17–61 (1960)
18. Xu, A.W.: A fast and exact algorithm for the median of three problem: a graph decomposition approach. J. Comput. Biol. 16(10), 1369–1381 (2009)
19. Hickey, G., Paten, B., Earl, D., Zerbino, D., Haussler, D.: HAL: a hierarchical format for storing and analyzing multiple genome alignments. Bioinformatics 29(10), 1341–1342 (2013)

CSAX: Characterizing Systematic Anomalies in eXpression Data

Keith Noto[1,*], Carla Brodley[1], Saeed Majidi[1],
Diana W. Bianchi[2,3], and Donna K. Slonim[1,2]

[1] Tufts University, Medford, MA, U.S.A.
noto@cs.tufts.edu
[2] Tufts University School of Medicine, Boston, MA, U.S.A.
[3] Tufts Medical Center, Boston, MA, U.S.A.

Abstract. Methods for translating gene expression signatures into clinically relevant information have typically relied upon having many samples from patients with similar molecular phenotypes. Here, we address the question of what can be done when it is relatively easy to obtain healthy patient samples, but when abnormalities corresponding to disease states may be rare and one-of-a-kind. The associated computational challenge, anomaly detection, is a well-studied machine learning problem. However, due to the dimensionality and variability of expression data, existing methods based on feature space analysis or individual anomalously-expressed genes are insufficient. We present a novel approach, CSAX, that identifies pathways in an individual sample in which the normal expression relationships are disrupted. To evaluate our approach, we have compiled and released a compendium of public microarray data sets, reformulated to create a testbed for anomaly detection. We demonstrate the accuracy of CSAX on the data sets in our compendium, compare it to other leading anomaly-detection methods, and show that CSAX aids both in identifying anomalies and in explaining their underlying biology. We note the potential for the use of such methods in identifying subclasses of disease. We also describe an approach to characterizing the difficulty of specific expression anomaly detection tasks and discuss how one can estimate the feasibility of a specific task. Our approach provides an important step towards identification of individual disease patterns in the era of personalized medicine.

1 Introduction

The development in the 1990's of techniques for genome-wide monitoring of expression data [1,2] has had a dramatic impact on the field of molecular medicine. Personalized diagnostics based on expression array signatures are increasingly moving into the clinic [3, 4]. However, methods for designing microarray-based diagnostics or discovering disease subtypes require a reasonable number of samples representing each patient class [5]. There are techniques for boosting the

---

* Author's current address: AncestryDNA (Ancestry.com), 153 Townsend St., Suite 800, San Francisco, CA 94107.

R. Sharan (Ed.): RECOMB 2014, LNBI 8394, pp. 222–236, 2014.
© Springer International Publishing Switzerland 2014

signal when relatively few samples are available [6, 7], but these don't eliminate the need for a representative set of samples that fully characterize the molecular variability underlying the phenotypes of interest. For rare diseases or genetically heterogeneous disorders, another analysis paradigm is needed. Here we demonstrate that by characterizing the expression patterns of "normal" samples, it is often possible to identify abnormal samples even when each abnormality is one-of-a-kind.

The problem of determining which samples to flag as abnormal, given only normal training data, is related to the computational field of *anomaly detection*, sometimes called *outlier detection*. Anomaly detection is an active research area in both statistics and data mining [8]. It is regularly applied to such problems as spam detection, identifying potential credit-card theft, verifying online identities, and correcting errors in census data.

There have been several previous efforts that explicitly apply anomaly detection methods to bioinformatics problems, including correction in genome annotation [9] and identifying changes in the steady-state behavior of stochastic gene regulatory networks [10]. A related approach by Torkamani and Schork [11] identifies genes whose expression pattern is unusual in a given cellular context. With respect to gene expression microarray data in particular, the task of identifying differentially expressed genes has been viewed in the framework of outlier detection [12–15], as has the problem of identifying array artifacts [16]. Perhaps the closest approach to that we describe here is that of Tomlins, *et al.* [17], who use outlier detection to identify common translocations in cancer, but the outliers still refer to individual genes rather than samples.

The underlying machine learning problem, that of identifying "abnormal" samples given only "normal" samples as training data, is a challenging one. Microarray data is particularly ill-suited for anomaly detection, as for many other machine learning problems, because of its noise level, the dimensionality of a typical data set (hundreds of samples but tens of thousands of genes), and the expectation that only a small fraction of those genes may provide any information about the classification of the samples.

Fortunately, other characteristics make the problem potentially tractable. We expect meaningful expression changes to reflect unusual regulation in specific functional pathways. We can therefore use prior knowledge about the relationships between genes to identify anomalous examples. Such information has the added advantage that it may provide hints to the underlying cause of the detected anomalies.

To evaluate the utility of such an approach, we created a compendium of data sets for anomaly detection from published microarray classification data sets. On this compendium, we compare several state-of-the-art methods for anomaly detection, including CSAX, a novel approach that we designed to boost the signal from the prior method most robust to irrelevant features [18] while identifying the gene sets that best distinguish each anomalous sample. Our results show that in many cases, anomaly detection can both identify unusual samples and produce meaningful information about the nature of the anomalous data.

There is also a question of what abnormalities we can expect to identify. For example, a single abnormal sample characterized by abnormal expression of a single gene could not possibly be detected by any method – the data are simply too noisy. Our method is applicable when a sizable number of genes' expression levels are sufficiently different. We characterize the classes of anomalies we can expect to detect, and we discuss how clinical intuition might be applied to identify anomalies suitable for these methods.

2   Data and Methods

2.1   Compendium of Microarray Anomaly Detection Data Sets

We assembled a compendium of 28 microarray anomaly detection tasks from published classification studies that involve at least two classes of samples (*e.g.*, healthy vs. disease, or multiple tissue types). We created an anomaly detection task from a microarray data set by (i) designating one class (usually the least-abundant or the least physiologically-normal) as the "anomalous" class, and all samples from the other classes as the "normal" class, which may therefore be quite heterogeneous; (ii) creating a training set from some of the normal microarrays, chosen at random, and (iii) creating a test set from the rest of the normal microarrays and all of the samples in the anomalous class (see Figure 1). We created our compendium from suitably-sized data sets (with several human microarray samples from at least two clearly- defined classes) found by browsing GEO (www.ncbi.nlm.nih.gov/geo) or from publications known to us, combined with a testbed of expression classification data sets assembled elsewhere for the development of computational methods (see Acknowledgements). The compendium includes all the data sets with which we experimented that had previouly been released publicly. Details and data can be found at bcb.cs.tufts.edu/csax.

In most envisioned applications, such as diagnosing rare diseases from blood samples, anomalies are likely to be one of a kind. However, in each data set in this compendium, we have a collection of relatively similar anomalies and we know which samples we should expect to identify as anomalous. We can therefore use this compendium as a "gold standard" data set to evaluate the accuracy of our methods.

2.2   A New Method for Expression Anomaly Detection

There are many existing computational methods for anomaly detection in generic high-dimensional data sets. The most successful general approaches include density-based methods such as the Local Outlier Factor (LOF) [19], which identifies outliers by comparing their distances from their nearest neighbors to the typical distances between nearby training examples, and one-class support vector machines (SVMs) [20]. However, neither of these methods is especially well suited for handling the dimensions of expression microarray data.

We recently developed an anomaly-detection method called *Feature Regression and Classification*, or FRaC [18]. When applied to expression data, FRaC

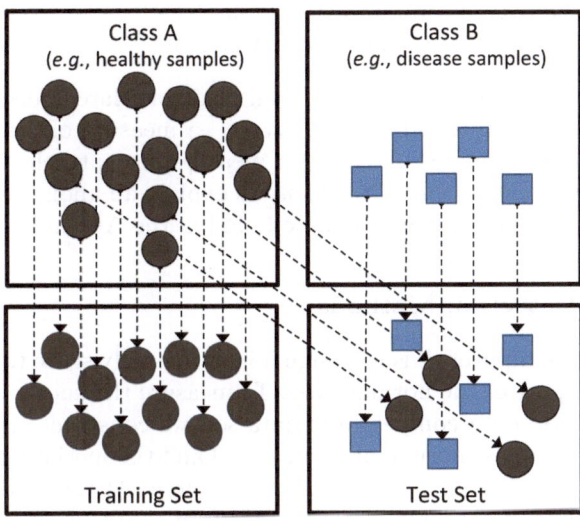

Fig. 1. To create an anomaly detection task from a microarray study with two classes of samples, A and B, we randomly select a portion of the A samples for the test set and use the remainder for training. The task is then to identify samples from class B after training on samples from class A alone.

predicts the expression level of each gene based on the levels of the others. A gene's expression level is considered "surprising" if a reliable relationship learned from the training data is violated in the test sample, as measured by the so-called "suprisal" [21] (or log-loss) score from the field of information theory. Anomaly scores for samples are computed by summing surprisal scores over all genes. FRaC is known to be robust to large numbers of irrelevant variables [18], making it well suited for identifying outliers in microarray data.

However, FRaC simply classifies samples as outliers without explicitly providing information about the nature of each anomaly. Instead, we would like to identify gene sets or pathways in which the sample is particularly anomalous. We therefore also developed a robust method, "Characterizing Systematic Anomalies in eXpression data" (CSAX), for doing so that uses the FRaC surprisal scores for each gene. Briefly, CSAX incorporates the following steps:

1. Using FRaC, measure the extent to which the expression levels of individual genes in the test sample are surprising, given the training data.
2. Measure the extent to which genes with surprising expression levels are involved in the same pathway or biological process. To do this, we rank the genes by their surprisal scores and look for gene sets with many high-surprisal genes using GSEA [22, 23]. One can use any collection of pre-defined gene sets for this step. We chose the Reactome gene sets (www.reactome.org) here, as a reasonable-sized yet sufficiently broad collection.
3. Measure the variance of the gene sets selected in the previous step to improve the robustness of the predictions via bootstrap aggregation, or *bagging*, a common machine learning technique. We find that (i) the gene sets that appear enriched using FRaC and GSEA alone are not necessarily the same

as those consistently enriched over several iterations of bagging, and (ii) the anomaly detection accuracy is noticeably higher using bagging.[1]

Given a training set, an unlabeled microarray, and a collection of related gene sets, the output of CSAX is (i) a real-valued *anomaly score*—a measure of our confidence that the microarray is anomalous, and (ii) a measurement attached to each gene set, indicating the extent to which it is likely to explain an anomaly in the microarray data. Further details of CSAX can be found in the Appendix.

2.3   Evaluating Anomaly Detection Methods

The goal of any anomaly detection approach is to assign a high anomaly score to unlabeled samples that are not part of the normal class. To measure the success of each anomaly detection method on a compendium data set, we construct an ROC curve [24] from the test set labels (normal or anomalous) and the method's predicted anomaly scores, and we calculate the area under the curve (AUC). The AUC can be viewed as the likelihood that an anomaly detector assigns a higher anomaly score to a test set anomaly than it does to a test set normal. Thus, higher AUC scores are better, the best possible score being 1.0. The AUC is a common performance measure that is independent of both the number and the proportion of anomalies in the test set.

For each expression data set in our compendium, we create an anomaly detection task (see Section 2.1 and Figure 1) by randomly selecting 75% of the normal microarrays for training. The remaining 25% of the normal microarrays and all of the anomalous microarrays are put in random order to make up the test set. We repeat this process 20 times and report an average AUC for each anomaly detection method for each data set.

We compare the performance of CSAX to that of two top-performing anomaly detection methods: computing the local outlier factor (LOF) [19], and using one-class support vector machines (SVMs) [20]. We also compare against the performance of our own method, FRaC [18].

3   Results

3.1   Detection and Characterization of Anomalous Samples

The AUC scores of CSAX, FRaC, LOF and SVMs on our compendium data sets are shown in Table 1. Each of the methods has the best performance on some data set. If we average the AUC scores over all the data sets, FRaC and CSAX are tied for the best performance. Overall, FRaC has the highest AUC (or is tied for highest AUC) on the largest number of data sets (16). Yet none of FRaC, SVMs, or LOF directly implicates specific gene sets as contributing to the identification of anomalous samples. CSAX, which was created to take advantage

---

[1] The anomaly ranking with bagging is more accurate on 20 of the 28 data sets in our compendium (data in online supplementary materials).

| | Average AUC | | | |
Task	SVM	LOF	FRaC	CSAX
atrt	**1.00**	**1.00**	**1.00**	0.99
bcat	**0.97**	0.95	**0.97**	0.87
bild	0.78	0.77	0.88	**1.00**
biomarkers	0.59	0.93	**0.95**	0.94
breast.basal	0.75	0.69	**0.76**	0.73
breast.er	0.65	0.78	**0.83**	0.82
desmoplastic	0.43	0.41	0.43	**0.53**
diabetes	**0.49**	0.45	**0.49**	0.44
downs	0.65	**0.66**	0.64	0.58
ethnic	0.59	0.65	0.66	**0.67**
gender	0.85	0.65	0.83	**0.98**
hematopoiesis	0.69	0.79	0.89	**0.92**
leukemia	**0.93**	0.88	**0.93**	0.92
lymphomas	0.59	0.84	**0.87**	0.81
meningiomas	0.55	0.61	0.65	**0.69**
meta.1.2	**1.00**	0.94	0.98	0.87
mind.body	0.43	**0.61**	0.54	0.53
multitumor	**1.00**	0.88	**1.00**	0.99
revlimid	0.64	0.47	0.56	**0.66**
ross2	0.96	0.91	**0.98**	**0.98**
ross3	0.98	**1.00**	**1.00**	**1.00**
roth07	0.63	0.59	**0.67**	0.65
sepsis	0.61	0.64	**0.68**	0.64
shakes	0.43	**0.45**	**0.45**	0.43
smokers	**0.65**	0.62	0.58	0.60
smokers2	0.55	0.63	0.72	**0.73**
survey	0.61	0.67	**0.88**	0.86
tzd	0.57	0.54	**0.60**	0.59
Best AUC	7	5	**16**	10
Avg AUC	0.699	0.751	**0.765**	**0.765**

Table 1. The average AUC over 20 replicate experiments of four anomaly detection methods on the tasks in our compendium: One-class SVMs [20], LOF [19], FRaC [18], and CSAX. A different random subset of the normal class is chosen as the training data for each replicate. "Best AUC" shows a count of the number of data sets in which the method has the highest AUC of the four (or is tied for the highest). "Average AUC" averages the AUC scores over all the data sets for that method.

of FRaC's strong performance while offering a useful functional characterization of the anomalous samples, also performs well and identifies the gene sets that are most surprisingly dysregulated. These can provide valuable information about the pathways disrupted in the anomalous samples.

For example, the "bild" data set consists of human mammary epithelial cells in which exogenous oncogenes (either *myc, ras, E2F3, β- catenin*, or *src*) are expressed. The *src* pathway was selected as the anomalous class for the compendium because it had the fewest samples. The most anomalous pathway across all of the *src* samples according to CSAX is "NCAM signaling for neurite outgrowth," with a median rank of 2, meaning that this pathway was ranked either first or second in at least half of the bagging trials. The next two top pathways, both with median rank 3, were "Signaling by FGFR" and "Downstream signaling of activated FGFR." *FGFR* signaling is mediated by *src* [25]. *NCAM* binds to *FGFR-1* and its role in cell migration depends on both *FGFR-1* and *src* activation [26], showing that *src* activation in the anomalous samples produces anomalous gene sets reflecting the direct effects of *src* expression. These low median rank scores suggest that there is remarkable consistency across the different test samples, which of course need not be the case for all envisioned applications.

As another example, the "leukemia" data set distinguishes between acute myeloid ("normal") and acute lymphoblastic leukemia ("anomalous") samples. The top gene set identified by CSAX, with a median rank of 5, is "Regulation of signaling by CBL." *CBL*, an oncogene known to be translocated or mutated in many acute myeloid leukemias, has more recently been discovered to play a broader role in many myeloid neoplasms [27].

We note that these gene sets do not indicate *differential expression* in the genes in these sets between normal and anomalous samples, in the way that typical gene set analyses do. Rather, they emphasize that *relationships* between the expression levels of genes in the indicated sets (to other genes in the genome) that are reliably established in the training samples are broken in the anomalous samples.

3.2   Characterizing Heterogeneity through Anomaly Detection

CSAX is also valuable for characterizing heterogeneity among anomalous samples. The *hematopoiesis* data set is the compendium data set with the most well-characterized heterogeneity in the anomalous class (cell lines of lymphoid origin): it includes B-cells, pre-B-cells, natural killer cells, and T-cells, all of which are collectively to be distinguished from hematopoietic cells of myeloid lineage. Figure 2A plots the first two principal components in a principal component analysis (PCA) of the normalized gene expression data from the lymphoid samples. While a subset of the T-cells stands out, the rest of the samples are rather jumbled together. Adding the third component (not shown) does little to improve these class distinctions. However, a similar plot (Figure 2B) on a matrix of the CSAX output (rows represent gene sets, columns represent samples, and the data are the median ranks of the gene sets in those samples) on the same input data does a much nicer job of separating the four sub-classes. Coincidentally, the top two principal components in both of these analyses explain approximately 55% of the variance in their data sets (54.92 and 54.88 respectively). Adding in the third principal component (Figure 2C) makes it clear that there appear to be two distinct T-cell subclasses.

Several prior methods such as PADGE [13], COPA [17,28], and GTI [12] that identify outlier expression of *genes* were intended to solve this problem of finding subgroups of samples grouped by which genes show unusual expression. However, CSAX may be more powerful at characterizing heterogeneity because it relies not on unusual expression of any particular genes, but on detecting a breakdown in the expected *relationships* in gene expression in a single sample, and identifying pathways enriched for such broken relationships. Accordingly, even if no two samples in a data set have the same dysregulated genes, CSAX may be able to find common patterns, allowing characterization of the heterogeneous anomalies into functionally-related subclasses.

3.3   How Hard Is an Anomaly Detection Task?

The variation in performance across the compendium seems to depend strongly on characteristics of the individual data sets. For example, on the "leukemia" data, where differential expression is known to be substantial across a large number of genes, all four methods perform quite well, while on the "diabetes"

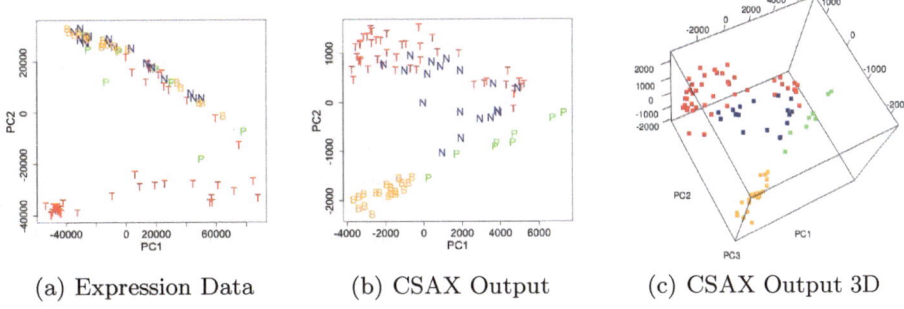

(a) Expression Data (b) CSAX Output (c) CSAX Output 3D

Fig. 2. Using anomaly detection to characterize heterogeneity. **a:** Plot of the first two principal components of the gene expression data for all anomalous samples in the *hematopoiesis* data set. Red "T"s correspond to T-cells, gold "B"s to B-cells, green "P"s to pre-B-cells, and blue "N"s to natural killer cells. **b:** Plot of the first two principal components of the results of CSAX on the hematopoiesis data set. Labels are the same as in part **a. c:** Adding the third principal component to the results from part **b**, in a three-dimensional plot, shows the clear separation of the different types of samples in the CSAX results. Furthermore, two distinct subgroups of T-cells emerge. Colors are the same as in parts **a** and **b**.

data set, which is known to have only subtle expression differences between the normal and anomalous classes [23], all the methods perform poorly. We would like to be able to predict what sorts of anomalous expression patterns should be detectable by these methods. Accordingly, we sought to characterize the difficulty of each of the compendium data sets.

We found no reliable way to characterize the difficulty of a data set using only the training data. Given our compendium of gold-standard data, however, we can still learn about characteristics of solvable problems using what we know about the test data sets. We can then apply this information to help us predict the utility of anomaly detection in applications where we don't know the right answer.

We discovered that the ratio between the median distance separating the training data from an anomalous example and the median distance between the training data and a test-set normal example is an excellent predictor of the eventual performance (measured by AUC) of an anomaly detector, regardless of which anomaly detector we use. We refer to this measure as the *relative anomaly aggregate distance* (RAAD).

We define *RAAD* as:

$$RAAD = \frac{\underset{x \in \mathcal{X}, q \in \mathcal{Q}^a}{\text{median}} |x - q|}{\underset{x \in \mathcal{X}, q \in \mathcal{Q}^n}{\text{median}} |x - q|} \tag{1}$$

where $\mathcal{X}$ is the training set, $\mathcal{Q}^a$ are the test set anomalous instances, $\mathcal{Q}^n$ are the test set normal instances, and $|x - q|$ indicates the vector distance between a training and test set instance (*i.e.,* each gene's expression is one component of

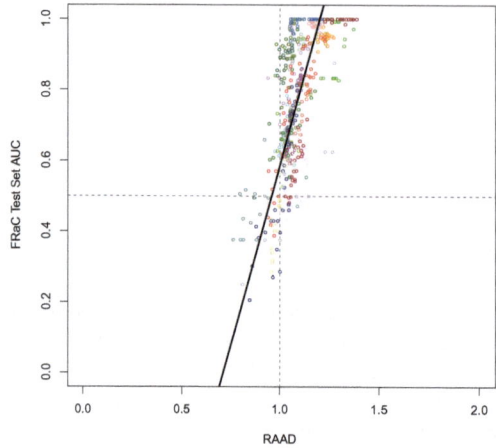

Fig. 3. A scatterplot comparing *RAAD* with test set performance over the 28 compendium data sets (each shown in its own color). Points show 20 replicates (random selection of training/test samples) for each data set. Horizontal and vertical lines show an AUC of 0.5 (random guessing) and a *RAAD* of 1.0 (normal and anomalous instances equidistant), respectively. (Scatter plots showing individual data sets are available in supplementary material.)

a high-dimensional vector). We use the l_1 norm ("Manhattan" distance) when computing vector distance because it is intuitive–the total distance is the sum of differences in gene expression between two microarrays.

The scatterplot in Figure 3 shows the relationship between *RAAD* and performance using FRaC.

In real applications, where the test data labels are unknown, clinicians' intuition about the degree of expression variation one might expect among the normal class and among the types of envisioned anomalies can be used to estimate whether anomaly detection methods are likely to be helpful. Further, a small test set with some heterogeneous anomalies can be used to estimate the RAAD score, helping to determine the value of obtaining and analyzing additional samples.

4   Discussion

We have shown that it is often possible to detect and characterize anomalous expression data given training data from normal samples only, and that the two methods designed with expression data in mind perform best, albeit with different strengths. FRaC learns reliable relationships between genes' expression patterns from the training data, and identifies anomalies when these patterns break down. This method is therefore entirely data-driven; it does not rely on prior knowledge about gene sets or relationships. Yet it makes sense that many clinically-important anomalies would be characterized by a breakdown in the expected relationships between genes' expression patterns. So it is perhaps not surprising that FRaC is particularly effective on these large data sets.

On the average, CSAX is about as accurate as FRaC (Table 1), but it has two very important differences: (i) it identifies gene sets that may help to explain the nature of the anomaly, and (ii) it uses fewer gene expression features in its predictions (because of the discount in Equation 3 in the Appendix), instead gaining its power from using known gene sets to integrate prior knowledge about gene relationships into the anomaly detection process.

The identification of gene sets with known functional roles associated with an anomaly is one of the primary goals of our work. It provides insight into the nature of the anomaly, and allows experts to follow up by ordering relevant tests. The fact that the performance of CSAX is comparable to that of FRaC while using fewer genes is evidence that the gene sets identified by our method are indeed relevant, because in general the use of fewer features hurts performance.

We also observed that performance depends less on the computational method used than on the difficulty of the data set itself. While the RAAD score is useful for characterizing the difficulty of anomalies whose classification is already known, in most cases such data will not be available. Thus, clinical intuition about the nature of anticipated anomalies will need to come into play. If the anomalous samples are likely to be no more different from the normal samples than the normal samples are from each other, no method is likely to succeed. Prior knowledge about expression variability and heterogeneity of the samples under consideration is expected to be helpful here.

This paradigm is also particularly useful for handling heterogeneity in expression data. If the anomalous samples are sufficiently heterogeneous, traditional classification methods are likely to struggle. The anomaly detection paradigm, however, can provide information about the particular characteristics of each sample, allowing clustering of the samples into classes. Potential applications include diseases such as lung cancer that are suspected to represent heterogeneous collections of related disorders, but where traditional approaches to molecular classification have not been sufficiently effective. Demonstrating the strength of CSAX for this application will be an important next step.

We note that as the cost of sequencing continues to decrease, genome-wide studies of expression will continue shifting away from microarrays towards RNA-seq approaches. However, there is no reason that CSAX cannot be applied in an RNA-seq setting. Indeed, it might be even more powerful, as changes in the relative abundance of different isoforms can be integrated into the analysis. Future work should therefore include demonstrating CSAX's power to identify systematic anomalies in RNA-seq data.

Overall, we have demonstrated that in many biomedically-interesting cases, it is indeed possible to identify and characterize individual anomalous samples from their expression patterns. Such one-of-a-kind analyses are crucial steps as we move towards the era of personalized medicine.

Acknowledgments. This research was supported by award number R01-HD-058880 from the Eunice Kennedy Shriver National Institute of Child Health and Human Development. The content is solely the responsibility of the authors and does not necessarily reflect the views of the NICHD or the National Institutes of Health. Additional support was provided by a Senior Faculty Research Fellowship from Tufts University. The authors thank Pablo Tamayo, George

Steinhardt, and Arthur Liberzon of the Broad Institue for access to their expression classification benchmark data, which led to many of the compendium data sets, and Stan Letovsky and Jill Mesirov for their feedback on earlier versions of this work.

Availability. Details, references, data, and source code are available at http://bcb.cs.tufts.edu/csax. FRaC source code is available at http://bcb.cs.tufts.edu/frac/.

References

1. Lockhart, D., Dong, H., Byrne, M., Follettie, M., Gallo, M., Chee, M., Mittmann, M., Wang, C., Kobayashi, M., Horton, H., Brown, E.: Expression monitoring by hybridization to high-density oligonucleotide arrays. Nature Biotech. 14, 1675–1680 (1996)
2. Shalon, D., Smith, S., Brown, P.: A DNA micro-array system for analyzing complex DNA samples using two-color fluorescent probe hybridization. Gen. Res. 6, 639–645 (1996)
3. Mehta, R., Jain, R., Badve, S.: Personalized medicine: the road ahead. Clin. Breast Cancer 11(1), 20–26 (2011)
4. Glas, A.M., Floore, A., Delahaye, L.J., Witteveen, A.T., Pover, R.C., Bakx, N., Lahti-Domenici, J.S., Bruinsma, T.J., Warmoes, M.O., Bernards, R., Wessels, L.F., Van't Veer, L.J.: Converting a breast cancer microarray signature into a high-throughput diagnostic test. BMC Genomics 7, 278 (2006)
5. Slonim, D.: From patterns to pathways: gene expression data analysis comes of age. Nature Genetics 32(suppl.), 502–508 (2002)
6. Tusher, V., Tibshirani, R., Chu, G.: Significance analysis of microarrays applied to the ionizing radiation response. PNAS 98(9), 5116–5121 (2001)
7. Dougherty, E.: Small sample issues for microarray-based classification. Comp. Funct. Genomics 2(1), 28–34 (2001)
8. Chandola, V., Banerjee, A., Kumar, V.: Anomaly detection: A survey. ACM Comput. Surv. 41(3), 15:1–15:58 (2009)
9. Mikkelsen, T., Galagan, J., Mesirov, J.: Improving genome annotations using phylogenetic profile anomaly detection. Bioinformatics 21(4), 464–470 (2005)
10. Kim, H., Gelenbe, E.: Anomaly detection in gene expression via stochastic models of gene regulatory networks. BMC Genomics 10(S3), S26 (2009)
11. Torkamani, A., Schork, N.: Prestige centrality-based functional outlier detection in gene expression analysis. Bioinformatics 25(17), 2222–2228 (2009)
12. Mpindi, J.P., Sara, H., Haapa-Paananen, S., Kilpinen, S., Pisto, T., Bucher, E., Ojala, K., Iljin, K., Vainio, P., Bjorkman, M., Gupta, S., Kohonen, P., Nees, M., Kallioniemi, O.: GTI: a novel algorithm for identifying outlier gene expression profiles from integrated microarray datasets. PLoS One 6(2), e17259 (2011)
13. Li, L., Chaudhuri, A., Chant, J., Tang, Z.: PADGE: analysis of heterogeneous patterns of differential gene expression. Physiol. Genomics 32(1), 154–159 (2007)
14. Ghosh, D.: Discrete nonparametric algorithms for outlier detection with genomic data. J. Biopharm. Stat. 20(2), 193–208 (2010)
15. Karrila, S., Lee, J., Tucker-Kellogg, G.: A comparison of methods for data-driven cancer outlier discovery, and an application scheme to semisupervised predictive biomarker discovery. Cancer Inform. 10, 109–120 (2011)

16. Sauer, U., Preininger, C., Hany-Schmatzberger, R.: Quick and simple: quality control of microarray data. Bioinformatics 21, 1572–1578 (2005)
17. Tomlins, S., Rhodes, D., Perner, S., Dhanasekaran, S., Mehra, R., Sun, X., Varambally, S., Cao, X., Tchinda, J., Kuefer, R., Lee, C., Montie, J., Shah, R., Pienta, K., Rubin, M., Chinnaiyan, A.: Recurrent fusion of TMPRSS2 and ETS transcription factor genes in prostate cancer. Science 310, 644–648 (2005)
18. Noto, K., Brodley, C., Slonim, D.: FRaC: A feature-modeling approach for semi-supervised and unsupervised anomaly detection. Data Mining and Knowledge Discovery 25, 109–133 (2011)
19. Breunig, M., Kriegel, H., Ng, R., Sander, J.: LOF: identifying density-based local outliers. ACM SIGMOD Record 29(2), 93–104 (2000)
20. Schölkopf, B., Smola, A.J., Williamson, R.C., Bartlett, P.L.: New support vector algorithms. Neural Computation 12(5), 1207–1245 (2000)
21. Tribus, M.: Thermodynamics and Thermostatics: An Introduction to Energy, Information and States of Matter, with Engineering Applications. D. Van Nostrand Company Inc., New York (1961)
22. Subramanian, A., Tamayo, P., Mootha, V.K., Mukherjee, S., Ebert, B., Gillette, M., Paulovich, A., Pomeroy, S., Golub, T., Lander, E., Mesirov, J.: Gene set enrichment analysis: A knowledge-based approach for interpreting genome-wide expression profiles. Proceedings of the National Academy of Sciences 102(43), 15545–15550 (2005)
23. Mootha, V., Lindgren, C., Eriksson, K.-F., Subramanian, A., Sihag, S., Lehar, J., Puigserver, P., Carlsson, E., Ridderstråle, M., Laurila, E., Houstis, N., Daly, M., Patterson, N., Mesirov, J., Golub, T.R., Tamayo, P., Spiegelman, B., Lander, E.S., Hirschhorn, J.N., Altshuler, D., Groop, L.C.: PGC-1alpha-responsive genes involved in oxidative phosphorylation are coordinately downregulated in human diabetes. Nature Genetics 34(3), 267–273 (2003)
24. Spackman, K.A.: Signal detection theory: Valuable tools for evaluating inductive learning. In: Proceedings of the Sixth International Workshop on Machine Learning, pp. 160–163. Morgan Kaufmann Publishers Inc., San Francisco (1989)
25. Sandilands, E., Akbarzadeh, S., Vecchione, A., McEwan, D., Frame, M., Heath, J.: Src kinase modulates the activation, transport and signalling dynamics of fibroblast growth factor receptors. EMBO Reports 8, 1162–1169 (2007)
26. Francavilla, C., Cattaneo, P., Berezin, V., Bock, E., Ami, D., de Marco, A., Chrisofori, G., Cavallaro, U.: The binding of ncam to fgfr1 induces a specific cellular response mediated by receptor trafficking. J. Cell. Biol. 187(7), 1101 (2009)
27. Kales, S., Ryan, P., Nau, M., Lipkowitz, S.: Cbl and human myeloid neoplasms: the Cbl oncogene comes of age. Cancer Res. 70(12), 4789–4794 (2010)
28. MacDonald, J.W., Ghosh, D.: COPA–cancer outlier profile analysis. Bioinformatics 22(23), 2950–2951 (2006)
29. Noto, K., Brodley, C., Slonim, D.: Anomaly detection using an ensemble of feature models. In: Proceedings of the 10th IEEE International Conference on Data Mining (ICDM 2010). IEEE Computer Society Press (2010)
30. Croft, D., O'Kelly, G., Wu, G., Haw, R., Gillespie, M., Matthews, L., Caudy, M., Garapati, P., Gopinath, G., Jassal, B., Jupe, S., Kalatskaya, I., Mahajan, S., May, B., Ndegwa, N., Schmidt, E., Shamovsky, V., Yung, C., Birney, E., Hermjakob, H., D'Eustachio, P., Stein, L.: Reactome: a database of reactions, pathways and biological processes. Nucleic Acids Research 39, D691–D697 (2011)
31. Chang, C.-C., Lin, C.-J.: LIBSVM: A library for support vector machines (2001) Software available at, http://www.csie.ntu.edu.tw/~cjlin/libsvm

Appendix: Methods Details

FRaC

In 2010, we developed *feature regression and classification* (FRaC [18,29]). Given a training set and an unlabeled example, the steps involved in FRaC are for all genes i:

1. Infer a predictive model C_i of the expression of gene i from some of the training data. The model will use the expression of some of the other genes to make its predictions. For this step, we use an ϵ-SVR (support vector regression) model with a linear kernel, the ϵ parameter (in the loss function) set to zero, and the C parameter (for regularization) set to 1.0. Preliminary experiments with microarray anomaly detection showed that FRaC is not very sensitive to these choices, and these settings prove to work well (data not shown).
2. Use held-aside training data (*i.e.*, not used in the previous step) to estimate the accuracy of the model; *e.g.*, to build a model E_i of the predictive error. We use leave-one-out cross-validation to sample the predictive error and we model E_i as a normal distribution $\mathcal{N}(\mu, \sigma)$, where μ and σ are set to the sample mean and standard deviation, respectively.
3. Use the predictive model C_i to predict the expression of gene i in the unlabeled example.
4. Compute the likelihood of the error of the prediction using the error model E_i.
5. The anomaly score we assign to gene i is the log loss, or *surprisal*, of the likelihood computed in the previous step.

The input to FRaC is a training set and a test set. The output of FRaC is a table of anomaly scores, one for each gene and for each test instance. When running FRaC by itself, the anomaly score is computed as the sum of the anomaly scores for each gene. When running FRaC for the purpose of generating input data for GSEA (*i.e.*, CSAX), aggregating gene anomaly scores is not necessary. In either case, we follow the method outlined above using our own implementation. Source code and documentation are available at `bcb.cs.tufts.edu/frac/`.

CSAX

CSAX uses FRaC (as described in the previous section) to compute an anomaly score for each gene, and uses GSEA to see which gene sets are associated with genes whose expression is particularly surprising.

The input to GSEA is (i) a list of genes, ranked by some measure of differential expression between two classes, and (ii) a collection of gene sets. The output of GSEA is a list of the gene sets, ranked by their "enrichment," a measure of their significance in distinguishing between the two classes of microarray. GSEA is implemented in Java and available at `http://www.broadinstitute.org/gsea/downloads.jsp`. We use the Java archive (gsea2-2.07.jar) and run the "pre-ranked" version of GSEA (`xtools.gsea.GseaPreranked`) on gene sets with at least seven and at most 500 genes, with 1000 permutations and a weighted scoring scheme (see GSEA documentation). The output of GSEA is a table, listing each gene set, its *enrichment score* (see [22, 23] for details), and other statistics, including a normalized version of the enrichment score that considers the size of each gene set.

This approach has the important advantage of identifying the gene sets that may best explain an anomaly, which is one of the primary goals of our research. However, applying this method to test set microarrays that come from the normal class will also identify gene sets that that are statistically enriched, even though these sets are effectively random and depend on how accurately the training set represents the true distribution of the normal class. Specifically, when the training set is too small to capture the full diversity of the normal sample space, there will be false positive results. For the envisioned applications, we need to better distinguish the results characterizing normal test samples from those characterizing anomalies.

We therefore use bagging to address this effect: over multiple iterations, we take a random subset of the training set and run FRaC and GSEA on it. This process produces multiple GSEA enrichment rankings for each gene set. The gene sets that best explain a true difference between an unlabeled microarray and the training set will appear at or near the top of GSEA's ranked list over multiple iterations of bagging, whereas the gene sets that are only enriched because their genes are misrepresented in the training set (because of the small sample size) are less likely to do so.

The challenge that remains is how to select the informative gene sets from the GSEA output lists, and how to combine their enrichment into a single anomaly score. A single gene set may not be enough by itself to fully characterize an anomaly, so we must consider multiple gene sets, but we only want the most informative ones—those that are ranked highly in GSEA's output tables. The method that we use is to first look at the collection of rankings for each gene set and to compute its median. For example, if a gene set appears in the #1 position more often than not, its median rank will be 1.

Formally, let $\mathcal{G}$ be our collection of G gene sets, and B be the number of bagging iterations. Let $r_b(\mathbf{g})$ be the ranking (*i.e.*, $1, 2, ...$) of gene set $\mathbf{g}$ in the bth iteration of bagging, when ordered by enrichment score. Let $V(\mathbf{g})$ be the median of all the rankings of gene set $\mathbf{g}$, *i.e.*,

$$V(\mathbf{g}) = \underset{b \in 1...B}{\text{median}} [r_b(\mathbf{g})]. \tag{2}$$

We consider the gene sets with the best median rankings (lowest values of V) to be the most informative ones. Let $\mathcal{M}$ be the G gene sets in $\mathcal{G}$, ordered by their median ranking V, *i.e.*, $\mathcal{M}_1$ is the gene set g with the lowest $V(\mathbf{g})$, and $V(\mathcal{M}_i) \leq V(\mathcal{M}_{i+1})$ for all $i \in [1, 2, ..., G-1]$.

We also run FRaC and GSEA on the entire training set (as opposed to the iterations of bagging). Let $ES(\mathbf{g})$ be the enrichment score calculated by GSEA for gene set $\mathbf{g}$ on the full training set. We define $ES(\mathbf{g}) = 0$ for any gene set $\mathbf{g}$ that does not appear in the table. Note that GSEA normalizes its enrichment score for gene set size. This normalization depends on the enrichment scores and is helpful when comparing against other gene sets on the same microarray, but is not helpful when comparing across microarrays. Therefore, we use normalized enrichment score for ranking, but ES refers to the raw enrichment score.

To compute an anomaly score, we combine the enrichment scores of each gene set, discounted by their position in $\mathcal{M}$ as

$$\text{anomaly score} = \sum_{i=1}^{G} \gamma^{i-1} \times ES(\mathcal{M}_i). \tag{3}$$

The single parameter γ controls how many of the highest-ranking gene sets are included in the computation of the anomaly score, and, by extension, how many genes influence the predictions.

In our experiments, we set $\gamma = 0.95$. Overall, we observe similar performance for different values of γ (results in supplementary materials online). In our experiments, we perform $B = 40$ iterations of bagging and use the $G = 1,079$ Reactome pathways gene sets [30]. Source code and documentation for CSAX can be found at http://bcb.cs.tufts.edu/csax/.

One-class Support Vector Machines

To compare our approach to one-class support vector machines [20], we use the LIBSVM [31] implementation with default settings. Preliminary investigation showed the AUC scores are not sensitive to a wide range of parameter settings (data not shown).

LOF

To compare our approach to LOF [19], we use our own implementation. LOF requires the specification of a single parameter, MinPts, which is the size of the neighborhood of microarrays. Following a suggestion in the original presentation of LOF [19], we compute the LOF using all possible values of MinPts, and take the maximum LOF. Source code and documentation for our implementation can be found at http://bcb.cs.tufts.edu/csax/.

WhatsHap: Haplotype Assembly for Future-Generation Sequencing Reads

Murray Patterson[1,2,*], Tobias Marschall[1,*], Nadia Pisanti[3,4], Leo van Iersel[1], Leen Stougie[1,5], Gunnar W. Klau[1,5,**], and Alexander Schönhuth[1,**]

[1] Life Sciences, CWI Amsterdam, The Netherlands
{m.d.patterson,t.marschall,gunnar.klau,a.schoenhuth}@cwi.nl
[2] LBBE, CNRS and Université de Lyon 1, Villeurbanne, France
[3] Department of Computer Science, University of Pisa, Italy
[4] LIACS, Leiden University, The Netherlands
[5] VU University Amsterdam, The Netherlands

Abstract. The human genome is diploid, that is each of its chromosomes comes in two copies. This requires to *phase* the *single nucleotide polymorphisms (SNPs)*, that is, to assign them to the two copies, beyond just detecting them. The resulting haplotypes, lists of SNPs belonging to each copy, are crucial for downstream analyses in population genetics. Currently, statistical approaches, which avoid making use of direct read information, constitute the state-of-the-art. *Haplotype assembly*, which addresses phasing directly from sequencing reads, suffers from the fact that sequencing reads of the current generation are too short to serve the purposes of genome-wide phasing.

Future sequencing technologies, however, bear the promise to generate reads of lengths and error rates that allow to bridge all SNP positions in the genome at sufficient amounts of SNPs per read. Existing haplotype assembly approaches, however, profit precisely, in terms of computational complexity, from the limited length of current-generation reads, because their runtime is usually exponential in the number of SNPs per sequencing read. This implies that such approaches will not be able to exploit the benefits of long enough, future-generation reads.

Here, we suggest WhatsHap, a novel dynamic programming approach to haplotype assembly. It is the first approach that yields provably optimal solutions to the *weighted minimum error correction (wMEC)* problem in runtime linear in the number of SNPs per sequencing read, making it suitable for future-generation reads. WhatsHap is a *fixed parameter tractable (FPT)* approach with coverage as the parameter. We demonstrate that WhatsHap can handle datasets of coverage up to 20x, processing chromosomes on standard workstations in only 1-2 hours. Our simulation study shows that the quality of haplotypes assembled by WhatsHap significantly improves with increasing read length, both in terms of genome coverage as well as in terms of switch errors. The switch error rates we achieve in our simulations are superior to those obtained by state-of-the-art statistical phasers.

---

[*] Joint first authorship.
[**] Joint last authorship.

R. Sharan (Ed.): RECOMB 2014, LNBI 8394, pp. 237–249, 2014.
© Springer International Publishing Switzerland 2014

1   Introduction

The human genome is *diploid*, that is, each of its chromosomes comes in two copies (except for sex chromosomes in males), one from the mother and one from the father. These parental copies are affected by different *single nucleotide polymorphisms (SNPs)*, and assigning the variants to the copies is an important step towards the full characterization of an individual genome. The corresponding assignment process is referred to as *phasing* and the resulting groups of SNPs are called *haplotypes*. Phasing SNPs in population studies allows to, for example, identify selective pressures and subpopulations, and to link possibly disease-causing SNPs with one another [13]. This explains that phasing SNPs has been an instrumental step in many human whole-genome projects [5,28]. In the meantime, globally concerted efforts have generated *reference panels* of haplotypes, for various populations, which may serve corresponding downstream analyses [29,30].

There are two major approaches to phasing variants. The first class of approaches relies on *genotypes* as input, which are lists of SNP alleles, together with their zygosity status. While *homozygous* alleles show on both chromosomal copies, and obviously apply for both haplotypes, *heterozygous* alleles show on only one of the copies, and have to be partitioned into two groups. If m is the number of heterozygous SNP positions, there are 2^m many possible haplotypes. This illustrates that directly phasing from genotype data is a hard computational problem. The corresponding approaches are usually statistical in nature, and they integrate existing reference panels. The underlying assumption is that the haplotypes to be computed are a mosaic of reference haplotype blocks that arises from recombination during meiosis. The output is the statistically most likely mosaic, given the observed genotypes. Most prevalent approaches are based on latent variable modeling [17,21,26]. Other approaches use Markov chain Monte Carlo techniques [23].

The other class of approaches makes direct usage of sequencing read data. Such approaches virtually assemble reads from identical chromosomal copies and are referred to as *haplotype assembly* approaches. Following the parsimony principle, the goal is to compute two haplotypes to which one can assign all reads with the least amount of sequencing errors to be corrected and/or erroneous reads to be removed. Among such formulations, the *minimum error correction (MEC)* problem has gained most of the recent attention. The MEC problem, which we will formally define in Section 2, consists of finding the minimum number of corrections to the SNP values to be made to the input in order to be able to arrange the reads into two haplotypes without conflicts. A major advantage of MEC is that it can be easily adapted to a weighted version (wMEC), in order to deal with phred-based error probabilities. Such error schemes are common in particular for *next-generation sequencing (NGS)* data. An optimal solution for the wMEC problem then translates to a maximum likelihood scenario relative to the errors to be corrected.

In tera-scale sequencing projects, e.g., [5,28], ever increasing read length and decreasing sequencing cost make it clearly desirable to phase directly from read data. However, statistical approaches are still the methodology of choice because: (i) most NGS reads are still too short to bridge so-called *variant deserts*. Successful read-based phasing, however, requires that all pairs of neighboring heterozygous SNP alleles are covered; and (ii) the MEC problem is $\mathcal{NP}$-hard, and so are all other similar problem formulations.

Most advanced existing algorithmic solutions to MEC [6,16] take time exponential in the number of variants per read, and, ironically, often benefit precisely from variant deserts, because these allow to decompose a problem instance into independent parts. A major motivation behind read-based approaches, however, is to handle long reads that cover as many variants as possible, thereby bridging all variant deserts. Hence, the current perception of haplotype assembly is often that it underlies theoretical limitations that are too hard to overcome.

Here, we present a *fixed parameter tractable (FPT)* approach to wMEC where coverage, that is the number of fragments that cover a SNP position, is the only parameter. Hence, the runtime of our approach is, for the first time, *polynomial (in fact: linear) in the number of SNPs per read*, which addresses the future sequencing technologies that will generate reads of several tens of thousands of base pairs (bp) in length, and that the currently existing approaches are not suitable for processing such data. A carefully engineered implementation of our algorithm allows the treatment of whole-genome datasets of maximum coverage up to 20x on the order of hours on a standard workstation. For datasets of higher coverage, we provide a technique for choosing a reasonable selection of reads. We demonstrate that the resulting haplotypes suffer from only minor amounts of errors, even on high-coverage datasets, while we provide a provably optimal solution to the wMEC problem on bounded-coverage datasets. To do so, we test against a long-read benchmark dataset that we produced. Such a dataset will be useful for future tools that leverage long reads.

2   The Minimum Error Correction (MEC) Problem

The input to the MEC problem is a matrix $\mathcal{F}$ with entries in $\{0, 1, -\}$. Each row of $\mathcal{F}$ corresponds to a fragment/read. Each column of $\mathcal{F}$ corresponds to a SNP position. The "$-$" symbol, which is referred to as a *hole*, is used when a fragment does not contain any information at the corresponding SNP position. This can be either because the SNP position is not covered by the read, or because the read gives no accurate information at that position. Let n be the number of rows (or fragments) of $\mathcal{F}$ and m the number of columns (or SNP positions).

A *haplotype* can formally be defined as a string of length m consisting of 0's and 1's. If h is a haplotype, then the i-th row of $\mathcal{F}$ is said to *conflict* with h if there is some SNP position j for which $h(j) \neq \mathcal{F}(i, j)$ while $\mathcal{F}(i, j) \neq -$. We say that $\mathcal{F}$ is *conflict free* if there exist two haplotypes h_1, h_2 such that each row of $\mathcal{F}$ does not conflict with at least one of h_1 and h_2. Under the *all-heterozygous assumption*, where all columns correspond to heterozygous sites, h_1 must be the complement of h_2.

The goal of MEC is to make $\mathcal{F}$ conflict free by flipping a minimum number of entries of $\mathcal{F}$ from 0 to 1 or vice versa. The weighted variant of MEC, denoted wMEC, has an additional weight function w as input. This weight function assigns a non-negative weight $w(i, j)$ to each entry $\mathcal{F}(i, j)$ of $\mathcal{F}$. This weight can reflect the relative confidence that the entry is correctly sequenced. The goal of wMEC is to make $\mathcal{F}$ conflict free by flipping entries in $\mathcal{F}$ with a minimum total weight.

The MEC problem, which is also called *minimum letter flip*, was introduced by Lippert *et al.* [22]. Cilibrasi *et al.* [7] showed that this problem is NP-hard even if each fragment is "gapless", i.e., if it consists of a consecutive sequence of 0's and 1's with holes to the left and to the right. Panconesi and Sozio [25] were the first to propose a practical heuristic for solving MEC. An exact branch and bound algorithm and a heuristic genetic algorithm were presented by Wang *et al.* [31]. Levy *et al.* [19] designed a greedy heuristic to assemble the haplotype of the genome of J. Craig Venter. Bansal *et al.* [4] developed an MCMC method to sample a set of likely haplotypes. In a follow-up, some of the authors proposed a much faster MAX-CUT-based heuristic algorithm called HAPCUT [3], which they show to outperform [25,19], while showing similar accuracy to [4] in shorter running time. Very recently, Selvaraj *et al.* [27] combine the HAPCUT [3] algorithm with proximity-ligation, which exploits information from "chromosome territories", to develop a method which reports good results on whole-genome haplotype reconstruction. In another recent paper, He *et al.* [16] proposed an exact dynamic programming algorithm. However, their algorithm depends exponentially on the length of the longest read, which means that for practical data this method has to ignore all long reads.

The weighted variant of MEC was first suggested by Greenberg *et al.* [12]. Zhao *et al.* [34] propose a heuristic for a special case of wMEC and present experiments showing that wMEC is more accurate than MEC.

More recently, in 2012, Aguiar and Istrail [2,1] propose a different heuristic approach for MEC which they show to perform well compared to previous methods. Exact *integer linear programming (ILP)* based approaches were also proposed very recently by Fouilhoux and Mahjoub [11] and Chen *et al.* [6]. Both methods have difficulties solving practical instances optimally. For this reason, Chen *et al.* also propose a heuristic for solving difficult subproblems.

3   A Dynamic Programming Algorithm for wMEC

We now present the WHATSHAP algorithm for solving wMEC. WHATSHAP is an exact dynamic programming approach that solves wMEC instances in linear time if we assume bounded coverage.

Consider the input matrix $\mathcal{F}$ of the wMEC problem. Each entry $\mathcal{F}(i, j) \neq -$ is associated with a confidence degree $w(i, j)$ telling how likely it is that $\mathcal{F}(i, j)$ is correctly sequenced and that its fragment i is correctly mapped to location j. We use such values as a weight for the correction we need to minimize in the wMEC model. When these weights are log-likelihoods, summing them up corresponds to multiplying probabilities and, thus, finding a minimum weight solution corresponds to finding a maximum likelihood bipartition of the reads/fragments.

Our *dynamic programming (DP)* formulation is based on the observation that, for each column, only *active* fragments need to be considered; a fragment i is said to be active in every column j that lies in between its leftmost non-hole entry and its rightmost non-hole entry. Thus, paired-end reads remain active in the "internal segment" between the two reads. Let $F(j)$ be the set of fragments that are active at SNP position j and let F be the set of all fragments. The aim is to find a bipartition (R^*, S^*) of F such that the changes in R^* and S^* to make $\mathcal{F}$ conflict free have minimum total weight.

Proceeding columnwise from SNP position 1 to m, our approach computes a DP table column $C(j, \cdot)$ for each $j \in \{1, \dots, m\}$. We say that a bipartition $B' = (R', S')$ of all fragments F *extends* bipartition $B = (R, S)$ of $F(j)$, if $R \subseteq R'$ and $S \subseteq S'$. We define $\mathcal{B}(X)$ to be the set of all bipartitions of X. Given a bipartition (R, S), we denote $\mathcal{B}(X \mid (R, S))$ the set of all bipartitions of X that extend (R, S), that is,

$$\mathcal{B}(X \mid (R, S)) := \left\{ (R', S') \in \mathcal{B}(X) \mid R \subseteq R' \text{ and } S \subseteq S' \right\} .$$

The basic idea of our dynamic program is as follows: for every bipartition $B = (R, S)$ of $F(j)$, entry $C(j, B)$ gives the minimum cost of a bipartition of all fragments F that renders positions $1, \dots, j$ conflict free *and* which extends B. By definition of $C(j, B)$, the cost of an optimal solution to the wMEC problem then equals $\min_{B \in F(m)} C(m, B)$. An optimal bipartition of the fragments can be obtained by backtracking along the columns of the DP table up to the first SNP position in $\mathcal{F}$.

To compute the contribution $\Delta_C(j, (R, S))$ of column j to the cost $C(j, (R, S))$ of bipartition (R, S), we define the following quantities.

Definition 1. *For a position j and a set R of fragment indices in $F(j)$, let $W^0(j, R)$ (resp. $W^1(j, R)$) denote the cost of setting position j on all fragments of R to 0 (resp. 1), flipping if required: i.e.,*

$$W^0(j, R) = \sum_{\substack{i \in R \\ \mathcal{F}(i, j) = 1}} w(i, j) \quad \text{and} \quad W^1(j, R) = \sum_{\substack{i \in R \\ \mathcal{F}(i, j) = 0}} w(i, j) .$$

Hence, given a bipartition (R, S) of $F(j)$, the minimum cost to make position j conflict free is

$$\Delta_C(j, (R, S)) := \min\{W^0(j, R), W^1(j, R)\} + \min\{W^0(j, S), W^1(j, S)\} .$$

Notice that, under the *all heterozygous assumption*, where one wants to enforce all SNPs to be heterozygous, the equation becomes

$$\Delta_C(j, (R, S)) := \min\{W^0(j, R) + W^1(j, S), W^1(j, R) + W^0(j, S)\} .$$

In both cases, we only need the four values $W^0(j, R)$, $W^1(j, R)$, $W^0(j, S)$, and $W^1(j, S)$ to compute $\Delta_C(j, (R, S))$. We now proceed to state in detail our DP formulation.

	1	2
0	0_5	– ···
1	1_3	0_2 ···
2	1_6	1_1 ···
3	–	0_2 ···

Fig. 1. WHATSHAP toy example. The small numbers next to the matrix of entries $\mathcal{F}$ denote the flipping weights.

Initialization. The first column $C(1, \cdot)$ of C is initialized to $\Delta_C(1, \cdot)$ as defined above.

Example. Assume $F(1) = \{f_0, f_1, f_2\}$ with $\mathcal{F}(0, 1) = 0, \mathcal{F}(1, 1) = 1,$ and $\mathcal{F}(2, 1) = 1$. Moreover, let $w(0, 1) = 5, w(1, 1) = 3,$ and $w(2, 1) = 6$. See Figure 1. Then $C(1, \cdot)$ is filled in as follows:

$$C\big(1, (\{f_0, f_1, f_2\}, \emptyset)\big) = \min\{9, 5\} + \min\{0, 0\} = 5$$
$$C\big(1, (\{f_0, f_1\}, \{f_2\})\big) = \min\{3, 5\} + \min\{6, 0\} = 3$$
$$C\big(1, (\{f_0, f_2\}, \{f_1\})\big) = \min\{6, 5\} + \min\{3, 0\} = 5$$
$$C\big(1, (\{f_1, f_2\}, \{f_0\})\big) = \min\{9, 0\} + \min\{0, 5\} = 0$$

Note that we need consider only half of the $2^{|F(1)|}$ bipartitions, because $C(j, (R, S)) = C(j, (S, R))$ for every bipartition $B = (R, S)$ and every SNP position j.

Recurrence. We compute $C(j + 1, \cdot)$ from $C(j, \cdot)$ as follows. When computing costs of bipartitions for $F(j + 1)$ we need only to keep track of the effect that this has on the bipartition of $F(j)$ through their intersection, which we denote by $F_{j+1}^{\cap} = F(j) \cap F(j + 1)$. For a bipartition (R, S) of $F(j + 1)$ we define $R_{j+1}^{\cap} = R \cap F_{j+1}^{\cap}$ and $S_{j+1}^{\cap} = S \cap F_{j+1}^{\cap}$. The recursion then becomes:

$$C(j + 1, (R, S)) = \Delta_C(j + 1, (R, S)) + \min_{B \in \mathcal{B}(F(j) \mid (R_{j+1}^{\cap}, S_{j+1}^{\cap}))} C(j, B) . \quad (1)$$

The first term accounts for the cost of the current SNP position, while the second term accounts for costs incurred at previous SNP positions. The minimum selects the best score with respect to the first j positions over all partitions that extend (R, S).

Example (continued). We extend the example with a second SNP position. Assume $F(2) = \{f_1, f_2, f_3\}$ with $\mathcal{F}(1, 2) = 0, \mathcal{F}(2, 2) = 1,$ and $\mathcal{F}(3, 1) = 0$. Moreover, let $w(1, 2) = 2, w(2, 2) = 1,$ and $w(3, 1) = 2$. See Figure 1. Then $C(2, \cdot)$ is filled in as follows:

$$C\big(2, (\{f_1, f_2, f_3\}, \emptyset)\big) = \min\{4, 1\} + \min\{0, 0\} +$$
$$\min\big\{C\big(1, (\{f_0, f_1, f_2\}, \emptyset)\big), C\big(1, (\{f_1, f_2\}, \{f_0\})\big)\big\}$$
$$= 4 + 0 + \min\{5, 0\} = 4$$

$$C\big(2, (\{f_1, f_2\}, \{f_3\})\big) = \min\{1, 2\} + \min\{0, 2\} +$$
$$\min\big\{C\big(1, (\{f_0, f_1, f_2\}, \emptyset)\big), C\big(1, (\{f_1, f_2\}, \{f_0\})\big)\big\}$$
$$= 1 + 0 + \min\{5, 0\} = 1$$

$$C\big(2, (\{f_1, f_3\}, \{f_2\})\big) = \min\{0, 4\} + \min\{1, 0\} + \min\big\{C\big(1, (\{f_0, f_1\}, \{f_2\})\big)\big\}$$
$$= 0 + 0 + 3 = 3$$

$$C\big(2, (\{f_2, f_3\}, \{f_1\})\big) = \min\{1, 2\} + \min\{0, 2\} + \min\big\{C\big(1, (\{f_0, f_2\}, \{f_1\})\big)\big\}$$
$$= 1 + 0 + 5 = 6$$

Algorithm Engineering. To compute a column, say j, of the DP table, we have to go through all bipartitions of the active fragments $F(j) = \{f_0, \ldots, f_{|F(j)|-1}\}$ at SNP position j. Because of the observed symmetry it is sufficient to store $2^{|F(j)|-1}$ entries in column j. We order these entries by a mapping of indices $k \in \{0, \ldots, 2^{|F(j)|-1} - 1\}$ to bipartitions, using a binary encoding such that each bit k_ℓ in the binary representation of k tells whether fragment f_ℓ is in the first or in the second part of the bipartition. We break the above mentioned symmetry by assigning $f_{|F(j)|-1}$ always to the first set. Formally, this results in the mapping:

$$B : k \mapsto \big(\{f_{|F(j)|-1}\} \cup \{f_\ell \mid k_\ell = 0\}, \{f_\ell \mid k_\ell = 1\} \mid \ell < |F(j)| - 1\big)$$

for all $k \in \{0, 1\}^{|F(j)|-1}$.

Example. Assume there is a SNP position j for which $F(j) = \{f_0, f_1, f_2\}$. Then $k \in \{0, 1, 2, 3\}$ and thus $C(p, \cdot)$ has four entries each one being encoded in two bits as follows. $00 \mapsto (\{f_0, f_1, f_2\}, \emptyset)$, $01 \mapsto (\{f_0, f_2\}, \{f_1\})$, $11 \mapsto (\{f_2\}, \{f_0, f_1\})$, $10 \mapsto (\{f_1, f_2\}, \{f_0\})$. Notice that $f_{|F(p)|-1} = f_2$, as a sort of *pivot*, is always in the first part of the bipartition.

For an efficient computation of $\Delta_C\big(j, B_j(k)\big)$, we enumerate all bipartitions $k \in \{0, \ldots, 2^{|F(j)|-1} - 1\}$ in *Gray code* order. This ensures that at most one bit is flipped between two consecutive bipartitions. Therefore, in moving from one bipartition to the next, only one fragment swaps sides and updating the four values $W^0(j, R)$, $W^1(j, R)$, $W^0(j, S)$, and $W^1(j, S)$ can be done in constant time. As $\Delta_C\big(j, (R, S)\big)$ can be computed from these values in constant time, and moving from one Gray code to the next can be done in (amortized) constant time using the algorithm from [24], we conclude that $\Delta_C(j, \cdot)$ can be computed in $O(2^{\mathrm{cov}(j)-1})$ time, where $\mathrm{cov}(j) = |F(j)|$ denotes the physical coverage at SNP position j.

To efficiently implement the DP recursion, one can compute an *intermediate projection column* as follows. For all $B \in \mathcal{B}(F_{j+1}^\cap)$, store

$$\overline{C}(j, B) = \min_{B' \in \mathcal{B}(F(j) \mid B)} C(j, B') \ .$$

Table $\overline{C}(j, \cdot)$ can be filled while computing $C(j, \cdot)$ without any additional (asymptotic) runtime expense. Using this precomputed table, Recursion (1) can be written as

$$C(j+1, (R, S)) = \Delta_C(j+1, (R, S)) + \overline{C}(j, (R_{j+1}^{\cap}, S_{j+1}^{\cap})) \ .$$

The algorithm has a runtime of $O(2^{k-1}m)$, where k is the maximum value of $\text{cov}(\cdot)$, and m is the number of SNP positions. Note that the runtime is independent of read length.

An optimal bipartition can be obtained by backtracking. To do this efficiently, we store tables $\overline{D}(j, \cdot)$ that store the indices of the partitions that define the minima in $\overline{C}(j, \cdot)$. Formally,

$$\overline{D}(j, B) = \underset{B' \in \mathcal{B}(F(j) \mid B)}{\arg\min} C(j, B') \ .$$

Using these auxiliary tables, the sets of fragments that are assigned to each allele can be reconstructed in $O(km)$ time. To backtrace an optimal bipartition, we need to store the rightmost DP column $C(m, \cdot)$ and the backtracking tables $\overline{D}(j, B)$ for $j \in \{1, \ldots, m-1\}$, which takes total space $O(2^{k-1}m)$. This leads to a dramatically reduced memory footprint in practice compared to storing the whole DP table C.

Backtracking gives us optimal fragment bipartitions (R_j^*, S_j^*) for each position j. It is then straightforward to derive the two haplotypes h_1 and h_2 from this as follows:

$$h_1(j) = \begin{cases} 0 & \text{if } W^0(j, R_j^*) < W^1(j, R_j^*) \\ 1 & \text{otherwise} \ , \end{cases} \quad \text{and}$$

$$h_2(j) = \begin{cases} 0 & \text{if } W^0(j, S_j^*) < W^1(j, S_j^*) \\ 1 & \text{otherwise} \ . \end{cases}$$

4   Experimental Results

The focus of the present paper is on very long reads and the promise they hold for read-based phasing. Since such data sets are not available today, we perform a simulation study. We use all variants, that is SNPs, deletions, insertions, and inversions, reported by [19] to be present in Venter's genome. These variants were introduced into the reference genome (hg18) to create a reconstructed diploid human genome with fully known variants and phasings. Using the read simulator SimSeq [10], we simulated a variety of data sets, that reflect current technology as well as possible future developments. Regarding the former, we used HiSeq and MiSeq error profiles to generate a 2x100 bp and a 2x250 bp paired-end data set, respectively. The distribution of the internal segment size (i.e., fragment size minus size of read ends) was chosen to be 100 bp and 250 bp, respectively, which reflects current library preparation protocols. Furthermore, we created an additional MiSeq data set with 2.5 kbp internal segment size resembling mate-pair sequencing. Longer reads with 1 000 bp, 5 000 bp, 10 000 bp, and 50 000 bp were simulated with two different uniform error rates of 1 % and 5 %. All data sets were created to have 30x average coverage and were mapped to the human genome using BWA MEM [20].

To not confound results by considering positions of (possibly) wrongly called SNPs, we always used the set of true positions of heterozygous SNPs that were introduced into the genome. We extracted all reads that covered at least two such SNP positions to be used for phasing. Next, we pruned the data sets to target coverages of 5x, 10x, 15x, and 20x by removing (randomly selected) reads that violated the coverage constraints until no more such reads exist. The resulting problem instances were then solved to optimality using WHATSHAP, the DP algorithm described above.

To our knowledge, no other methods exist that can solve instances of wMEC with very long reads to optimality in practice. The DP approach of He et al. [16] has a worst-case complexity linear in 2^r where r is the length of the longest read (in terms of the number of SNPs covered). For coverage pruned at 15x and read length 5 000, r equals 30. For read length 50 000, r reaches a value of 147, which is clearly too large to run He et al.'s approach. In the ILP approach of Chen et al. [6], the key to solving MEC to optimality is to decompose the problem into independent blocks. Such a decomposition becomes less and less possible for longer reads, thus rendering such an approach infeasible for very long reads. After submission of this article, we found the similar yet independently developed DP approach of [9] that, however, addresses only the unweighted MEC. Moreover, due to our careful algorithm engineering, we have a lower asymptotic run-time, and can practically manage coverage up to 20x rather than 12x. A detailed performance analysis against all of these tools will appear in the full version of this work.

Our approach solved any problem instance with 15x coverage or below in less than 10 minutes on a single core (of an Intel Xeon E5-2620 CPU). For coverage 20x no problem instance took longer than 2.5 hours. The accuracy performance is summarized in Figure 2. There, the percentage of chromosome 1 that could be phased (y-axis) is plotted against the percentage of errors in the predicted haplotypes (x-axis) for different read lengths and coverages. A SNP position is *unphasable* if it is not covered by any read that also covered another SNP. Furthermore, we report an unphasable position whenever one of the two haplotypes contains no read at that position. Among those positions that *are* phasable of the reported haplotypes, we compute the number of errors, which is the sum of *zygosity errors* and *switch errors*, by comparing to the true haplotypes. A zygosity error occurs when a position is reported to be homozygous when it is truly heterozygous (and vice versa). A switch operation at position t on a binary string s is defined to result in the binary string $s[1 \ldots t]\bar{s}[t+1 \ldots |s|]$, where the $\bar{\ }$ operation flips all bits in a binary string. The switch error is now defined as the minimum number of such operations needed to transform the predicted haplotype (after all positions with zygosity errors have been removed)

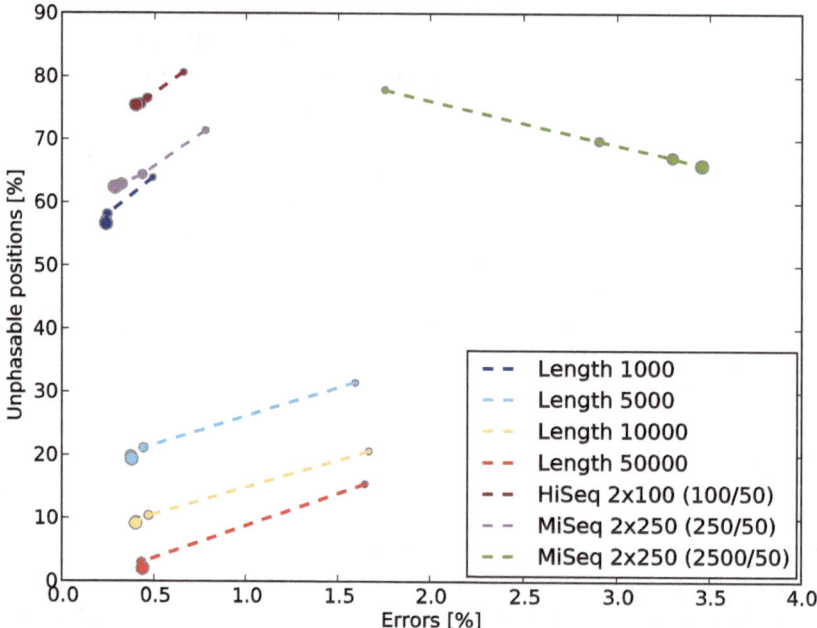

Fig. 2. Performance of phasing human chromosome 1 with 68 184 heterozygous SNPs in total using different simulated data sets and different coverages. The *unphasable positions* percentage (y-axis) gives the fraction of the SNP positions that could not be phased due to not being covered by reads that span more than one SNP position. Length 1 000, 5 000, 10 000, and 50 000 refer to reads of this length from a hypothetical sequencer with an error rate of 1%. HiSeq/MiSeq refers to using error profiles specific to these instruments during read sampling; in parentheses: mean/standard deviation of internal segment size (i.e., fragment size minus length of read ends). Data sets are pruned to four different target coverages (5x, 10x, 15x, 20x) encoded by circle diameter in the plot (larger means more coverage).

into the true haplotype. Note that the number of switch errors can increase when the number of unphasable positions goes down (see MiSeq experiments in Figure 2, for instance), as less gaps mean more contiguous fragments where such errors can be made.

Figure 2 clearly shows that long reads will indeed facilitate read-based phasing. For short reads of current HiSeq or MiSeq instruments, large portions of chromosome 1 cannot be phased. This was to be expected since short reads cannot span SNP deserts and, in general, rarely contain many SNPs. For the HiSeq data set, we found a paired-end read to cover only 2.2 SNPs on average. For the long read data sets (i.e., non-HiSeq/MiSeq) shown in Figure 2, an error rate of 1% was used. Repeating the experiments with an error rate of 5% yields nearly identical results. This exemplifies that errors are indeed corrected by solving the wMEC problem on sufficiently long reads.

Interestingly, the importance of a high coverage seems to be limited, especially for long reads. For 10x, 15x, and 20x, the corresponding circles in Figure 2 are often close together. On the other hand, the influence of read length was much more drastic, highlighting that our approach, which is able to handle long reads but only limited coverage, is well suited for future data sets that allow for read based phasing.

5   Conclusions and Further Work

We have presented WHATSHAP, a dynamic programming approach for haplotype assembly. WHATSHAP is the first exact approach for the wMEC problem with runtime linear in the number of SNPs. WHATSHAP is thus ready to benefit from increasing read lengths, which will boost the quality of haplotype assembly-based predictions.

While our approach handles datasets with possibly long reads, it can only deal with limited coverage. Although WHATSHAP can handle coverage as large as 20x on a standard workstation, and larger coverage does not seem to significantly improve the quality of the predicted haplotypes as shown in our simulation study, a number of possible ways to cope with higher coverage are under investigation. A first possibility is a divide and conquer heuristic approach that operates on high coverage portions of the matrix by (i) (randomly/suitably) splitting the fragments into as many subsets as necessary to make each one of them a *slice* of limited coverage, (ii) solving each slice separately using the dynamic programming approach, and finally (iii) merging the resulting *super-reads* and applying iteratively the method again. Another possibility is to just properly select reads up to the manageable coverage and to discard the rest.

In the literature there are several graph representations of haplotype data (the *fragment conflict graph* defined in [18] and many of its variants), and consequently the optimization problems we have mentioned are seen there as finding the minimum number of graph editing operations that make the graph bipartite. In particular, for the conflict graph variant used in [11], the MEC problem turns out to be equivalent to finding the *Maximum Induced Bipartite Subgraph (MIBS)*. It follows that our dynamic programming approach for MEC can be generalized to a FPT approach for MIBS where the parameter is the pathwidth of the graph.

In this work we have concentrated on assembling SNP haplotypes from reads of a sequenced genome. As a next step we will integrate predictions from statistical phasers into our approach. In some sense, the *super-read* obtained from a *slice*, mentioned above, can be viewed as a reference haplotype from a *reference panel* for an existing population. Hence, reference haplotypes can be seamlessly integrated into this merging step (iii) for a hybrid approach. Hybrid methods are the future of sequencing data analysis, and the field is already moving quickly in this direction [8,15,14,27,32,33].

In addition, haplotyping mostly refers to only SNPs for historical reasons [29,30]. To fully characterize an individual genome, however, haplotyping must produce exhaustive lists of both SNPs and non-SNPs, that is, larger variants. This has become an essential ingredient of many human whole-genome projects

[5,28]. In this paper we focused on SNPs variants and we identify the integration of non-SNP-variants as a challenging future research direction.

Lastly, we used prior knowledge of the true SNP positions in the genome in our simulation study. But since our method only scales linearly in the number of SNP positions, one could conceivably use as input the raw read input, to produce a "de novo" haplotype. Since SNPs comprise roughly 5% of positions, and the runtime of our method is on the order of 10 minutes on average (for sufficient 15x coverage), such a de novo haplotype could be generated in about 3 hours. The heterozygous sites of this constructed haplotype then correspond to the SNP positions. It hence follows that this tool could be used for SNP discovery, and perhaps for larger variants as well.

Acknowledgments. Murray Patterson was funded by a Marie Curie ABCDE Fellowship of ERCIM. Leo van Iersel was funded by a Veni, and Alexander Schönhuth was funded by a Vidi grant of the Netherlands Organisation for Scientific Research (NWO).

References

1. Aguiar, D., Istrail, S.: Hapcompass: A fast cycle basis algorithm for accurate haplotype assembly of sequence data. J. of Comp. Biol. 19(6), 577–590 (2012)
2. Aguiar, D., Istrail, S.: Haplotype assembly in polyploid genomes and identical by descent shared tracts. Bioinformatics 360, i352–i360 (2013)
3. Bansal, V., Bafna, V.: HapCUT: an efficient and accurate algorithm for the haplotype assembly problem. Bioinformatics 24(16), i153–i159 (2008)
4. Bansal, V., et al.: An MCMC algorithm for haplotype assembly from whole-genome sequence data. Genome Research 18(8), 1336–1346 (2008)
5. Boomsma, D.I., et al.: The Genome of the Netherlands: design, and project goals. European Journal of Human Genetics (2013), doi:10.1038/ejhg.2013.118
6. Chen, Z.Z., Deng, F., Wang, L.: Exact algorithms for haplotype assembly from whole-genome sequence data. Bioinformatics 29(16), 1938–1945 (2013)
7. Cilibrasi, R., van Iersel, L., Kelk, S., Tromp, J.: On the complexity of several haplotyping problems. In: Casadio, R., Myers, G. (eds.) WABI 2005. LNCS (LNBI), vol. 3692, pp. 128–139. Springer, Heidelberg (2005)
8. Delaneau, O., Howie, B., Cox, A., Zagury, J., Marchini, J.: Haplotype estimation using sequencing reads. Am. J. of Human Genetics 93(4), 687–696 (2013)
9. Deng, F., Cui, W., Wang, L.: A highly accurate heuristic algorithm for the haplotype assembly problem. BMC Genomics 14(suppl. 2), S2 (2013)
10. Earl, D.A., et al.: Assemblathon 1: A competitive assessment of de novo short read assembly methods. Genome Research (2011), doi:10.1101/gr.126599.111
11. Fouilhoux, P., Mahjoub, A.: Solving VLSI design and DNA sequencing problems using bipartization of graphs. Comp. Optim. and Appl. 51(2), 749–781 (2012)
12. Greenberg, H., Hart, W., Lancia, G.: Opportunities for combinatorial optimization in computational biology. Informs J. on Computing 16(3), 211–231 (2004)
13. Hartl, D., Clark, A.: Principles of Population Genetics. Sinauer Associates, Inc., Sunderland (2007)
14. He, D., Eskin, E.: Hap-seqX: expedite algorithm for haplotype phasing with imputataion using sequence data. Gene. 518(1), 2–6 (2013)

15. He, D., Han, B., Eskin, E.: Hap-seq: an optimal algorithm for haplotype phasing with imputation using sequencing data. J. Comp. Biol. 20(2), 80–92 (2013)

16. He, D., et al.: Optimal algorithms for haplotype assembly from whole-genome sequence data. Bioinformatics 26(12), i183–i190 (2010)

17. Howie, B., Donnelly, P., Marchini, J.: A flexible and accurate genotype imputation method for the next generation of genome-wide association studies. PLoS Genetics 5(6), e1000529 (2009)

18. Lancia, G., Bafna, V., Istrail, S., Lippert, R., Schwartz, R.: SNPs problems, complexity and algorithms. In: Meyer auf der Heide, F. (ed.) ESA 2001. LNCS, vol. 2161, pp. 182–193. Springer, Heidelberg (2001)

19. Levy, S., et al.: The diploid genome sequence of an individual human. PLoS Bio. (2007), doi:10.1371/journal.pbio.0050254

20. Li, H.: Aligning sequence reads, clone sequences and assembly contigs with BWA-MEM. arXiv (1303.3997) (2013)

21. Li, Y., et al.: MaCH: using sequence and genotype data to estimate haplotypes and unobserved genotypes. Genet. Epidemiol. 34, 816–834 (2010)

22. Lippert, R., et al.: Algorithmic strategies for the single nucleotide polymorphism haplotype assembly problem. Briefings in Bioinformatics 3(1), 23–31 (2002)

23. Menelaou, A., Marchini, J.: Genotype calling and phasing using next-generation sequencing reads and a haplotype scaffold. Bioinformatics 29(1), 84–91 (2013)

24. Mossige, S.: An algorithm for Gray codes. Computing 18, 89–92 (1977)

25. Panconesi, A., Sozio, M.: Fast hare: A fast heuristic for single individual SNP haplotype reconstruction. In: Jonassen, I., Kim, J. (eds.) WABI 2004. LNCS (LNBI), vol. 3240, pp. 266–277. Springer, Heidelberg (2004)

26. Scheet, P., Stephens, M.: A fast and flexible statistical model for large-scale population genotype data: Applications to inferring missing genotypes and haplotypic phase. American Journal of Human Genetics 78, 629–644 (2006)

27. Selvaraj, S., et al.: Whole-genome haplotype reconstruction using proximity-ligation and shotgun sequencing. Nature Biotechnology 31, 1111–1118 (2013)

28. The 1000 Genomes Project Consortium: A map of human genome variation from population-scale sequencing. Nature 467(7319), 1061–1073 (2010)

29. The International HapMap Consortium: A second generation human haplotype map of over 3.1 million SNPs. Nature 449, 851–861 (2007)

30. The International HapMap Consortium: Integrating common and rare genetic variation in diverse human populations. Nature 467, 52–58 (2010)

31. Wang, R.S., Wu, L.Y., Li, Z.P., Zhang, X.S.: Haplotype reconstruction from SNP fragments by minimum error correction. Bioinformatics 21(10), 2456–2462 (2005)

32. Yang, W.Y., Hormozdiari, F., Wang, Z., He, D., Pasaniuc, B., Eskin, E.: Leveraging reads that span multiple single nucleotide polymorphisms for haplotype inference from sequencing data. Bioinformatics 29(18), 2245–2252 (2013)

33. Zhang, Y.: A dynamic Bayesian Markov model for phasing and characterizing haplotypes in next-generation sequencing. Bioinformatics 29(7), 878–885 (2013)

34. Zhao, Y.T., Wu, L.Y., Zhang, J.H., Wang, R.S., Zhang, X.S.: Haplotype assembly from aligned weighted SNP fragments. Computational Biology and Chemistry 29, 281–287 (2005)

Simultaneous Inference of Cancer Pathways and Tumor Progression from Cross-Sectional Mutation Data*

Benjamin J. Raphael[1] and Fabio Vandin[1,2,**]

[1] Department of Computer Science and Center for Computational Molecular Biology,
Brown University, Providence, RI, USA
[2] Department of Mathematics and Computer Science, University of Southern
Denmark, Odense, Denmark
braphael@cs.brown.edu, vandinfa@imada.sdu.dk

Abstract. Recent cancer sequencing studies provide a wealth of somatic mutation data from a large number of patients. One of the most intriguing and challenging questions arising from this data is to determine whether the temporal order of somatic mutations in a cancer follows any common progression. Since we usually obtain only one sample from a patient, such inferences are commonly made from cross-sectional data from different patients. This analysis is complicated by the extensive variation in the somatic mutations across different patients, variation that is reduced by examining combinations of mutations in various pathways. Thus far, methods to reconstruction tumor progression at the pathway level have restricted attention to known, *a priori* defined pathways.

In this work we show how to *simultaneously* infer pathways and the temporal order of their mutations from cross-sectional data, leveraging on the exclusivity property of driver mutations within a pathway. We define the Pathway Linear Progression Model, and derive a combinatorial formulation for the problem of finding the optimal model from mutation data. We show that while this problem is NP-hard, with enough samples its optimal solution uniquely identifies the correct model with high probability even when errors are present in the mutation data. We then formulate the problem as an integer linear program (ILP), which allows the analysis of datasets from recent studies with large number of samples. We use our algorithm to analyze somatic mutation data from three cancer studies, including two studies from The Cancer Genome Atlas (TCGA) on large number of samples on colorectal cancer and glioblastoma. The models reconstructed with our method capture most of the current knowledge of the progression of somatic mutations in these cancer types, while also providing new insights on the tumor progression at the pathway level.

---

* This work is supported by NIH grant R01HG007069-01 and by NSF grant IIS-1247581.
** Corresponding author.

R. Sharan (Ed.): RECOMB 2014, LNBI 8394, pp. 250–264, 2014.
© Springer International Publishing Switzerland 2014

1   Introduction

Cancer is a disease caused by the accumulation of somatic mutations, changes in the genome that appear during the lifetime of an individual. High-throughput DNA sequencing technologies are now measuring these mutations in thousands of cancer genomes through projects such as The Cancer Genome Atlas (TCGA) [26], International Cancer Genome Consortium (ICGC) [34] and many others. In the analysis of somatic mutations in cancer, two important questions arise. First, which mutations are the *driver* mutations responsible for cancer and which are merely random, *passenger* mutations? Second, is there any temporal order to the driver mutations in a single cancer patient? While the first question might be addressed in part by comparing the observed frequencies of mutations across different individuals [10, 20], the second question is much more difficult to address from *cross-sectional* data, sequencing data taken from single time-points across different individuals. Answering the question about temporal progression, and more specifically determining what mutations occur early in the progression of cancer, is essential for both a basic understanding of cancer biology and for developing targeted treatments. The ideal dataset to determine temporal progression is a *longitudinal* dataset consisting of measurements of somatic mutations from multiple time-points in a single individual. However, such datasets are nearly impossible to obtain from human tumors: it is difficult to obtain multiple samples from the same patient without also perturbing the tumor with surgery, chemotherapy, or other treatments.

A number of methods for inferring temporal progression of mutations from cross-sectional data have been introduced [12, 11, 6, 5, 23, 28, 18, 16, 2–4, 17, 24, 25] (see Section 1.1). These methods consider models of increasing complexity for cancer progression: trees, mixtures of trees, and Bayesian network models with different constraints. However, such approaches infer progression at the level of individual mutations, or individual genes. The difficulty with this approach is that cancers exhibit extensive mutational heterogeneity: the somatic mutations, including driver mutations, vary widely across individuals with the same cancer. Thus, the main signal used to infer temporal order, co-occurrence of mutations in different samples, is very weak. A major reason for this mutational heterogeneity is that somatic mutations perturb various signaling, regulatory, and metabolic pathways [31]. Thus, different individuals may harbor driver mutations in different genes within the same pathway. Since driver mutations target pathways, it is possible that the order in which mutations arises is at the pathway level, not at the gene level. There has been some initial work in inferring pathway order [17, 8]. These approaches demonstrated some advantages over gene based approaches, but restricted attention to known, annotated pathways. Most annotated pathways are large and overlap with other pathways, thus creating problems for the discovery of mutation progression in smaller sets of interacting genes.

An alternative to known pathways is to examine sets of genes or mutations *de novo*. However, the large number of such combinations will quickly overwhelm such an exhaustive approach. Recently, it has been observed that driver mutations in pathways tend to be mutually exclusive meaning that an individual rarely has more than one driver mutation in a pathway [33] and this observation has been used to successfully identify pathways in cancer datasets [22, 29, 9, 21]. Mutual exclusivity is a powerful signal to constrain the combinations of genes and mutations to examine. In this paper, we design an algorithm to infer *simultaneously* pathways and their temporal order from cross-sectional data, leveraging on the expected exclusivity of mutations within pathways. We apply this algorithm to simulated and real sequencing data from colorectal and glioblastoma cancers. The progression models produced by our method are in agreement with the current knowledge of the progression of mutations in these cancer types, and also propose some novel hypothesis. The sets identified by our method mostly correspond to known pathways or sets of interacting genes, showing the ability of our approach to simultaneously identify cancer pathways and the tumor progression they define.

1.1   Previous Work

After the seminal work of Fearon and Vogelstein [14] that proposed a model for progression of mutations in colorectal cancer, a number of computational methods have been designed to reconstruct the progression of genetic events leading to cancer from cross-sectional data assuming that the order is at the gene level. These methods consider models of increasing complexity. The model of Fearon and Vogelstein [14] describes a *linear sequence* (or *path*) on genes. Desper et. al [12, 11] considered *trees* on the genes. A number of works [6, 5, 23, 28], have proposed the inference of *mixture of trees* on the genes to model the progression of cancer. While providing a first advance in understanding cancer progression, these methods assume that cancer progresses through disjoint paths, with no possible convergence of different paths, that is a stringent constraint on the model. More general methods that include convergence describe the model in terms of probabilistic directed acyclic graphs, or Bayesian networks [18, 16, 2–4, 17, 24, 25]. These methods impose different restrictions on the model to limit the search space and to represent possible features of cancer progression. In practice these methods can include at most a dozen of genes in their analysis; the exception is the method presented in [25] that uses a mixed integer linear program (MILP) to infer the best (constrained) Bayesian network. However they consider only models (networks) in which the number of parents of a node is bounded by a small value k (in [25] values of $k \leq 4$ are used).

We note that the model we are interested in could be defined as a Bayesian network in which the genes in a pathway are the parents of all the genes in the next pathway in the progression, and the probability model should reflect the exclusivity among mutations in the parents of a particular node. None of the methods above has considered the exclusivity among genes (or mutations) in their model, and to the best of our knowledge no such model has been proposed in the machine learning literature.

Two recent works [17, 8] have considered the inference of the progression model at the pathway level, where the assignment of genes into pathways was defined *a priori* and provided in input to the method. As pointed out in the introduction, the *a priori* assignment of genes into pathways is complicated by the fact that such pathways are large and moreover often a gene is assigned to multiple pathways, that limit the ability to detect smaller sets of interacting genes related to tumor progression.

1.2 Contributions

This work initiates the study of the simultaneous identification of cancer pathways and their mutation order in tumor progression from cross-sectional data, without relying on the *a priori* definition of pathways, and contains the following contributions. First, we formalize the Pathway Linear Progression Model for tumor progression, in which mutations within each pathway are mutually exclusive, while they satisfy a linear progression across pathways. We show that the computational problem, that we call Pathway Linear Progression Reconstruction problem, of identifying the model that provides the best (in terms of number of errors) explanation of the observed data is NP-hard. Moreover, we prove that under reasonable assumptions and with enough samples, the correct progression model is uniquely identified by the optimal solution of the Pathway Linear Progression Reconstruction problem even when the data contains errors.

Second, we formulate the Pathway Linear Progression Reconstruction problem as an Integer Linear Program (ILP), providing an exact solution for datasets of realistic size. Using simulated data we show that the correct progression model is identified by our algorithm under realistic assumptions for the error model when enough samples are considered. We also show that when genes that are not correlated with tumor progression are included in the analysis, our algorithm identifies the correct order among genes in pathways driving the progression.

Third, we run our algorithm on somatic mutation data from The Cancer Genome Atlas (TCGA) studies on colorectal and glioblastoma cancers, and on a different colorectal cancer study. We show that the progression models produced by our method recapitulate most of the current knowledge of the progression of mutations in colorectal cancer, and also propose novel hypothesis for the progression driving colorectal and glioblastoma cancer. In particular, on somatic mutation data from 224 TCGA samples of colorectal cancer, our algorithm identifies models that are in agreement with current knowledge of the progression of mutations in this cancer type. Moreover our method groups members of the Raf-Ras pathways and SMADs and interacting genes in different sets. On somatic mutation data from 251 glioblastoma multiforme samples, our method defines a model with gene sets corresponding to part of the Rb1, PI3K, and p53 pathway.

2 Methods and Algorithms

2.1 Model and Problem Definition

We are given mutation data from m samples $s_1, s_2, \ldots, s_m$, consisting of the mutation status of each of n genes $g_1, g_2, \ldots, g_n$. This data is represented by a

$m \times n$ binary mutation matrix M, with samples on the rows and genes on the columns, where $M_{i,j} = 1$ if g_j is mutated in sample s_i, and $M_{i,j} = 0$ otherwise. We use r_i to denote the ith row of M, and c_j to denote the jth column of M. We now define a model in which the mutation data comes from a linear progression on sets of genes (see Figure 1a).

Pathway Linear Progression Model (PLPM). A $m \times n$ mutation matrix M satisfies the Pathway Linear Progression Model $PLPM(K)$ with parameter $K > 1$ if there exists a partition $\mathcal{P} = \{P_1, P_2, \ldots, P_K\}$ of the columns $\{c_1, \ldots, c_n\}$ of M into K sets such that:

1. for each row r_i of M, 1's within each set P_k are *mutually exclusive*, that is: for all $1 \le k \le K$ we have $|\{c_j \in P_k : M_{i,j} = 1\}| \le 1$;
2. each row r_i of M satisfies the *progression* on the sets $P_1, \ldots, P_K$, that is: for all $1 < k \le K$, if $|\{c_j \in P_k : M_{i,j} = 1\}| > 0$ then $|\{c_j \in P_{k-1} : M_{i,j} = 1\}| > 0$.

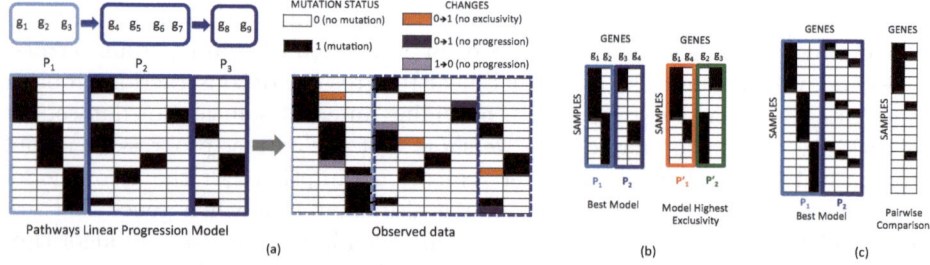

Fig. 1. The Pathway Linear Progression Model(PLPM). (a) A linear progression on gene sets (pathways) generates a mutation matrix with exclusive mutation within each set, and a progression of mutations across the sets. In real data errors that disrupt the exclusivity or the progression are present. (b-c) Problems of considering only exclusivity or only progression in reconstructing a PLPM. (b) Left: the correct progression model. Right: the (incorrect) partition that is inferred by maximizing the (total) exclusivity of sets. Since the correct model does not show perfect exclusivity (due to errors, etc., present in real data), maximizing the exclusivity does not lead to recover the correct model. (c) Left: the correct progression model. Right: genes pairwise comparison reveals no information about the progression. Progression at the genes level needs to appear as significant co-occurrence between 1's in two columns, while in the example for each pair of columns in $P_1 \times P_2$ the number of samples in which they are both 1 is exactly the expected number under the independence of the two columns. An arbitrary partition, most probably not correct, would then be reported by considering only the progression signal among genes.

Given m and $\mathcal{P}$, let $PLPM(\mathcal{P})$ be the set of $m \times n$ mutation matrices M for which the constraints 1 and 2 are satisfied. We denote a mutation matrix $M \in PLPM(\mathcal{P})$ by saying that M *satisfies* the PLPM defined by $\mathcal{P}$. Therefore a mutation matrix M satisfies the Pathway Linear Progression Model $PLPM(K)$ if there exists a partition $\mathcal{P} = \{P_1, P_2, \ldots, P_K\}$ such that $M \in PLPM(\mathcal{P})$. Each set P_k in a partition $\mathcal{P}$ defines a set of genes, or *pathway*, that by their interaction perform a certain function or process in the cell. When at least one of the columns of P_k has value 1 in r_i, we say that P_k is *mutated* in r_i.

Due to a number of factors (passenger mutations in driver genes, false positives and false negatives in mutations detection, etc.), a mutation matrix M is a noisy observation from the PLPM, and M may therefore no satisfy $PLPM(K)$, or equivalently there is no partition $\mathcal{P}$ for which $M \in PLPM(\mathcal{P})$. For a given K, we are thus interested in finding a partition $\mathcal{P} = \{P_1, P_2, \ldots, P_K\}$ such that M is *close* to satisfy the Pathway Linear Progression Model having $\mathcal{P}$ as as partition. In particular, we look for a partition $\mathcal{P}^*$ that minimizes the number of entries of M that must be changed (*flips* of $0 \to 1$ or $1 \to 0$) so that the resulting mutation matrix $M' \in PLPM(\mathcal{P}^*)$. More formally, given two binary $m \times n$ matrices M, M', let $d(M, M') = \sum_{i=1}^{m} \sum_{j=1}^{n} |M_{i,j} - M'_{i,j}|$. For a mutation matrix M and a partition $\mathcal{P}$, we define $f(M, \mathcal{P}) = \min_{M' \in PLPM(\mathcal{P})} d(M, M')$. Let $\mathcal{P}(K)$ be the set of all possible partitions (of the columns of M) into K sets.

Pathway Linear Progression Reconstruction Problem. Given a $m \times n$ mutation matrix M and an integer value $K > 1$, find $\mathcal{P}^* = \arg\min_{P \in \mathcal{P}(K)} f(M, \mathcal{P})$.

In our formulation, there are two requirements that a partition has to satisfy: the exclusivity of mutations within each set of the partition, and the progression across the sets. One may think that considering only one of the two requirements for the optimization is enough. The examples of Figure 1b show that this is not true.

We therefore need to identify the best partition $\mathcal{P}^*$ by simultaneously considering both exclusivity and progression. We have the following result. (Due to space constraints, proofs are omitted from this manuscript.)

Theorem 1. *The Pathway Linear Progression Reconstruction problem is NP-hard for any value of K.*

2.2   Conditions for Reconstruction with Errors

Real mutation data has various sources of error that result in both false positive and false negative mutations. Thus, rather than observing a mutation matrix M satisfying $PLPM(K)$, we observe a perturbed mutation matrix $\tilde{M}$. A natural question is to determine conditions under which the partition $\mathcal{P}$ that defines M can be recovered as a solution of the Pathway Linear Progression Reconstruction problem when either M or $\tilde{M}$ is given as input. In this section, we prove that if the number m of samples is large enough, then with bounded error probability $\mathcal{P}^*$ is the unique solution to Pathway Linear Progression Reconstruction under two different models for generating M and $\tilde{M}$, respectively.

First, given a partition $\mathcal{P}$ of K sets $P_1, P_2, \ldots, P_K$, we define a *uniform* generative model $UPLPM(\mathcal{P})$ for an $m \times n$ mutation matrix M as follows. For each row r_i, $i = 1, \ldots, m$, we select the positions for the 1's by: i) choose the *stage in the progression* of r_i, that is a value $t \in \{1, 2, \ldots, K\}$, uniformly at random; ii) for each $j, 1 \leq j \leq t$, choose one of the columns in P_j uniformly at random to be 1 in r_i. We have the following.

Theorem 2. *Let M be a $m \times n$ mutation matrix generated from $UPLPM(\mathcal{P})$. If $m \geq Kn^2 \ln \frac{2n^2}{\delta}$, then $\mathcal{P}$ is the* unique *optimal solution to the Pathway Linear Progression Reconstruction problem with probability $\geq 1 - \delta$.*

Next, we consider the case of a mutation matrix $\tilde{M}$ that is a perturbation of an $m \times n$ mutation matrix M generated from $UPLPM(\mathcal{P})$. We generate such an $\tilde{M}$ as follows. We assume that for each $P \in \mathcal{P} : |P| = \frac{n}{K}$, and in each row one entry chosen uniformly at random has been flipped with probability q, independently for each row. We call the set of such matrices $UPLPM(\mathcal{P}, q)$. We prove the following.

Theorem 3. *Let $q = \frac{K^2 n(K - \varepsilon n^2)}{n^3(K-1)^2 + n^3 K - 2Kn^2 + 2K^3} \geq 0$ for some $\varepsilon > 0$, and let $\tilde{M}$ be an $m \times n$ mutation matrix from $UPLPM(\mathcal{P}, q)$. If $m \geq \frac{8}{\varepsilon^2} \ln \frac{2n^2}{\delta}$, then $\mathcal{P}$ is the unique optimal solution to the Pathway Linear Progression Reconstruction problem with probability $\geq 1 - \delta$.*

2.3   Algorithm

We now formulate the Pathway Linear Progression Reconstruction problem as an integer linear program (ILP). For a partition $\mathcal{P}$, let $p_{j,k}$ be a 0-1 variable with $p_{j,k} = 1$ if column c_j is assigned to set P_k, and $p_{j,k} = 0$ otherwise. Let $a_{i,k}$ be a 0-1 variable with $a_{i,k} = 1$ if the set P_k is considered mutated (after required flips are made) in row r_i, and $a_{i,k} = 0$ otherwise. We also define auxiliary 0-1 variables $f_{i,k}$ for $1 \leq i \leq m$ and $1 \leq k \leq K$, where intuitively $f_{i,k} = 1$ if we need to flip one of the entries of columns in P_k for P_k to be mutated in row r_i, and 0 otherwise. A valid solution to our problem then satisfies the following constraints:

- each column is assigned to exactly one set: for $1 \leq j \leq n$, $\sum_{k=1}^{K} p_{j,k} = 1$;
- for each set P_k, at least one column is assigned to it: for $1 \leq k \leq K$, $\sum_{j=1}^{n} p_{j,k} \geq 1$;
- for each sample the progression model is satisfied: for $1 \leq i \leq m$ and $1 \leq k \leq K - 1$, $a_{i,k} \geq a_{i,k+1}$;
- for each row r_i, the set P_k is considered mutated if it has a 1 in r_i or if one of its entries in row r_i is flipped to make it mutated (i.e., $f_{i,k} = 1$): for $1 \leq k \leq K$ and $1 \leq i \leq m$, $\sum_{j=1}^{n} M_{i,j} p_{j,k} + f_{i,k} \geq a_{i,k}$.

For a particular partition $\mathcal{P}$, the value of the objective function is the minimum number of entries of M that we need to flip to satisfy the constraints defined by $\mathcal{P}$. If we consider a given sample r_i and a given set P_k, after variables $p_{j,k}, a_{i,k},$ and $f_{i,k}$ have been fixed, the number of entries of P_k that are flipped in r_i is given by $\sum_{j=1}^{n} M_{i,j} p_{j,k} - a_{i,k} + 2f_{i,k}$. Since we want to minimize the total number of entries that are flipped, the objective function is:

$$\min \sum_{i=1}^{m} \sum_{k=1}^{K} \left(\sum_{j=1}^{n} M_{i,j} p_{j,k} - a_{i,k} + 2f_{i,k} \right).$$

The contribution of one row to the objective function is interpreted as follows. The term $\sum_{j=1}^{n} M_{i,j} p_{j,k}$ counts the number of observed 1's in r_i for set P_k. Assume that $\sum_{j=1}^{n} M_{i,j} p_{j,k} > 0$: if we consider set P_k mutated in r_i ($a_i = 1$), the number of entries of M that we need to flip to satisfy the progression model is $\sum_{j=1}^{n} M_{i,j} p_{j,k} - 1$, and it is $\sum_{j=1}^{n} M_{i,j} p_{j,k}$ otherwise ($a_{i,k} = 0$). If instead $\sum_{j=1}^{n} M_{i,j} p_{j,k} = 0$, if we do not consider P_k mutated ($a_{i,k} = 0$) then the number of entries to be flipped is 0, while if we consider P_k mutated ($a_{i,k} = 1$) then the number of flips is 1, obtained by having $f_{i,k} = 1$ as enforced by the last constraint above. Note that this reasoning assumes that $f_{i,k} = 0$ whenever $\sum_{j=1}^{n} M_{i,j} p_{j,k} > 0$ or $a_{i,k} = 0$, that is not forced by the constraints above but is obtained when the objective function is minimized.

3   Experimental Results

In this section we present the results of our experimental analysis on simulated data, and on data from cancer studies. In all cases we solved the ILP using CPLEX v12.3 with default parameters.

3.1   Simulated Data

We performed a number of experiments using simulated data to assess the robustness of our method to different levels of noise. We considered data coming according to a progression model $\mathcal{P}$ to which noise was added. In particular, we considered a progression model with $K = 5$ stages, each containing 5 genes, and generated 100 datasets with m samples from this model, adding noise by flipping each entry of the corresponding mutation matrix with probability p. Note that this error model is more complex and more realistic than the one we analyzed in Theorem 3. The progression stage for each sample was chosen uniformly at random (between 1 and 5), and for a sample the mutated gene in a stage is chosen uniformly at random. We considered values of $m = 50, 100, 500, 1000,$ and $p = 0.001, 0.01, 0.05,$ that are values in the expected range for passenger mutation probability given the background mutation rate and the length of the genes [29, 30]. (Note that when $p = 0.05$, the expected number of errors per sample is > 1.)

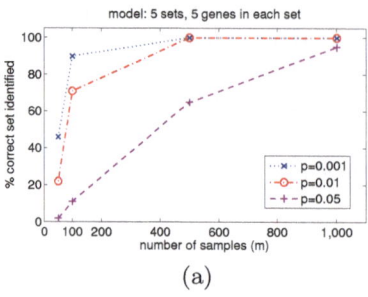 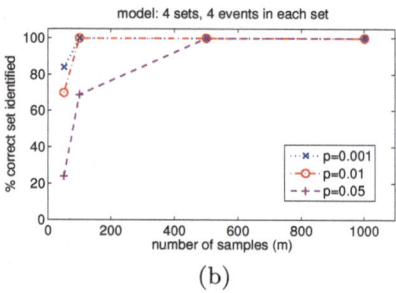

(a)   (b)

Fig. 2. Fraction of times (over 100 trials) the entire correct order is identified by the ILP on m samples where the mutation matrix M comes from a progression model with K sets each containing ℓ genes. Each entry of M is flipped with probability p; the results for different values of p are shown. (a) Results for $K = 5, \ell = 5$. (b) Results for $K = 4, \ell = 4$.

For each combination m, p we recorded the fraction of times in which the optimal solution identified by solving the ILP (fed with the correct value for K) corresponded to $\mathcal{P}$ (Figure 2a). As we can see, when $m = 50$ samples are included in the analysis, most of the time the correct progression model is not identified, even when the error probability is not very high ($p = 0.001$). However, for $p \leq 0.01$, when 100 samples are analyzed the correct progression model is reported most of the times, and when 500 samples are analyzed the correct model is reported every time. In contrast, when $p = 0.05$, with 500 samples the correct model is reported only 65% of the times, and it is reported 95% of the times when 1000 samples are considered. These results show that while data from reasonably sized cancer studies can sometimes be used to infer the correct progression model, studies of size larger then currently available may be required to identify the correct progression model if the noise level (i.e., p) is high.

To understand how the complexity (i.e., number of sets, number of genes in each set) of $\mathcal{P}$ impact the number of samples required to reliably identify the correct model, we considered a "simpler" model, consisting of $K = 4$ stages of progression, each including 4 genes (Figure 2b). Mutations from this model were generated as for the model above, and we considered the same values for m and p. In this case, for a given pair (m, p), the fraction of times the correct model is reported is always greater or equal to the fraction of times the correct model with 5 sets and 5 events in each set was identified for the same pair (m, p). For example, when $p \leq 0.01$ and $m = 50$ samples are considered, the correct model is reported at least 70% of the times, while with 500 samples the correct model is reported every time even when $p = 0.05$.

These results show that the number of samples required to identify the correct model is sensitive to the parameters of the progression model, confirming and extending the analytical results of Section 2.2, and moreover show that currently available cancer studies have sufficient samples sizes to identify progression models where the number of sets and the number of genes in each set is not too high, while more samples may be required to correctly identify models with a large number of sets (stages) and a large number of genes, or when the probability of false positives and false negatives is very high.

We also used simulations to assess the impact of the inclusion of genes not related to the progression on the accuracy of our method. We considered the progression model with $K = 5$ stages, each consisting of 5 genes related to the progression, described above, and also included mutations for 25 additional genes, not related to the progression, each mutated in 5% of the samples independently of all other events. We generated 100 datasets from this model for each of the values $m =$ 50, 100 500, fixing $p = 0.001$. For $m = 50$, the inferred model never corresponded to the correct model on the 25 genes related to progression; for $m = 100$, the inferred model on the 25 genes related to the progression was reported 41% of the times, while for $m = 500$ this happened 100% of the times. This shows that even when genes not associated with the progression model are included in the analysis, our method is able to correctly reconstruct the relationship between the genes associated with the progression when the number of samples is sufficiently high. Our analysis also shows that spurious associations are more likely to be observed in late stages of the inferred progression (data not shown).

3.2   Cancer Data

We used our ILP to analyze somatic mutation data from published cancer studies. We first analyzed the dataset from a colorectal cancer study [32] considered in [17]. We then analyzed two large datasets from The Cancer Genome Atlas (TCGA) studies on colorectal cancer [27] and glioblastoma multiforme [7].

For all these datasets we used the ILP to identify the set $\mathcal{P}_K^*$ of cardinality K of minimum weight for $K = 2, \ldots, 8$, and then considered the best progression model to be the set $\mathcal{P}^*$ of minimum weight among the different solutions obtained: $\mathcal{P}^* = \arg\min_{K \in \{2,\ldots,8\}} f(\mathcal{P}_K^*)$. To assess the statistical significance of our observation we computed a p-value using a permutation test, estimating the probability of obtaining a set of size K of weight less or equal to $\mathcal{P}^*$ when the mutations are place independently in the samples preserving the mutation frequency of the genes. For each gene we also computed the fraction of times it is reported in a particular stage of the progression using *bootstrap* datasets [13]; this measures the stability to random fluctuations in the samples population of the assignment of a particular gene to a stage in the progression.

Colorectal Cancer. We analyzed the mutations reported from the 95 samples considered in [32], for the 8 genes mutated with frequency above 5%: APC, EPHA3, EVC2, FBXW7, KRAS, PIK3CA, TCF7L2, TP53. The PLPM of minimum weight is shown in Figure 3a. The progression model inferred with our method shares some similarities with the one inferred in [17] (Figure 3b), and is consistent with the proposed linear order of mutations in colorectal cancer: mutations in APC occur early in the progression, while KRAS mutations appear later.

Interestingly, in [17] the order of TP53 mutations were reported as independent of other mutations and mutations in PIK3CA were reported as independent of KRAS mutations, while in our model mutations in TP53 and PIK3CA are reported to appear after APC mutations, but before KRAS mutations. TP53 mutations have been reported to appear after APC in [1], while TP53 mutations are considered to appear after KRAS mutations [14, 15]. TP53 mutations and PIK3CA mutations are significantly exclusive in this dataset ($p < 0.008$ by Fisher exact test), and are therefore potentially related. Of the 71 samples that contain a TP53 mutation or a PIK3CA mutation (or

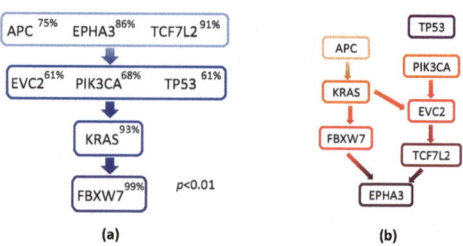

Fig. 3. Progression models for colorectal data [32]. (a) PLPM inferred using our method. The p-value from the permutation test is reported. For each gene, the fraction of times it appears in the same stage out of 100 bootstrap dataset is shown. (b) Model from [17]. In [17] all parents of a node must be mutated for a gene to be mutated. TP53 is independent of other genes.

both - 1 sample), 51 also present a KRAS mutation, while only 8 samples with a KRAS mutation do not have a TP53 mutation or a PIK3CA mutation. Therefore, the most reasonable explanation, assuming a linear order among pathways, is that KRAS mutations come after TP53/PIK3CA mutations.

Since the model inferred in [17] considers TP53 mutations as independent of the other mutations, we assessed how well the data is described by the two models when TP53 is ignored. In particular, we found that 12 samples have mutations that (ignoring TP53 and assuming no errors) do not conform to the PLPM model in Figure 3a, while 22 samples have mutations that (ignoring TP53 and assuming no errors) do not conform to the model of [17]. For example, none of the 5 samples with EPHA3 mutations come from the model of [17], while only 1 such sample does not come from the PLPM model in Figure 3a. Therefore our model provides a better explanation of the colorectal cancer data from [32].

TCGA Colorectal Cancer. We analyzed 224 colorectal samples from the TCGA study on this cancer type. We download mutation data from the Broad GDAC Firehose[1], including single nucleotide variants and indels. We restricted our analysis to the 14 genes identified as recurrently mutated by MutSigCV [20].

The progression model inferred by our method is shown in Figure 4a. Interestingly, the progression model restricted to the genes APC, TP53, PIK3CA, and KRAS is the same we identify from the smaller dataset of [32]. Moreover, the bootstrap analysis reveals that these genes and NRAS have the most stable assignments to the different stages of the progression.

---

[1] https://confluence.broadinstitute.org/display/GDAC/Home

As noted before, TP53 mutations are usually reported as appearing after KRAS mutations. However, even considering only (TP53, PIK3CA) in the second stage of the progression model, and considering (BRAF, NRAS, KRAS) in the third stage, we have that 58 samples contain a mutation in the set (TP53, PIK3CA) and not in (BRAF, NRAS, KRAS), while 48 samples contain a mutation in (BRAF, NRAS, KRAS) and not in (TP53, PIK3CA), therefore is more reasonable to assume that mutations in (TP53, PIK3CA) (that show again significant exclusivity of mutations - $p < 0.0032$ by Fisher exact test) appear before mutations in (BRAF, NRAS, KRAS). Moreover, a recent analysis [19] suggested

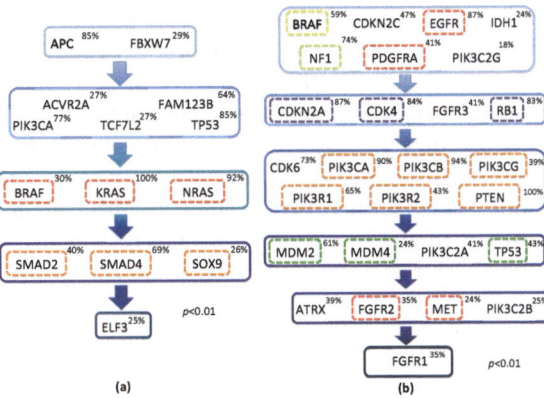

Fig. 4. PLPM models from TCGA cancer datasets. (a) PLPM for TCGA colorectal data [27]. Dashed boxes identify genes in the same pathway, with different colors for different pathways. (b) PLPM for TCGA glioblastoma multiforme data [7]. Dashed boxes identify genes that are in a set with at least another gene in the same pathway (as annotated in [7]), with different colors for different pathways. For each PLPM, the p-value from the permutation test is reported, and the fraction of times genes appear in the same stage out of 100 bootstrap dataset is shown.

that in 3 other cancer types mutations in TP53 appear early during tumorigenesis, while KRAS mutations appear later in the tumor development.

Two sets in our model contain genes all in the same pathway or interacting. In particular, (BRAF, KRAS, NRAS) is part of the Ras-Raf pathway, and SOX9 interacts with SMAD2, that interacts with SMAD4. (For both these sets, the probability that these genes are assigned to the same set in the partition under a random assignment is < 0.05.) This shows that our method identifies sets that correspond to pathways or sets of interacting genes without any *a priori* information about the interactions among genes and their assignment to pathways.

TCGA Glioblastoma Multiforme. We analyzed 251 samples from a recent TCGA study on this cancer type [7]. We restricted our analysis to the 27 genes reported in [7] as part of the landscape of pathway alterations in GBM, mostly obtained from manual curation.

For each gene, we considered single nucleotide variants, indels, and copy number aberrations consistent with the report in [7] for these genes. The progression model inferred by our method is shown in Figure 4b. In 4 of the 6 sets in the progression model inferred by our method, at least 50% of the genes are part of the same pathway (as annotated in [7]), and each set has such genes coming from a different pathway; for 3 of these sets, all but 1 gene are part of the same pathway. In particular, the second set in the progression contains mostly genes from the Rb1 pathway, the third set contains mostly genes in the PI3K pathway, and the fourth set in the progression contains mostly genes in the p53 pathway. Moreover, 5 of the 6 sets in the model (i.e., all sets with at least 2 genes) contain at a least a pair of genes that are in the same pathway. The bootstrap analysis reveals that on average the assignment (in terms of progression stage) of genes that are in a set with at least another gene in the same pathway is more stable than the assignment of the other genes. For the first 4 sets in the progression model, this is true also considering only the genes in the specific set. This show that the model reported by our method identifies pathway relations among genes in the different stages.

4   Conclusions

In this paper we study the problem of the simultaneous identification of cancer pathways and the tumor progression from cross-sectional mutation data. We formally define a model in which mutations within each pathway are exclusive, while they satisfy a linear progression at the pathway level. We prove that the problem of reconstructing the best model is NP-hard, and provide an ILP formulation to solve the problem for reasonably sized datasets. Moreover we show, analytically and with synthetic data, that under reasonable assumptions on the progression model and on the errors occurring in real data the optimal solution provided by our method captures the correct progression model when enough samples are considered.

We analyze somatic mutations data from three cancer studies, and show that most of the current knowledge of the mutation progression in these studies is captured by the models produced by our method. Most of the sets in the models obtained from these datasets correspond to interacting genes or part of known pathways, showing the ability of our method to correctly infer cancer pathways while inferring the progression of genetic events leading to cancer.

There are many directions in which our work can be extended. In certain cases more information about the probability of false positives and false negatives is known, even for each single gene. Our ILP formulation can easily incorporate such such information whenever available. Moreover, while our current formulation requires all the genes to be included in the model, the inclusion in the analysis of all measured somatic mutations from whole-exome or whole-genome sequencing requires to explicitly model the fact that some genes have mutations not associated with tumor progression; our model can be modified to not include all the genes in the cancer pathways defining the progression. For example, this can obtained by relaxing the constraint that each gene appears in exactly one set of the model and including in the objective function a penalization term for mutations that are not included in the model. Finally, more complex models at the pathway level could be considered. However, the challenges of simultaneous reconstruction of cancer pathways and complex models among them from a finite number of samples imply that these generalizations, and the comparison of the different models one can obtain, will not be straightforward.

References

1. Attolini, C.S.-O., Cheng, Y.-K., Beroukhim, R., Getz, G., Abdel-Wahab, O., et al.: A mathematical framework to determine the temporal sequence of somatic genetic events in cancer. Proc. Natl. Acad. Sci. U S A 107(41), 17604–17609 (2010)
2. Beerenwinkel, N., Sullivant, S.: Markov models for accumulating mutations. Biometrika 96(3), 645–661 (2009)
3. Beerenwinkel, N., Eriksson, N., Sturmfels, B.: Evolution on distributive lattices. J. Theor. Biol. 242(2), 409–420 (2006)
4. Beerenwinkel, N., Eriksson, N., Sturmfels, B.: Conjunctive bayesian networks. Bernoulli 13(4), 893–909 (2007)
5. Beerenwinkel, N., Rahnenführer, J., Däumer, M., Hoffmann, D., Kaiser, R., et al.: Learning multiple evolutionary pathways from cross-sectional data. J. Comput. Biol. 12(6), 584–598 (2005)
6. Beerenwinkel, N., Rahnenführer, J., Kaiser, R., Hoffmann, D., Selbig, J., et al.: Mtreemix: a software package for learning and using mixture models of mutagenetic trees. Bioinformatics 21(9), 2106–2107 (2005)
7. Brennan, C.W., Verhaak, R.G.W., McKenna, A., Campos, B., Noushmehr, H., et al.: The somatic genomic landscape of glioblastoma. Cell 155(2), 462–477 (2013)
8. Cheng, Y.-K., Beroukhim, R., Levine, R.L., Mellinghoff, I.K., Holland, E.C., et al.: A mathematical methodology for determining the temporal order of pathway alterations arising during gliomagenesis. PLoS Comput. Biol. 8(1), e1002337 (2012)
9. Ciriello, G., Cerami, E., Sander, C., Schultz, N.: Mutual exclusivity analysis identifies oncogenic network modules. Genome Res. 22(2), 398–406 (2012)
10. Dees, N.D., Zhang, Q., Kandoth, C., Wendl, M.C., Schierding, W., et al.: Music: identifying mutational significance in cancer genomes. Genome Res. 22(8), 1589–1598 (2012)
11. Desper, R., Jiang, F., Kallioniemi, O.P., Moch, H., Papadimitriou, C.H., et al.: Inferring tree models for oncogenesis from comparative genome hybridization data. J. Comput. Biol. 6(1), 37–51 (1999)
12. Desper, R., Jiang, F., Kallioniemi, O.P., Moch, H., Papadimitriou, C.H., et al.: Distance-based reconstruction of tree models for oncogenesis. J. Comput. Biol. 7(6), 789–803 (2000)
13. Efron, B., Tibshirani, R.: An introduction to the bootstrap, 1st edn. Chapman and Hall (1994)
14. Fearon, E.R., Vogelstein, B.: A genetic model for colorectal tumorigenesis. Cell 61(5), 759–767 (1990)
15. Fearon, E.R.: Molecular genetics of colorectal cancer. Annu. Rev. Pathol. 6, 479–507 (2011)
16. Gerstung, M., Baudis, M., Moch, H., Beerenwinkel, N.: Quantifying cancer progression with conjunctive bayesian networks. Bioinformatics 25(21), 2809–2815 (2009)
17. Gerstung, M., Eriksson, N., Lin, J., Vogelstein, B., Beerenwinkel, N.: The temporal order of genetic and pathway alterations in tumorigenesis. PLoS One 6(11), e27136 (2011)
18. Hjelm, M., Höglund, M., Lagergren, J.: New probabilistic network models and algorithms for oncogenesis. J. Comput. Biol. 13(4), 853–865 (2006)
19. Kandoth, C., McLellan, M.D., Vandin, F., Ye, K., Niu, B., et al.: Mutational landscape and significance across 12 major cancer types. Nature 502(7471), 333–339 (2013)

20. Lawrence, M.S., Stojanov, P., Polak, P., Kryukov, G.V., Cibulskis, K., et al.: Mutational heterogeneity in cancer and the search for new cancer-associated genes. Nature 499(7457), 214–218 (2013)
21. Leiserson, M.D.M., Blokh, D., Sharan, R., Raphael, B.J.: Simultaneous identification of multiple driver pathways in cancer. PLoS Comput. Biol. 9(5), e1003054 (2013)
22. Miller, C.A., Settle, S.H., Sulman, E.P., Aldape, K.D., Milosavljevic, A.: Discovering functional modules by identifying recurrent and mutually exclusive mutational patterns in tumors. BMC Med. Genomics 4, 34 (2011)
23. Rahnenführer, J., Beerenwinkel, N., Schulz, W.A., Hartmann, C., von Deimling, A., et al.: Estimating cancer survival and clinical outcome based on genetic tumor progression scores. Bioinformatics 21(10), 2438–2446 (2005)
24. Sakoparnig, T., Beerenwinkel, N.: Efficient sampling for bayesian inference of conjunctive bayesian networks. Bioinformatics 28(18), 2318–2324 (2012)
25. Shahrabi Farahani, H., Lagergren, J.: Learning oncogenetic networks by reducing to mixed integer linear programming. PLoS One 8(6), e65773 (2013)
26. The Cancer Genome Atlas Network. Comprehensive genomic characterization defines human glioblastoma genes and core pathways. Nature 455(7216), 1061–1068 (2008)
27. The Cancer Genome Atlas Network. Comprehensive molecular characterization of human colon and rectal cancer. Nature 487(7407), 330–337 (2012)
28. Tofigh, A., Sjölund, E., Höglund, M., Lagergren, J.: A global structural em algorithm for a model of cancer progression. Advances in Neural Information Processing Systems 24, 163–171 (2011)
29. Vandin, F., Upfal, E., Raphael, B.J.: De novo discovery of mutated driver pathways in cancer. Genome Res. 22(2), 375–385 (2012)
30. Vandin, F., Upfal, E., Raphael, B.J.: Finding driver pathways in cancer: models and algorithms. Algorithms Mol. Biol. 7(1), 23 (2012)
31. Vogelstein, B., Papadopoulos, N., Velculescu, V.E., Zhou, S., Diaz Jr., L.A., et al.: Cancer genome landscapes. Science 339(6127), 1546–1558 (2013)
32. Wood, L.D., Parsons, D.W., Jones, S., Lin, J., Sjöblom, T., et al.: The genomic landscapes of human breast and colorectal cancers. Science 318(5853), 1108–1113 (2007)
33. Yeang, C.-H., McCormick, F., Levine, A.: Combinatorial patterns of somatic gene mutations in cancer. FASEB J. 22(8), 2605–2622 (2008)
34. Zhang, J., Baran, J., Cros, A., Guberman, J.M., Haider, S., et al.: International Cancer Genome Consortium Data Portal–a one-stop shop for cancer genomics data. Database (2011)

DIPSPADES: Assembler for Highly Polymorphic Diploid Genomes

Yana Safonova[1], Anton Bankevich[1,2], and Pavel A. Pevzner[1,3]

[1] Algorithmic Biology Laboratory, St. Petersburg Academic University,
Russian Academy of Sciences, St. Petersburg, Russia
yana.safonova@ablab.spbau.ru, anton.bankevich@gmail.com
[2] Theodosius Dobzhansky Center for Genome Bioinformatics,
St. Petersburg State University, St. Petersburg, Russia
[3] Dept. of Computer Science and Engineering, University of California, San Diego,
La Jolla, CA, USA

Abstract. While the number of sequenced diploid genomes of interest have been steadily increasing in the last few years, assembly of highly polymorphic (HP) diploid genomes remains challenging. As a result, there is shortage of tools for assembling HP genomes from NGS data. The initial approaches to assembling HP genomes were proposed in the pre-NGS era and are not well suited for NGS projects. We present the first de Bruijn graph assembler DIPSPADES for HP genomes and demonstrate that it significantly improves on the state-of-the-art in the HP genome assembly.

Keywords: genome assembly, polymorphism, de Bruijn graph, SPAdes.

1 Introduction

Assembly of highly polymorphic (HP) diploid genomes is a complex computational problem. When two haplomes are very similar, e.g., as human haplomes that differ from each other by only $\approx 0.1\%$ of nucleotides, both haplomes are usually assembled as a single reference genome (with further analysis of SNPs). Assembling found SNPs into human haplomes is a difficult but well studied problem (Aguiar and Istrail, 2012 [1], Xie et al., 2008 [18], He et al., 2010 [11], Zhao et al., 2005 [20], Bansal et al., 2008 [4]).

This paper addresses an even more challenging problem of assembling haplomes that differ from each other by 0.4–10% (e.g., like in HP sea squirt genomes). The standard assembly approaches fail to reconstruct individual haplomes in HP genomes; moreover, it is not clear whether the algorithms proposed for human haplome assembly can contribute to assembling HP genomes.

Assembly of a diploid genome can result in two types of contigs: *haplocontigs* (contigs representing both haplomes) and *consensus contigs* (representing a consensus of both haplomes for the orthologous regions) (Fig. 1). Consensus contigs do not adequately represent haplomes but are rather a mosaic of segments from both haplomes. Thus, in each polymorphic site of a diploid genome, the alleles

R. Sharan (Ed.): RECOMB 2014, LNBI 8394, pp. 265–279, 2014.
© Springer International Publishing Switzerland 2014

present in a consensus contig are somewhat randomly chosen from one of haplomes. In practice, since some regions of HP genome are less polymorphic than others, conventional assemblers generate a mixture of haplocontigs and consensus contigs while assembling HP genomes. We also define *double contigs*, a pair of haplocontigs representing both haplomes for the same genomic regions (Fig. 1).

Two approaches were proposed for assembling HP genomes (referred to as *HP assemblies* below). The first approach for HP assembles (applied to fish *F. rubripes*) was proposed in the pre-NGS era and was based on generating consensus contigs by intentionally ignoring differences between haplomes. To achieve this goal, Aparicio et al., 2002 [2] constructed the overlap graph of Sanger reads while allowing large differences in overlaps between reads. A similar approach was applied to genome assembly of the sea squirt *S. intestinalis* (Dehal et al., 2002 [8]). The resulting assembly was further used as a reference to align reads and restore both haplomes. This approach, while feasible with Sanger reads, is not very practical in the case of NGS reads that are more amenable to the de Bruijn graph approaches.

The second approach (Huang et al., 2012 [12], Vinson et al., 2005 [17]) is to generate haplocontigs using a conventional assembly algorithm and to further reconstruct allelic relationships between haplotypes based on pairwise contig alignments. In reality, such approaches generate a mixture of haplocontigs and consensus contigs since the degree of polymorphism varies along the HP genomes. As the result, assemblies generated by this approach tend to be fragmented since they represent a mosaic of consensus and haplocontigs. Recently, Donmez and Brudno, 2011 [9] proposed advanced methods of generating haplocontigs based on the overlap graph approach.

We present DIPSPADES, a new algorithm for assembling HP genomes. DIPSPADES uses the de Bruijn graph constructed by SPAdes assembler [3] to generate both consensus and haplocontigs. Instead of analyzing contigs or long read alignments (as in the previous approaches), we use the de Bruijn graph to mask polymorphism in contigs and to produce a more comprehensive representation of the genome by both consensus contigs and haplocontigs. The question how accurate are these assemblies constructed in the pre-NGS era remains open since there is no gold standard for checking the validity of HP assemblies. Such benchmarking of HP assemblies is an important goal of this paper.

To provide the first comprehensive benchamarking of HP assemblies, we took advantage of a unique dataset generated in the course of a recent massive effort to sequence 37 genomes of *S. commune* conducted in Dr. Alexey Kondrashov laboratory at Moscow State University (see below for description of all datasets that were analyzed in this paper).

S. commune is a model organism (wood-degrading mushroom) whose genome is ideally suited for benchmarking HP genome assemblers. The unique feature of the widely distributed haploid *S. commune* is that two different organisms differ by $7 - 12\%$ even if collected on the same continent (and up to 25% on different continents). Thus, combining reads from two *S. commune* genomes perfectly models an HP genome, yet allowing one to test the quality of assembly, the bottleneck in previous studies of assembly algorithms for diploid genomes.

Benchmarking of DIPSPADES on both simulated and real fungi datasets (with polymorphism rate varying from 0.4 to 10 percent) demonstrated that DIPSPADES significantly improves assemblies of HP genomes. DIPSPADES is also an excellent *comparative assembler* that can be used to generate a consensus assembly of multiple similar genomes (due to lack of space this result will be described elsewhere).

Fig. 1 shows DIPSPADES pipeline (green arrows) that allows to achieve the goal of constructing double contigs (black arrow).

2 Definitions

Let $DB(\text{GENOME}, k)$ be the de Bruijn graph [7] of a genome GENOME and its reverse complement GENOME$'$, where vertices and edges correspond to $(k-1)$-mers and k-mers, respectively. Each chromosome in GENOME and GENOME$'$ corresponds to a path in this graph; a set of these paths represents the *genome traversal* of the graph. In this paper, we will work with *condensed de Bruijn graphs* [3], where each edge is assigned a *length* (in k-mers) and the length of a path is the sum of its edge lengths (rather than the number of edges in the condensed de Bruijn graph). Let $DB(\text{READS}, k)$ be the de Bruijn graph constructed from a set READS of reads from GENOME and their reverse complements. For simplicity we first consider an idealized case with full coverage of GENOME and error-free reads. In this case, the graphs $DB(\text{READS}, k)$ and $DB(\text{GENOME}, k)$ coincide. In reality, DIPSPADES analyzes error-prone reads and gaps in coverage.

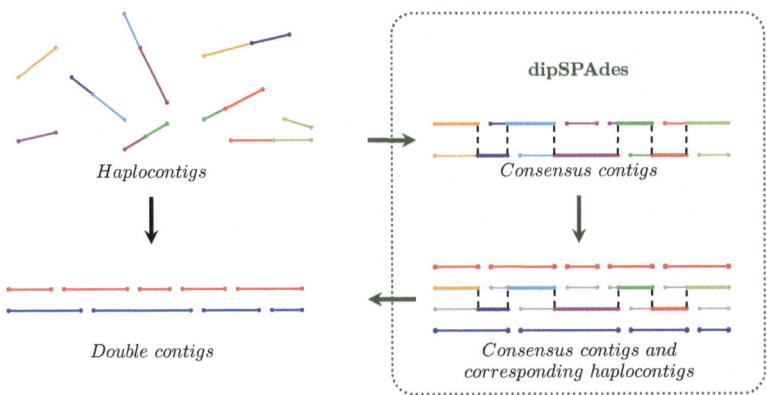

Fig. 1. The DIPSPADES pipeline (green arrows). Conventional assemblers generate haplocontigs that represent both haplomes shown in red and blue. However, in practice colors of haplocontigs are unknown. DIPSPADES uses the de Bruijn graph to generate consensus contigs by combining and extending haplocontigs. Afterwards, DIPSPADES restores allelic relations using alignment of haplocontigs to the consensus contigs.

A diploid genome GENOME = (GENOME$_1$ ∪ GENOME$_2$) can be viewed as two similar double-stranded haplomes GENOME$_1$ and GENOME$_2$. Typically, differences between haplomes are represented as a collection of SNPs and short indels. Given a pairwise alignment, we use the notion of *percent identity* (percent of matches among all columns of the alignment) to measure similarity between sequences. Correspondingly *divergence* is 100 minus percent identity. For example, the distribution of the divergence in the alignment of two fungi genomes from *Schizophyllum commune* demonstrates that ≈ 88 percent of genome length have divergence below 20%.

3   Methods

3.1   Motivation

Consider an imaginary genome GENOME = $aRbRcRd$ with a perfectly conserved long repeat R of multiplicity 3 and four unique regions a, b, c, d (Fig. 2(a)). The de Bruijn graph DB(GENOME, k) has five edges: R, a, b, c, d (Fig. 2(b)). Now imagine that GENOME evolved into two haplomes in such a way that in the first haplome GENOME$_1$, 1st and 3rd copy of repeat R significantly diverged resulting in unique regions R_1 and R_3. Similarly, in the second haplome GENOME$_2$, the 2nd copy of repeat R significantly diverged resulting in a unique region R_2. The resulting haplomes can be represented as aR_1bRcR_3d and $aRbR_2cRd$ (Fig. 2(c)). The de Bruijn graph DB(GENOME$_1$ ∪ GENOME$_2$, k) is shown in Fig. 2(d). While the genomic traversal in the assembly graph is unknown, SPAdes and other assemblers generate a set of subpaths of this traversal referred to as contigs. For example, consider an edge R_1 in DB(GENOME$_1$ ∪ GENOME$_2$, k). Note that the only edge that can follow R_1 in genome traversal is b. Similarly, the only edge that can precede R_1 in genome traversal is a. Thus, aR_1b and similarly bR_2c and cR_3d are substrings of either GENOME$_1$ or GENOME$_2$. Moreover, analysis of these contigs reveals that aR_1b overlaps with bR_2c that in turn overlaps with cR_3d and thus would lead to the assembly of the entire genome into a consensus contig $aR_1bR_2cR_3d$ (Fig. 2(e)).

In other words, the endpoints of haplocontigs from different haplomes do not match since breakpoints of haplocontigs often happen at different positions on different haplomes. As a result, overlaps between these haplocontigs allow one to assemble them into longer sequences. The example above illustrates how divergence in diploid genomes helps to improve the assembly of HP genomes but presents a highly idealized case. In practice, this approach will not work for a variety of reasons, e.g., fragment b in two haplomes may be highly diverged preventing one from detecting an overlap between aR_1b and bR_2c. To address this problem, DIPSPADES uses a *polymorphism masking algorithm* described below that essentially suppresses differences between b in aR_1b and b in bR_2c (Subsection 3.2). We refer to the resulting contigs as *masked haplocontigs* and acknowledge that such polymorphism masking may produce a version of b that belongs to neither GENOME$_1$ nor GENOME$_2$.

Thus, DIPSPADES consists of two parts. First, it masks polymorphisms to reveal overlaps between contigs in graph $DB(GENOME_1 \cup GENOME_2, k)$ (Subsection 3.2). Second, it searches for overlaps in masked contigs and extends them, thus improving the quality of assembly (Subsection 3.3). However, at the last stage of DIPSPADES, we reconstruct double contigs in both haplomes (Subsection 3.4).

3.2   Polymorphism Masking

De Bruijn Graphs of Diploid Genomes. Since haplomes $GENOME_1$ and $GENOME_2$ are similar, the de Bruijn graphs $DB(GENOME_1, k)$ and $DB(GENOME_2, k)$ are also similar. Fig. 3(a) and 3(b) show two imaginary haplomes with low and high polymorphism rates, respectively. Polymorphic sites are shown by red in one haplome and blue in another. We color a k-mer (edge in the de Bruijn graph) as red, blue, or black, depending on whether it belongs only to the first genome, only to the second genome or to both genomes, respectively. Red/blue edges in $DB(GENOME_1 \cup GENOME_2, k)$ often aggregate into red/blue paths as shown in Fig. 3(c), 3(d). A red and a blue path between the same vertices form a *bulge*. We refer to paths forming a bulge as *alternative paths*.

The average bulge length depends on the polymorphism rate: bulges are short in the case of low polymorphism rate (Fig. 3(e)) and long in the case of of high polymorphism rate (Fig. 3(f)). Distributions of bulge lengths for *C. albicans* (low polymorphism rate) and *A. protococcarum* (high polymorphism rate) genomes show that average bulge lengths (for k-mers size 56) are ≈ 139 nt and ≈ 833 nt, respectively. Bulges are very prominent in HP genomes, e.g., in the de Bruijn graph of *A. protococcarum* genome, 99.2% of the total length of edges in the graph belongs to bulges of length less than 25000 (for k-mers size 56).

We distinguish between bulges caused by sequencing errors and bulges caused by polymorphisms. The former type of bulges are artifacts that are removed by existing fragment assembly algorithms while the later type of bulges are important for HP genome assembly. We assume that bulges caused by errors in reads have been removed by SPAdes (e.g., by removing alternative paths with lower coverage) before DIPSPADES even starts analyzing later type of bulges.

Haplocontigs versus Consensus Contigs. HP genomes represent a mosaic of regions with varying degrees of polymorphisms. Conventional assemblers consider the non-polymorphic regions of HP genomes as repeats, attempt to resolve them (e.g., by using read-pairs), and output the resulting haplocontigs.

As was advocated in [3], for regions with low polymorphism rate, it makes sense to intentionally collapse short bulges rather than retain information about polymorphisms. The resulting consensus contigs represent a mixture of haplomes $GENOME_1$ and $GENOME_2$ since parts of genome that contain deleted polymorphic variations are not represented in the assembly. We refer to such a mixture of haplomes as CONSENSUSGENOME, i.e. in each polymorphic site of a diploid genome, the alleles present in the CONSENSUSGENOME are somewhat randomly

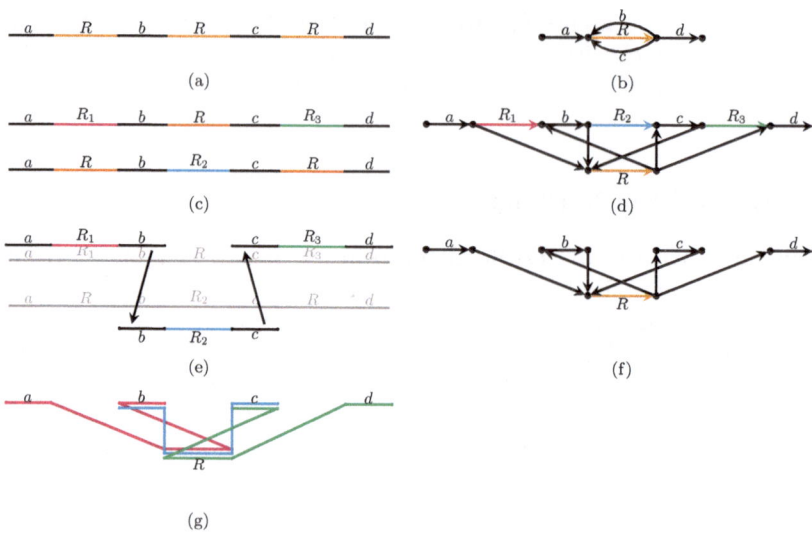

Fig. 2. Fig. (a) shows a genome GENOME that contains a repeat R with multiplicity 3. Fig. (b) shows the de Bruijn graph DB(GENOME, k). Fig. (c) illustrates highly polymorphic haplomes GENOME$_1$ and GENOME$_2$. In the first haplome, the 1st and the 3rd copy of the repeat R diverged, resulting in a unique regions R_1 and R_3. In the second haplome, the 2nd copy of the repeat R diverged, resulting in a unique region R_2. Fig. (d) shows the de Bruijn graph DB(GENOME$_1 \cup$ GENOME$_2$, k). Fig. (e) shows overlaps between contigs obtained from DB(GENOME$_1 \cup$ GENOME$_2$, k) allowing one to construct the consensus as a single contig. Fig. (f) shows CONSENSUSGRAPH. Fig. (g) shows contigs (red, blue, and green paths) that map to graph in Fig. (f).

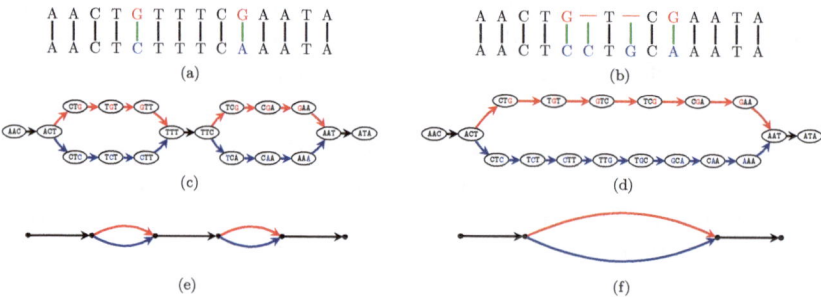

Fig. 3. Low (left) and high (right) rates of polymorphisms result in short and long bulges in the de Bruijn graphs. Genomes ((a), (b)), uncondensed de Bruijn graphs ((c), (d)), and condensed de Bruijn graph ((e), (f)).

chosen from one of haplomes (Fig. 4(a)). For the sake of simplicity, we assume that there are no rearrangements between GENOME$_1$ and GENOME$_2$. In practice, DIPSPADES considers micro-inversions and micro-transpositions (inversions and transpositions with short span) as polymorphisms. The breakpoints of genome rearrangements with longer span usually turn into breakpoints within contigs output by DIPSPADES.

Polymorphism Masking Algorithm. As we mentioned above, polymorphisms in a diploid genome are represented by bulges of varying lengths in the de Bruijn graph DB(GENOME$_1$ ∪ GENOME$_2$, k). DIPSPADES uses a different and more complex bulge finding approach than existing assemblers that typically deal with rather short bulges (e.g. ≈ 250 nt long). In contrast, DIPSPADES collapses bulges that may be two order of magnitude longer since bulges formed by highly diverged regions tend to be very long. Here we describe an algorithm for finding and masking polymorphisms by collapsing bulges (compare with aggressive bulge collapsing in [3]).

For a bulge (P_1, P_2) formed by alternative paths P_1 and P_2 in the de Bruijn graph, we define $divergence(P_1, P_2)$ as the divergence between these paths. We collapse bulge (P_1, P_2) (COLLAPSEBULGE procedure in the pseudocode below) if $divergence(P_1, P_2)$ is below a fixed threshold $maxDivergence$. Distributions of divergence in bulges for *C. albicans* (low polymorphic rate) and *A. proto-coccarum* (high polymorphic rate) illustrate that the vast majority of bulges have divergence below 20% for both low and high polymorphism rates. We used the divergence threshold $maxDivergence = 20\%$ for the tests described in the Results section.

We refer to the graph after the aggressive bulge collapsing as the CONSENSUSGRAPH and represent the CONSENSUSGENOME as a traversal in this graph. Fig. 4(b) shows two bulges corresponding to two polymorphic sites. When these bulges are collapsed (Fig. 4(c)) parts of the genome that correspond to edges 2 and 6 are no longer present in the CONSENSUSGRAPH.

Note that this procedure sometimes collapses bulges that do not represent orthologous regions of a diploid genome but instead collapses non-orthologous regions with spurious similarities. On the other hand, it does not collapse some highly diverged orthologous regions. Also some subgraphs of the de Bruijn graph are so tangled that it is difficult to find and collapse bulges in these subgraphs. Effective algorithms for bulge finding will be described elsewhere. For simplicity, we further assume that all orthologous regions were collapsed correctly.

Polymorphism Masking in Contigs. Most fragment assembly algorithms discard information about the alternative paths removed during the bulge collapsing. In contrast, DIPSPADES capitalizes on a unique feature of SPAdes and *projects* k-mers from the removed alternative path to the retained alternative paths (see [3]). Thus, the projection procedure defines mapping of every path in DB(GENOME$_1$ ∪ GENOME$_2$, k) to a path in CONSENSUSGRAPH. E.g., one can map haplocontigs and even haplomes GENOME$_1$ and GENOME$_2$ (if they were

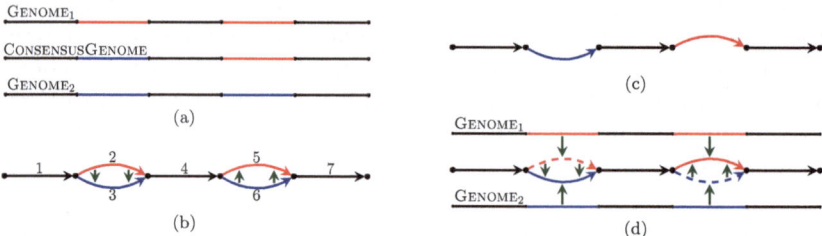

Fig. 4. Fig. (a) illustrates a possible CONSENSUSGENOME of haplomes GENOME₁ and GENOME₂. Fig. (b) shows the de Bruijn graph DB(GENOME₁∪GENOME₂, k). The directions of green arrows match the direction of the bulge collapsing in the polymorphism masking algorithm. Fig. (c) shows CONSENSUSGRAPH generated by the polymorphism masking algorithm. Fig. (d) shows haplocontigs representing GENOME₁ and GENOME₂ that map to the same paths in CONSENSUSGRAPH.

known) to CONSENSUSGRAPH. Both GENOME₁ and GENOME₂ map to the same path that corresponds to CONSENSUSGENOME (Fig. 4(d)).

We apply the described polymorphism masking procedure to haplocontigs to obtain *masked haplocontigs*. Since haplocontigs are substrings of either GENOME₁ or GENOME₂, masked haplocontigs are substrings of CONSENSUSGENOME. It is much easier to analyze overlaps of masked haplocontigs (as compared to overlaps of haplocontigs) since in most cases the overlapping segments of masked haplocontigs are 100% similar and thus, are easy to detect.

Consider again the example shown in Fig. 2(f). After extracting contigs aR_1b, bR_2c, and cR_3d from the de Bruijn graph, we collapse bulges eventually transforming it into a graph shown in Fig. 2(d). For the sake of simplicity, Fig. 2(f) assumes an unrealistic case when regions a, b, c, d have not diverged at all. However, even if these regions diverged, DIPSPADES would mask polymorphisms between these regions. We remark that while the graph in Fig. 2(b) is merely a different drawing of the graph in Fig. 2(g), it also contains information about contigs that map to this graph (red, blue, and green paths in Fig. 2(d)). In particular, the contigs aR_1b, bR_2c, and cR_3d in Fig. 2(f) have collapsed to paths aRb, bRc, and cRd in Fig. 2(d). As a result, the collapsed bulges in Fig. 2(d) turn into *superpaths* in Fig. 2(f) (see [15]). This transformation allows us to extend the described approach from the idealized (Fig. 2(e)) to real and complex de Bruijn graph. Below we describe how DIPSPADES benefits from superpaths we generate at this stage.

3.3   Consensus Overlap Graph

Edges of the CONSENSUSGRAPH represent the consensus contigs that are typically rather fragmented. Moreover, even when we use paired reads to resolve repeats in this graph (e.g., using the read-pair analysis tool in SPAdes), the resulting assembly remains fragmented (see Results). Below we show how divergence between haplomes can be used to improve the consensus assembly.

In the previous subsection we described how to mask polymorphisms in haplocontigs. Though the masked haplocontigs represent substrings of CONSENSUSGENOME and cover CONSENSUSGENOME, each position in the CONSENSUSGENOME is covered twice by masked haplocontigs (e.g., in Fig. 4(d), each edge of the CONSENSUSGRAPH is covered by two haplocontigs with masked polymorphisms). In such an assembly, some masked haplocontigs may become *redundant*, (i.e., contained within other masked haplocontigs), and thus can be removed. In Fig. 4(d), contigs corresponding to GENOME$_1$ and GENOME$_2$ are the same and one contig from each pair can be safely removed.

DIPSPADES generates the consensus contigs that significantly improve the consensus assembly. This method utilizes superpaths and uses a strategy illustrated in Fig. 2. We apply the overlap graph assembly technique [6] to utilize information about the overlaps between the masked haplocontigs.

The Overlap Graph of Masked Haplocontigs. DIPSPADES constructs an overlap graph on the set of masked haplocontigs. A set of strings STRINGS is called *proper* if no string in this set is a substring of another string. Given a proper set of strings STRINGS, we construct the *overlap graph* as follows. The vertices of the overlap graph are strings from STRINGS. We connect vertices $string_1$ and $string_2$ by a directed edge if a sufficiently long suffix of $string_1$ equals to a prefix of $string_2$ (the default overlap length threshold is $minOverlap = 1500$ nucleotides). The overlap graph is obtained from this graph after removing all *transitive edges*, i.e., we remove an edge $(string_1, string_2)$ if there is an alternative directed path from $string_1$ to $string_2$ in the graph. The non-branching paths in the overlap graph are reported as *consensus contigs*.

3.4   Haplotype Assembly

Each consensus contig C corresponds to regions C_1 and C_2 in haplomes that are similar to C. Below we describe how DIPSPADES reconstructs sequences of C_1 and C_2 based on the projection of haplocontigs to the consensus contigs. Since each haplocontig is projected to a masked haplocontig and masked haplocontigs are assembled into consensus contigs, each haplocontig is aligned to a substring of a consensus contig.

For a consensus contig C, let HAPLO(C) denote the set of haplocontigs that are projected to C. Ideally each contig from HAPLO(C) is a substring of either C_1 or C_2. Thus, our goal is to split HAPLO(C) into disjoint subsets D_1 and D_2 such that contigs from D_1 correspond to C_1 and contigs from D_2 correspond to C_2. Afterwards, we reconstruct C_1 and C_2. Due to lack of space algorithms for splitting HAPLO(C) and for reconstruction of C_1 and C_2 will be described elsewhere.

3.5   DIPSPADES Algorithm

The pseudocode below computes haplocontigs using procedure COMPUTE-CONTIGS. In DIPSPADES, it is implemented using SPAdes assembly pipeline. DIPSPADES can also use haplocontigs computed by other tools.

Algorithm. DIPSPADES workflow

1: **procedure** DIPSPADES(READS, k, $maxDivergence$, $minOverlap$)
2:    GRAPH ← DEBRUIJNGRAPH(READS, k)
3:    HAPLOCONTIGS ← CONTIGS(GRAPH, READS)
4:    **for each** bulge formed by alternative paths P_1 and P_2 in GRAPH
5:        **if** DIVERGENCE(P_1, P_2) ≤ $maxDivergence$
6:            COLLAPSEBULGE(P_1, P_2)
7:    CONSENSUSGRAPH ← GRAPH
8:    MASKEDHAPLOCONTIGS ← MASKPOLYMORPHISMS(CONSENSUSGRAPH, HAPLOCONTIGS)
9:    CONSENSUSCONTIGS ← CONSENSUSCONTIGS(MASKEDHAPLOCONTIGS, $minOverlap$)
10:    DOUBLECONTIGS ← HAPLOTYPEASSEMBLY(CONSENSUSCONTIGS, MASKEDHAPLOCONTIGS)
11:    **return** CONSENSUSCONTIGS, DOUBLECONTIGS
---

4   Results

Datasets. We benchmarked DIPSPADES and other assembly tools on simulated and real datasets (Table 1). To simulate a diploid genome we used a single haplome of the diploid fungus *Candida dubliniensis* (haplome size 14.6 Mbp). We generated a polymorphic copy of each chromosome with divergence 10% using the uniform random distribution of SNPs and indels and simulated error-free paired-end reads with read length 100, insert size 270 and 11× coverage. It results in perfect coverage of genome by 56-mers.

We also analyzed three real datasets of Illumina reads for genomes of *Schizophyllum commune*, *Candida albicans* and recently sequenced (unpublished) fungus *Amoeboaphelidium protococcarum*, strain *X-5*. Sequencing data for *S. commune* and *A. protococcarum* were provided by the Laboratory of Evolutionary Genomics at Moscow State University directed by Dr. Alexey Kondrashov. Sequencing data for *Candida albicans* were obtained from NCBI (accession number SRX113442).

S. commune [14] is a haploid fungus, but divergence between genomes of any two *S. commune* organisms is high (≥ 7%). *A. protococcarum* is a diploid fungus with extremely high rate of divergence between haplomes (≈ 10%). Finally we analyzed diploid fungus *Candida albicans* (divergence ≈ 0.4%) to estimate the performance of various algorithms for low-polymorphism data.

Assembly Tools. Currently, HaploMerger is the only available (and practical) tool for assembling HP genomes from short reads. Thus, we benchmarked the performance of DIPSPADES and HaploMerger (build 20120810) [12] in generating consensus contigs. We also benchmarked the performance of SPAdes 2.5.1 [3] and Velvet 1.2.10 [19] in generating haplocontigs. In addition, we benchmarked SPAdes* (aggresive bulge collapsing in DIPSPADES followed by the repeat resolution algorithm implemented in SPAdes 2.5.1) in generating consensus contigs. We introduced SPAdes* in our benchmarking to illustrate advantages of

DIPSPADES as compared to standard assemblers run in the mode of the aggressive bulge collapsing.

We ran Velvet and HaploMerger with default parameters. While DIPSPADES and HaploMerger can be run on any set of haplocontigs, in this study we applied them to SPAdes haplocontigs (as shown in a recent independent benchmarking study by Magoc at al., 2013 [13], SPAdes typically improves on other assemblers in the case of relatively small genomes). To generate results for SPAdes and SPAdes*, we turned off the bulge removal procedure in SPAdes and removed all edges with low coverage instead. We expect Velvet and SPAdes to produce assemblies with total length close to the double length of a haplome, and SPAdes*, DIPSPADES, and HaploMerger to produce assemblies with total length close to the length of haplome. In each row of the benchmarking tables we highlighted the entries with the best results. In all tables, only contigs of length ≥ 500 bp were used.

Analysis of the assembly errors is a non-trivial task in the case of HP genomes. While there are excellent tools for analyzing assemblies (e.g., QUAST [10], GAGE [16], and others), neither of them is designed for HP genomes. For example, QUAST often reports misassemblies that represent alignment artifacts specific for HP genomes rather than assembly errors. We thus resorted to manual analysis of all misassemblies reported by QUAST in assemblies of simulated HP diploid genome. This manual analysis revealed 2, 1, 1, and 0 misassemblies for HaploMerger, SPAdes, DIPSPADES, and Velvet, respectively. We also investigated assembly errors in consensus assembly of *S. commune* where also very few potential misassemblies were found. This analysis will be presented elsewhere.

Benchmarking. Table 2 shows results of assembly of simulated HP diploid genome based on *C. dubliniensis* haplome. We mixed two libraries of *S. commune* and computed contigs from the graph that was constructed from the resulting mixed library. Such assembly tends to be very fragmented due to the very tangled de Bruijn graph structure (see Table). In the case of *A. protococcarum*, we had 2 Illumina paired-end libraries (see Table 1). DIPSPADES allows one to mix contigs from assemblies of different libraries. To obtain haplocontigs we assembled two libraries separately and mixed computed contigs. Tables 4 and 5 present benchmarking results for *A. protococcarum* and *C. albicans*, respectively.

Summary. Table 2 reveals some limitations of HaploMerger: for simulated *C.dubliniensis* data it results in N50 equal to only ≈ 6 Kbp, (as compared to ≈ 116 Kbp for DIPSPADES). This disappointing performance reflects limitations of HaploMerger in the case when the original assembly is fragmented. One can see that in this case, HaploMerger hardly improved on the original fragmented assembly by SPAdes (slightly increasing N50 from 5392 to 5876). This limitation (acknowledged in [12]) often prevents application of HaploMerger for NGS data where N50 is typically small.

Table 3 illustrates the difficulties of the diploid assembly in a real setting. As expected, SPAdes generates rather short contigs (N50 = 3598) since *S. commune*

Table 1. Information about genomes and sequencing data. For *S. commune*, the coverage of 4 strains varied from 15 to 34. All datasets were sequenced using Illumina technology with read length 100 bp.

Dataset name	Estimated genome size (Mbp)	Average divergence (%)	Insert size (bp)	Average coverage
Simulated HP diploid genome	14.6 × 2	10.0	270	11
S. commune (4 strains)	38.9	7.0	233	34
A. protococcarum (1st library)	11.0 × 2	10.0	270	667
A. protococcarum (2nd library)	11.0 × 2	10.0	170	166
C. albicans	14.5 × 2	0.4	196	43

Table 2. Assembly of simulated HP diploid genome based on *C. dubliniensis* haplome.

	Velvet	SPAdes	HaploMerger	SPAdes*	DipSPAdes
Expected total length (Mbp)	29.2	29.2	14.6	14.6	14.6
Total length (Mbp)	28.41	28.11	**14.64**	16.3	14.45
# contigs	7626	7973	3739	525	**433**
Largest contig	29486	29488	29488	**491990**	**491969**
N50	5378	5392	5876	88380	**116291**
N75	3085	3094	3345	40720	**50879**

Table 3. Assembly of two *S. commune* (A8 and B3) genomes. Haplocontigs were obtained from the assembly graph that was constructed from a mixed library of A8 and B3 reads. HaploMerger failed to produce results on these haplocontigs since it typically requires contigs with N50 exceeding tens of Kbp.

	Velvet	SPAdes	HaploMerger	SPAdes*	DipSPAdes
Expected total length (Mbp)	77.8	77.8	38.9	38.9	38.9
Total length (Mbp)	39.15	60.33	N/A	45.91	**38.85**
# contigs	34406	26820	N/A	5721	**3147**
Largest contig	37580	44596	N/A	**231443**	181171
N50	1219	3598	N/A	24931	**27625**
N75	761	1694	N/A	8477	**14065**

Table 4. Assembly of *A. protococcarum*. Columns "SPAdes (IS = 170)" and "SPAdes (IS = 270)" illustrate results of assemblies of libraries with corresponding insert sizes. For obtaining consensus contigs we used mixture of contigs from these runs.

	Velvet	SPAdes (IS=170)	SPAdes (IS=270)	Haplo-Merger	SPAdes*	DipSPAdes
Expected total length (Mbp)	22.0	22.0	22.0	11.0	11.0	11.0
Total length (Mbp)	19.57	23.25	24.45	12.24	16.5	**11.15**
# contigs	13926	4620	1902	742	1490	**230**
Largest contig	90937	138704	200276	200276	205337	**435986**
N50	1656	8760	28942	38265	30393	**130702**
N75	994	4689	14470	19842	12787	**62530**

Table 5. Assembly of *C. albicans*.

	Velvet	SPAdes	HaploMerger	SPAdes*	DipSPAdes
Expected total length (Mbp)	29.0	29.0	14.5	14.5	14.5
Total length (Mbp)	11.28	17.37	2.84	**14.85**	13.93
Number of contigs	6731	4007	**337**	1540	1174
Largest contig	34870	112388	92126	**116985**	**116985**
N50	2276	8788	23529	25691	**27961**
N75	1155	3300	8115	10639	**12456**

strains represent a mosaic of highly diverged and conserved regions. As a result, HaploMerger failed to produce results on this dataset. In contrast, DIPSPADES was able to significantly improve on SPAdes (N50 = 27625). Table 3 also shows some shortcomings of SPAdes* that generated assembly with excessive total length due to its inability to collapse some polymorphisms.

Table 4 illustrates performance of DIPSPADES and other tools on real HP diploid genome. One can see that DIPSPADES greatly improved assembly (from N50 = 28942 to N50 = 130702).

Table 5 illustrates that even in the case of the low polymorphism rate ($\approx$ 14 times lower than in *S. commune* but $\approx$ 5 times higher than in human). DIPSPADES allows one to significantly improve assembly (from N50 = 8788 to N50 = 27961). In contrast, HaploMerger generated very poor assembly that covers only $\approx$ 20% of the genome.

Discussion. While the number of sequenced diploid genomes of interest have been steadily increasing in the last few years, assembly of polymorphic diploid genomes remains challenging. The lion's share of diploid genomes (probably most) feature much higher polymorphism rates than the human genome ($\approx$ 0.1%). Since assembly of HP diploid genomes is challenging, inbreeding is often a necessary step to enable high-quality assemblies [5]. This strategy allows one to breed organisms with $\approx$ 10–fold reduction in polymorphism rates after sufficient number of generations. However, the inbreeding approach is time-consuming and often fails to generate viable offspring due to the high death rates of inbred organisms [5].

While HaploMerger remains the only practical NGS assembler for HP genomes, it relies on standard assembly tools and is primarily designed for analyzing long contigs produced by these tools. However, the reality is that, in the case of HP genomes, the contigs produced by these tools are often short and are not well suited for further analysis by HaploMerger. Moverover HaploMerger is optimized for genomes with very high rate of polymorphisms. For example, we observed that HaploMerger produces low-quality incomplete assemblies of genomes with relatively low polymorphism rates (Table 5). Thus, there is plenty of room for further improvement of assemblies of HP genomes. DIPSPADES is the first de Bruijn graph assembly tool for NGS data that is optimized for HP genomes (both medium and extremely high divergence). We have shown that DIPSPADES generates consensus assemblies that significantly improve on the state-of-the-art tools.

Acknowledgments. We are indebted to Drs. Alexey Kondrashov, Sergey Naumenko, Maria Baranova, and Yegor Bazykin for many helpful discussions. The work of YS, AB, and PAP was supported by the Government of the Russian Federation, grant 11.G34.31.0018. The work of AB was supported by grant 11.G34.31.0068 to Dr. Stephen J. O'Brien. PAP was partially supported by the National Institutes of Health, grant 3P41RR024851-02S1. We would like to thank Dr. Alexey Kondrashev for providing sequencing data of *S. commune* and

A. protococcarum that were generated with support of Government of the Russian Federation, grant 11.G34.31.0008. We are grateful to Dr. Shengfeng Huang and Dr. Anton Korobeynikov for help with HaploMerger benchmarking. We are also grateful to Dr. Alla Lapidus for help with preparing this paper.

References

1. Aguiar, D., Istrail, S.: Hapcompass: a fast cycle basis algorithm for accurate haplotype assembly of sequence data. Journal of Computational Biology 19, 577–590 (2012)
2. Aparicio, S., et al.: Whole-genome shotgun assembly and analysis of the genome of Fugu rubripes. Science 297, 1301–1310 (2002)
3. Bankevich, A., Nurk, S., Antipov, D., Gurevich, A.A., Dvorkin, M., Kulikov, A.S., Lesin, V.M., Nikolenko, S.I., Pham, S., Prjibelski, A.D., Pyshkin, A.V., Sirotkin, A.V., Vyahhi, N., Tesler, G., Alekseyev, M.A., Pevzner, P.A.: SPAdes: A New Genome Assembly Algorithm and Its Applications to Single-Cell Sequencing. Journal of Computational Biology 19, 455–477 (2012)
4. Bansal, V., Halpern, A.L., Axelrod, N., Bafna, V.: An MCMC algorithm for haplotype assembly from whole-genome sequence data. Genome Research 18, 1336–1346 (2008)
5. Barriere, A., Yang, S., Pekarek, E., Thomas, C., Haag, E., Ruvinsky, I.: Detecting heterozygosity in shotgun genome assemblies: Lessons from obligately outcrossing nematodes. Genome Research 19, 470–480 (2009)
6. Batzoglou, S., Jaffe, D., Stanley, K., Butler, J., Gnerre, S., Mauceli, E., Berger, B., Mesirov, J., Lander, E.: Arachne: a whole-genome shotgun assembler. Genome Research 12, 177–189 (2002)
7. Compeau, F., Pevzner, P., Tesler, G.: How to apply de bruijn graphs to genome assembly. Nature Biotechnology 29, 987–991 (2011)
8. Dehal, P., et al.: The draft genome of Ciona intestinalis: Insights into chordate and vertebrate origins. Science 298, 2157–2167 (2002)
9. Donmez, N., Brudno, M.: Hapsembler: An Assembler for Highly Polymorphic Genomes. In: Bafna, V., Sahinalp, S.C. (eds.) RECOMB 2011. LNCS, vol. 6577, pp. 38–52. Springer, Heidelberg (2011)
10. Gurevich, A., Saveliev, V., Vyahhi, N., Tesler, G.: QUAST: Quality Assessment Tool for Genome Assemblies. Bioinformatics 29, 1072–1075 (2013)
11. He, D., Choi, A., Pipatsrisawat, K., Darwiche, A., Eskin, E.: Optimal algorithms for haplotype assembly from whole-genome sequence data. Bioinformatics 26, i183–i190 (2010)
12. Huang, S., Chen, Z., Huang, G., Yu, T., Yang, P., Li, J., Fu, Y., Yuan, S., Chen, S., Xu, A.: HaploMerger: reconstructing allelic relationships for polymorphic diploid genome assemblies. Genome Research 22, 1581–1588 (2012)
13. Magoc, T., Pabinger, S., Canzar, S., Liu, X., Su, Q., Puiu, D., Tallon, L.J., Salzberg, S.L.: GAGE-B: An evaluation of genome assemblers for bacterial organiss. Bioinformatics 29, 1718–1725 (2013)
14. Ohm, R.A., et al.: Genome sequence of the model mushroom Schizophyllum commune. Nature 28, 957–963 (2010)
15. Pevzner, P., Tang, H., Waterman, M.: An Eulerian path approach to DNA fragment assembly. Proc. Natl. Acad. Sci. U S A 98, 9748–9753 (2001)

16. Salzberg, S.L., et al.: GAGE: A critical evaluation of genome assemblies and assembly algorithms. Genome Research 22, 557–567 (2012)
17. Vinson, J.P., Jaffe, D.B., O'Neill, K., Karlsson, E.K., Stange-Thomann, N., Anderson, S., Mesirov, J.P., Satoh, N., Satou, Y., Nusbaum, C., Birren, B., Galagan, J.E., Lander, E.S.: Assembly of polymorphic genomes: algorithms and application to Ciona savignyi. Genome Research 15, 1127–1135 (2005)
18. Xie, M., Wang, J., Chen, J.: A model of higher accuracy for the individual haplotyping problem based on weighted SNP fragments and genotype with errors. Bioinformatics 24, i105–i113 (2008)
19. Zerbino, D., Birney, E.: Velvet: algorithms for de novo short read assembly using de Bruijn graphs. Genome Res. 18, 821–829 (2008)
20. Zhao, Y.Y., Wu, L.Y., Zhang, J.H., Wang, R.S., Zhang, X.S.: Haplotype assembly from aligned weighted SNP fragments. Computational Biology and Chemistry 29, 281–287 (2005)

An Exact Algorithm to Compute the DCJ Distance for Genomes with Duplicate Genes

Mingfu Shao[1], Yu Lin[1,2], and Bernard Moret[1]

[1] Laboratory for Computational Biology and Bioinformatics, EPFL, Lausanne, Switzerland
[2] Department of Computer Science and Engineering, University of California, San Diego, La Jolla, California
{mingfu.shao,yu.lin,bernard.moret}@epfl.ch

Abstract. Computing the edit distance between two genomes is a basic problem in the study of genome evolution. The double-cut-and-join (DCJ) model has formed the basis for most algorithmic research on rearrangements over the last few years. The edit distance under the DCJ model can be computed in linear time for genomes without duplicate genes, while the problem becomes **NP**-hard in the presence of duplicate genes. In this paper, we propose an ILP (integer linear programming) formulation to compute the DCJ distance between two genomes with duplicate genes. We also provide an efficient preprocessing approach to simplify the ILP formulation while preserving optimality. Comparison on simulated genomes demonstrates that our method outperforms MSOAR in computing the edit distance, especially when the genomes contain long duplicated segments. We also apply our method to assign orthologous gene pairs among human, mouse and rat genomes, where once again our method outperforms MSOAR.

Keywords: DCJ distance, adjacency graph, maximum cycle decomposition, orthology assignment.

1 Introduction

The combinatorics and algorithmics of genomic rearrangements have been the subject of much research since the problem was formulated in the 1990s [1]. The advent of whole-genome sequencing has provided us with masses of data on which to study genomic rearrangements. Genomic rearrangements include inversions, transpositions, circularizations, and linearizations, all of which act on a single chromosome, and translocations, fusions, and fissions, which act on two chromosomes. These operations can all be described in terms of the single double-cut-and-join (DCJ) operation [2, 3], which has formed the basis for most algorithmic research on rearrangements over the last few years [4–8]. A DCJ operation makes two cuts in the genome, either in the same chromosome or in two different chromosomes, producing four cut ends, then rejoins the four cut ends in a different order.

A basic problem in genome rearrangements is to compute the edit distance between two genomes, i.e., the minimum number of operations needed to transform one genome into another. Under the inversion model, Hannenhalli and Pevzner gave the first polynomial-time algorithm to compute the edit distance for unichromosomal genomes

R. Sharan (Ed.): RECOMB 2014, LNBI 8394, pp. 280–292, 2014.
© Springer International Publishing Switzerland 2014

[9], which was later improved to linear time [10]. As for the multichromosomal genomes, the edit distance under the Hannenhalli-Pevzner model (inversions and translocations) has been studied through a series of papers [9, 11–13], culminating in a fairly complex linear-time algorithm [4]. Under the DCJ model, the edit distance can be computed in linear time for two multichromosomal genomes in a simple and elegant way [2].

All of these algorithms assume genomes contain no duplicate genes. However, gene duplications are widespread events and have long been recognized as a major driving force of evolution [14, 15]. For example, in human genomes segmental duplications are hotspots for non-allelic homologous recombination leading to genomic disorders, copy-number polymorphisms, and gene and transcript innovations [16]. The problem of computing the inversion distance for genomes in the presence of duplicate genes has been proved **NP**-hard [17]. Suksawatchon *et al.* proposed a heuristic for this problem using binary integer programming [18], which was later extended to handle gene deletion [19]. Chen *et al.* decomposed this problem into two new optimization problems, called the *minimum common partition* and the *maximum cycle decomposition*, for which efficient heuristics were designed [17]. They packaged the whole algorithms into the SOAR software system, and applied SOAR to assign orthologs on a genome-wide scale. Later, they extended SOAR to unite rearrangements and single-gene duplications as a new software package, called MSOAR, which can be applied to detect inparalogs in addition to orthologs [20]. Recently, they incorporated tandem duplications into their model, and demonstrated that the new system achieved a better sensitivity and specificity than MSOAR [21].

In this paper, we focus on the problem of computing the edit distance for two genomes with duplicate genes under the DCJ model. This problem is also **NP**-hard, which can be proved by a reduction from the **NP**-hard problem of *breakpoint graph decomposition* [22]. We first reduce this problem to the problem of finding the optimal consistent decomposition of the corresponding adjacency graph, then formulate the latter problem as an integer linear program. We also provide an efficient preprocessing approach to reduce the ILP formulation while preserving optimality. Finally, we compare our method with MSOAR on both simulated and biological datasets.

2 Problem Statement

We model one genome as a set of chromosomes, and each chromosome as a linear or circular list of genes. Homologous genes are grouped into *gene families*. In this paper, we study two genomes with the same gene content: each gene family has the same number of genes in both genomes. Assuming that two genomes G_1 and G_2 have the same gene content, we say a bijection between G_1 and G_2 is *valid* if it specifies n homologous gene pairs, where n is the number of genes in each genome. If G_1 and G_2 contain only singleton gene families (exactly one gene in each family in each genome), then there is a unique valid bijection between G_1 and G_2, and the DCJ distance between G_1 and G_2 can be computed in linear time [2]. If G_1 and G_2 contain gene families with multiple genes in each genome, then there are many valid bijections between G_1 and G_2. Different valid bijections define different one-to-one correspondences between homologous genes, yielding possibly different DCJ distances between

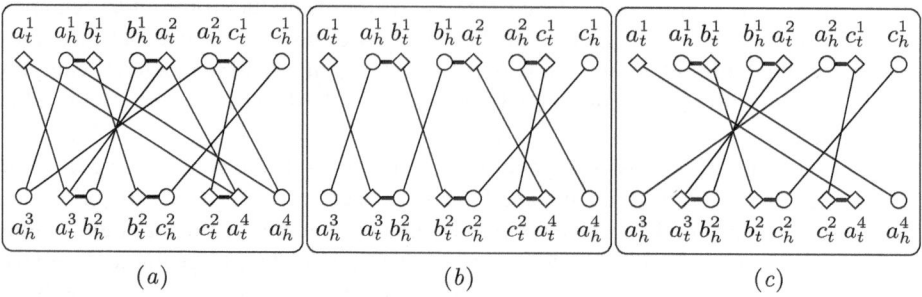

Fig. 1. An example of adjacency graph and its two consistent decompositions. Genome 1 contains one linear chromosome, (a^1, b^1, a^2, c^1), and genome 2 also contains one linear chromosome $(-a^3, -b^2, -c^2, a^4)$. Genes in the same gene family are represented by the same label, and distinguished by different superscripts. All black edges are represented by long thin lines, and all grey edges are represented by short thick lines. (a) The corresponding adjacency graph, in which head extremities are represented by circles, while tail extremities are represented by diamonds. (b) A consistent decomposition with 2 odd-length paths, whose corresponding valid bijection maps a^1 to a^3 and a^2 to a^4. (c) Another consistent decomposition with 2 odd-length paths and 1 cycle, whose corresponding valid bijection maps a^1 to a^4 and a^2 to a^3.

G_1 and G_2. In this paper, we study the following *generalized DCJ distance problem*: given two genomes G_1 and G_2 with the same gene content, find a valid bijection between G_1 and G_2 that minimizes the DCJ distance. We denote the generalized DCJ distance between G_1 and G_2 as $d(G_1, G_2)$.

We use the notation introduced by Bergeron *et al.* [2] for gene orders. The two ends of a gene g are called *extremities*, the head as g_h and the tail as g_t. If genes f and g are homologous, its corresponding extremities (f_h and g_h, f_t and g_t) are also *homologous*. Two consecutive genes a and b can be connected by one *adjacency*, which is represented by a set of two extremities; thus adjacencies come in four types: $\{a_t, b_t\}$, $\{a_h, b_t\}$, $\{a_t, b_h\}$, and $\{a_h, b_h\}$. If gene g lies at one end of a linear chromosome, then this end can be represented by a set of one extremity, $\{g_t\}$ or $\{g_h\}$, called a *telomere*. The set of all extremities of a genome is called the *extremity set*.

Let G_1 and G_2 be two genomes with the same gene content, and let S_1 and S_2 be the extremity sets of G_1 and G_2, respectively. The *adjacency graph* with respect to G_1 and G_2 can be written as $AG = (V, E)$, with $V = S_1 \cup S_2$ and where E is composed of two types of edges, *black edges* and *grey edges*. Two extremities in different extremity sets (one is in S_1 and the other one is in S_2) are connected by one black edge if they are homologous, and two extremities in the same extremity set are connected by one grey edge if they form an existing adjacency. Figure 1a gives an example.

We say that a cycle (or path) in the adjacency graph is *alternating* if any two adjacent edges in this cycle (or path) consist of one black edge and one grey edge. The *length* of a cycle (or path) is defined as the number of its black edges. A *decomposition* of the adjacency graph is a set of vertex-disjoint alternating cycles and paths that cover all vertices and all grey edges. We say a decomposition is *consistent* if for any two homologous genes f and g, either both (f_h, g_h) and (f_t, g_t) are in this decomposition, or neither of them is in this decomposition. Figure 1*b* and 1*c* give two examples of consistent decompositions.

Given two genomes G_1 and G_2 with the same gene content, there is a natural one-to-one correspondence between the set of all possible valid bijections from G_1 to G_2 and the set of all possible consistent decompositions of the adjacency graph with respect to G_1 and G_2. In fact, if one valid bijection is given, which maps gene f in G_1 to a homologous gene g in G_2, then we can keep the black edges (f_h, g_h) and (f_t, g_t) in the decomposition. We do the same thing for every pair of genes specified by this valid bijection; this process culminates in a consistent decomposition. On the other hand, if we are given a consistent decomposition of the corresponding adjacency graph, we can collect all homologous gene pairs (f, g) indicated by black edges (f_h, g_h) and (f_t, g_t), which form a valid bijection from G_1 to G_2. Given a consistent decomposition with c cycles and o odd-length paths, exactly $(|V|/4 - c - o/2)$ DCJ operations are needed to transform G_1 into G_2 [2]. Thus, we can write $d(G_1, G_2) = \min_{D \in \mathcal{D}}(|V|/4 - c_D - o_D/2) = |V|/4 - \max_{D \in \mathcal{D}}(c_D + o_D/2)$, where $\mathcal{D}$ is the space of all consistent decompositions, and c_D and o_D are the numbers of cycles and odd-length paths in a decomposition D, respectively. This formula transforms the generalized DCJ distance problem into the *maximum cycle decomposition problem*, which asks for a consistent decomposition of the adjacency graph such that the number of cycles plus half the number of odd-length paths in this decomposition is maximized.

3   ILP for the Maximum Cycle Decomposition Problem

In [23], we described a capping method to remove telomeres by introducing *null extremities*. All null extremities are homologous to each other, but none is homologous to any other extremity. Let $AG = (V = S_1 \cup S_2, E)$ be the adjacency graph with respect to two given genomes G_1 and G_2. Suppose that G_1 and G_2 contain $2 \cdot k_1$ and $2 \cdot k_2$ telomeres respectively. The "telomere removal" proceeds as follows (see Figure 2 for an example). For each extremity $u \in S_1$ coming from each telomere in G_1, we add one null extremity τ to S_1 and add one grey edge to E that connects u and τ. Similarly, for each extremity $v \in S_2$ coming from each telomere in G_2, we add one null extremity τ to S_2 and add one grey edge to E that connects v and τ. If we additionally have $k_1 < k_2$, we then add $(k_2 - k_1)$ pairs of null extremities to S_1, each of which is connected by one more grey edge added to E. We finally add black edges connecting all possible pairs of null extremities between S_1 and S_2. We can prove that this telomere removal process does not change $d(G_1, G_2)$ using the same argument as in [7, 23]. In the following we assume that each vertex is adjacent to exactly one grey edge in the adjacency graph, and that the consistent decompositions consist of only cycles.

Now we formulate the maximum cycle decomposition problem as an integer linear program. Let $AG = (V, E)$ be the adjacency graph with respect to two given genomes G_1 and G_2 with the same gene content. For each edge $e \in E$, we create binary variable x_e to indicate whether e will be in the final decomposition. First, we require that all grey edges be in the final decomposition:

$$x_e = 1, \quad \forall e \text{ that are grey}$$

Second, we require that the final decomposition be consistent:

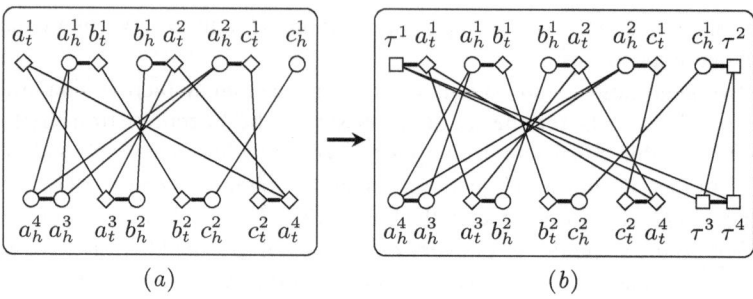

Fig. 2. An example of the telomere removal. Genome 1 contains one linear chromosome, (a^1, b^1, a^2, c^1), and genome 2 contains one circular chromosome $(-a^3, -b^2, -c^2, a^4)$. (a) The corresponding adjacency graph. (b) The adjacency graph after the telomere removal, in which null extremities are represented by squares.

$$x_{(f_h, g_h)} = x_{(f_t, g_t)}, \quad \forall f \in G_1 \text{ and } \forall g \in G_2 \text{ that are homologous}$$

Third, we require that for each vertex exactly one adjacent black edge adjacent be chosen:

$$\sum_{(u,v) \in E, \ v \in S_2} x_{(u,v)} = 1, \quad \forall u \in S_1$$

$$\sum_{(u,v) \in E, \ u \in S_1} x_{(u,v)} = 1, \quad \forall v \in S_2$$

These three groups of constraints guarantee that all selected edges form a consistent decomposition.

Now we count the number of cycles. We first index the vertices arbitrarily, $V = \{v_1, v_2, \cdots, v_{|V|}\}$. For each vertex v_i, we create variable y_i to indicate the *label* of v_i. We set a distinct positive bound i for each y_i:

$$0 \le y_i \le i, \quad 1 \le i \le |V|$$

We require that all vertices in the same cycle in the final decomposition have the same label, which can be guaranteed by requiring that, for each selected edge, the two adjacent vertices have the same label:

$$y_i \le y_j + i \cdot (1 - x_e), \quad \forall e = (v_i, v_j) \in E$$
$$y_j \le y_i + j \cdot (1 - x_e), \quad \forall e = (v_i, v_j) \in E$$

Then, for each vertex v_i, we create binary variable z_i to indicate whether y_i is equal to its upper bound i:

$$i \cdot z_i \le y_i, \quad 1 \le i \le |V|$$

Since all vertices in the same cycle have the same label and all upper bounds are distinct, there is exactly one vertex in each cycle whose label can be equal to its upper bound. Finally, we set the objective to

$$\max \sum_{1 \leq i \leq |V|} z_i,$$

which is equal to the number of cycles.

There are $O(|E|)$ variables and $O(|E|)$ constraints in this ILP formulation.

4   Fixing Cycles of Length Two

A cycle of length two in the adjacency graph indicates one shared adjacency. The following theorem gives a sufficient condition to fix this cycle while preserving optimality, which can be used to narrow the search for an optimal bijection.

Theorem 1. *Given an adjacency graph $AG = (V, E)$, if a length-two cycle C contains some vertex with total degree 2, then there exists an optimal consistent decomposition of AG that contains C.*

Proof. Let $\{a_h^1, b_h^1, a_h^2, b_h^2\}$ be the four vertices of C, where a_h^1 and b_h^1 form an adjacency in G_1 while a_h^2 and b_h^2 form an adjacency in G_2, and (a_h^1, a_h^2) and (b_h^1, b_h^2) are the two black edges of C. Let a_h^1 be the vertex of total degree 2; then the gene family of $\{a^1, a^2\}$ is a singleton family, and thus edge (a_h^1, a_h^2) appears in every consistent decomposition. Now we prove the theorem by contradiction. Suppose that edge (b_h^1, b_h^2) is not in any optimal consistent decomposition. Take any optimal consistent decomposition D, in which b_h^1 is linked to b_h^4 and b_h^2 is linked to b_h^3. Since D is consistent, we know that edges (b_t^1, b_t^4) and (b_t^2, b_t^3) are also in D. We now transform D into a new decomposition D'' that contains edge (b_h^1, b_h^2) by exchanging two pairs of edges. Figure 3 illustrates this process. First, we remove edges (b_h^1, b_h^4) and (b_h^3, b_h^2) from D and add edges (b_h^1, b_h^2) and (b_h^3, b_h^4); denote this inconsistent decomposition by D'. Since in this step one cycle is split into two small cycles, we have that $c_{D'} = c_D + 1$. Now, we remove edges (b_t^1, b_t^4) and (b_t^3, b_t^2) from D' and add edges (b_t^1, b_t^2) and (b_t^3, b_t^4) to obtain the consistent decomposition D''. This step involves at most two cycles of D', and merges these two

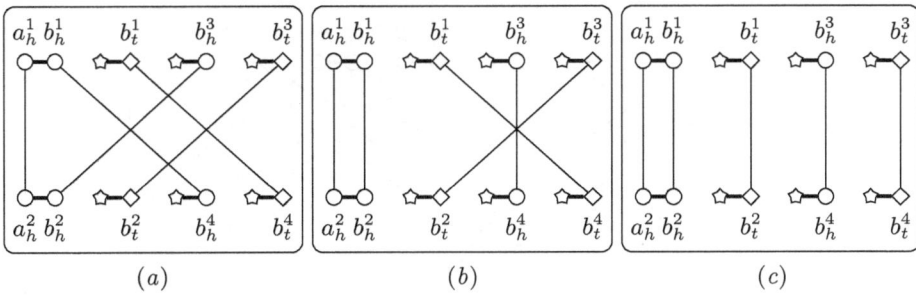

Fig. 3. The process of building a new optimal consistent decomposition that contains edge (b_h^1, b_h^2). (a) One optimal consistent decomposition D without edge (b_h^1, b_h^2). Star represents unrelated extremities. (b) The inconsistent decomposition D'. (c) The consistent decomposition D''.

cycles together in the worst case. Thus, we have $c_{D''} \geq c_{D'} - 1$. Overall, we have that $c_{D''} \geq c_D$, which means D'' is also an optimal consistent decomposition—the desired contradiction.   □

If all four vertices in a cycle of length two have degree larger than 2, then it is possible that this cycle is not part of any optimal consistent decomposition. Figure 4 gives such an example. Moreover, this example also shows that if a shared adjacency appears exactly once in each genome, it is still possible that the corresponding cycle of length two is not part of any optimal consistent decomposition.

5   Experimental Results

We compare our method with MSOAR on both simulated and biological datasets. The input for both methods is two genomes with the same gene content, and the output is a bijection between the two genomes, plus the DCJ distance calculated as $n - c - o/2$, where n is the number of genes in each genome, and c and o are the numbers of cycles and odd-length paths in the adjacency graph induced by the bijection. We use both the accuracy of the bijection, which is defined as the percentage of correct gene pairs (compared with a reference bijection), and the deviation from the true evolutionary distances, to evaluate the performance of the two methods.

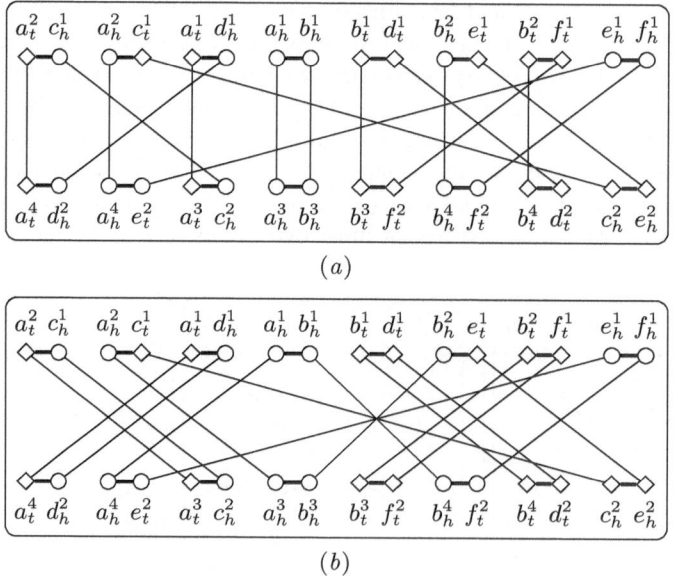

Fig. 4. An example of a cycle of length two that is not part of any optimal consistent decomposition. (a) A consistent decomposition with 4 cycles that contains the cycle of length two of $\{a_h^1, b_h^1, a_h^3, b_h^3\}$. (b) An optimal consistent decomposition with 5 cycles.

For our method, given two genomes, we first build the adjacency graph and then employ the telomere removal technique to obtain a new adjacency graph without telomeres. Then we apply Theorem 1 to fix possible cycles of length two, and finally invoke GUROBI [24] to solve the ILP formulation. Since the ILP solver might take a long time, we set a time limit of two hours for each instance in our experiments—the best solution will be returned if the ILP solver does not terminate in two hours. For MSOAR, we run its binary version downloaded from `http://msoar.cs.ucr.edu/`. We compare our method with MSOAR, rather than the latest version MSOAR 2.0, because we focus on genomes with the same gene content, which implicitly requires that, after the speciation event, only DCJ operations are involved. Compared with MSOAR, MSOAR 2.0 aims to identify tandem duplications of genes *after* the speciation. Thus, under our evolutionary model that does not contain postspeciation duplications, MSOAR and MSOAR 2.0 are equivalent.

5.1 Simulation Results

We simulate artificial genomes under an evolutionary model including segmental duplications and DCJs. We introduce duplicated genes through segmental duplications. For each segmental duplication, we uniformly select a position to start duplicating a segment of the genome and place the new copy to a new position. Since the average copy number of each gene in human, mouse and rat genomes, are 1.46, 1.55 and 1.28, respectively, we set the average copy number to 1.5 in our simulation. From a genome of 1,000 distinct genes, we generate an ancestor genome with 1,500 genes, by randomly performing $500/L$ segmental duplications of length L (in terms of the number of genes in the segment). We then simulate two extant genomes from the ancestor by randomly performing N DCJs (in terms of inversions) independently. Thus, the true evolutionary distance between the two extant genomes is $2 \cdot N$. The reference bijection consists of those gene pairs that correspond to the same gene in the ancestor. We test three different lengths for segmental duplications ($L = 1, 2, 5$); results illustrate the trends and capabilities of the two methods in handling genomes with duplicated segments. We also vary the number of DCJs over a broad range ($N = 200, 210, \cdots, 500$) that reaches beyond the saturation point. For each setting, we randomly simulate 5 independent instances, and calculate the average accuracy of the bijection and the average deviation from the true evolutionary distances over these 5 instances for both methods.

Figure 5 shows the deviation from the true evolutionary distances for both methods. The first observation is that saturation starts occurring for a true evolutionary distance of 720: the DCJ distance obtained from the reference bijection is smaller than the true evolutionary distance, and the gap increases along with the increase of the true evolutionary distance. Second, when the true evolutionary distance is less than 720, our method obtains very accurate DCJ distances while MSOAR usually overestimates the DCJ distance. The difference is particularly pronounced for $L \geq 2$: in such cases, there exist identical segments in each genome, a situation that creates problems when MSOAR tries to partition each genome into a minimum number of common segments [17].

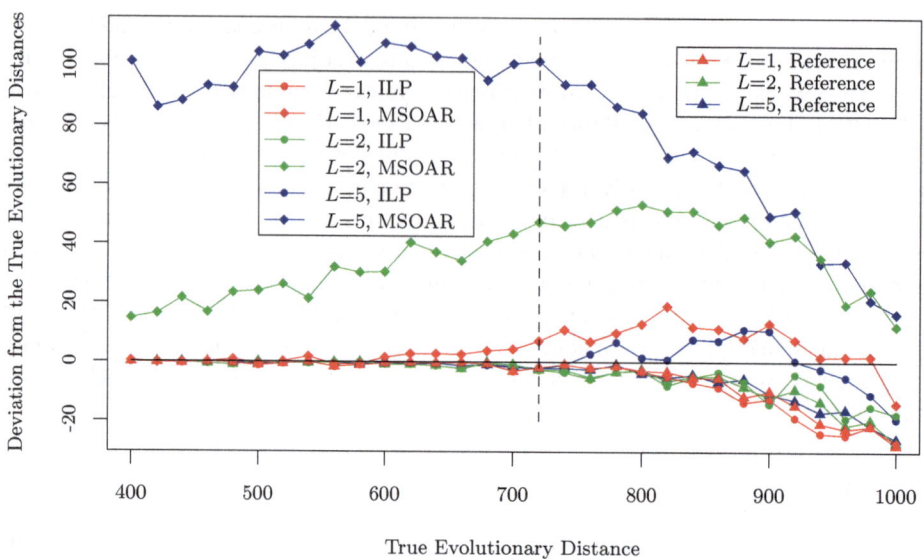

Fig. 5. Deviation from the true evolutionary distances on simulation data. Diamonds track MSOAR, circles track our method, and triangles track the reference bijection.

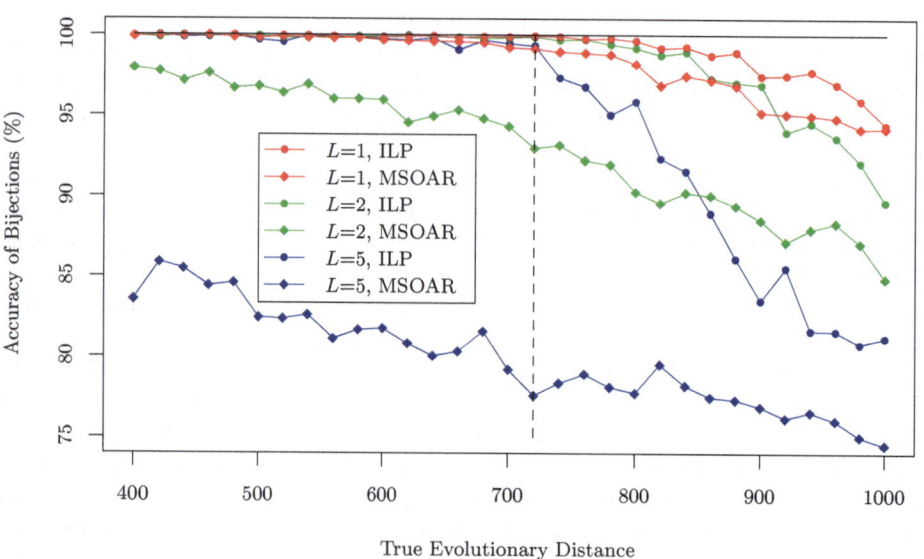

Fig. 6. The accuracy of the bijections on simulation data. Diamonds track MSOAR, while circles track our method.

Figure 6 shows the the accuracy of the bijections for both methods. For $L = 1$, both methods can correctly identify most gene pairs. For $L \geq 2$, our method significantly outperforms MSOAR. For large L, the accuracy of our method decreases rapidly beyond saturation, but continues to dominate MSOAR.

The running time of MSOAR grows slowly as the the true evolutionary distance increases. For the most complicated case of $L = 5$ and the true evolutionary distance is 1000, MSOAR can finish in less than 2000 seconds. Regarding our method, when the true evolutionary distance is relatively small (≤ 640 when $L = 5$, ≤ 740 when $L = 2$, and ≤ 820 when $L = 1$), the preprocessing method can fix a considerable portion of the adjacency graph, leaving a small ILP instance that can be solved very quickly (even faster than MSOAR). When the true evolutionary distance is relatively large, the ILP solver cannot terminate in two hours and a sub-optimal solution is obtained. Usually, this solution is equal or very close to the optimal solution, because the ILP solver can find the optimal solution very quickly, but must spend more time to verify that it is optimal. This observation is also verified by the very high accuracy before the saturation point shown in Figure 6.

5.2 Application to Orthology Assignment

Under a parsimonious evolutionary scenario, the optimal valid bijection between two genomes with the same gene content minimizes the number of DCJs after speciation, and thus infers the orthologous gene pairs [17]. We test both methods for assigning orthologous genes between pairs of genomes. Human, mouse, and rat genomes are well annotated, so we chose them to evaluate the performance of the two methods. For each species, we downloaded the information for all protein-coding genes from Ensembl (http://www.ensembl.org), including gene family names, positions on the chromosomes and gene symbols. If a gene has multiple alternative products, we keep its longest isoform. Two genes are considered homologous if they have the same Ensembl gene family name; they are considered orthologous if they have the same gene symbol. (Note that two orthologous genes are necessarily homologous, but two homologous genes need not be orthologous.) For a pair of genomes, we keep only orthologous gene pairs, thereby obtaining two genomes with the same gene content; our reference bijection is then defined by these orthologous gene pairs. For both methods, we use gene family and position information to infer orthologous relationships and compare them to the reference bijection.

The results of comparing these three genomes are shown in Table 1. Both methods mostly agree with annotation, indicating that the parsimonious model is appropriate

Table 1. Comparison of human, mouse and rat genomes

species pairs	gene pairs	accuracy of bijection (%)		DCJ distance	
		MSOAR	our method	MSOAR	our method
human mouse	14876	98.63	99.18	933	894
human rat	12971	98.79	99.28	1320	1294
mouse rat	13525	98.60	99.26	968	916

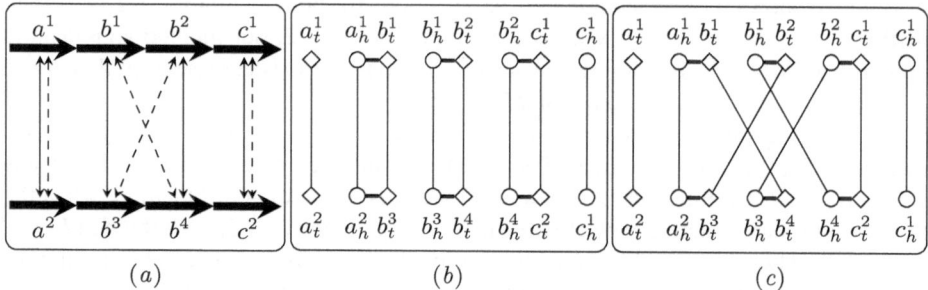

Fig. 7. Comparison of the reference bijection with our bijection. (*a*) Two identical segments. Our bijection is shown by solid lines while reference bijection is shown by dashed lines. (*b*) The adjacency graph corresponding to the our bijection, in which there are 3 cycles. (*c*) The adjacency graph corresponding to the reference bijection, in which there is only 1 cycle.

when comparing these genomes; our method obtains slightly better accuracy. On human and mouse for example, our bijection has 122 different gene pairs compared with the reference bijection. Among these pairs, 34 of them can be explained by a simple structure, illustrated in Figure 7. For two identical segments, our method outputs a sequential bijection for which no DCJ operation is needed, while the reference bijection contains a crossover, for which at least two DCJ operations are needed. The other 87 pairs can be explained by 32 pairs of segments, for each of which our bijection needs fewer DCJ operations than the reference bijection. On the comparison of the DCJ distance, our method gets fewer DCJ operations than MSOAR in all three pairs of genomes.

6   Conclusion

We formulated the maximum cycle decomposition problem as an integer linear program. We proved a theorem that can be used to reduce the complexity while preserving optimality. The combination of the two gives a practical method to compute the exact DCJ distance for genomes with duplicate genes. Such a method is crucial for comparative genomics, since duplicate genes are commonly observed in most species.

The ILP formulation can be extended in various ways. First, we can use the relaxed LP (linear programming) techniques to design possible approximation algorithms. Second, when we apply it to do orthology assignment, we can also take the sequence similarity information into account, by adding a term of the form $\sum_{e \in E} w_e \cdot x_e$ to the objective function, where w_e can be set to the similarity of the two genes. How to combine sequence similarity and DCJ distances remains an unexplored problem, but our ILP formulation provides a first step by allowing us to study linear combinations of the two.

We assumed that, after a speciation event, only DCJ operations are involved. This assumption is clearly unrealistic—it was made to simplify the problem and enable us to devise a first exact solution. However, now that our ILP method has proved successful, we can combine it with our previous work [23] to include single-gene deletion and single-gene insertion in the model.

References

1. Fertin, G., Labarre, A., Rusu, I., Tannier, E., Vialette, S.: Combinatorics of Genome Rearrangements. MIT Press (2009)
2. Bergeron, A., Mixtacki, J., Stoye, J.: A unifying view of genome rearrangements. In: Bücher, P., Moret, B.M.E. (eds.) WABI 2006. LNCS (LNBI), vol. 4175, pp. 163–173. Springer, Heidelberg (2006)
3. Yancopoulos, S., Attie, O., Friedberg, R.: Efficient sorting of genomic permutations by translocation, inversion and block interchange. Bioinformatics 21(16), 3340–3346 (2005)
4. Bergeron, A., Mixtacki, J., Stoye, J.: A new linear-time algorithm to compute the genomic distance via the double cut and join distance. Theor. Comput. Sci. 410(51), 5300–5316 (2009)
5. Chen, X.: On sorting permutations by double-cut-and-joins. In: Thai, M.T., Sahni, S. (eds.) COCOON 2010. LNCS, vol. 6196, pp. 439–448. Springer, Heidelberg (2010)
6. Chen, X., Sun, R., Yu, J.: Approximating the double-cut-and-join distance between unsigned genomes. BMC Bioinformatics 12(suppl. 9), S17 (2011)
7. Yancopoulos, S., Friedberg, R.: Sorting genomes with insertions, deletions and duplications by DCJ. In: Nelson, C.E., Vialette, S. (eds.) RECOMB-CG 2008. LNCS (LNBI), vol. 5267, pp. 170–183. Springer, Heidelberg (2008)
8. Moret, B.M.E., Lin, Y., Tang, J.: Rearrangements in phylogenetic inference: Compare, model, or encode? In: Chauve, C., El-Mabrouk, N., Tannier, E. (eds.) Models and Algorithms for Genome Evolution. Computational Biology, vol. 19, pp. 147–172. Springer, Berlin (2013)
9. Hannenhalli, S., Pevzner, P.A.: Transforming cabbage into turnip (polynomial algorithm for sorting signed permutations by reversals). In: Proc. 27th Ann. ACM Symp. Theory of Comput. (STOC 1995), pp. 178–189. ACM Press, New York (1995)
10. Bader, D.A., Moret, B.M.E., Yan, M.: A fast linear-time algorithm for inversion distance with an experimental comparison. J. Comput. Biol. 8(5), 483–491 (2001)
11. Jean, G., Nikolski, M.: Genome rearrangements: a correct algorithm for optimal capping. Inf. Proc. Letters 104(1), 14–20 (2007)
12. Ozery-Flato, M., Shamir, R.: Two notes on genome rearrangement. J. Bioinf. Comp. Bio. 1(1), 71–94 (2003)
13. Tesler, G.: Efficient algorithms for multichromosomal genome rearrangements. J. Comput. Syst. Sci. 65(3), 587–609 (2002)
14. Bailey, J.A., Eichler, E.E.: Primate segmental duplications: crucibles of evolution, diversity and disease. Nature Reviews Genetics 7(7), 552–564 (2006)
15. Lynch, M.: The Origins of Genome Architecture. Sinauer (2007)
16. Jiang, Z., Tang, H., Ventura, M., Cardone, M.F., Marques-Bonet, T., She, X., Pevzner, P.A., Eichler, E.E.: Ancestral reconstruction of segmental duplications reveals punctuated cores of human genome evolution. Nature Genetics 39(11), 1361–1368 (2007)
17. Chen, X., Zheng, J., Fu, Z., Nan, P., Zhong, Y., Lonardi, S., Jiang, T.: Assignment of orthologous genes via genome rearrangement. ACM/IEEE Trans. on Comput. Bio. & Bioinf. 2(4), 302–315 (2005)
18. Suksawatchon, J., Lursinsap, C., Bodén, M.: Computing the reversal distance between genomes in the presence of multi-gene families via binary integer programming. Journal of Bioinformatics and Computational Biology 5(1), 117–133 (2007)
19. Laohakiat, S., Lursinsap, C., Suksawatchon, J.: Duplicated genes reversal distance under gene deletion constraint by integer programming. Bioinformatics and Biomedical Engineering, 527–530 (2008)

20. Fu, Z., Chen, X., Vacic, V., Nan, P., Zhong, Y., Jiang, T.: MSOAR: a high-throughput or- tholog assignment system based on genome rearrangement. Journal of Computational Biol- ogy 14(9), 1160–1175 (2007)
21. Shi, G., Zhang, L., Jiang, T.: MSOAR 2.0: Incorporating tandem duplications into ortholog assignment based on genome rearrangement. BMC Bioinformatics 11(1), 10 (2010)
22. Kececioglu, J., Sankoff, D.: Exact and approximation algorithms for sorting by reversals, with application to genome rearrangement. Algorithmica 13(1), 180–210 (1995)
23. Shao, M., Lin, Y.: Approximating the edit distance for genomes with duplicate genes under DCJ, insertion and deletion. BMC Bioinformatics 13(suppl. 19), S13 (2012)
24. Gurobi Optimization Inc. Gurobi optimizer reference manual (2013)

HIT'nDRIVE: Multi-driver Gene Prioritization Based on Hitting Time

Raunak Shrestha[1,2,*], Ermin Hodzic[3,*], Jake Yeung[2,4,*], Kendric Wang[2],
Thomas Sauerwald[5], Phuong Dao[6], Shawn Anderson[2], Himisha Beltran[7],
Mark A. Rubin[7], Colin C. Collins[2,8], Gholamreza Haffari[9], and S. Cenk Sahinalp[3,10]

[1] CIHR Bioinformatics Training Program, University of British Columbia,
Vancouver, BC, Canada
[2] Laboratory for Advanced Genome Analysis, Vancouver Prostate Centre,
Vancouver, BC, Canada
[3] School of Computing Science, Simon Fraser University,
Burnaby, BC, Canada
[4] Genome Science and Technology Program, University of British Columbia,
Vancouver, BC, Canada
[5] Computer Laboratory, University of Cambridge, Cambridge, United Kingdom
[6] National Center for Biotechnology Information, NLM, NIH, Bethesda, MD, USA
[7] Weill Cornell Cancer Center, New York, NY, USA
[8] Department of Urologic Sciences, University of British Columbia, Vancouver, BC, Canada
[9] Faculty of Information Technology, Monash University, Melbourne, VIC, Australia
[10] School of Informatics and Computing, Indiana University, Bloomington, IN, USA
cenk@sfu.ca

Abstract. A key challenge in cancer genomics is the identification and prioritization of genomic aberrations that potentially act as drivers of cancer. In this paper we introduce HIT'nDRIVE, a combinatorial method to identify aberrant genes that can collectively influence possibly distant "outlier" genes based on what we call the "random-walk facility location" (RWFL) problem on an interaction network. RWFL differs from the standard facility location problem by its use of "multi-hitting time", the expected minimum number of hops in a random walk originating from any aberrant gene to reach an outlier. HIT'nDRIVE thus aims to find the smallest set of aberrant genes from which one can reach outliers within a desired multi-hitting time. For that it estimates multi-hitting time based on the independent hitting times from the drivers to any given outlier and reduces the RWFL to a weighted multi-set cover problem, which it solves as an integer linear program (ILP). We apply HIT'nDRIVE to identify aberrant genes that potentially act as drivers in a cancer data set and make phenotype predictions using only the potential drivers - more accurately than alternative approaches.

1 Introduction

Over the past decade, high-throughput sequencing efforts have revealed the importance of genomic aberrations in the progression of cancer [1]. During the time course of cancer evolution, tumor cells accumulate numerous genomic aberrations, however only a

---

* These authors contributed equally to this work.

R. Sharan (Ed.): RECOMB 2014, LNBI 8394, pp. 293–306, 2014.
© Springer International Publishing Switzerland 2014

few "driver aberrations" are expected to confer crucial growth advantage - and have potential to be used as therapeutic targets. The identification of these driver aberrations and the specific genes they alter poses a significant challenge as they are greatly outnumbered by functionally inconsequential "passenger" aberrations which contribute further towards cancer heterogeneity [1, 2].

While several methods for finding drivers of cancer have been described previously, most of them rely on the recurrence frequency of single nucleotide variants with respect to the background mutation rate in a population of tumors [3, 4]. These approaches are restricted to identifying only highly recurrent mutations as driver events. However, recent whole-genome studies have revealed that important genes may be recurrently mutated in only a small fraction of the tumor cohort under study, and can be subtype-specific [5–7]. Furthermore, personalized rare drivers are likely to arise during later stages of tumor evolution and be isolated to a small fraction of tumor cells [8, 9].

Perhaps the first computational method to consider large scale genomic variants as driver events is by Akavia *et al.* [10], which correlates genes with highly recurrent copy number alterations with variation in gene expression profiles within a Bayesian network. Similarly, Masica and Karchin [11] correlate gene mutation information with expression profile changes in other genes, again with no prior knowledge of pathways or protein interactions. Another approach, (Multi) Dendrix [12] aims to simultaneously identify multiple driver pathways, assuming mutual exclusivity of mutated genes among patients, using either a Markov chain Monte Carlo algorithm or integer linear programming (ILP). Finally, MEMo by Ciriello *et al.* [13], identifies sets of proximally-located genes from interaction networks, which are also recurrently altered and exhibit patterns of mutual exclusivity across the patient population. To the best of our knowledge, the first method to link copy number alterations to expression profile changes within an interaction network is by Kim *et al.* [14] which connects specific "causal" aberrant genes with potential targets in a protein interaction network. Similarly, method, PARADIGM [15], computes gene-specific inferences using factor graphs to integrate various genomic data to infer pathways altered in a patient. A more recent tool, HotNet by Vandin *et al.* [16], was the first to use a network diffusion approach to compute a pairwise influence measure between the genes in the (gene interaction) network and identify subnetworks enriched for mutations. TieDIE [17] also uses the diffusion model to identify a collection of pathways and subnetworks that associate a fixed set of driver genes to expression profile changes in other genes. Briefly, the network diffusion approach aims to measure the influence of one node over another by calculating the stationary proportion of a "flow" originating from the starting node, that ends up in the destination node. Since it is based on the stationary distribution, the inferences that can be made by the diffusion model are time independent. In that sense, the diffusion approach is very similar to Rooted PageRank, the stationary probability of a random walk originating at a source node, being at a given destination node. A final method, DriverNet by Bashashati *et al* [18], also aims to correlate single nucleotide alterations with target genes expression profile changes, but only among direct interaction partners. The novel feature of DriverNet is that it aims to find the "minimum" number of potential drivers that can "cover" targets.

Our Contributions. In this paper we present a novel integrative method that considers potential driver events at the genomic level, i.e. single nucleotide mutations, structural or copy number changes. Our contributions are as follows:

1. We present HIT'nDRIVE, an algorithm that aims to identify "the most parsimonious" set of patient specific driver genes which have sufficient "influence" over a large proportion of outlier genes. HIT'nDRIVE formulates this as a "random-walk facility location" problem (RWFL), a combinatorial optimization problem, which, to the best of our knowledge, has not been explored earlier. RWFL differs from the standard facility location problem by its use of "multi-source hitting time" (or multi-hitting time) as an alternative distance measure between a set of aberrant genes (potential drivers) and an outlier gene. Multi-hitting time generalizes the notion of hitting time [19]: we define it as the expected minimum number of hops in which a random walk originating from any aberrant gene reaches the outlier for the first time (in the human gene or protein interaction network). RWFL problem thus asks to find the smallest (the most parsimonious) set of aberrant genes from which one can reach (at least a given fraction of) all outliers within a user defined multi-hitting time. We believe that applications of RWFL problem may extend beyond its application to driver gene identification - to influence analysis in social networks, disease networks, etc.

2. Since RWFL problem is NP-hard, we estimate the multi-hitting time based on the independent hitting times of the drivers to an outlier, which provides an upper bound on the multi-hitting time. Our experiments show that this estimate works well for the human protein interaction network.

3. More importantly, our estimate enables us to reduce the RWFL problem to a weighted multi-set cover problem, for which we give an ILP formulation. For the specific problem instances we consider, our ILP formulation is solvable exactly by CPLEX in less than two days on a standard PC.

4. Note that hitting time as a measure for influence of one potential driver on an outlier gene is quite different from the diffusion-based measures or the Rooted PageRank: hitting time essentially measures the expected distance/time between a source node and a destination node in a random walk. We argue that hitting time is a better measure to capture the influence of one (driver) node over another as it is (i) parameter free (diffusion model introduces at least one additional parameter - the proportion of incoming flow "consumed" at a node in each time step), (ii) it is time dependent (while the diffusion model and PageRank measures the stationary behavior) and (iii) it is more robust (w.r.t. small perturbations in the network; see [20]).

5. We also show that, by a simple Monte Carlo method, the hitting time in networks with n nodes that have constant average degree and small diameter (as per the human protein interaction network) can be estimated in $\tilde{O}(n^2)$ time. For computing the hitting time in general networks, alternative methods [21] require to perform a complete matrix inversion, which takes $O(n^{2+c})$ time for some $c > 0.37$.

6. We have applied HIT'nDRIVE to identify genes subject to somatic mutation and copy number changes that potentially act as drivers in glioblastoma cancer. We then used the identified potential drivers to perform phenotype prediction on the cancer data set, solely based on gene expression profiles of small subnetworks "seeded"

by the drivers. For that we extended the OptDis method [22] by focusing only on driver-seeded subnetworks and achieved a higher accuracy than the alternative approaches.

2   HIT'nDRIVE Framework

HIT'nDRIVE naturally integrates genome and transcriptome data from a number of tumor samples for identifying and prioritizing aberrated genes as potential drivers. It "links" aberrations at the genomic level to gene expression profile alterations through a gene or protein interaction network. For that, it aims to find the *smallest* set of aberrated genes that can "explain" most of the observed gene expression alterations in the cohort. In other words, HIT'nDRIVE identifies the minimum number of potential drivers which can "cause" a user-defined proportion of the downstream expression effects observed.

HIT'nDRIVE uses a particular "influence" value of a potential driver gene on other (possibly distant) genes based on the (gene or protein) interaction network in use. In order to capture the uncertainty of interaction of genes with their neighbours, it considers a random walk process which propagates the effect of sequence alteration in one gene to the remainder of the genes through the network. As a result, the influence is defined to be the inverse of hitting-time, the expected length (number of hops) of a random walk which starts at a given potential driver gene, and "hits" a given target gene the first time in a (protein or gene) interaction network. More specifically, for any two nodes $u, v \in V$ of an undirected, connected graph $G = (V, E)$, let the random variable $\tau_{u,v}$ denote the number of hops in a random walk starting from u to visit v for the first time. The hitting-time $H_{u,v}$, thus is defined as $H_{u,v} = E[\tau_{u,v}]$ [23].

In order to capture synthetic lethality like scenarios, HIT'nDRIVE also considers multiple aberrated genes as potential drivers. For that, we define the influence value (of a set of potential driver genes on a target) as the inverse of multi(source)-hitting time, i.e., the expectation of the smallest number of hops in one of the random walk processes, simultaneously starting at each one of the potential drivers and ending at a given outlier for the first time. More specifically, let $U \subseteq V$ be a subset of nodes of G and $v \in (V - \{U\})$ be a single node. We thus define the multi(source)-hitting time $H_{U,v}$ as $H_{U,v} = E[\min_{u \in U} \tau_{u,v}]$.

HIT'nDRIVE formulates the process of potential driver gene discovery in terms of the "random-walk facility location" (RWFL) problem, which, for a single patient can be described as follows.

Let X be a set of potential driver genes and $\mathcal{Y}$ be a set of expression altered (outlier) genes. Then, for a user defined k, HIT'nDRIVE can aim to return k potential driver genes as solution to the following optimization problem:

$$\arg\min_{X \subseteq \mathcal{X}, |X|=k} \max_{y \in \mathcal{Y}} H_{X,y}$$

where $H_{X,y}$ denotes the multi-hitting time from the gene set X to the gene y.
RWFL problem resembles the standard (minimax) "facility location" problem in which one seeks a set of nodes as facilities in a graph such that the maximum distance from any node in the graph to its closest facility is minimized. RWFL differs from standard

facility location by its use of $H_{X,y}$ as a distance measure between a collection of nodes to any other node, which aims to capture the uncertainty in molecular interactions during the propagation of one or more signals, by random walks starting from one or more origins (reminiscent of the underlying Brownian motion).

Since the standard facility location is an NP-hard problem, RWFL problem is NP hard as well. As shown in the next section, we overcome this difficulty by introducing a good estimate on the multi-hitting time that helps us to reduce RWFL problem to the weighted multi-set cover problem, which we solve through an ILP formulation in Section 3. (Although the use of set-cover for representing the most parsimonious solution in a bioinformatics context is not new [24], to the best of our knowledge this is the first use of the multi-set cover formulation for maximum parsimony.) In this formulation, we use a slightly different objective: given a user defined upper bound on the maximum multi-hitting time, we now aim to minimize the number of potential drivers that can "cover" (a user defined proportion of) the outlier genes. For more than one patient, we minimize the number of drivers that can "cover" (a user defined proportion of) patient-specific outliers such that each such outlier is covered by potential drivers that are aberrant in that patient.

2.1   Estimating Hitting Time on a Protein-Protein Interaction (PPI) Network

As mentioned before, HIT'nDRIVE estimates the multi-hitting time $H(U, v)$ between a set of nodes U and a single node v, as a function of independent hitting times $H(u, v)$ for all $u \in U$ - as will be shown later. However, even computing $H(u, v)$ is not a trivial task in a general graph $G = (V, E)$ as it requires a solution to a system of $|V|$ linear equations with $|V|$ variables. Below we show how to efficiently calculate $H(u, v)$ for all $u, v \in V$ for a graph $G = (V, E)$ with constant average degree and small diameter - as per the available human protein interaction network (or any small world network).

Let $H_{max} = \max_{u,v}\{H_{u,v}\}$. Our aim is to estimate $H_{u,v}$ empirically by performing independent random walks and taking the average of the observed hitting times. More formally, for any given number of iterations $m > 1$ and pair $u, v \in V$, let $X_1, X_2, ..., X_m$ be a sequence of independent random variables which have the same distribution as $\tau_{u,v}$ for every $1 \leq i \leq m$. Then the empirical hitting time is defined as $\tilde{H}_{u,v} = \frac{1}{m} \cdot \sum_{i=1}^{m} X_i$. The following theorem shows how fast $\tilde{H}_{u,v}$ converges to $H_{u,v}$.

Theorem 1. *Assume that G is a graph such that the maximum hitting time satisfies $H_{max} \leq Cn$ for some constant $C > 0$ and let u, v be an arbitrary pair of nodes. Then for any $\varepsilon \in [\frac{1}{n^4}, 1]$, after $m = (128C)^2(1/\varepsilon)^2(\log_2 n)^3$ iterations, the returned estimate $\tilde{H}_{u,v}$ satisfies*

$$\Pr\left[|\tilde{H}_{u,v} - H_{u,v}| \leq \varepsilon n\right] \geq 1 - n^{-3}.$$

Moreover, with probability at least $1 - n^{-7}$, the total number of random walk hops made is at most $m \cdot 32Cn\log_2 n = O((1/\varepsilon)^2 n \log^4 n)$.

We provide the proof of Theorem 1 in the Supplements. To obtain the empirical estimates of all n^2 hitting times $H_{u,v}$ efficiently, observe that taking a single random walk starting from u until all nodes are visited gives an estimate for all n hitting times $H_{u,v}$ with $v \in V$. Since for fixed $v \in V$, all m estimates for $H_{u,v}$ (coming from m iterations) are independent, we conclude by the first statement of Theorem 1 and the union bound that with probability at least $1 - n^{-2}$, for fixed $u \in V$ all n estimates $\tilde{H}_{u,v}$ approximate $H_{u,v}$ up to an additive error of εn. Similarly, the total number of random walk hops to obtain all these n approximations is $O((1/\varepsilon)^2 n \log^4 n)$ with probability at least $1 - n^{-6}$. Finally, we do the above procedure for all n possible starting vertices $u \in V$, so that with probability at least $1 - n^{-1}$, we have an εn-additive approximation for each of the n^2 hitting times, and the total number of random walk hops is $O((1/\varepsilon)^2 n^2 \log^4 n)$ with probability at least $1 - n^{-5}$.

2.2   Estimating Multi-source Hitting Time via Single-Source Hitting Times

Given $U = \{u_1, u_2, \ldots, u_k\}$, we now show how to estimate $H_{U,v}$ by a function of independent pairwise hitting times $H_{u_i,v}$ for all $u_i \in U$. A natural estimate is

$$H_{U,v} \approx \frac{1}{\sum_{i=1}^{k} \frac{1}{H_{u_i,v}}} \tag{1}$$

Let the conductance of graph G be defined as $\Phi(G) = \min_{0 \subsetneq S \subsetneq V} \frac{|E(S, V\setminus S)|}{\min\{\text{vol}(S), \text{vol}(V\setminus S)\}}$. Many real-world networks including preferential attachment graphs are known to have large conductance [25]. For such graphs, our next theorem provides mathematical evidence for the accuracy of our estimate in (1).

Theorem 2. *Let $G = (V, E)$ be any graph with constant conductance $\Phi > 0$. Then there is an integer $C = C(\Phi) > 0$ such that, given an integer k, a set of nodes $U = \{u_1, u_2, \ldots, u_k\}$ and node $v \in V$ satisfying $\frac{1}{k \cdot \frac{\deg(v)}{2|E|}} \geq \log^{1.5} n$, the following inequality holds:*

$$H_{U,v} \leq C \cdot \frac{1}{\sum_{i=1}^{k} \frac{\deg(v)}{2|E|}}.$$

In particular, for any pair of nodes u, v with $\deg(v) \leq \frac{2|E|}{\log^{1.5} n}$ we have $H_{u,v} = O(\frac{|E|}{\deg(v)})$.

We provide the proof in the Supplements. Note that the bound in Theorem 2 differs from our estimate in equation (1) in that $\frac{1}{H_{u_i,v}}$ is replaced by $\frac{\deg(v)}{2|E|}$. However, for graphs with constant conductance, we have $H_{u,v} \leq H_{\pi,v} + O(\log n)$, where $H_{\pi,v}$ is the hitting time for a random walk starting according to the stationary distribution π, given by $\pi(w) = \frac{\deg(w)}{2|E|}$ for every $w \in V$. Hence $\frac{2|E|}{\deg(v)} = H_{v,v} \leq H_{\pi,v} + O(\log n)$. Since $H_{\pi,v} = \sum_{u \in U} \pi(u) \cdot H_{u,v}$, it follows that, given any fixed node v, it holds for "most nodes" u that $\tilde{H}_{u,v}$ is not much smaller than $\frac{2|E|}{\deg(v)} - O(\log n)$.

3   Reformulation of RWFL as a Weighted Multi-set Cover Problem

Since RWFL is NP-hard we reduce it to the weighted set cover problem, which we solve via an ILP formulation. This formulation also generalizes RWFL to allow patient-specific drivers and outlier genes. Consider a bipartite graph $G_{bip}(\mathcal{X}, \mathcal{Y}, \mathcal{E})$ where $\mathcal{X}$ is

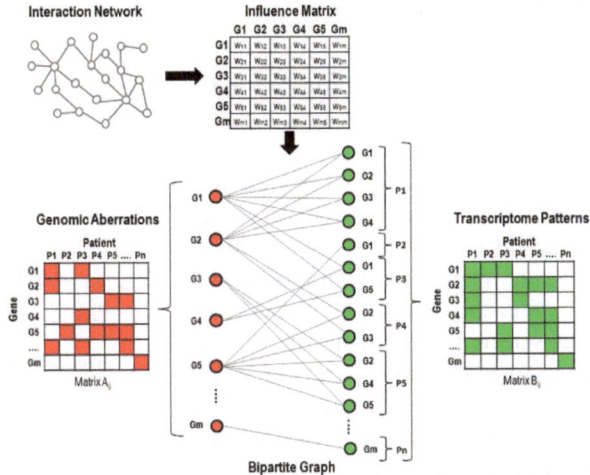

Fig. 1. Schematic overview of construction of bipartite graph in HIT'nDRIVE. The influence matrix derived from the interaction network contains the *inverse* hitting time between every pair of genes. *A* and *B* are gene-patient matrices showing the genomic abberations and expression alteration events, respectively. The red color in *A* indicates the aberration status of a gene in a patient. Similarly, the green color in *B* indicate expression altered genes in a patient. The edges in the bipartite graph are weighted by the inverse hitting time within the PPI network.

the set of aberrant genes, $\mathcal{Y}$ is the set of patient-specific expression altered genes, and $\mathcal{E}$ is the set of edges. If gene g_i is mutated in a patient p, we set edges between g_i and all of the expression altered genes in the same patient (g_j, p) where the edges are weighted by the inverse pairwise-hitting times $w_{i,j} := H_{g_i,g_j}^{-1}$; see the Figure 1 for more details.

We now define a minimum weighted multi-set cover (WMSC) problem on G_{bip}, whose solution provides an exact solution to RWFL problem, provided our estimate of the multi-hitting times are accurate, i.e.

$$\arg\min_{X \subseteq \mathcal{X}} |X| \quad \text{such that} \quad \max_{y \in \mathcal{Y}} H_{X,y} \leq \varDelta \tag{2}$$

where $\varDelta$ is the maximum allowed multi-hitting time from the drivers to any expression altered gene.

WMSC asks to compute as the potential driver gene set, the smallest set which "sufficiently" covers "most" of the patient specific expression altered genes:

$$\arg\min_{X \in \mathcal{X}} \min_{Y \subseteq \mathcal{Y}, |Y| \geq \alpha |\mathcal{Y}|} |X| \quad \text{such that} \quad \forall y \in Y : \sum_{x \in X} w_{x,y} \geq \gamma_y \tag{3}$$

where $0 < \alpha \leq 1$ represents the fraction of patient-specific expression altered genes that we believe are causally linked to the potential drivers. The left-hand-side of the constraints in (2) and (3) are related by $H_{X,y}^{-1} \approx \sum_{x \in X} H_{x,y}^{-1}$, as mentioned in Section 2.2. The introduction of γ_y makes it possible to control the minimum amount of "coverage" needed for *individual* expression alteration events (each patient potentially indicates a unique expression alteration event for each gene).

3.1 An ILP Formulation for WMSC

We formulate WMSC as an ILP and solve it using an off-the-shelf ILP solver. The ILP formulation for our combinatorial optimization problem is as Figure 2 where there is a binary variable x_i, y_j, e_{ij}, respectively, for each potential driver, expression alteration event, and edge in the bipartite graph. The first constraint ensures that a selected driver contributes to the coverage of each of the expression alteration events it is connected to - in each patient. The second constraint ensures that selected (patient-specific) driver genes cover at least a (γ) fraction of the sum of all incoming edge weights to each expression alteration event. This constraint corresponds to setting a lower

$$\min_{x_1,...,x_{|X|}} \sum_i x_i$$
s.t.
$$\forall i, j : x_i = e_{ij}$$
$$\forall j : \sum_i e_{ij} w_{ij} \geq y_j \gamma \sum_i w_{ij}$$
$$\sum_j y_j \geq \alpha |\mathcal{Y}|$$
$$x_i, e_{ij}, y_j \in \{0, 1\}$$

Fig. 2. ILP formulation

bound on the joint influence (i.e. our estimate on the inverse of multi-hitting time) of selected (patient specific) drivers on an expression alteration event. The third constraint ensures that the selected driver genes collectively cover at least an α fraction of the set of expression alteration events.

4 Evaluation Framework

Evaluating computational methods for predicting cancer drivers is challenging in the absence of the ground truth (i.e. follow-up biological experiments). We refer to previous studies [18] that observe the overlap between predicted driver genes and known cancer genes compiled in public resources such as the Cancer Gene Census (CGC) database [26] or the Catalogue of Somatic Mutations in Cancer (COSMIC) database [27] and we provide those numbers as well. However, we mainly focused on testing whether our predictions provide insight into the cancer phenotype and improve classification accuracy on an independent cancer dataset. The classifiers we evaluate are based on network "modules", a set of functionally related genes (e.g. in a signaling pathway), which are connected in an interaction network and include at least one potential driver. They then use module features, such as the average expression of genes in the module, for phenotype classification.

Using such module features, we hope that the classifier in use does not *overfit* on rare drivers and is able to *generalize* the signal coming from rare drivers to new patients.

For classification purposes we primarily use OptDis [22] for *de novo* identification of modules which include (i.e. are seeded by) at least one predicted driver gene. In general, OptDis performs supervised dimensionality reduction on the set of connected subnetworks.

It projects the high dimensional space of all connected subnetworks to a user-specified lower dimensional space of subnetworks such that, in the new space, the samples belonging to the same (different) class are closer (respectively, more distant) to each other with respect to a normalized distance measure (typically L_1).

Since the human PPI network has a small diameter, there is significant overlap between many modules seeded by potential driver genes. In order to limit the number of overlapping modules (and achieve further dimensionality reduction) we first compute the top 10 modules seeded by each driver gene that have the best individual "discriminative scores" (a linear combination of the average in-class distance and out-class distance [22]). The modules seeded by all potential drivers are then collectively sorted based on their discriminative score. Among these modules, we greedily pick a subset in a way that the ith module is added to our result subset R if its maximum pairwise node overlap with any module already in R is no more than a user-defined threshold.

5   Experiments

We use a publicly available cancer dataset representing matched genomic aberration (somatic mutation, copy-number aberration) and transcriptomic patterns (gene-expression data) of 156 Glioblastoma Multiforme (GBM) samples [5] from The Cancer Genome Atlas (TCGA). We make use of a global network of protein-protein interaction (PPI) from the Human Protein Reference Database (HPRD) version April 2010 [28] to derive the influence values based on the hitting time. We use the same PPI Network for module identification using our modification to OptDis. We ran HIT'nDRIVE with different combinations of values for the variables α and γ as given in Figure 3-A. For a fixed γ, the number of selected driver genes increased linearly with the value of α. The increase in the number of drivers is expected as more drivers are required to cover larger fraction of abnormal expression events.

Evaluation Based on CGC and COSMIC Databases. To assess whether the genes identified by HIT'nDRIVE are essential players in cancer, we first analyzed the concordance of the predicted drivers with the genes annotated in CGC and COSMIC database. Gene sets resulting with the parameters $\gamma = 0.7$ and $\alpha = \{0.1, 0.2, ..., 0.9\}$ were analyzed (Figure 3-B). The fraction of driver genes affiliated with cancer in the CGC and COSMIC databases increase with increasing values of α. The remainder of results are obtained for parameter values $\gamma = 0.7$ and $\alpha = 0.9$ this results in 107 driver genes covering the majority (22933) of outlier genes in 156 patients.

Phenotype Classification Using Dysregulated Modules Seeded with the Predicted Drivers. We evaluated the driver genes identified by HIT'nDRIVE using phenotype classification (as described in Section 4 and results are shown in Figure 4). Briefly, drivers identified from the TCGA dataset were used as seeds for discovering discriminative subnetwork modules. The module expression profiles were used to classify normal vs. glioblastoma samples through repeated cross-validation on the validation dataset. First, HIT'nDRIVE using hitting time based influence values, was compared against DriverNet, which greedily identifies driver genes using direct gene interactions from the HPRD network. Across the appreciable range of discriminative modules discovered by OptDis, HIT'nDRIVE demonstrates better accuracy in classifying the cancer phenotype, with a maximum accuracy of 96.9% and a mean accuracy of 93.4% (Figure 4). Next, comparing the HIT'nDRIVE deduced drivers against a comparable number of genes with the highest node-degrees in the PPI network reveals a clear advantage to HITnDRIVE. This trend was observed when genes were used as individual classification features (blue vs. orange plots) as well as when they were used as seeds for

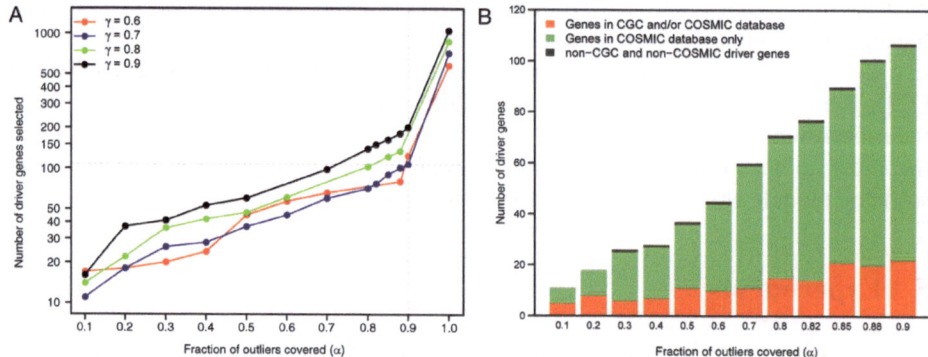

Fig. 3. Behavior of HIT'nDRIVE as a function of α and γ. (A) The number of selected drivers and covered outliers as α increases for various values of γ. Note that some of the data points are missing for the problems which could not be solved within 48 hours. (B) Concordance of GBM driver genes with that of COSMIC and Cancer Gene Census database for $\gamma = 0.7$.

module-based features (red vs. brown plots). Comparing the classification accuracy of HITnDRIVE deduced drivers against 107 genes randomly selected from the entire list of aberrant genes (red vs. grey plots) provides additional support for the relevance of drivers selected by HITnDRIVE. This is also confirmed by comparing the performance of hitting-time based influence values against those derived from the diffusion model [16] (red vs. black plots) both employed by HITnDRIVE.

Sensitivity of HIT'nDRIVE to Small Perturbations of the PPI Network. We perturbed the PPI network by swapping endpoints at random of 20% edges and recalculated pairwise hitting times. We observed that almost all changes are less than 10% relative to the original values, most of them being between 1% and 5%. However, impact on accuracy of classification using HIT'nDRIVE output can be noticed in Figure 4.

Prediction of Frequent and Rare Drivers. The 107 driver genes nominated by HIT'nDRIVE are aberrated at varying frequencies in the tumor population (Figure 5-A). CHEK2 and EGFR are the two most frequently aberrated drivers (at 46.8% and 42.3% respectively), followed by CDKN2A (31.4%), MTAP (30.1%) and CDKN2B (29.5%). Some of these frequent drivers harbour different types of genomic aberrations in different patients. For example, EGFR shows somatic mutation and high copy-number gain in 14.2% and 32.7% of the patients, respectively. Similarly, PTEN harbours somatic mutation in 12.8% and homozygous deletion in 3.9% of the patients. Amplification in EGFR, PDGFRA, mutations in CHEK2, TP53, PTEN, RB1, and deletions in CDKN2A have been previously associated with GBM [5, 29, 30]. HIT'nDRIVE also identified infrequent drivers, which we defined as genes that are genomically aberrant in at most 2% of the cases. Out of 27 (25.2%) rare driver genes identified, four genes (FLI1, BMPR1A, MYST4, and BRCA2) were implicated in the CGC database. Despite being aberrant in a small fraction of patients, the rare drivers are specifically associated with

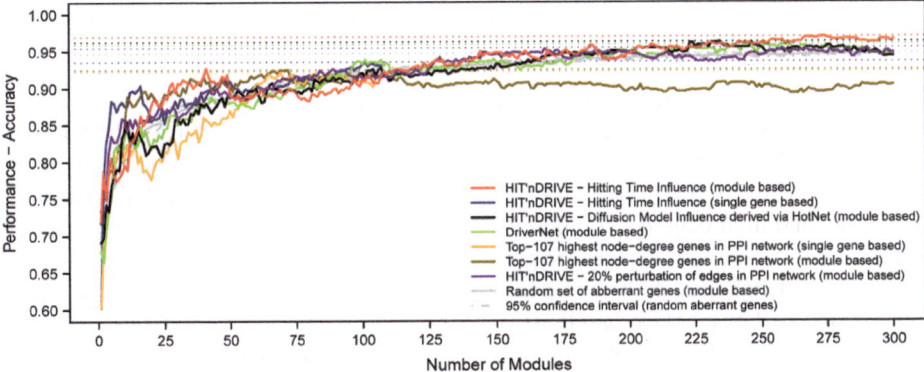

Fig. 4. Phenotype classification using the identified drivers obtained by various methods. The dysregulated sets of modules seeded by the 107 chosen drivers are used to predict phenotype in the validation dataset using using k-nearest neighbour classifier with k=1. We used the HPRD-PPI Network for module identification using our modification to OptDis.

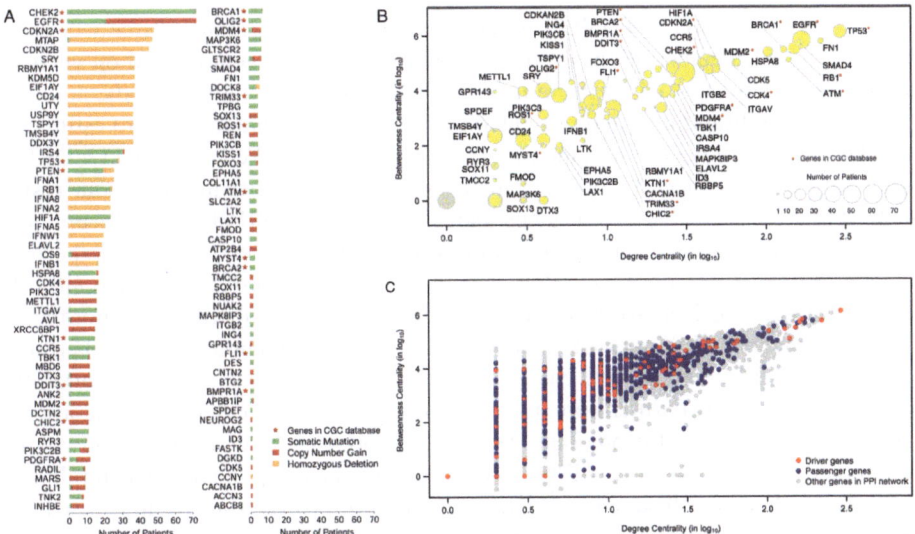

Fig. 5. Characteristics of driver genes of GBM predicted by HIT'nDRIVE. (A) Recurrence frequency of the aberration in the driver genes predicted by HIT'nDRIVE. (B) The centrality of the predicted drivers in the PPI network. The size of the circles is proportional to the recurrence frequency of the genomic aberration of the gene. (C) Centrality of the "driver" and "passenger" genes is colored by red and blue dots respectively; all other nodes in the PPI network apart from the driver and passenger genes are represented as grey dots.

cancer development, DNA repair, cell growth and migration, cell death and survival. Some rare drivers like MAG and BMPR1A have also been closely linked with GBM progression [31, 32].

Prediction of Low-degree and High-degree Drivers. The drivers predicted by HIT'nDRIVE include a number of well-known high-degree "hubs" such as TP53, EGFR, RB1 and BRCA1, which occupy the central position (with high degree and high betweenness, i.e. the proportion of shortest paths between all pairs of nodes that go through that node, and high degree - computed by the igraph [33] R package.) in the PPI network (Figure 5-B). If these genes are perturbed, they dysregulate several other genes and the associated signaling pathways. Moreover, HIT'nDRIVE also identified low-degree genes (such as IFNA2, UTY, and RYR3) that reside in the periphery of the PPI network. Some of these low-degree genes are only aberrant in a small fraction of patients. Since driver genes and passenger genes display similar network characteristics (Figure 5-C), and identified driver genes have both low and high degrees in the network, HIT'nDRIVE likely selects drivers irrespective of known network biases.

6 Conclusion and Future Work

We have presented HIT'nDRIVE, a combinatorial method to capture the collective effects of driver gene aberrations on possibly distant "outlier" genes based on what we call the "random-walk facility location" (RWFL) problem. We introduced the notion of "multi-source hitting time" and presented efficient and accurate methods to estimate it based on single-source hitting time in large-scale networks. We applied HIT'nDRIVE to identify genes subject to somatic mutation and copy number in GBM. Our results showed that the predicted driver genes identified by HIT'nDRIVE are well-supported in databases of important cancer genes. Furthermore, these drivers were able to perform phenotype predictions more accurately than the alternative approaches. Importantly, the discovery of these drivers were not biased by the frequency of aberration and/or the degree of a gene in the PPI network. Our approach can easily integrate various aberration types such as single nucleotide changes, copy number changes, structural variations, and splice variations. Furthermore, it can be straightforwardly extended to incorporate epigenome and/or gene-fusions data. As gene networks increase in density and volume of interaction, HIT'nDRIVE will be able to capture such improvements naturally. Finally our method is well suited to identify patient-specific driver-aberrations which can potentially be used as therapeutic targets.

Supplements: All supplementary material can be found and downloaded at http://compbio.cs.sfu.ca/software-hitndrive.

References

1. Stratton, M.R., Campbell, P.J., Futreal, P.A.: The cancer genome. Nature 458(7239), 719–724 (2009)
2. Greenman, C., Stephens, P., Smith, R., Dalgliesh, G.L., Hunter, C., et al.: Patterns of somatic mutation in human cancer genomes. Nature 446(7132), 153–158 (2007)

3. Greenman, C., Wooster, R., Futreal, P.A., Stratton, M.R., Easton, D.F.: Statistical analysis of pathogenicity of somatic mutations in cancer. Genetics 173(4), 2187–2198 (2006)
4. Youn, A., Simon, R.: Identifying cancer driver genes in tumor genome sequencing studies. Bioinformatics 27(2), 175–181 (2011)
5. Parsons, D.W., Jones, S., Zhang, X., Lin, J.C.H., Leary, R.J., Angenendt, P., et al.: An integrated genomic analysis of human glioblastoma multiforme. Science 321(5897), 1807–1812 (2008)
6. Cancer Genome Atlas Network: Integrated genomic analyses of ovarian carcinoma. Nature 474(7353), 609–615 (2011)
7. Cancer Genome Atlas Network: Comprehensive molecular characterization of human colon and rectal cancer. Nature 487(7407), 330–337 (2012)
8. Greaves, M., Maley, C.C.: Clonal evolution in cancer. Nature 481(7381), 306–313 (2012)
9. Ding, L., Ley, T.J., Larson, D.E., Miller, C.A., Koboldt, D.C., et al.: Clonal evolution in relapsed acute myeloid leukaemia revealed by whole-genome sequencing. Nature 481(7382), 506–510 (2012)
10. Akavia, U.D., Litvin, O., Kim, J., Sanchez-Garcia, F., Kotliar, D., et al.: An integrated approach to uncover drivers of cancer. Cell 143(6), 1005–1017 (2010)
11. Masica, D.L., Karchin, R.: Correlation of somatic mutation and expression identifies genes important in human glioblastoma progression and survival. Cancer Research 71(13), 4550–4561 (2011)
12. Leiserson, M.D.M., Blokh, D., Sharan, R., Raphael, B.J.: Simultaneous identification of multiple driver pathways in cancer. PLoS Computational Biology 9(5), e1003054 (2013)
13. Ciriello, G., Cerami, E., Sander, C., Schultz, N.: Mutual exclusivity analysis identifies oncogenic network modules. Genome Research 22(2), 398–406 (2012)
14. Kim, Y.A., Wuchty, S., Przytycka, T.M.: Identifying causal genes and dysregulated pathways in complex diseases. PLoS Computational Biology 7(3), e1001095 (2011)
15. Vaske, C.J., Benz, S.C., Sanborn, J.Z., Earl, D., Szeto, C., et al.: Inference of patient-specific pathway activities from multi-dimensional cancer genomics data using PARADIGM. Bioinformatics 26(12), i237–i245 (2010)
16. Vandin, F., Upfal, E., Raphael, B.J.: Algorithms for detecting significantly mutated pathways in cancer. Journal of Computational Biology: a Journal of Computational Molecular Cell Biology 18(3), 507–522 (2011)
17. Paull, E.O., Carlin, D.E., Niepel, M., Sorger, P.K., Haussler, D., et al.: Discovering causal pathways linking genomic events to transcriptional states using Tied Diffusion Through Interacting Events (TieDIE). Bioinformatics, 1–8 (2013)
18. Bashashati, A., Haffari, G., Ding, J., Ha, G., Lui, K., et al.: DriverNet: uncovering the impact of somatic driver mutations on transcriptional networks in cancer. Genome Biology 13(12), R124 (2012)
19. Liben-Nowell, D., Kleinberg, J.: The link-prediction problem for social networks. Journal of the American Society for Information Science and Technology 58(7), 1019–1031 (2007)
20. Hopcroft, J., Sheldon, D.: Manipulation-resistant reputations using hitting time. In: Bonato, A., Chung, F.R.K. (eds.) WAW 2007. LNCS, vol. 4863, pp. 68–81. Springer, Heidelberg (2007)
21. Tetali, P.: Design of on-line algorithms using hitting times. SIAM J. Comput. 28(4), 1232–1246 (1999)
22. Dao, P., Wang, K., Collins, C., Ester, M., Lapuk, A., Sahinalp, S.C.: Optimally discriminative subnetwork markers predict response to chemotherapy. Bioinformatics 27(13) (July 2011)
23. Levin, D.A., Peres, Y., Wilmer, E.L.: Markov Chains and Mixing Times. American Mathematical Society (2008)

24. Hormozdiari, F., Alkan, C., Eichler, E.E., Sahinalp, S.C.: Combinatorial algorithms for structural variation detection in high-throughput sequenced genomes. Genome Research 19(7), 1270–1278 (2009)
25. Mihail, M., Papadimitriou, C.H., Saberi, A.: On certain connectivity properties of the internet topology. J. Comput. Syst. Sci. 72(2), 239–251 (2006)
26. Futreal, P.A., Coin, L., Marshall, M., Down, T., Hubbard, T., et al.: A census of human cancer genes. Nature reviews. Cancer 4(3), 177–183 (2004)
27. Forbes, S.A., Bindal, N., Bamford, S., Cole, C., Kok, C.Y., et al.: COSMIC: mining complete cancer genomes in the Catalogue of Somatic Mutations in Cancer. Nucleic Acids Research 39(database issue), D945–D950 (2011)
28. Prasad, T.S.K., Kandasamy, K., Pandey, A.: Human Protein Reference Database and Human Proteinpedia as discovery tools for systems biology. Methods in Molecular Biology 577, 67–79 (2009)
29. Cancer Genome Atlas Network: Comprehensive genomic characterization defines human glioblastoma genes and core pathways. Nature 455(7216), 1061–1068 (2008)
30. Verhaak, R.G.W., Hoadley, K.A., Purdom, E., Wang, V., Qi, Y., et al.: Integrated genomic analysis identifies clinically relevant subtypes of glioblastoma characterized by abnormalities in PDGFRA, IDH1, EGFR, and NF1. Cancer Cell 17(1), 98–110 (2010)
31. McKerracher, L., David, S., Jackson, D.L., Kottis, V., Dunn, R.J., et al.: Identification of myelin-associated glycoprotein as a major myelin-derived inhibitor of neurite growth. Neuron 13(4), 805–811 (1994)
32. Piccirillo, S.G.M., Reynolds, B.A., Zanetti, N., Lamorte, G., Binda, E., et al.: Bone morphogenetic proteins inhibit the tumorigenic potential of human brain tumour-initiating cells. Nature 444(7120), 761–765 (2006)
33. Csardi, G., Nepusz, T.: The igraph software package for complex network research. InterJournal Complex Systems, 1695 (2006)

Modeling Mutual Exclusivity
of Cancer Mutations

Ewa Szczurek and Niko Beerenwinkel

Department of Biosystems Science and Engineering, ETH Zurich, Mattenstrasse 26,
4058 Basel, Switzerland, SIB Swiss Institute of Bioinformatics
{ewa.szczurek,niko.beerenwinkel}@bsse.ethz.ch

Recent years in cancer research were characterized by both accumulation of data and growing awareness of its overwhelming complexity. Consortia like The Cancer Genome Atlas [1] generated large collections of tumor samples, recording presence or absence of genomic alterations, such as somatic point mutations, amplifications, or deletions of genes. One of the basic tasks in the analysis of tumor genomic data is to elucidate sets of genes involved in a common oncogenic pathway. A *de novo* approach to this task is to search for mutually exclusive patterns in cancer genomic data [2, 3, 4, 5], where these alterations tend not to occur together in the same patient. Such patterns are commonly evaluated and ranked by their coverage and impurity. Coverage is defined as the number of patient samples in which at least one alteration occurred, while impurity refers to non-exclusive, additional alterations that violate strict mutual exclusivity. Mutually exclusive patterns have frequently been observed in cancer data, and were associated with functional pathways [6].

Previous *de novo* approaches identified mutually exclusive patterns either with an online learning approach [2], or by maximizing a mutual exclusivity weight, which increases with coverage and decreases with impurity [3, 4, 5]. However, in the absence of a statistical model of the data, the definition of the weight, although intuitively reasonable, remains arbitrary. None of the existing approaches deal with the problem of errors in the data, which may arise due to measurement noise, as well as uncertainty in mutation calling and interpretation. We show that ignoring errors in the data, particularly false positives, may lead to false ranking of patterns.

To address these limitations, we propose a probabilistic, generative model of mutually exclusive patterns in the data. The model contains coverage as well as impurity as parameters, together with false positive and false negative rates. We show analytically that the model parameters are identifiable, and propose efficient algorithms for their estimation, as well as for pattern evaluation and ranking. Via comparison of the mutual exclusivity model to the null model assuming independent alterations of genes, our approach allows statistical testing for mutual exclusivity, both in the presence and absence of errors.

We first evaluate the performance of our approach in the case when, as it is done in the literature, the data is assumed to record no false positive or negative alterations. On simulated patterns our mutual exclusivity test proves more powerful than the weight-based permutation test applied previously.

R. Sharan (Ed.): RECOMB 2014, LNBI 8394, pp. 307–308, 2014.
© Springer International Publishing Switzerland 2014

In glioblastoma multiforme data [7], we find novel, biologically relevant patterns, which are not detected by the permutation test. Next, we examine the bias introduced in pattern ranking by ignorance of errors, especially false positives, and show that when the error rates are known, our approach is able to accurately estimate the true coverage and impurity and rank the patterns accordingly. Finally, we analyze the practical limits of accurate parameter estimation in the most difficult, but also most realistic case, where the data contains errors occurring at unknown rates. We apply our approach to a large, pan-cancer collection of 3299 tumor samples from twelve tumor types [8], for which the model accounting for the presence of false positives can accurately be estimated. This model is shown to be more flexible than the model assuming no errors in the data, and is applied to identify universal, significant mutual exclusivity patterns. Both extensive simulations, as well as application to glioblastoma and pan-cancer data show that our statistical approach to mutual exclusivity provides increased flexibility and power to detect cancer pathways from tumor data in the presence of noise.

References

[1] TCGA: Comprehensive genomic characterization defines human glioblastoma genes and core pathways. Nature 455(7216), 1061–1068 (2008)
[2] Miller, C., Settle, S., Sulman, E., Aldape, K., Milosavljevic, A.: Discovering functional modules by identifying recurrent and mutually exclusive mutational patterns in tumors. BMC Medical Genomics 4(1), 34+ (2011)
[3] Vandin, F., Upfal, E., Raphael, B.J.: *De Novo* discovery of mutated driver pathways in cancer. Genome Res. 22(2), 375–385 (2012)
[4] Leiserson, M.D.M., Blokh, D., Sharan, R., Raphael, B.J.: Simultaneous Identification of Multiple Driver Pathways in Cancer. PLoS Comput. Biol. 9(5), e1003054+ (2013)
[5] Zhao, J., Zhang, S., Wu, L.Y., Zhang, X.S.: Efficient methods for identifying mutated driver pathways in cancer. Bioinformatics 28(22), 2940–2947 (2012)
[6] Yeang, C.H., Mccormick, F., Levine, A.: Combinatorial patterns of somatic gene mutations in cancer. FASEB J. 22(8), 2605–2622 (2008)
[7] Brennan, C.W., Verhaak, R.G.W., McKenna, A., Campos, B., Noushmehr, H., Salama, S.R., Zheng, S., Chakravarty, D., Sanborn, J.Z., Berman, S.H., Beroukhim, R., Bernard, B., Wu, C.J., Genovese, G., Shmulevich, I., Barnholtz-Sloan, J., Zou, L., Vegesna, R., Shukla, S.A., Ciriello, G., Yung, W.K., Zhang, W., Sougnez, C., Mikkelsen, T., Aldape, K., Bigner, D.D., Van Meir, E.G., Prados, M., Sloan, A., Black, K.L., Eschbacher, J., Finocchiaro, G., Friedman, W., Andrews, D.W., Guha, A., Iacocca, M., O'Neill, B.P., Foltz, G., Myers, J., Weisenberger, D.J., Penny, R., Kucherlapati, R., Perou, C.M., Hayes, D.N., Gibbs, R., Marra, M., Mills, G.B., Lander, E., Spellman, P., Wilson, R., Sander, C., Weinstein, J., Meyerson, M., Gabriel, S., Laird, P.W., Haussler, D., Getz, G., Chin, L.: The Somatic Genomic Landscape of Glioblastoma. Cell 155(2), 462–477 (2013)
[8] Ciriello, G., Miller, M.L., Aksoy, B.A., Senbabaoglu, Y., Schultz, N., Sander, C.: Emerging landscape of oncogenic signatures across human cancers. Nat. Genet. 45(10), 1127–1133 (2013)

Viral Quasispecies Assembly
via Maximal Clique Enumeration

Armin Töpfer[1,2], Tobias Marschall[3], Rowena A. Bull[4], Fabio Luciani[4],
Alexander Schönhuth[3,*], and Niko Beerenwinkel[1,2,*,**]

[1] Department of Biosystems Science and Engineering,
ETH Zurich, Basel, Switzerland
[2] SIB Swiss Institute of Bioinformatics, Switzerland
[3] Centrum Wiskunde & Informatica, Amsterdam, The Netherlands
[4] Inflammation and Infection Research Centre, School of Medical Sciences, UNSW,
Sydney, Australia
niko.beerenwinkel@bsse.ethz.ch

Genetic variability of virus populations within individual hosts is a key determinant of pathogenesis, virulence, and treatment outcome. It is of clinical importance to identify and quantify the intra-host ensemble of viral haplotypes, called viral quasispecies. Ultra-deep next-generation sequencing (NGS) of mixed samples is currently the only efficient way to probe genetic diversity of virus populations in greater detail. Major challenges with this bulk sequencing approach are (i) to distinguish genetic diversity from sequencing errors, (ii) to assemble an unknown number of different, unknown, haplotype sequences over a genomic region larger than the average read length, (iii) to estimate their frequency distribution, and (iv) to detect structural variants, such as large insertions and deletions (indels) that are due to erroneous replication or alternative splicing. Even though NGS is currently introduced in clinical diagnostics, the *de-facto* standard procedure to assess the quasispecies structure is still single-nucleotide variant (SNV) calling. Viral phenotypes cannot be predicted solely from individual SNVs, as epistatic interactions are abundant in RNA viruses. Therefore, reconstruction of long-range viral haplotypes has the potential to be adopted, as data is already available.

We present HaploClique, a computational method that combines a probabilistic model of sequence similarity and structural similarity with a graph theoretical method to reconstruct viral quasispecies from NGS paired-end data. We define a read alignment graph, in which nodes correspond to single-end and paired-end alignments (Figure 1A right). We draw an edge between nodes if alignments (i) have sufficient overlap, (ii) are compatible in the insert size (Figure 1A left), defined as the unsequenced fragment between read pairs, and (iii) show that sequences are sufficiently similar. Taken together these criteria ensure that both reads are likely to stem from the same haplotype (Figure 1B). If alignments stem from the same haplotype, their sequences are identical up to sequencing errors in the intervals of overlap. The corresponding probability is computed

---

* Equal contributions.
** Corresponding author.

R. Sharan (Ed.): RECOMB 2014, LNBI 8394, pp. 309–310, 2014.
© Springer International Publishing Switzerland 2014

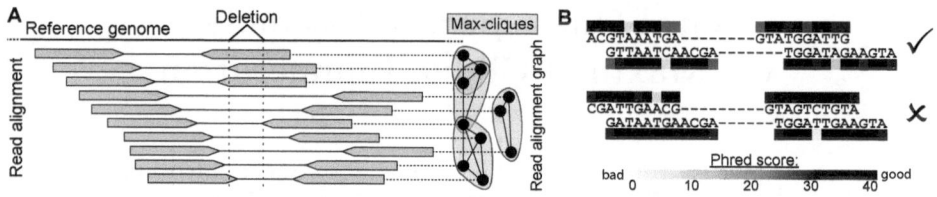

Fig. 1. (A) Paired-end read alignment with a deletion harboring haplotype and the corresponding read alignment graph with max-cliques of minimal size three, based on the insert size compatibility. (B) Phred scores of paired-end reads are used to assess sequence similarity. The unsequenced fragment is indicated by the gap symbol '-'. Top example is sequence compatible, as reads differ only in sequencing errors, bases with low phred scores. Bottom example is not sequence compatible, because reads differ in bases with high pred scores.

using the base calling quality scores (phred scores). In addition, we compute the probability that the non-overlapping alignment sequences are identical.

We develop a maximal clique (max-clique) enumeration approach (Figure 1A right) to cluster NGS reads. Max-cliques are fully connected subgraphs that cannot be extended and consist of reads with mutually compatible alignments. We use max-cliques to reconstruct haplotype sequences and detect indels. In detail, the structural similarity of all reads in a max-clique and its deviation from the empirical insert size distribution allows to detect large indels. The consensus sequence of each max-clique, called super-read, is the predicted error-corrected local haplotype. By iterating read alignment graph construction and max-clique enumeration of super-reads, haplotype fragments grow in length and possibly allow full-length reconstruction. Haplotype abundance estimation is performed by counting original reads that participated in a super-read.

In extensive simulation studies, we benchmarked the accuracy and robustness of estimating haplotype frequencies, the error correction performance, and the minimal distance of two haplotypes to be perfectly distinguishable. We showed that HaploClique outperforms the state-of-the-art tools ShoRAH, PredictHaplo, and QuRe on a simulated dataset of five well known HIV strains with a low coverage of 600x. HaploClique successfully reconstructed one haplotype at its full length and the other strains are covered with reconstructed haplotypes of sizes 5-6 kb, where the original strain lengths are 9-10 kb. The structure prediction accuracy and robustness was assessed for varying deletion sizes between 100 bp and 1 kb. We applied HaploClique to a clinical hepatitis C virus infected sample and detected a novel deletion of size 357 ± 167 bp, which was validated by two independent long-read sequencing experiments. HaploClique is able to predict large indels that cannot be detected by current computational methods and reconstruct full-length haplotypes from low coverage samples. HaploClique's implementation is available at **https://github.com/armintoepfer/haploclique**.

Correlated Protein Function Prediction via Maximization of Data-Knowledge Consistency*

Hua Wang[1], Heng Huang[2,**], and Chris Ding[2]

[1] Department of Electrical Engineering and Computer Science
Colorado School of Mines, Golden, Colorado 80401, USA
[2] Department of Computer Science and Engineering
University of Texas at Arlington, Arlington, Texas 76019, USA
huawangcs@gmail.com, {heng,chqding}@uta.edu

Abstract. Protein function prediction in conventional computational approaches is usually conducted one function at a time, fundamentally. As a result, the functions are treated as separate target classes. However, biological processes are highly correlated, which makes functions assigned to proteins are not independent. Therefore, it would be beneficial to make use of function category correlations in predicting protein function. We propose a novel Maximization of Data-Knowledge Consistency (MDKC) approach to exploit function category correlations for protein function prediction. Our approach banks on the assumption that two proteins are likely to have large overlap in their annotated functions if they are highly similar according to certain experimental data. We first establish a new pairwise protein similarity using protein annotations from knowledge perspective. Then by maximizing the consistency between the established *knowledge similarity* upon annotations and the *data similarity* upon biological experiments, putative functions are assigned to unannotated proteins. Most importantly, function category correlations are elegantly incorporated through the knowledge similarity. Comprehensive experimental evaluations on *Saccharomyces cerevisiae* data demonstrate promising results that validate the performance of our methods.

Keywords: Protein Function Prediction, Mutli-Label Classification, Symmetric Nonnegative Matrix Factorization.

1 Introduction

Due to its significant importance in post-genomic era, protein function prediction has been extensively studied and many computational approaches have been proposed in the past decade. Among numerous existing algorithms, graph based approaches and data integration based approaches have demonstrated effectiveness due to their clear connections to biological facts.

---

* This work was partially supported by US NSF IIS-1117965, IIS-1302675, IIS- 1344152.
** Corresponding author.

R. Sharan (Ed.): RECOMB 2014, LNBI 8394, pp. 311–325, 2014.
© Springer International Publishing Switzerland 2014

Since many biological experimental data can be readily represented as networks, graph-based approaches are the most natural way to predict protein function [1]. Neighborhood-based methods [2–5] assign functions to a protein based on the most frequent functions within a neighborhood of the protein, and they mainly differ in how the "neighborhood" of a protein is defined. Network diffusion based methods [6, 7] view the interaction network as a flow network, on which protein functions are diffused from annotated proteins to their neighbors in various ways. Other function prediction approaches via biological networks include graph cut based approaches [8, 9], and those derived from kernel methods [10]. More recently, the authors developed a graph-based protein function prediction method [11] using PPI graph to take advantage of the function-function correlations by considering protein function prediction as a multi-label classification problem, which takes the same perspective as this work. Experimental data from one single source often incomplete and sometimes even misleading [12], therefore predicting protein function using multiple biological data has attracted increased attention. [13] proposed a kernel-based data fusion approach to integrate multiple experimental data via a hybrid kernel and use support vector machine (SVM) for classification. [14] presented a locally constrained diffusion kernel approach to combine multiple types of biological networks. Artificial neural network is employed in [15] for the integration of different protein interaction data.

Most existing computational approaches usually consider protein function prediction as a standard classification problem [13, 16, 17]. Typically, these approaches make prediction one function at a time, fundamentally, *i.e.*, the classification for each functional category is conducted independently. However, in reality most biological functions are highly correlated, and protein functions can be inferred from one another through their interrelatedness [11, 18]. These function category correlations, albeit useful, are seldom utilized in predicting protein function. In this study, we explore this special characteristic of the protein functional categories and make use of the function-function correlations to improve the overall predictive accuracy of protein functions.

1.1 Multi-label Correlated Protein Function Prediction

Because a protein is usually observed to play several functional roles in different biological processes within the same organism, it is natural to annotate it with multiple functions. Therefore, protein function prediction is a *multi-label classification* problem [19, 11, 20–22, 18]. Multi-label data, such as those used in protein function prediction, present a new opportunity to improve classification accuracy through label correlations, which are absent in single-label data. For example, when applying Functional Catalogue (FunCat) annotation scheme (version 2.1) [23] on yeast genome, we observe that there is a big overlap between the proteins annotated to function "Cell Fate" (ID: 40) and those annotated to "Cell Type Differentiation" (ID: 43). As shown in the left panel of Figure 1, among 268 proteins annotated with function "Cell Fate" in yeast genome, 168 proteins are also annotated with function "Cell Type Differentiation", whereas the average number of proteins annotated with other functions is only about 51. As a result, we reasonably speculate that these two functions are statistically correlated in a stronger way. As a result, if a protein is known to be annotated with function "Cell Fate" by either experimental or computational evidences, we have high confidence to annotate the same protein with function "Cell Type Differentiation" as well.

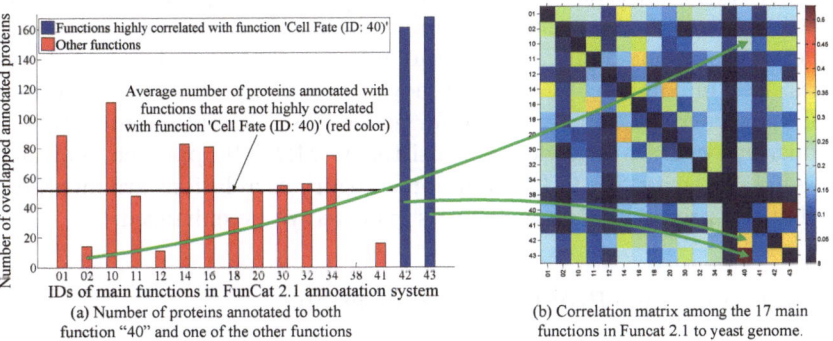

(a) Number of proteins annotated to both function "40" and one of the other functions

(b) Correlation matrix among the 17 main functions in Funcat 2.1 to yeast genome.

Fig. 1. Left: number of proteins annotated to both function 40 and one of the other functions. **Right:** visualization of the correlation values defined by Eq. (1) among the 17 main functions in FunCat 2.1 to yeast genome.

1.2   Data-Knowledge Consistency and Our Motivations

In protein function prediction, we need both experimental data and biological knowledge. Here we refer to *data* as original experimental measurements or results, such as protein sequences, protein-protein interaction (PPI) networks measured by yeast two-hybrid screening, gene expression profiles, *etc*. On the other hand, *knowledge* refers to human-curated research findings recorded in well structured databases or documented in biomedical literatures, such as human-encoded annotation databases, ontologies, *etc*.

In most existing approaches for protein function prediction, knowledge are routinely used as supervision in the classification tasks, *i.e.*, protein annotations are interpreted as labels assigned to data points. In this study, we employ knowledge information from a new perspective. Motivated by the observation that label indications in a multi-label classification task (*i.e.*, protein function annotations in protein function prediction problems) convey important attribute information [21], we use the function annotations of a protein as its description, and assess pairwise protein similarities upon such descriptions. The key assumption of our work is that two proteins are likely to have large overlap in their annotated functions if they are highly similar according to experimental data. More precisely, let $\mathbf{x}_i$ and $\mathbf{x}_j$ be descriptions of two proteins abstracted from experimental data, and $\mathbf{f}_i$ and $\mathbf{f}_j$ be the labeling vectors that encode the annotated functions of the same two proteins respectively, we evaluate the similarity between the two proteins in the following two different ways. The first one is based upon experimental data and denoted as $\mathcal{S}_D(\mathbf{x}_i, \mathbf{x}_j)$, while the second one is based upon biological knowledge and denoted as $\mathcal{S}_K(\mathbf{f}_i, \mathbf{f}_j)$. If functions $\mathbf{f}_i$ and $\mathbf{f}_j$ are annotated appropriately to proteins $\mathbf{x}_i$ and $\mathbf{x}_j$, *i.e.*, the data and the knowledge are consistent, we would expect that the two similarity measurements should be close given that they are normalized to the same scale, *i.e.*, $\mathcal{S}_D(\mathbf{x}_i, \mathbf{x}_j) \approx \mathcal{S}_K(\mathbf{f}_i, \mathbf{f}_j)$. With this assumption, we may determine the optimal function assignments to unannotated proteins by minimizing the difference between the two sets of similarities, *i.e.*, maximizing the consistency between experimental data and biological knowledge. In this paper, we formalize this assumption and propose our Maximization of Data-Knowledge Consistency (MDKC) approach. Through the knowledge similarity $\mathcal{S}_K(\mathbf{f}_i, \mathbf{f}_j)$, function category correlations are incorporated, such that the predictive performance is expected to be enhanced.

1.3   Notations and Problem Formalization

In protein function prediction, we are given K biological functions and n proteins. Without losing generality, we assume the first l proteins are annotated, our goal is to predict functions for the rest $n - l$ unannotated proteins.

Let $\mathbf{x}_i \in \mathbb{R}^p$ denote a protein, which is a vector description of the i-th protein constructed from certain biological experimental data, such as the amino acid histogram of a protein sequence. The pairwise similarities among the proteins are modeled as a symmetric matrix $W \in \mathbb{R}^{n \times n}$, where W_{ij} measures how similar proteins $\mathbf{x}_i$ and $\mathbf{x}_j$ are. W is usually seen as edge weight matrix of a graph where proteins correspond to vertices. In the simplest case of a PPI network, $W_{ij} = 1$ if protein $\mathbf{x}_i$ and protein $\mathbf{x}_j$ interact, and 0 otherwise. Every protein is assigned with a number of biological functions, which are described by a function annotation vector $\mathbf{y}_i \in \{0, 1\}^K$, such that $\mathbf{y}_i(k) = 1$ if protein $\mathbf{x}_i$ is annotated with the k-th function, $\mathbf{y}_i(k) = 0$ if it is not annotated with the k-th function or unannotated. $\{\mathbf{y}_i\}_{i=1}^l$ for the first l annotated proteins are known, and our objective is to learn $\{\mathbf{y}_i\}_{i=l+1}^n$ for the $n - l$ unannotated proteins. We write $Y = [\mathbf{y}_1, \ldots, \mathbf{y}_n]^T = [\mathbf{y}^{(1)}, \ldots, \mathbf{y}^{(K)}]$, where $\mathbf{y}^{(k)} \in \mathbb{R}^n$ is a class-wise function annotation vector. Besides the ground truth function assignment matrix Y, we also define $F = [\mathbf{f}_1, \ldots, \mathbf{f}_n]^T \in \mathbb{R}^{n \times K}$ as the predicted function assignment matrix, where $F_{ik} = \mathbf{f}_i(k)$ for $l + 1 \le i \le n$ indicates our confidence to assign the k-th function to an unannotated protein $\mathbf{x}_i$.

2   Formulation of Function Category Correlations

Before we proceed to the algorithm development of our new approach, we first explore and formalize the function category correlations, as they are one of our most important mechanism to boost protein function prediction performance.

As shown in the left panel of Figure 1, proteins assigned to two different functions may overlap. Statistically, the bigger the overlap is, the more closely the two functions are related. Therefore, functions assigned to a protein are no longer independent, but can be inferred from one another. In the extreme case, such as in parent-child hierarchy of protein function annotation systems, once we know a protein is annotated to a child function, we can immediately annotate all the ancestor functions to the same protein.

Using cosine similarity, we define a function category correlation matrix, $C \in \mathbb{R}^{K \times K}$, where C_{kl} captures the correlation between the k-th and l-th functions as following:

$$C_{kl} = \cos(\mathbf{y}^{(k)}, \mathbf{y}^{(l)}) = \frac{\langle \mathbf{y}^{(k)}, \mathbf{y}^{(l)} \rangle}{\|\mathbf{y}^{(k)}\| \, \|\mathbf{y}^{(l)}\|}, \tag{1}$$

where $\langle \cdot, \cdot \rangle$ denotes the inner product of two vectors and $\| \cdot \|$ denotes the ℓ_2 norm of a vector.

Using FunCat annotation scheme on yeast genome, function correlations defined in Eq. (1) are illustrated in the right panel of Figure 1. The high correlation value between functions "Cell Fate" and "Cell Type Differentiation" shown in the figure implies that they are highly correlated, which agrees with the observations shown in the left panel. In addition, as can be seen in the right panel of Figure 1, some other function pairs are also highly correlated, such as "Transcription" and "Protein With Binding Function or Cofactor Requirement", "Regulation of Metabolism and Protein Function" and "Cellular Communication/Signal Transduction Mechanism", *etc.* All these observations strictly comply with the biological truths, which justifies the correctness of our formulation for function category correlations in Eq. (1) from biological perspective.

3   The Maximization of Data-Knowledge Consistency (MDKC) Approach

We assume that two proteins tend to have large overlap in their assigned functions if they are very similar in terms of some experimental data. In order to predict protein functions upon this assumption, we evaluate the similarity between two proteins in the following two ways, one by experimental data called as *data similarity*, and the other by biological knowledge called as *knowledge similarity*. We denote the former as $\mathcal{S}_D(\mathbf{x}_i, \mathbf{x}_j)$, and the latter as $\mathcal{S}_K(\mathbf{f}_i, \mathbf{f}_j)$. If the functions annotated to proteins are consistent with their experimental data, we would expect the data similarity is close to the knowledge similarity:

$$\min \sum_{i,j} [\mathcal{S}_D(\mathbf{x}_i, \mathbf{x}_j) - \mathcal{S}_K(\mathbf{f}_i, \mathbf{f}_j)]^2,$$

$$\text{s.t.} \quad \mathbf{f}_i = \mathbf{y}_i, \ \forall \, 1 \leq i \leq l, \tag{2}$$

where the constraint fixes the functions assigned to annotated proteins to be ground truth. The optimization objective in Eq. (2) minimizes the overall difference between the two types of similarities, which thereby maximizes the data-knowledge consistency.

3.1   Optimization Framework of the MDKC Approach

In protein function prediction, the data similarity is already known in a priori. Namely, $\mathcal{S}_D\left(\mathbf{x}_i, \mathbf{x}_j\right) = W$, and W depends on input experimental data. For example, when input data are a PPI network, W could be the adjacency matrix of the PPI graph in the simplest case or any derived topological similarity; when input data are protein sequences, W could be the inverse Euclidean distances of amino acid histogram vectors; *etc.* Because W is input dependent, we defer its detailed definitions to Section 4 according to the experimental data used in the respective empirical evaluations.

Now we consider knowledge similarity. The simplest method is to count the number of common annotated functions of two proteins, *i.e.* $\mathbf{f}_i^T \mathbf{f}_j$. However, the problem of this straightforward similarity measurement lies in that it considers all the biological functions to be independent and is unable to explore the correlations among them. In particular, it will give zero similarity whenever two proteins do not share any annotated functions, although they could be strongly related if their annotated functions are highly correlated. For example, given a pair of proteins, one annotated with function "Cell Fate" and the other annotated with function "Cell Type Differentiation", although they may not share any common functions, they may still have certain similarities, either biologically or statistically, as illustrated in Figure 1. In the extreme case, in the parent-child annotation system, such as FunCat scheme used in this work, if protein $\mathbf{x}_i$ is annotated with one of the ancestor function of protein $\mathbf{x}_j$'s annotated function, the two proteins are closely related even they do not share any common functions. Therefore, in order to capture correlations among different functions, instead of the simple dot product, we compute the knowledge similarity as following:

$$\mathcal{S}_K\left(\mathbf{f}_i, \mathbf{f}_j\right) = \mathbf{f}_i^T C^{-1} \mathbf{f}_j = \mathbf{f}_i^T A \mathbf{f}_j, \tag{3}$$

where, for notation simplicity, we denote $A = C^{-1}$ in the sequel.

Note that, compared to the dot product similarity defined by $\mathbf{f}_i^T \mathbf{f}_j$ based on the Euclidean distance, the knowledge similarity computed by Eq. (3) is based on the Mahalanobis distance, where C acts as the covariance matrix encoding the human-curated prior knowledge for the biological species of interest. Statistically speaking, because the Euclidean distance is independent of input data while the Mahalanobis distance captures the second-order statistics of the input data, the latter is able to better characterize the relationships between the data points of a given input data set when its distribution is known in a priori. In protein function prediction, the Euclidean distance based knowledge similarity is independent of the concerned biological species, whereas the Mahalanobis distance based knowledge similarity is specific to the biological species of interest thereby has increased statistical power. Most importantly, function-function correlations, the most important advantage of a multi-label data set over the traditional single-label data set, are exploited for the later protein annotations tasks, which is an important contribution of the proposed method.

Utilizing the knowledge similarity defined in Eq. (3), we can formalize the data-knowledge consistency assumption in Eq. (2) by the following optimization problem:

$$\underset{F}{\arg\min} \ \sum_{i,j=1}^{n} \left(W_{ij} - \sum_{k,l=1}^{K} F_{ik} A_{kl} F_{jl} \right)^2, \tag{4}$$

$$\text{s.t.} \quad F_{ik} = Y_{ik}, \ \forall \ 1 \leq i \leq l, \ 1 \leq k \leq K. \tag{5}$$

In standard classification problems in machine learning, F_{ik} $(1 \leq i \leq l)$ are fixed for labeled data points. Specifically, a big F_{ik} indicates that data point $\mathbf{x}_i$ belongs to the k-th class, while a small F_{ik} indicates that $\mathbf{x}_i$ does not belong the k-th class. However, this assumption does not hold in the problem of protein function prediction. For an annotated protein, its associated functions refer to those who have certain experimental supports for the associations between this protein and its associated functions. On the other hand, the non-association between a protein and a function only means that we currently do not have any biological or computational evidence for the corresponding association. In reality, however, the protein could have the concerned function. And the exact goal of computational methods for protein function prediction is to identify putative protein functions, which could work as the candidates for further experimental screening. As a result, instead of using the hard constraints in Eq. (5), it is reasonable to relax the confidence variables F_{ik} $(1 \leq i \leq l)$ for annotated proteins to be dynamic variables, which approximate the ground truth function assignments. The constraint in Eq. (5) hence can be written to minimize the following penalty function:

$$\alpha \sum_{i=1}^{l} \sum_{k=1}^{K} \left(Y_{ik} - F_{ik} \right)^2, \tag{6}$$

where $\alpha > 0$ controls the relative importance of the penalty. Following the experiences in graph-based semi-supervised learning, we empirically set $\alpha = 0.1$ in all our experiments.

Finally, we write our objective in a more compact way using matrices to minimize the following:

$$J_{\text{MDKC}}(F) = \|W - FAF^T\|_F^2 + 2\alpha \operatorname{\mathbf{tr}} \left((Y - F)^T V (Y - F) \right),$$

$$s.t. \quad F \geq 0, \tag{7}$$

where $\|\cdot\|_F$ denotes the Frobenius norm of a matrix and $\operatorname{\mathbf{tr}}(\cdot)$ denotes the trace of a matrix. Here $V \in \mathbb{R}^{n \times n}$ is a diagonal indicator matrix, whose diagonal entry $V_{ii} = 1$ if the i-th protein is an annotated protein, while $V_{ii} = 0$ indicates that the i-th protein is unannotated. In Eq. (7), the constraint $F \geq 0$ is naturally enforced because W is nonnegative by definition. Most importantly, with this nonnegative constraint Eq. (7) will be enriched with clustering interpretation as detailed soon later, which makes the mathematical formulation of the proposed method more biologically meaningful.

We call Eq. (7) as our proposed Maximization of Data-Knowledge Consistency (MDKC) approach. Upon the solution of Eq. (7), we assign putative functions to unannotated proteins.

3.2   Computational Algorithm of MDKC Approach

Mathematically, Eq. (7) is a regularized NMF problem [24–26]. Although the optimization techniques for the NMF problem and its variants have been extensively studied in literature [27, 28, 24–26, 29, 30], solving Eq. (7) is challenging. Most, if not all, existing algorithms to solve NMF problems are only able to deal with rectangle input matrices (the number of rows of a matrix is different from that of columns) or asymmetric square matrices, but not symmetric input matrices such as the one used in our objective in Eq. (7). This is because the latter involves a fourth-order term due to the symmetric usage of the factor matrix F, which inevitably complicates the problem (More detailed analyses can be found in our earlier works [31, 32]). Traditional solutions to symmetric NMF typically rely on heuristics [27, 33], thus we introduce Algorithm 1 to solve Eq. (7) in a principled way. Due to space limit, the proofs of its correctness and convergence will be provided in the extended journal version.

Algorithm 1. Algorithm to solve Eq. (7)

Data: 1. Data similarity matrix W.
2. Function-function correlations matrix C.
3. Indication matrix Y derived from labels of annotated proteins.
Result: Factor matrices F.
1. Computer $A = C^{-1}$.
2. Initialize F following [27].
repeat
     3. Compute $F_{ij} \leftarrow F_{ij}[\frac{(WFA+\alpha VY)_{ij}}{(FAF^TFA+\alpha VF)_{ij}}]^{\frac{1}{4}}$.
until *Converges*

---

4   Results and Discussion

We evaluate the proposed MDKC approach on *Saccharomyces cerevisiae* genome data. We apply the proposed method on protein sequence data, and an integration of protein sequence data and PPI network data, respectively.

We use MIPS Functional Catalogue (FunCat) system [23] to annotate proteins, which is an annotation scheme for the functional description of proteins from prokaryotes, unicellular eukaryotes, plants and animals. Taking into account the broad and highly diverse spectrum of known protein functions, FunCat (version 2.1) consists of 27 main functional categories that cover general fields such as cellular transport, metabolism, cellular communication, *etc.* 17 main function categories in FunCat annotation scheme are involved to annotate yeast genome, which are listed in Table 1.

Table 1. Main functional categories in FunCat annotation scheme (version 2.1) and the corresponding number of annotated proteins to yeast species

Function ID	Function Description	Size
01	Metabolism	1397
02	Energy	336
10	Cell Cycle and DNA Processing	981
11	Transcription	1009
12	Protein Synthesis	476
14	Protein Fate	1125
16	Protein with Binding Function	1019
18	Regulation of Metabolism and Protein Function	246
20	Transport Facilitation and Transport Routes	995
30	Cellular Communication and Signal Transduction	231
32	Cell Rescue, Defense and Virulence	515
34	Interaction with the Environment	446
38	Transposable Elements, Viral and Plasmid Proteins	59
40	Cell Fate	268
41	Development	67
42	Biogenesis of Cellular Components	827
43	Cell Type Differentiation	437

4.1   Evaluation on Protein Sequence Data

Because sequence is the most fundamental form to describe a protein, which contains important structural, characteristic and genetic information, we first evaluate the proposed MDKC approach using protein sequences. We compare the predictive accuracy of our approach against functional similarity weight (FS) approach [4] and kernel-based data fusion (KDF) approach [13]. We also report the performance of majority voting (MV) approach [2] as a baseline. We employ broadly used average precision and average F1 score [4] as performance metrics.

Adaptive Decision Boundary for Prediction. To predict specific putative functions for unannotated proteins we need a decision boundary (threshold) for learned ranking values, say $\mathbf{y}^{(k)}$, of each class. In many semi-supervised learning algorithms, the threshold for classification is usually selected as 0, which, however, is not necessary to be the best choice. We use an adaptive decision boundary to achieve better predictive performance, which is adjusted such that the weighted training errors on annotated proteins are minimized.

Considering the binary classification problem for the k-th functional category, we denote b_k as the decision boundary, S_+ and S_- as the sets of positive and negative samples for the k-th class, and $e_+(b_k)$ and $e_-(b_k)$ as the numbers of misclassified positive and negative training samples. The adaptive (optimal) decision boundary is given as following [19, 11]:

$$b_k^{\text{opt}} = \arg\min_{b_k} \left[\frac{e_+(b_k)}{|S_+|} + \frac{e_-(b_k)}{|S_-|} \right]. \tag{8}$$

And the decision rule to assign a function to protein $\mathbf{x}_i$ is given by:

$$\begin{cases} \mathbf{x}_i \text{ is annotated with the } k\text{-th function if } F_{ik}^* > b_k^{\text{opt}}; \\ \mathbf{x}_i \text{ is not annotated with the } k\text{-th function if } F_{ik}^* \leq b_k^{\text{opt}}; \end{cases} \qquad (9)$$

Data Preparation. We obtain sequence data from GenBank [34], and describe a protein sequence through one kind of its elementary constituents, *i.e.*, trimers of amino acids. Trimer, a type of k-mer (when $k = 3$) broadly used in sequence analysis, considers the statistics of one amino acid and its vicinal amino acids, and regards any three consecutive amino acids as a unit to preserve order information, *e.g.*, "ART" is one unit, and "MEK" is another one. The trimer histogram of a sequence hence can be used to characterize a protein $\mathbf{x}_i$, which is denoted as P_i. Because histogram indeed is a probability distribution, we use Kullback-Leibler (KL) divergence [35], a standard way to assess the difference between two probability distributions, to measure the distance between two proteins, which is defined as:

$$D_{\text{KL}} (P_i \parallel P_j) = \sum_k P_i (k) \log \frac{P_i (k)}{P_j (k)}, \qquad (10)$$

where k denotes the index of the k-th trimer. Because KL divergence is non-symmetric, *i.e.*, $D_{\text{KL}} (P_i \parallel P_j) \neq D_{\text{KL}} (P_j \parallel P_i)$, we use the symmetrized KL divergence as following:

$$D_{\text{s-KL}} (i, j) = \frac{D_{\text{KL}} (P_i \parallel P_j) + D_{\text{KL}} (P_j \parallel P_i)}{2}. \qquad (11)$$

Finally, the pairwise data similarity W is defined by converting the symmetrized KL divergences through the standard way:

$$\begin{aligned} W_{ij} &= D_{\text{s-KL}} (i, i) + D_{\text{s-KL}} (j, j) - 2D_{\text{s-KL}} (i, j) \\ &= - \left[D_{\text{KL}} (P_i \parallel P_j) + D_{\text{KL}} (P_j \parallel P_i) \right]. \end{aligned} \qquad (12)$$

Improved Predictive Capability. We perform standard 5-fold cross validation to evaluate the compared approaches and report the average performance of 5 trials in Table 2. For FS approach, because it does not supply a threshold, we use the one giving best F1 score to make prediction. We implement two versions of our method to evaluate the contributions of each of its components. First, we solve Eq. (7) by Algorithm 1, which is the proposed method. Second, we solve a degenerate version of the problem in Eq. (7) by not incorporating the correlations between functional categories. Specifically, we replace FAF^T in Eq. (7) by FF^T, which is denoted by MDKC-S.

The results in Table 2 show that the MDKC-S and MDKC approaches clearly outperform the other compared approaches, which concretely quantify the advantage of our approaches. The improvement on classification performance of MDKC approach over MDKC-S approach clearly justify the usefulness of function-function correlations in predict putative protein functions.

Table 2. Average precision and average F1 score by the compared approaches in 5-fold cross validation on the main functional categories of FunCat annotation scheme

Approaches	Average Precision	Average F1 score
FS	33.65%	22.78%
KDF	53.45%	38.10%
MV	32.07%	29.46%
MDKC-S	56.51%	39.04%
MDKC	61.38%	42.17%

4.2   Evaluation on Integrated Biological Data

As mentioned earlier, biological data from one single experimental source only convey information for a certain aspect, which are usually incomplete and sometimes misleading. For example, similar sequences do not always have similar functions. In the extreme case, proteins with 100% sequence identity could perform different functional roles [12]. Therefore, integration of different biological data is necessary for more robust and complete protein function inferences. In general, results learned from a combination of different types of data are likely to lead to a more coherent model by consolidating information on various aspects of the same biological process. In this subsection, we evaluate the predictive performance using the integrated data from both PPI networks and protein sequences.

Data Preparation. We download PPI data for *Saccharomyces cerevisiae* species from BioGRID (version 2.0.56) [36]. By removing the proteins connected by only one PPI, we end up with 4403 annotated proteins with 86167 PPIs. We represent the protein interaction network as a graph, with vertices corresponding to the proteins, and edges corresponding to PPIs. The adjacency matrix of the graph is denoted as $B \in \{0,1\}^{n \times n}$ where $n = 4403$, such that $B_{ij} = 1$ if proteins $\mathbf{x}_i$ and $\mathbf{x}_j$ interact, and 0 otherwise.

The adjacency matrix B itself measures the similarity among proteins in the sense that two proteins are related if they interact. However, two critical problems prevent us from directly using B as data similarity $\mathcal{S}_D(\mathbf{x}_i, \mathbf{x}_j)$ to predict protein function. First, B only measures the local connectivity of a graph, and contains no information for connections via more than one edge. Therefore the important information contained in the global topology is simply ignored. Second, PPI data suffer from high noise due to the nature of high-throughput technologies, *e.g.*, false positive rate in yeast two-hybrid experiments is estimated as high as 50% [37]. Therefore, we use the Topological Measurement (TM) method [38] to compute the data similarity matrix W_{PPI}, which takes into consideration paths with all possible lengths on a network and weights the influence of every path by its length. Specifically, $(W_{\mathrm{PPI}})_{ij}$ between proteins $\mathbf{x}_i$ and $\mathbf{x}_j$ is computed as:

$$(W_{\text{PPI}})_{ij} = \sum_{k=2}^{|V|-2} \text{PR}^k (i, j),$$

$$\text{PR}^k (i, j) = \frac{\text{PS}^k (i, j)}{\text{MaxPS}^k (i, j)}, \tag{13}$$

where $|V|$ is the number of vertices in the PPI graph, $\text{PR}^k (i, j)$ is the path ratio of the paths of length k between proteins $\mathbf{x}_i$ and $\mathbf{x}_j$, and $\text{PS}^k (i, j)$ and $\text{MaxPS}^k (i, j)$ are defined as following:

$$\text{PS}^k (i, j) = \left(A^k \right)_{ij}, \tag{14}$$

where $(\cdot)_{ij}$ denotes the ij-th entry of a matrix, and

$$\text{MaxPS}^k (i, j) = \begin{cases} \sqrt{d_i d_j}, & \text{if } k = 2, \\ d_i d_j, & \text{if } k = 3, \\ \sum_{k \in N(i), l \in N(j)} \text{MaxPS}^{k-2} (k, l), & \text{if } k > 3. \end{cases} \tag{15}$$

where $d_i = \sum_j B_{ij}$ is the degree of the i-th vertex, and $N(i)$ denotes its neighboring vertices. The detailed explanation of TM measurement can be referred to [38].

We compute the sequence data similarity following the same ways in Section 4.1, which is denoted as and W_{sequence} respectively. The integrated data similarity W is hence computed as following:

$$W = W_{\text{PPI}} + \gamma W_{\text{sequence}}, \tag{16}$$

where γ is a parameter to balance the two data sources and we empirically select it as:

$$\gamma = \frac{\sum_{i,,i \neq j} W_{\text{PPI}} (i, j)}{\sum_{i,,i \neq j} W_{\text{sequence}} (i, j)}. \tag{17}$$

We compare the predictive performance of our MDKC approach to two data integration based protein function prediction approaches: kernel-based data fusion (KDF) approach [13] and locally constrained diffusion kernel (LCDK) approach [14], and two baseline approaches: majority voting (MV) approach [2] and iterative majority voting (IMV) approach [8]. The function-wise prediction performance measured by average precision and average F1 score in standard 5-fold cross validation are reported in Figure 2.

From the results in Figure 2(a) and Figure 2(b), we can see that the proposed MDKC approach consistently better than other compared approaches, sometimes very significantly, which again demonstrate the superiority of our approach.

A more careful examination on the results in Figure 2 shows that, although our approach outperforms the compared approaches in most functional categories, but not always, e.g., the average precision for function "Transposable Elements,

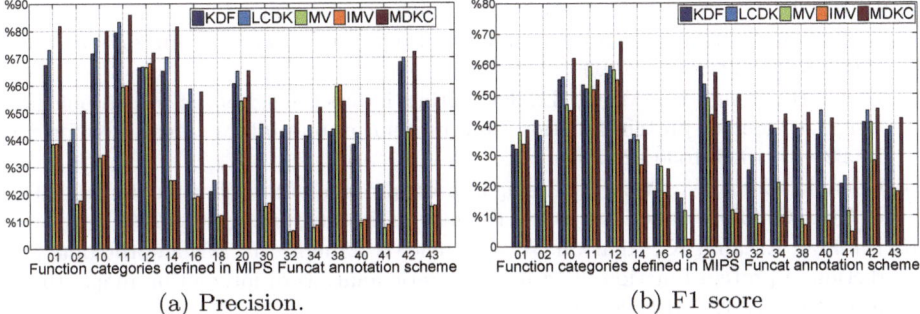

(a) Precision.   (b) F1 score

Fig. 2. Performance of 5-fold cross validation for the 17 main functional categories in FunCat annotation scheme (version 2.1) by KDF, LCDK, MV, GMV and the proposed MDKC approach

Viral and Plasmid Proteins" (ID: 38). By scrutinizing the function category correlations, defined in Eq. (1) and illustrated in the right panel of Figure 1, we can see the average correlation of function "Transposable Elements, Viral and Plasmid Proteins" with other functional categories is among the lowest. As a result, the presence/absence of this function category can not benefit from other functional categories, because it only has weak correlations with them. In contrast, prediction for the function categories with high correlations to others generally can benefit from our approach. This observation firmly testify the importance of function category correlations in predicting protein function.

5   Conclusions

In this paper, we presented a novel Maximization of Data-Knowledge Consistency (MDKC) approach to predict protein function, which attempts to make use of function category correlations to improve the predictive accuracy. Different from traditional approaches in predicting protein function, we employed annotation knowledge in a novelly different way to measure pairwise protein similarities. By maximizing consistency between the *knowledge similarity* computed from annotations and the *data similarity* computed from biological experimental data, optimal function assignments to unannotated proteins are obtained. Most importantly, function category correlations are incorporated in a natural way through the knowledge similarity. Comprehensive empirical evaluations have been conducted on *Saccharomyces cerevisiae* genome, promising results in the experiments justified our analysis and validated the performance of our methods.

References

1. Sharan, R., Ulitsky, I., Shamir, R.: Network-based prediction of protein function. Mol. System Biol. 3(1) (2007)
2. Schwikowski, B., Uetz, P., Fields, S.: A network of protein- protein interactions in yeast. Nat. Biotech. 18, 1257–1261 (2000)

3. Hishigaki, H., Nakai, K., Ono, T., Tanigami, A., Takagi, T.: Assessment of prediction accuracy of protein function from protein-protein interaction data. Yeast 18(6), 523–531 (2001)

4. Chua, H., Sung, W., Wong, L.: Exploiting indirect neighbours and topological weight to predict protein function from protein-protein interactions. Bioinformatics 22(13), 1623–1630 (2006)

5. Chua, H., Sung, W., Wong, L.: Using indirect protein interactions for the prediction of Gene Ontology functions. BMC Bioinformatics 8(suppl. 4), S8 (2007)

6. Nabieva, E., Jim, K., Agarwal, A., Chazelle, B., Singh, M.: Whole-proteome prediction of protein function via graph-theoretic analysis of interaction maps. Bioinformatics 21, 302–310 (2005)

7. Weston, J., Elisseeff, A., Zhou, D., Leslie, C., Noble, W.: Protein ranking: from local to global structure in the protein similarity network. Proc. Natl. Acad. Sci. USA 101(17), 6559 (2004)

8. Vazquez, A., Flammini, A., Maritan, A., Vespignani, A.: Global protein function prediction from protein-protein interaction networks. Nat. Biotechnol. 21, 697–700 (2003)

9. Karaoz, U., Murali, T., Letovsky, S., Zheng, Y., Ding, C., Cantor, C., Kasif, S.: Whole-genome annotation by using evidence integration in functional-linkage networks. Proc. Natl Acad. Sci. USA 101(9), 2888–2893 (2004)

10. Liang, S., Shuiwang, J., Jieping, Y.: Adaptive diffusion kernel learning from biological networks for protein function prediction. BMC Bioinformatics 9, 162 (2008)

11. Wang, H., Huang, H., Ding, C.: Function-function correlated multi-label protein function prediction over interaction networks. In: Chor, B. (ed.) RECOMB 2012. LNCS, vol. 7262, pp. 302–313. Springer, Heidelberg (2012)

12. Whisstock, J., Lesk, A.: Prediction of protein function from protein sequence and structure. Q. Rev. Biophysics 36(3), 307–340 (2004)

13. Lanckriet, G., Deng, M., Cristianini, N., Jordan, M., Noble, W.: Kernel-based data fusion and its application to protein function prediction in yeast. In: Proc. of Pacific Symp. on Biocomputing, vol. 9, pp. 300–311 (2004)

14. Tsuda, K., Noble, W.: Learning kernels from biological networks by maximizing entropy. Bioinformatics 20, 326–333 (2004)

15. Shi, L., Cho, Y., Zhang, A.: ANN Based Protein Function Prediction Using Integrated Protein-Protein Interaction Data. In: Proc. of International Joint Conf. on Bioinformatics, Systems Biol. and Intelligent Comp., pp. 271–277 (2009)

16. Shin, H., Lisewski, A., Lichtarge, O.: Graph sharpening plus graph integration: a synergy that improves protein functional classification. Bioinformatics 23(23), 3217 (2007)

17. Sun, L., Ji, S., Ye, J.: Adaptive diffusion kernel learning from biological networks for protein function prediction. BMC Bioinformatics 9(1), 162 (2008)

18. Wang, H., Huang, H., Ding, C.: Protein function prediction via laplacian network partitioning incorporating function category correlations. In: Proceedings of the Twenty-Third International Joint Conference on Artificial Intelligence, pp. 2049–2055. AAAI Press (2013)

19. Wang, H., Huang, H., Ding, C.: Image Annotation Using Multi-label Correlated Green's Function. In: Proc. of IEEE ICCV 2009, pp. 2029–2034 (2009)

20. Wang, H., Ding, C., Huang, H.: Multi-label linear discriminant analysis. In: Daniilidis, K., Maragos, P., Paragios, N. (eds.) ECCV 2010, Part VI. LNCS, vol. 6316, pp. 126–139. Springer, Heidelberg (2010)

21. Wang, H., Huang, H., Ding, C.: Multi-label feature transform for image classifications. In: Daniilidis, K., Maragos, P., Paragios, N. (eds.) ECCV 2010, Part IV. LNCS, vol. 6314, pp. 793–806. Springer, Heidelberg (2010)
22. Wang, H., Huang, H., Ding, C.: Image annotation using bi-relational graph of images and semantic labels. In: Proceedings of the IEEE Conference on Computer Vision and Pattern Recognition 2011 (CVPR 2011), pp. 793–800 (2011)
23. Mewes, H., Heumann, K., Kaps, A., Mayer, K., Pfeiffer, F., Stocker, S., Frishman, D.: MIPS: a database for genomes and protein sequences. Nucleic Acids Res. 27(1), 44 (1999)
24. Cai, D., He, X., Wu, X., Han, J.: Non-negative matrix factorization on manifold. In: Proc. of ICDM (2008)
25. Gu, Q., Zhou, J.: Co-clustering on manifolds. In: Proc. of SIGKDD (2009)
26. Cai, D., He, X., Han, J., Huang, T.S.: Graph regularized non-negative matrix factorization for data representation. IEEE Trans. Pattern Analysis Mach. Intell. 99 (2010)
27. Ding, C., Li, T., Peng, W., Park, H.: Orthogonal nonnegative matrix t-factorizations for clustering. In: SIGKDD (2006)
28. Ding, C., Li, T., Jordan, M.: Convex and semi-nonnegative matrix factorizations. IEEE Transactions on Pattern Analysis and Machine Intelligence 32(1), 45–55 (2010)
29. Wang, H., Nie, F., Huang, H., Makedon, F.: Fast nonnegative matrix tri-factorization for large-scale data co-clustering. In: Proceedings of the Twenty-Second International Joint Conference on Artificial Intelligence, vol. 2, pp. 1553–1558. AAAI Press (2011)
30. Wang, H., Nie, F., Huang, H., Ding, C.: Dyadic transfer learning for cross-domain image classification. In: Proc. of ICCV, pp. 551–556. IEEE (2011)
31. Wang, H., Nie, F., Huang, H., Ding, C.: Nonnegative matrix tri-factorization based high-order co-clustering and its fast implementation. In: Proceedings of ICDM (2011)
32. Wang, H., Huang, H., Ding, C., Nie, F.: Predicting protein-protein interactions from multimodal biological data sources via nonnegative matrix tri-factorization. In: Chor, B. (ed.) RECOMB 2012. LNCS, vol. 7262, pp. 314–325. Springer, Heidelberg (2012)
33. Li, T., Ding, C., Jordan, M.: Solving consensus and semi-supervised clustering problems using nonnegative matrix factorization. In: Proc. of ICDM (2007)
34. Benson, D., Karsch-Mizrachi, I., Lipman, D.: GenBank. Nucleic Acids Res. 34, D16–D20 (2006)
35. Kullback, S., Leibler, R.: On information and sufficiency. The Annals of Mathematical Statistics, 79–86 (1951)
36. Stark, C., Breitkreutz, B., Reguly, T., Boucher, L., Breitkreutz, A., Tyers, M.: BioGRID: a general repository for interaction datasets. Nucleic Acids Res. 34(database issue), D535 (2006)
37. Deane, C., Salwinski, L., Xenarios, I., Eisenberg, D.: Protein Interactions Two Methods for Assessment of the Reliability of High Throughput Observations. Mol. & Cellular Proteomics 1(5), 349–356 (2002)
38. Pei, P., Zhang, A.: A topological measurement for weighted protein interaction network. In: Proceedings of the 2005 IEEE Computational Systems Bioinformatics Conference, pp. 268–278 (2005)

Bayesian Multiple Protein Structure Alignment

Rui Wang and Scott C. Schmidler*

Program in Computational Biology & Bioinformatics,
Departments of Statistical Science and Computer Science,
Duke University, NC 27708, USA
rui.wang@duke.edu, schmidler@stat.duke.edu

Abstract. Multiple protein structure alignment is an important tool in computational biology, with numerous algorithms published in the past two decades. However, recently literature highlights a growing recognition of the inconsistencies among alignments from different algorithms, and the instability of alignments obtained by individual algorithms under small fluctuations of the input structures. Here we present a probabilistic model-based approach to the problem of multiple structure alignment, using an explicit statistical model. The resulting algorithm produces a Bayesian posterior distribution over alignments which accounts for alignment uncertainty arising from evolutionary variability, experimental noise, and thermal fluctuation, as well as sensitivity to alignment algorithm parameters. We demonstrate the robustness of this approach on alignments identified previously in the literature as "difficult" for existing algorithms. We also show the potential for significant stabilization of tree reconstruction in structural phylogenetics.

Keywords: protein structure alignment, Bayesian statistics, MCMC.

1 Introduction

Uncertainty in biological sequence alignments has received considerable attention recently, particularly in regard to its effect on phylogeny reconstruction [42, 26]. Alignment uncertainty arises from multiple sources: the stochasticity of the underlying evolutionary model, the limited information contained in the pair or set of sequences to be aligned, and the sensitivity to input parameters of alignment algorithms. In this paper we consider uncertainty arising in the context of protein *structure* alignment. Structural alignment of proteins is a key tool for understanding protein function, mechanism, and evolution (see [12, 18] for reviews), and is commonly used as a "gold-standard" for evaluating or calibrating sequence alignments [25]. Uncertainties in pairwise structure alignments have recently been considered by [37, 38, 31, 8] and again arise from multiple sources. First, most current algorithms formulate structure alignment as an optimization problem with respect to some similarity metric, and different metrics weight various structural properties differently (often with tunable weights),

---

* To whom correspondence should be addressed.

R. Sharan (Ed.): RECOMB 2014, LNBI 8394, pp. 326–339, 2014.
© Springer International Publishing Switzerland 2014

leading to considerable subjective or empirical bias [15, 19]. In addition, 3D protein structures are intrinsically flexible and dynamic - variability exceeding 1Å is common, and conformational changes may vary over tens of angstroms [6] - but high resolution (X-ray) structures are static snapshots. Sub-angstrom structural variation can cause substantial inconsistencies and apparently incorrect alignments by existing methods [29]. Such shortcomings have led to calls for new approaches to the multiple structure alignment problem [4]. Here we present a probabilistic approach to multiple structure alignment which addresses these issues explicitly.

Probabilistic modeling is a natural framework for accounting for uncertainty arising from multiple sources. For biological sequence analysis, probabilistic modeling has yielded highly effective tools for global and local pairwise sequence alignment [44, 41], multiple sequence alignment [21, 3, 24, 23], secondary structure prediction of proteins [34–36] and RNA [11, 33], and protein tertiary structure prediction [39, 22, 43]. See [10] for an introduction emphasizing sequence alignment and RNA base pairing. For multiple sequence alignment, probabilistic models such as hidden Markov models (HMMs) and multinomial mixture models [24, 23] also have a computational advantage over optimizing all pairwise distances, which is NP-hard [40]. The HMM approach instead aligns each input sequence to a (albeit unknown) profile model, reducing the problem to one of estimating the common profile using standard statistical algorithms, then pairwise aligning each input sequence to the model. Thus formulating the problem in terms of an explicit probabilistic model yields practical algorithms.

Here we describe a probabilistic model-based approach to multiple *structural* alignment. As with multiple sequence alignment, we build on the machinery of HMMs. However, the application of HMMs to 3D structures requires significant generalization and algorithmic development, so such models have not been applied to structure alignment previously. (Alexandrov and Gerstein [1] use an HMM to represent the core residue profile in a multiple structure alignment obtained by other means, but this does not address the alignment problem itself.) Our approach directly accounts for multiple sources of uncertainty in the alignment process, using Bayesian inference to identify multiple possible alignments and their relative probabilities. The alignment is obtained by averaging over all remaining uncertainty in the model. This approach also handles unknown alignment parameters (such as gap penalties and thresholds) to be adaptively estimated from the data in a coherent statistical framework. We also address an unresolved problem in sequence alignment HMMs - choosing the length of the model - via Bayesian model averaging. The resulting algorithm is significantly more robust; replaces heuristic optimization criteria with clear, testable statistical assumptions; and results in structural alignments that lead to significantly more stable and accurate phylogenies.

2   A Probabilistic Model for Protein Structure Families

Our approach to multiple structure alignment constructs a probabilistic model of the underlying protein family. This generalizes the approach introduced by

[37, 38, 31] for *pairwise* protein structure alignment, and the closely related approach for matching residues of two protein active sites developed by [17]. Let X_j be an $n_j \times 3$ matrix containing the C_α coordinates of protein j, for $j = 1, \ldots, m$. Our stochastic model represents each input structure X_j as being generated from "mean" or model structure U, to which insertions are added or deletions made stochastically, random noise added to the coordinates, and then an arbitrary Euclidean transformation applied. In the special case that the noise is independent $\Sigma_\epsilon = \phi^{-1} I$, this can be written

$$X_j = Y_j R_j + \mathbf{1}^T v_j \qquad\qquad Y_j \sim \mathtt{HMM}(\Theta)$$

where R_j is a rotation (special orthogonal matrix), and v_j an arbitrary translation, applied to the coordinates Y_j of the jth protein. Here $\Theta = \{U, \phi, \mu_{\mathrm{I}}, \phi_{\mathrm{I}}, Q\}$ denotes the collection of profile HMM model parameters (described below), with $U = [\mu_1^{\mathrm{M}}, \ldots, \mu_n^{\mathrm{M}}]^T$ the matrix of mean structure coordinates, $\phi = \{\phi_i^{\mathrm{M}}\}_{i=1}^n$ the corresponding emission precisions, and $\mathbf{1}$ the column vector of ones.

This hierarchical model combines ideas from two distinct fields: probabilistic sequence analysis [10], and statistical shape analysis [9]. The additive error ϵ models the combined effects of thermal fluctuation and conformational variability, evolutionary drift, and experimental measurement error. The HMM consists of Match (**Mat**), Insert (**Ins**), and Delete (**Del**) states, organized as in profile HMM sequence alignment [21, 10]. However in our model the **Mat** and **Ins** states emit 3D coordinate vectors from multivariate Gaussian distributions $y_i \sim \mathcal{N}(\mu_i^{\mathrm{M}}, \Sigma_i^{\mathrm{M}})$ with state-specific mean positions, rather than letters of a nucleotide or amino acid sequence from discrete distributions. To simplify, we assume **Ins** states share a common mean position μ^{I} and that covariance matrices Σ^{I} and Σ^{M} are diagonal, i.e. $\Sigma_j = \phi_j^{-1} I_3$. Importantly however, we allow the precision parameters ϕ_j for each input structure $j \in \{1, \ldots, m\}$ to be distinct. As the ϕ_j's are estimated via Bayesian inference along with all other parameters (see below), this enables the algorithm to adaptively determine the precision of each input structure, allowing us to analyze structures of varying experimental resolution and/or having a wide range of evolutionary divergence times or rates. As demonstrated in Results, this provides significant benefits over existing algorithms that implicitly weight each input structure equally. It also helps determine the core conserved residues by evaluating fluctuations relative to the variance in each structure. In addition, it aids in computation by preventing kinetic trapping of the MCMC chain in regions having only a subset of proteins aligned. Lastly, the Markov transition matrix Q assigns probabilities to transitions between the three types of states. Since the transitions (**Ins** $\rightarrow$ **Del**) and (**Del** $\rightarrow$ **Ins**) yield the same alignment, we constrain $Q(\mathbf{Del} \rightarrow \mathbf{Ins}) = 0$ as commonly done [41, 31]. Note that the transition probabilities are not currently site-dependent, but could be made so in future versions of the model.

3   Bayesian Multiple Alignment and MCMC Sampling

In our probabilistic framework, multiple structure alignment is achieved by simultaneous Bayesian estimation of the model parameters (including "mean" structure U) and the alignments of each input structure to the model. Let $A = \{A_j\}_{j=1}^m$ with A_j denoting an adjacency matrix specifying the alignment of protein j to the model, and $(R, v) = \{R_j, v_j\}_{j=1}^m$ the corresponding rotations and translations. Let $\Phi = (A, R, v)$. We compute the posterior distribution:

$$\pi(\Theta, \Phi | X_1, \ldots, X_m) \propto$$
$$p_0(n)\pi_0(\Theta, \Phi | n) \prod_{j=0}^{m-1} f(X_j | R_j, v_j, A_j, n, \Theta) p(A_j | \Theta, n)\pi_0(R_j, v_j) \quad (1)$$

where $f()$ is the likelihood function obtained from the Gaussian emission distributions and $p(A_j \mid \Theta, n)$ the prior on alignments implied by the Markov chain indel process of the HMM. To compute (1) we construct a Markov chain Monte Carlo (MCMC) algorithm [14] using Gibbs sampling and Metropolis-Hastings steps to sample the alignments, translations, rotations and the models from their joint posterior distribution. Here we emphasize non-standard steps in the sampling requiring specialized solutions, especially the sampling of rotations and changes to model dimension.

Let $C_j = (X_j - v_j)R_j^{-1} = [c_{j1}, \ldots, c_{jn_j}]$ denote the coordinate matrix after inverting the Euclidean transformation. The likelihood is then

$$f(X_j | R_j, v_j, A_j, n, \Theta) \propto \exp(-\frac{1}{2}s^2)(\phi^{\mathrm{I}})^{\frac{1}{2}\eta_I}(\phi_j^{\mathrm{M}})^{\frac{1}{2}\eta_{jM}}$$

where $\eta_{jM} = \Sigma_l\Sigma_k\delta_{jkl}^{\mathrm{M}}$, $\eta_I = \Sigma_k\delta_{jk}^{\mathrm{I}}$, and $s^2 = \Sigma_k\delta_{jk}^{\mathrm{I}}\phi^{\mathrm{I}}\|c_{jk}-\mu^{\mathrm{I}}\|^2 + \Sigma_k\Sigma_l\delta_{jkl}^{\mathrm{M}}\phi_j^{\mathrm{M}} \cdot \|c_{jk} - \mu_l^{\mathrm{M}}\|^2$. Here $\delta_{jkl}^{\mathrm{M}}$ equals 1 if c_{jk} is emitted from the lth **Mat** state in alignment A_j, and zero otherwise, and δ_{jk}^{I} equals 1 if c_{jk} is emitted from any **Ins** state.

A key distinction between sequence and structure alignment is the need for *invariance under Euclidean transformations*. Although an HMM that emits 3D coordinates is easily defined, alignment of 3D coordinate sequences cannot be done by straightforward application of sequence HMM techniques because each matching implies a distinct (distribution of) rotation and translation which depends on the matching globally. This requires joint evaluation of the likelihood (emission probability) simultaneously rather than locally, and this global dependence destroys the conditional independence structure required for efficient forward/backward recursions in HMMs.

Some structural alignment algorithms simultaneously optimize over matchings and rotation or translation transformations by iteratively maximizing the alignment given the superposition, and then the superposition given the alignment (reviewed in [12]). This suggests an iterative *sampling* scheme for probabilistic inference, a type of MCMC algorithm known as a Gibbs sampler [14], whereby having defined a joint distribution $\pi(\Theta, \Phi \mid X)$ over alignments and superpositions, one iteratively samples from the *conditional* distributions:

$$\pi(\Theta \mid \Phi, X), \quad \pi(A \mid \Theta, R, v, X), \quad \text{and} \quad \pi(R, v \mid \Theta, A, X)$$

Given Θ and (R, v), the alignments A_j are conditionally independent and can be sampled using standard stochastic dynamic programming recursions well-known in the sequence alignment HMM literature [10].

Although natural, such alternation of alignments and superpositions (Gibbs sampling) does not necessarily yield an efficient sampling method. Alignment and superposition are strongly correlated, and the convergence rate of the Gibbs sampler is determined by strength of that correlation. Thus is is important to develop efficient sampling steps which modify both simultaneously.

Rotations. A random-walk Metropolis proposal was constructed on the space $SO(3)$ of 3D rotations using a unit quaternion parametrization:

$$q = [q_0, q] = [\cos(\theta/2), v \sin \theta/2]$$

where θ is an angle around unit vector $v \in \mathbb{R}_3$. New rotations are proposed by independently sampling a vector v' uniformly on the unit sphere $\mathbb{S}_2$, and a small angle of rotation around that vector $\theta' \sim \Gamma(1, 40)$ to form a rotation $q' = [\cos(\theta'/2), v' \sin \theta'/2]$. The proposed rotation is then obtained by composition of q' with the current rotation q via quaternion multiplication: $q^* = q' \cdot q$. This proposal yields a symmetric random walk since the inverse rotation $(\theta, \mathbf{v})^{-1} = (\theta, -\mathbf{v})$. This approach performs much better than a random walk on $\mathbf{q}$ itself.

Because a change in rotation affects all atoms and can dramatically increase the RMSD, the rotations are sampled jointly with alignments. Conditional on the proposed rotation R', a new alignment A'_j is drawn by dynamic programming, and the pair (R'_j, A'_j) are accepted or rejected together. This overcomes the strong dependency between R_j and A_j that is problematic for a Gibbs sampler updating $R_j \mid A_j$ and $A_j \mid R_j$. Since this proposal is symmetric, the joint acceptance probability is given by

$$\alpha\left((R_j, A_j), (R'_j, A'_j)\right) = \min(1, \frac{f(X_j | R'_j, v_j, \Theta, n)}{f(X_j | R_j, v_j, \Theta, n)})$$

Sampling of translations is achieved in an analogous manner, proposing $v'_j \sim N(0, \sigma_v^2 I_3)$ (in practice $\sigma^2 = .1$ works well), then proposing a new alignment $A'_j \mid v'_j$ and accepting or rejecting jointly.

The above moves involve only local perturbations to the rotation/translation. For strongly multimodal posteriors, we previously developed a "library sampling" technique [31] to allow jumps between significantly different rot/trans pairs. We did not find this necessary here, perhaps due to additional mixing achieved by the transdimensional moves below. For strongly multimodal examples (e.g. matching a single domain to a homo-dimer), this may be necessary.

Transdimensional Moves. Because the number of "core" positions in a protein family is unknown *a priori*, the number of states n in the HMM cannot be fixed in advanced, and is subject to inference. The dimensions of the parameter vector Θ and alignments $\{A_j\}_{j=1}^m$ depend on n, and we use a reversible-jump step

[16] to allow n to vary. We update $(n, \Theta, \{A_j\})$ jointly by first sampling a new (n', Θ'), and then new alignments $\{A'_j\}$ conditional on (n', Θ'). n is proposed to increase or decrease by 1 with equal probability:

- $n \to n + 1$: Insert a new position of three states (**Mat,Del,Ins**)
- $n \to n - 1$: Delete an existing position of three states

The proposed location $i \in (1, \ldots, n + 1)$ for an inserted position is randomly sampled with probability proportional to $\lambda^{\min(m, k_{Ii})}$ where k_{Ii} is the total number of C_α's in all proteins emitted from the ith **Ins** state and $\lambda > 1$ is a constant set at 1.2 to achieve a reasonable acceptance rate. This tends to propose new positions into the model in locations where there are many insertions. The mean of the proposed **Mat** state is sampled as follows:

- When the **Ins** state of the ith position has no emissions we set $\boldsymbol{\mu}^M_{i,\text{new}} = \boldsymbol{z} + (\boldsymbol{\mu}^M_{i-1} + \boldsymbol{\mu}^M_i)/2$ for $\boldsymbol{z} \sim \mathcal{N}(\boldsymbol{0}, \sigma^2_\epsilon I_3)$
- Otherwise we set $\boldsymbol{\mu}^M_{i,\text{new}} = \hat{\omega} + \boldsymbol{z}$ for $\boldsymbol{z} \sim \mathcal{N}(\boldsymbol{0}, \hat{\sigma}^2 I_3)$, where $\hat{\omega}$ is the sample mean and $\hat{\sigma}$ the sample s.d. of the coordinates emitted from the ith **Ins** state

In practice setting $\sigma_\epsilon = 6$ achieves reasonable acceptance rates. New alignments $\{A'_j\}$ are then sampled from their conditional distributions $P(A_j | \Theta', R_j, \boldsymbol{v}_j, X_j)$ as above. Deletions are proposed by sampling $i \in (1, \ldots n)$ with weight $\lambda^{-\min(m, k^M_i)}$ where k^M_i is the number emissions from the ith **Mat** state, and then sampling the A_j's from their conditional distributions. The Metropolis-Hastings acceptance ratio for these transdimensional moves is: $\alpha^{\text{Ins}} = \min(1, \gamma)$ for

$$\gamma = \frac{\pi(\Theta', A', R, v, | X)p(u'_1) \prod_{j=0}^{m-1} p(A_j | \Theta, R_j, \boldsymbol{v}_j, X_j)}{\pi(\Theta, A, R, v | X)p(u_1)p(u_2|u_1) \prod_{j=0}^{m-1} p(A'_j | \Theta, R_j, \boldsymbol{v}_j, X_j)}$$

since $\left| \frac{\partial \Theta'}{\partial(\Theta, u_2)} \right| = 1$. Conversely, when deleting a layer $\alpha^{\text{Del}} = \min(1, \gamma^{-1})$.

For all examples in this paper we run at least three independent MCMC chains, each using a different input structure to initialize the profile HMM model and pre-aligning other input structures to the model with the pairwise structural alignment program FAST [45]. We monitor convergence using the Gelman-Rubin diagnostic [7].

Prior Distributions. Prior distributions for model parameters are taken as follows. Markov transition probability vectors Q(Mat → ·) and Q(Ins → ·) are given Dir(α, α, α) priors with $\alpha = 1$, and Q(Del → ·) is given Dir$(\alpha, 0, \alpha)$ to enforce the alignment uniqueness constraint discussed previously. Prior distributions for means of the **Mat** and **Ins** state emission distributions were constructed from quartiles (q_{a1}, q_{a2}, q_{a3}) of the input structure C_α atom coordinates along each axis $a \in \{x, y, z\}$. Prior distributions for μ^M_{ia}'s are independently normal with mean q_{a2} (median) and variance $1.5(q_{a3} - q_{a1})$ (interquartile range). Priors for the precisions of these states' emission distributions are taken to be $Ga(.1, .01)$. The profile model length n is given uniform prior distribution over the range

$[\lceil 0.5 n_{\min} \rceil, \lceil 1.5 n_{\max} \rceil]$, where $n_{\min}$ and $n_{\max}$ are the shortest and longest in-put protein lengths, respectively. Independent uniform prior distributions over rotations and translations are assigned to each structure, so $p(R_j, \boldsymbol{v}_j) \propto 1$.

4   Results

Alignment of Difficult Cases

We first test our algorithm on two example sets of protein structures that are difficult to automatically align due to structural variability. The first is a set of eight KH-domain type I structures taken from the SISYPHUS database (align-ment: AL00054790, pdb ids: 1ec6a, 1dt4a, 1viga, 2fmra, 1j5ka, 1j4wa1, 1j4wa2, 1k1ga), which was previously identified as a difficult case for multiple alignment algorithms [2]. Figures 1a and 1b highlight two regions in one of the structures (scop id: d1k1ga_) with great uncertainties about their structural conservation based on alignments from MUSTANG, MATT and POSA (recently evaluated as the most accurate among current alignment algorithms [4]). Also shown is the manual alignment from SISYPHUS.

Alignments from these algorithms give a binary assignment to each position, either conserved or not conserved, in some cases jumping back and forth in an evolutionarily implausible manner. In contrast, our algorithm computes a smooth posterior probability of inclusion, reflecting the uncertainty from mul-tiple possible good alignments. This avoids the instability of arbitrary cut-offs and indicates to the user where the alignment is ambiguous. We see that these two regions are assigned intermediate values which vary smoothly along the se-quence; for example our method assigns 40% probability match to 'G' where Mustang includes and Matt/POSA exclude. Moreover, those positions with no uncertainty (probability of conservation essentially equal to one) provide the only alignment that is identical to the manual alignment in both regions.

In the second example, we aligned six GroES proteins (1p3hn, 1aono, 1aonp, 1aons, 1pcqo, 1pf9o), five from *E. coli* and one from *M. tuberculosis*. The five *E. coli* GroES proteins have highly conserved structures (overall RMSD within 0.5Å) including their mobile loops, but GroEs from M. tuberculosis differs signifi-cantly in the mobile loop (Figure 1c). Pirovano *et al* [29] show that this structural variation of *M. tuberculosis* causes all of several popular algorithms considered (DALI, CE and even flexible alignment algorithms, MATT and FATCAT) to fail to align the 5 *E. coli* structures in this region, as well as in two other regions (boxes in Fig. 1d)) containing an insertion and a deletion, respectively. In sharp contrast, our alignment is shown in Fig. 1d. We use upper and lower-case letters to denote residues from **Mat** or **Ins** states for a sampled alignment, and color red positions with marginal posterior probability 95% of being insertions. It is seen that our algorithm successfully aligns all five *E. coli* structures perfectly, identifying the mobile loop as one large insertion in *M. tuberculosis*, and also identifying the single insertion and single deletion.

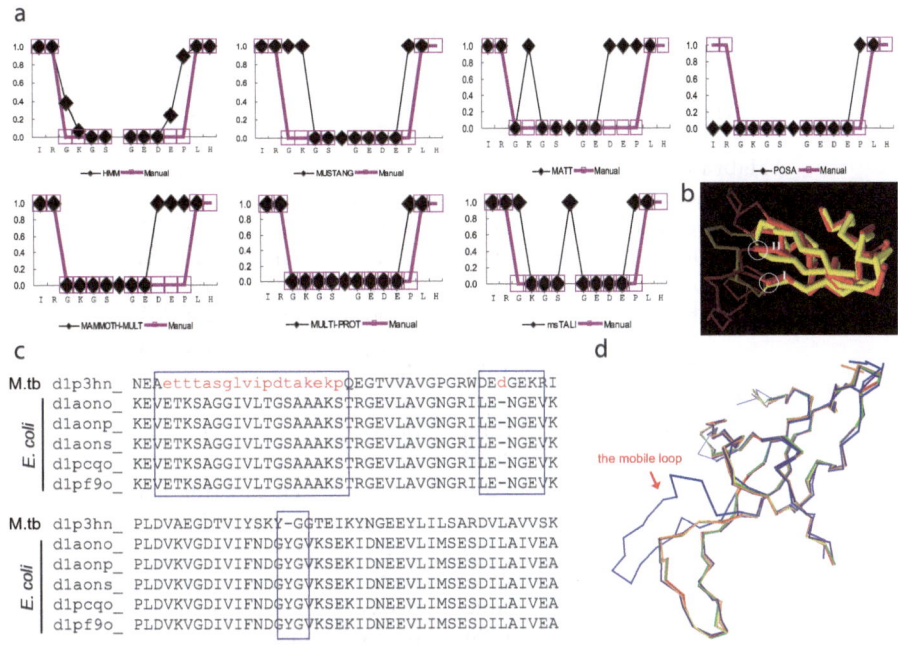

Fig. 1. Examples of protein structures that are previously difficult to align due to structural variability. (a) The manually curated alignment (AL00054790) from SISY-PHUS database is compared with the alignments obtained from our algorithm (HMM) and 6 other popular multiple alignment algorithms by looking at the two regions of aligned proteins. X-axis: the representative protein's sequences; Y-axis: the probability (for HMM algorithm) or assignment (1 or 0 for other algorithms) of a residue being conserved. (b) The representative protein structure d1k1ga_ (red) superposed by d1j5ka_ based on SISYPHUS manual alignment. Region I and II corresponding to the starting residues of "IRGKGS" and "GEDEPLH" in (a), respectively. (c) Visualization of the aligned GroES proteins from *E. coli* and *M. tuberculosis*, adopted from the Figure 4c in [29].(d) Our alignment of the six GroES proteins. Residues in lower case: insertions. Red: identified insertions/deletions (including the mobile loop) *M. tuberculosis*. Boxes: Regions poorly aligned by existing algorithms [29].

Hemoglobin Isoform Evolution

For the second test, we aligned 25 structures of vertebrate hemoglobin α subunit (Table 1) from the SCOP database [27]. The appropriateness of this alignment is assessed by visual inspection of the tree built from the RMSD matrix derived from the alignment, and compared with those trees by the multiple structural alignment algorithms SSM and MUSTANG, as well as the pairwise structural alignment algorithm Mammoth, denoted as Pairwise tree, MUSTANG tree and SSM tree, respectively. In particular the Pairwise Tree is constructed as follows: Mammoth was used to obtain a pairwise alignment of each pair of input structures, and the resulting pairwise RMSD distance matrix was used to build a Neighbor Joining (NJ) tree [32] with ClustalX.

Table 1. Hemoglobin subunits used in the analysis

species name	PDB id	α	β	Å	type
Aldabra Giant Tortoise	1wmu	deoxy	deoxy	1.65	D
Bar-headed goose	1c40	aquo-met	aquo-met	2.3	A
Bar-headed goose	1hv4	deoxy	deoxy	2.8	A
Bluefin tuna	1v4u	cmo	cmo	2	A
Bluefin tuna	1v4x	deoxy	deoxy	1.6	A
Bovine	1fsx	cmo	cmo	2.1	A
Bovine	1hda	deoxy	deoxy	2.2	A
Chicken	1hbr	deoxy	deoxy	2.3	D
Dusky rockcod	1la6	cmo	deoxy	2	A
Emerald rockcod	1hbh	deoxy	deoxy	2.2	A
Emerald rockcod	2h8d	deoxy	deoxy	1.78	A
Horse	1g0b	cmo	cmo	1.9	A
Horse	2dhb	deoxy	deoxy	2.8	A
Human	2dn3	cmo	cmo	1.25	A
Human	1ird	cmo	cmo	1.25	A
Human	2dn2	deoxy	deoxy	1.25	A
Human	1a3n	deoxy	deoxy	1.8	A
Rainbow trout	1ouu	cmo	cmo	2.5	I
Rainbow trout	1out	deoxy	deoxy	2.3	I
Red Stingray	1cg8	cmo	cmo	1.9	A
Red Stingray	1cg5	deoxy	deoxy	1.6	A
Spot	1spg	cmo	cmo	1.95	A
Hound shark	1gcw	cmo	cmo	2	A
Hound shark	1gcv	deoxy	deoxy	2	A
Yellow perch	1xq5	met	met	1.9	A

As can be seen in Fig. 2a, all three structure-based trees show aberrant branches (dashed circles): cmo-bound 1la6 clustered with deoxy proteins (Pairwise Tree); cmo-bound mammalian proteins clustered with deoxy avian proteins (MUSTANG Tree and SSM Tree), which deviate greatly from the sequence-based tree. For example, the SSM tree fails to identify the boundary between fishes and non-fishes and proposes a mixed monophyly consisting of cartilaginous fishes, reptiles/birds and mammals. The MUSTANG tree has the same topology as ours for unliganded proteins, but clusters the liganded subunits of horse and cow with the goose subunits. In contrast, all such these inconsistencies are resolved in our MAP tree (Figure 1b), correctly revealing three major monophylies: mammals/reptile/birds, cartilaginous fishes (sharks/rays), and bony fishes. This agrees with hemoglobins' highly conserved function in vertebrates. Note that the alignments obtained by the three multiple structural alignment methods are of similar length and quality, as measured by mean pairwise RMSD (number residues aligned): 1.02 (136) for SSM, 1.08 (140.6) for MUSTANG, and 0.918±0.004 (136.8±0.2) for our algorithm (posterior means).

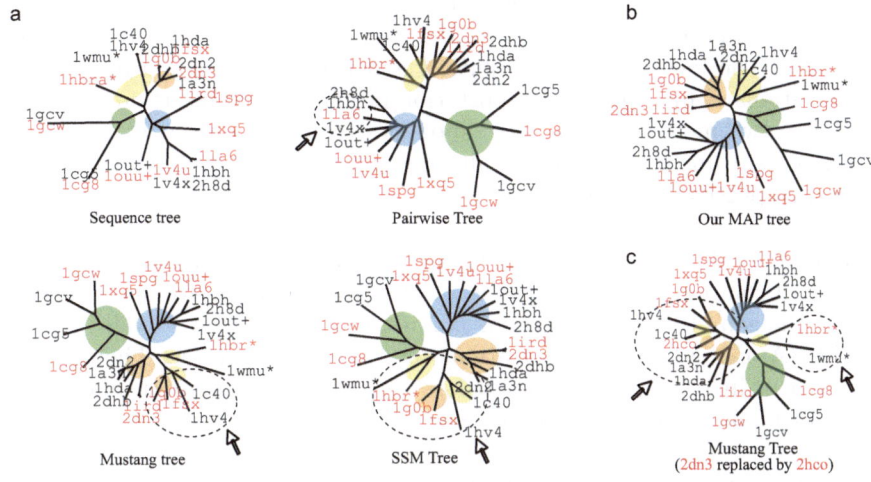

Fig. 2. Analysis of Hgb α subunits. (a) trees built from sequence alignment by ClustalX, and three structural alignments by SSM, Mustang and our algorithm. (b) trees based on multiple structural alignment by our algorithm and Mustang, after replacing the human CMO-bound Hgb with a lower resolution structure. Fonts in red: liganded; black: unliganded; ellipses in blue: bony fishes; green: cartilagineous fishes; yellow: reptiles and birds; brown: mammals. '*' Hgb D; '+' Hgb I; otherwise Hgb A.

Thus the improvement comes not from finding a *better* alignment, but in *averaging over alignment uncertainty* to construct a more stable and accurate tree.

As emphasized above, our algorithm produces not just a single multiple alignment, but a posterior distribution over *all* alignments, allowing it to account for uncertainty. We can take advantage of this to obtain a probabilistic estimate of the NJ tree as follows: For each alignment sampled from the posterior, we calculate the pairwise RMSD for every pair of input structures, and use them construct a NJ tree. Trees with zero nodal distance [5] are combined. This calculation gives the Bayesian marginal posterior distribution over (equivalence classes of) tree, averaging over all possible alignments weighted according to their respective posterior probabilities. This identifies two NJ trees which account for > 98% of the posterior probability (78% (Figure 2a) and 20%, respectively). They are 95.8% similar in topology [28], differing only in placement of chicken and tortoise hemoglobins: the MAP (78%) tree places these with the two goose proteins, while the less probable (20%) tree places them with the shark proteins.

To demonstrate the robustness of our algorithm to input noise, we repeated the analysis replacing one of the human CMO-bound structures (2dn3, 1.25 Å) by a lower resolution predecessor (2hco, 2.7 Å). This effected a topological change in trees obtained from Mustang alignments (Fig 2c), placing 2hco in a separate branch from another human cmo-Hgb (1ird), and grouping the D isoforms (1wmu,1hbr) with cartilagineous fish. In contrast, our algorithm retains the sistership of the human cmo-Hgbs (1ird and 2hco) with posterior probability $> 80\%$, but estimates their divergence (0.4) to be much larger than that of 1ird-2dn3 (0.08). The low- and high-resolution structures differ significantly in only a few positions, which have high posterior probability of insertion under our model. The two most probable tree topologies (the only ones with probability $> 20\%$) remain unchanged, simply altering their relative probabilities to 26% and 39%, respectively.

Running Time: The algorithm was deployed on Duke Shared Cluster Resource in a parallel computing fashion. For a set of m structures of average length 100 residues, the sampler takes 3 seconds per iteration using $m + 1$ nodes (3G Hz, leq512M RAM), and 150-200k iterations (5-7 days) to converge. The computation time can increase significantly if fewer ($< m/2$) nodes are available, or if the size of input structures is large (e.g. > 300 aa), or the structures contain higher percentage of flexible regions ($> 30\%$). Thus the computational expense of the current implementation prohibits large-scale comparisons with other multiple alignment algorithms at present, and we currently view the algorithm as a tool for in-depth analysis of individual protein families, rather than for database screening.

5   Discussion

We have described a probabilistic approach to multiple protein structure alignment based on an explicit statistical model of variability in protein families. The Bayesian approach avoids sensitivity to alignment parameters by using statistical inference to adaptively learn parameters. As described in [31] for pairwise alignments, this model generalizes many existing structural alignment algorithms, which correspond to MAP alignments under varying choices of prior and noise distributions. In addition, our results indicate that averaging over alignment uncertainty makes phylogenetic tree reconstruction significantly more robust. The model may be extended to treat tree topology as an additional parameter for Bayesian inference, using algorithms for sampling phylogenetic trees from their conditional posterior distributions. This would allow methods for simultaneous sequence alignment/phylogeny reconstruction [42, 26] to incorporate structural information, which is known to be conserved over much longer evolutionary time scales than sequence.

Acknowledgements. This work was partially supported by NSF grant DMS-0605141 and NIH grant R01GM090201.

References

1. Alexandrov, V., Gerstein, M.: Using 3D hidden Markov models that explicitly represent spatial coordinates to model and compare protein structures. BMC Bioinformatics 5 (2004)

2. Andreeva, A., Prlic, A., Hubbard, T.J.P., Murzin, A.G.: SISYPHUS - structural alignments for proteins with non-trivial relationships. Nucleic Acids Research 35, D253–D259 (2007)
3. Baldi, P., Chauvin, Y., Hunkapiller, T., McClure, M.A.: Hidden Markov-models of biological primary sequence information. Proceedings of the National Academy of Sciences U.S.A 91(3), 1059–1063 (1994)
4. Berbalk, C., Schwaiger, C.S., Lackner, P.: Accuracy analysis of multiple structure alignments. Protein Science 18(10), 2027–2035 (2009)
5. Bluis, J., Shin, D.: Nodal distance algorithm: Calculating a phylogenetic tree comparison metric. In: Proc. 3rd IEEE Symposium on Bioinformatics and Bioengineering, pp. 87–94 (2003)
6. Burra, P.V., Zhang, Y., Godzik, A., Stec, B.: Global distribution of conformational states derived from redundant models in the PDB points to non-uniqueness of the protein structure. Proceedings of the National Academy of Sciences U.S.A 106(26), 10505–10510 (2009)
7. Castelloe, J.M., Zimmerman, D.L.: Convergence assessment for reversible jump MCMC samplers. Technical Report 313, University of Iowa, Dept. of Statistics and Actuarial Science (2002)
8. Challis, C., Schmidler, S.C.: A Stochastic Evolutionary Model for Protein Structure Alignment and Phylogeny. Molecular Biology and Evolution 29(11), 3375–3387 (2012)
9. Mardia, K.V., Dryden, I.L.: Statistical Shape Analysis. Wiley (1998)
10. Durbin, R., Eddy, S., Krogh, A., Mitchison, G.: Biological Sequence Analysis: Probabilistic Models of Proteins and Nucleic Acids. Cambridge University Press (1998)
11. Eddy, S.R., Durbin, R.: RNA sequence analysis using covariance models. Nucleic Acids Research 22(11), 2079–2088 (1994)
12. Eidhammer, I., Jonassen, I., Taylor, W.R.: Structure comparison and structure patterns. Journal of Computational Biology 7(5), 685–716 (2000)
13. Geisbrecht, B.V., Dowd, K.A., Barfield, R.W., Longo, P.A., Leahy, D.J.: Netrin binds discrete subdomains of DCC and UNC5 and mediates interactions between DCC and heparin. Journal of Biological Chemistry 278(35), 32561–32568 (2003)
14. Spiegelhalter, D.J., Gilks, W.R., Richardson, S. (eds.): Markov Chain Monte Carlo in Practice. Chapman & Hall (1996)
15. Godzik, A.: The structural alignment between two proteins: Is there a unique answer? Protein Science 5(7), 1325–1338 (1996)
16. Green, P.J.: Reversible jump Markov chain Monte Carlo computation and Bayesian model determination. Biometrika 82(4), 711–732 (1995)
17. Green, P.J., Mardia, K.V.: Bayesian alignment using hierarchical models, with applications in protein bioinformatics. Biometrika 93(2), 235–254 (2006)
18. Hasegawa, H., Holm, L.: Advances and pitfalls of protein structural alignment. Current Opinion in Structural Biology 19(3), 341–348 (2009)
19. Kolodny, R., Petrey, D., Honig, B.: Protein structure comparison: implications for the nature of 'fold space', and structure and function prediction. Current Opinion in Structural Biology 16(3), 393–398 (2006)
20. Krissinel, E., Henrick, K.: Secondary-structure matching (SSM), a new tool for fast protein structure alignment in three dimensions. Acta Crystallographica Section D-Biological Crystallography 60, 2256–2268 (2004)
21. Krogh, A., Brown, M., Mian, I.S., Sjolander, K., Haussler, D.: Hidden Markov-models in computational biology - applications to protein modeling. Journal of Molecular Biology 235(5), 1501–1531 (1994)

22. Lathrop, R., Rogers, R., Smith, T., White, J.: A Bayes-optimal sequence-structure theory that unifies protein sequence-structure recognition and alignment. Bulletin of Mathematical Biology 60(6), 1039–1071 (1998)
23. Lawrence, C.E., Altschul, S.F., Boguski, M.S., Liu, J.S., Neuwald, A.F., Wootton, J.C.: Detecting subtle sequence signals: a Gibbs sampling strategy for multiple alignment. Science 262(5131), 208–214 (1993)
24. Lawrence, C.E., Reilly, A.A.: An expectation maximization (EM) algorithm for the identification and characterization of common sites in unaligned biopolymer sequences. Proteins: Structure, Function, and Bioinformatics 7(1), 41–51 (1990)
25. Levitt, M., Gerstein, M.: A unified statistical framework for sequence comparison and structure comparison. Proceedings of the National Academy of Sciences, U.S.A 95(11), 5913–5920 (1998)
26. Lunter, G., Rocco, A., Mimouni, N., Heger, A., Caldeira, A., Hein, J.: Uncertainty in homology inferences: Assessing and improving genomic sequence alignment. Genome Research 18(2), 298–309 (2008)
27. Murzin, A.G., Brenner, S.E., Hubbard, T., Chothia, C.: SCOP: A structural classification of proteins database for the investigation of sequences and structures. Journal of Molecular Biology 247(4), 536–540 (1995)
28. Nye, T.M.W., Lio, P., Gilks, W.R.: A novel algorithm and web-based tool for comparing two alternative phylogenetic trees. Bioinformatics 22(1), 117–119 (2006)
29. Pirovano, W., Feenstra, K.A., Heringa, J.: The meaning of alignment: lessons from structural diversity. BMC Bioinformatics 9 (2008)
30. Rajagopalan, S., Deitinghoff, L., Davis, D., Conrad, S., Skutella, T., Chedotal, A., Mueller, B.K., Strittmatter, S.M.: Neogenin mediates the action of repulsive guidance molecule. Nature Cell Biology 6(8), 756–762 (2004)
31. Rodriguez, A., Schmidler, S.C.: Bayesian protein structure alignment (under revision)
32. Saitou, N., Nei, M.: The neighbor-joining method - a new method for reconstructing phylogenetic trees. Molecular Biology and Evolution 4(4), 406–425 (1987)
33. Sakakibara, Y., Brown, M., Hughey, R., Mian, I.S., Sjolander, K., Underwood, R.C., Haussler, D.: Stochastic context-free grammers for tRNA modeling. Nucleic Acids Research 22(23), 5112–5120 (1994)
34. Schmidler, S.C., Liu, J.S., Brutlag, D.L.: Bayesian segmentation of protein secondary structure. Journal of Computational Biology 7(1-2), 233–248 (2000)
35. Schmidler, S.C., Liu, J.S., Brutlag, D.L.: Bayesian protein structure prediction. Case Studies in Bayesian Statistics 5, 363–378 (2001)
36. Schmidler, S.C.: Statistical Models and Monte Carlo Methods for Protein Structure Prediction. PhD thesis, Stanford University (2002)
37. Schmidler, S.C.: Fast Bayesian shape matching using geometric algorithms (with discussion). In: Bernardo, J.M., Bayarri, S., Berger, J.O., Dawid, A.P., Heckerman, D., Smith, A.F.M., West, M. (eds.) Bayesian Statistics, vol. 8, pp. 471–490. Oxford University Press, Oxford (2006)
38. Schmidler, S.C.: Bayesian flexible shape matching with applications to structural bioinformatics (submitted)
39. Simons, K.T., Kooperberg, C., Huang, E., Baker, D.: Assembly of protein tertiary structures from fragments with similar local sequences using simulated annealing and a Bayesian scoring function. Journal of Molecular Biology 268(1), 209–225 (1997)
40. Wang, L., Jiang, T.: On the complexity of multiple sequence alignment. Journal of Computational Biology 1(4), 12 (1994)

41. Webb, B.J.M., Liu, J.S., Lawrence, C.E.: BALSA: Bayesian algorithm for local sequence alignment. Nucleic Acids Research 30(5), 1268–1277 (2002)
42. Wong, K.M., Suchard, M.A., Huelsenbeck, J.P.: Alignment uncertainty and genomic analysis. Science 319(5862), 473–476 (2008)
43. Wouter, B., Mardia, K.V., Taylor, C.C., Ferkinghoff-Borg, J., Krogh, A., Hamelryck, T.: A generative, probabilistic model of local protein structure. Proceedings of the National Academy of Sciences, U.S.A 105(26), 8932–8937 (2008)
44. Zhu, J., Liu, J.S., Lawrence, C.E.: Bayesian adaptive sequence alignment algorithms. Bioinformatics 14(1), 25–39 (1998)
45. Zhu, J.H., Weng, Z.P.: FAST: A novel protein structure alignment algorithm. Proteins-Structure Function and Bioinformatics 58(3), 618–627 (2005)

Gene-Gene Interactions Detection
Using a Two-Stage Model

Zhanyong Wang[1], Jae-Hoon Sul[1], Sagi Snir[2], Jose A. Lozano[3],
and Eleazar Eskin[1,*]

[1] Computer Science Department, University of California Los Angeles, United States
[2] Institute of Evolution, Department of Evolutionary and Environmental Biology,
Faculty of Natural Sciences, University of Haifa, Israel
[3] Intelligent Systems Group, Department of Computer Science and Artificial
Intelligence, University of the Basque Country, Donostia, Spain
eeskin@cs.ucla.edu

Abstract. Genome wide association studies (GWAS) have discovered
numerous loci involved in genetic traits. Virtually all studies have reported associations between individual single nucleotide polymorphism
(SNP) and traits. However, it is likely that complex traits are influenced
by interaction of multiple SNPs. One approach to detect interactions
of SNPs is the brute force approach which performs a pairwise association test between a trait and each pair of SNPs. The brute force approach is often computationally infeasible because of the large number of
SNPs collected in current GWAS studies. We propose a two-stage model,
Threshold-based Efficient Pairwise Association Approach (TEPAA), to
reduce the number of tests needed while maintaining almost identical
power to the brute force approach. In the first stage, our method performs the single marker test on all SNPs and selects a subset of SNPs
that achieve a certain significance threshold. In the second stage, we
perform a pairwise association test between traits and pairs of the SNPs
selected from the first stage. The key insight of our approach is that we
derive the joint distribution between the association statistics of a single
SNP and the association statistics of pairs of SNPs. This joint distribution allows us to provide guarantees that the statistical power of our
approach will closely approximate the brute force approach. We applied
our approach to the Northern Finland Birth Cohort data and achieved
63 times speedup while maintaining 99% of the power of the brute force
approach.

1 Introduction

Genome-wide association studies (GWAS) attempt to discover genetic variation
associated with disease traits. To perform GWAS, studies collect genetic variation of individuals and their disease status or disease related traits. GWAS
studies typically collect single nucleotide polymorphisms (SNPs) because technologies allow for very cost efficient collection of SNPs. Since SNPs are so prevalent in the genome, they are likely to be correlated with other genetic variations.

---

* To whom correspondence should be addressed.

R. Sharan (Ed.): RECOMB 2014, LNBI 8394, pp. 340–355, 2014.
© Springer International Publishing Switzerland 2014

Current GWAS studies collect about a million SNPs in thousands of individuals. The standard approach for identifying associations between SNPs and traits is that for each SNP, we compare the average trait value of individuals who have one allele of a SNP and that of individuals who have the other allele of the SNP. If the difference between the two average trait values is above a certain threshold, we declare that the SNP is significant associated with the trait. We refer to computing the difference in the average trait values for each SNP as the "single marker test", and it has successfully identified many individual SNPs associated with several complex diseases [1, 2, 5, 6, 16].

Current studies on certain complex diseases have also suggested that some SNPs influence diseases through interactions [3, 19, 23]. In an extreme scenario, two SNPs may not have any effect on a disease independently, but they may affect the disease when both are present. To detect an interaction of SNPs, one needs to consider the association between a trait and a pair of SNPs. One approach to find such associations is to divide individuals into two groups: one group of individuals who have a certain combination of alleles for a pair of SNPs and the other group of individuals who have different combinations of alleles for the pair of SNPs. We then compute the difference in the average trait value between the two groups to determine whether the pair of SNPs is significantly associated with the trait. Finding an association between a trait and a pair of SNPs is called the "pairwise association test", and recently, several different methods have been proposed [7, 13, 14, 24, 25].

One major challenge in discovering pairs of SNPs associated with a trait is that it requires enormous computation. One needs to compute associations between a trait and $4 \times \binom{M}{2}$ pairs of SNPs where M is the number of SNPs available for testing. When M is close to one million as in current GWAS, an exhaustive pairwise search that considers all pairs of SNPs considers 2000 billion pairs of SNPs, which is a computationally challenging task. As the number of SNPs in GWAS keeps increasing with the improvement of technologies to collect SNPs, the exhaustive search becomes even more computationally infeasible.

In this paper, we present a Threshold-based Efficient Pairwise Association Approach (TEPAA) for detecting associations between traits and pairs of SNPs using a two-stage model. In the first stage, our method performs the single marker test on all individual SNPs and selects a subset of SNPs that exceed a certain significance threshold (called "the first stage threshold") for further consideration. In the second stage, individual SNPs that are selected in the first stage are paired with each other, and we perform the pairwise association test on those pairs. In this method, there exists a trade-off between the probability of the method detecting a pair of SNPs associated with a trait (called "statistical power of the method") and the computational burden (or cost). Intuitively, statistical power increases as we include more SNPs in the second stage, which means higher cost. The first stage threshold determines this trade-off, and we derive the analytical power of our method which allows us to determine the threshold and to control this trade-off. The key insight of our approach is that we derive the joint distribution between the association statistics of single SNP and the association statistics of pairs of SNPs. This joint distribution allows us to provide guarantees that the statistical power of our approach will closely approximate the brute force approach. We can accurately compute the analytical power of our two stage model at any first stage threshold and compare it to the power of the brute force approach. Hence, we are able to choose as few SNPs as possible in the first stage while achieving almost the same power as the brute force approach.

While recently developed methods such as TEAM [25, 26] significantly reduce the computational burden of searching for pairs of associated SNPs, to our knowledge very few methods are feasible to apply to full size human GWAS datasets. The SIXPAC method developed by Pe'er and Prabhu utilizes a novel randomization technique that requires $10\times$ to $100\times$ fewer tests than a brute-force approach to find long-range interactions using standard two-locus test [15]. However, their method only handles case-control data and can not apply to quantitative traits. Wan et al developed an approach BOOST, which designed a Boolean representation of data and used a screening stage to filter out most nonsignificant SNP interactions [22]. However, their method can not apply to quantitative traits either.

The only existing method that is feasible on a full size human GWAS dataset to detect SNP pairs associated with quantitative traits is FastEpistasis [17]. FastEpistasis is a brute-force approach which conducts pairwise associations for all pairs of SNPs, or SNP pairs specified by users. The advantage of FastEpistasis is that their method is parallelled and utilizes high-performance computer architectures with multiple cores. Our method utilizes a two-stage strategy and greatly reduced the number of pairwise association tests with little power loss.

We note that in this paper, we are only considering pairs of SNPs which are far apart from each other. There is another class of methods which consider multiple SNPs close to each other [12, 20, 21]. These problems are completely different and characterized by very different challenges. For example, the computational burden which is the focus of our paper is different because the number of pairs of SNPs near each other is significantly smaller than the total number of pairs of SNPs. In addition, neighboring SNPs are typically correlated with each other, referred to as in linkage disequilibrium (LD). Pairs of SNPs far from each other are typically independent or unlinked which is an observation that we leverage in our approach.

2   Results

2.1   Overview of the Two-Stage Model TEPAA

We present a two-stage model, TEPAA, for detecting associations between traits and pairs of SNPs. In this first stage, the association statistics for all SNPs are computed. Any SNPs which have a statistic higher than a pre-determined threshold then advance to the second stage in which all pairs of these SNPs are evaluated. The first stage threshold is important in determining power and cost of our method because it controls the number of SNPs to be selected in the first stage. For a truly associated pair of SNPs to be identified using our approach, both SNPs must advance to the second round and thus must have association statistics higher than the first stage threshold. Clearly, the more stringent the threshold, the smaller the number of SNPs in the second stage and the smaller number of pairs of SNPs which must be evaluated speeds up this method. On the other hand, more stringent thresholds increase the chance that at least one of the pair of truly associated SNPs will not be more significant than the first stage threshold and the pair will not be identified by the method. Hence, there is a trade-off between power and cost, which is determined by the first stage threshold.

Our method chooses the first stage thresholds such that the two-stage model loses only a small amount of power but increases computational efficiency dramatically compared to the exhaustive search. To find such thresholds, we first derive the analytical power and cost of both the brute force approach and the two-stage model. This analysis allows us to choose the threshold that yields the desired power and cost, and hence it allows us to control the trade-off between the two. To derive the analytical power of our two stage model, we use the framework of Multivariate Normal Distribution(MVN) to model the association statistics [8–10]. We use a MVN to approximate the joint distribution between the association statistic of single SNP and the association statistic of pairs of SNPs. The non-centrality parameters (NCPs) of statistics are considered to be the mean vector in the MVN and correlations among statistics as a covariance matrix in the MVN. The NCPs and correlations can be calculated from the data and thus we obtained all the parameters of the MVN. The details of the analysis are discussed in Section 3.4.

From our analysis, we observe that the thresholds which control the power loss of the two stage approach depend on the minor allele frequency (MAF) of the SNPs. In particular, more common SNPs can be filtered out with less significant thresholds than rare SNPs. In order to efficiently implement TEPAA using MAF dependent thresholds for each pair, we group the SNPs into bins based on their MAFs to apply the correct thresholds to each possible pair. After disregarding rare variants with $MAF < 0.05$, we categorize all common SNPs into 9 bins according to their MAF, with step size 0.05. Each pair of SNPs would have two thresholds, one for each SNP in the first stage. In total, we have $\binom{9}{2}+9$ categories of SNP pairs. We pre-compute the first stage thresholds for each combination of two MAFs in order to achieve 1% power loss, while achieving high cost savings. We sort the SNPs within each bin by their association statistics and use binary search to rapidly obtain the set of SNPs above a single threshold to efficiently implement the first stage of our method.

2.2   Application of TEPAA to the NFBC Data

We applied TEPAA to the Northern Finland Birth Cohort (NFBC) data to demonstrate the utility of our two stage model and the cost saving on a real data. The Northern Finland Birth Cohort Data contains $5,326$ individuals, and $331,476$ SNPs are genotyped. The histogram of all SNPs' MAFs is shown in Fig. 1(a). As described in detail in Section 3.5, we categorize all common SNPs

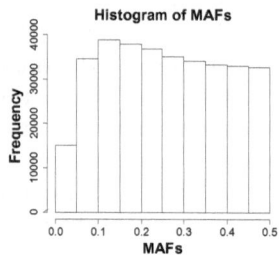

Histogram of MAFs

MAF of SNP B

MAF of SNP A	0.05	0.1	0.15	0.2	0.25	0.3	0.35	0.4	0.45	0.5
0.05	1.13	5.2	5.86	5.70	5.54	5.30	5.13	5.00	4.97	4.94
0.1	-	5.97	13.45	13.07	12.72	12.16	11.78	11.48	11.40	11.34
0.15	-	-	7.57	14.71	14.32	13.69	13.26	12.93	12.84	12.77
0.2	-	-	-	7.15	13.91	13.31	12.89	12.56	12.48	12.41
0.25	-	-	-	-	6.77	12.95	12.54	12.22	12.14	12.07
0.3	-	-	-	-	-	6.19	11.99	11.69	11.61	11.55
0.35	-	-	-	-	-	-	5.81	11.32	11.25	11.18
0.4	-	-	-	-	-	-	-	5.52	10.96	10.90
0.45	-	-	-	-	-	-	-	-	5.44	10.83
0.5	-	-	-	-	-	-	-	-	-	5.38

(a) The distribution of all SNPs' MAFs

(b) The number of SNP pairs in each category. Numbers are shown in factor of 100 millions.

Fig. 1. The Distribution of all SNPs' MAFs and number of SNP pairs in each category

into 9 bins according to their MAFs. The number of SNP pairs in each category is shown in Fig. 1(b). The first stage thresholds of TEPAA are pre-computed for each category in order to have the power loss at 1% using the methods described in Section 3.5. The cost saving for each category is summarized in Table 1. Based on Fig. 1(b) and Table 1, the estimated overall cost saving is 63.2 times, which is the ratio between total number of pairwise association tests in brute force approach and that of TEPAA.

For all SNPs in each bin, we calculate the association statistics and sort the SNPs in descending order of their statistics. We perform our analysis using the dominant model which is standard for analysis of epistatic interactions. We note that the basic approach of TEPAA can be extended to other models such as recessive or additive as well.

We compare the performance of the brute force approach and TEPAA to detect the SNP pairs associated with the phenotype "CRP" (C-reactive protein) on a machine with 2.3 GHz AMD Opteron Processor. Since it is impractical to run the brute force on the whole chromosome, the CPU time of the brute force approach is estimated from one single chromosome by scaling, which is estimated to be $1,542$ hours for phenotype "CRP". The CPU time of TEPAA is 24.5 hours for the same phenotype. We achieved 62.9 times of cost saving, which verifies our analysis of the cost savings of TEPAA when achieving 1% of power loss. However, both brute-force approach and two-stage model report no significant SNP interactions under the significance threshold 10^{-12}. This is understandable since this data set contains only $5,326$ individuals. In the next section, we show that the brute force approach and TEPAA have similar power when there exists significant SNP interactions.

2.3   TEPAA Controls Power Loss in Simulated Data

To demonstrate that TEPAA has only 1% power loss using the pre-computed first stage thresholds, we perform simulations where we implant a significant SNP-SNP interaction to the NFBC data and then detect the SNP pair using TEPAA.

We created phenotype data using the phenotype "CRP" (C-reactive protein) in the NFBC data as a starting point. To simulate the significant SNP pairs, we randomly sample the MAF of each SNP from $[0.05, 0.5)$. The alleles of each individuals at these two simulated SNPs are then sampled from the MAF. The phenotypes of the individuals with causal alleles at the SNP pairs are increased by a selected effect size so that the pairs has 50% power in the brute-force approach. Then we apply both the brute-force approach and the two-stage approach to the simulated dataset. The first stage significance thresholds in the two-stage approach are selected in order to obtain 1% power loss.

We generated $10,000$ simulated SNP pairs and applied both approaches. The power for each approach is calculated as the proportion of experiments that the approach detected the implanted SNP pairs among all 10000 experiments. The power of brute-force approach is 51% while the power of TEPPA is 50.8%. The practical power loss is 0.4%. We note that the power loss is lower than we expected because the thresholds are chosen for MAF frequency bins to be conservative and valid for all members of that bin.

3   Methods

3.1   Association Test between One SNP and Traits

We first illustrate the method to detect association between traits and one SNP. A traditional approach to identify the association is that for each SNP, we compare the average trait value of individuals who carry the causal allele at the SNP and that of the individuals who do not have the causal allele at the SNP of interest. If the difference between those two values is above a certain threshold, we declare that the investigated SNP has a significant correlation with the trait. This approach is referred to as "single marker test" and has been successful in many association studies. We analyze the power of the "single marker test" as follows.

Assume we are investigating SNP A, with minor allele frequency (MAF) to be p_A and the causal allele is the minor allele (for the case where the causal allele is the major allele, we have similar analysis). Let N be the number of individuals and y_i be the trait value of individual i. Then the number of individuals with the minor allele at SNP A can be denoted as $N_A = N \cdot p_A$ and the number of individuals without the minor allele at SNP A can be denoted as $N_{\neg A} = N \cdot p_{\neg A} = N \cdot (1 - p_A)$ We use x_i^A to denote the allele of individual i at SNP A. y_i is any real number and $x_i^A \in \{0, 1\}$. We set $x_i^A = 1$ when the allele of individual i at SNP A is the minor allele and $x_i^A = 0$ otherwise.

We assume that a trait value of individual i follows the normal distribution with a certain mean μ and a variance σ^2. If the minor allele affects the trait, the mean trait value (μ) of individuals with the minor allele will increase by a certain value β_A (effect size). Now, we can obtain the distribution of y_i as

$$y_i \sim N(\mu + x_i^A \beta_A, \sigma^2) \tag{1}$$

Let $\bar{Y}_A$ be the average trait value of individuals who have the causal allele at SNP A and $\bar{Y}_{\neg A}$ be the average trait value of individuals who do not carry the causal allele at SNP A. Then we can derive the distributions of $\bar{Y}_A$ and $\bar{Y}_{\neg A}$ as follows:

$$\bar{Y}_A = \frac{\sum_{i:x_i^A=1} y_i}{N_A} \sim N(\mu + \beta_A, \frac{\sigma^2}{N \cdot p_A}), \bar{Y}_{\neg A} = \frac{\sum_{i:x_i^A=0} y_i}{N_{\neg A}} \sim N(\mu, \frac{\sigma^2}{N \cdot p_{\neg A}}) \quad (2)$$

We normalize the difference between $\bar{Y}_A$ and $\bar{Y}_{\neg A}$ to obtain the following statistic S_A, which is normally distributed with mean $\lambda_A \sqrt{N}$ (the non-centrality parameter) and unit variance.

$$S_A = \frac{\bar{Y}_A - \bar{Y}_{\neg A}}{\sqrt{\frac{\sigma^2}{N \cdot p_A \cdot (1-p_A)}}} \sim N(\lambda_A \sqrt{N}, 1), \; where \; \lambda_A = \frac{\beta_A \sqrt{p_A(1-p_A)}}{\sigma} \quad (3)$$

Given the significance level α and the observed value of the test statistic S_A, the SNP is deemed as significant, or statistically associated with the trait, if $|S_A| \geq \Phi^{-1}(1 - \alpha/2)$, where Φ^{-1} is the quantile function of the standard normal distribution. For simplicity, we use the notation $T = \Phi^{-1}(1 - \alpha/2)$ as the per-SNP threshold.

We declare all those SNPs with statistic $|S_A| > T$ to be associated with trait. So the per-causal-SNP power of a putative causal SNP A, which is the probability of $|S_A| > T$, can be calculated as

$$P_1(A) = P(|S_A| > T) = \Phi\left(-T + \lambda_A \sqrt{N}\right) + 1 - \Phi\left(T + \lambda_A \sqrt{N}\right) \quad (4)$$

The average power $\overline{P_1}$ is obtained by averaging per-causal-SNP powers over all putative causal SNPs.

3.2 The Brute-Force Approach for Pairwise Association Test

Current studies on complex disease have also suggested that some SNPs influence traits in pairs. Only when both causal alleles appear on a pair of SNPs, the trait value is increased. To detect the interaction of SNPs that influence the trait, we need to consider the association between a trait and a pair of SNPs (pairwise association test). We analyze the power of the brute force approach which calculates the association between a trait and all pairs of SNPs as follows.

We assume there exists a SNP pair AB, composed of SNP A and SNP B, that influence a trait. Assume the causal alleles are minor alleles at both SNPs. Our statistic is the difference between the average trait value of individuals who have minor alleles on both SNPs and that of individuals who do not have minor allele on at least one of the two SNPs A and B. Here we assume the two SNPs have same (positive) direction of effect. We use the same notation as in section 3.1. The expected number of individuals who have minor alleles at both SNPs can be computed as $N_{AB} = N \cdot p_A \cdot p_B$ and the expected number of individuals who do not have minor alleles at both SNPs can be computed as $N_{\neg AB} = N \cdot (1 - p_A \cdot p_B)$. If an individual carries the causal alleles at both SNPs A and B, the mean of trait value is increased or decreased by the effect size of the SNP pairs, which is denoted as β_{AB}. Then we can write the distribution of y_i as

$$y_i \sim N(\mu + x_i^A x_i^B \beta_{AB}, \sigma^2) \tag{5}$$

Let $\bar{Y}_{AB}$ be the average trait value of individuals with causal alleles at both SNPs and let $\bar{Y}_{\neg AB}$ be the average trait value of individuals without causal alleles at both SNPs. For simplicity, let $\sum_{11}$ denote $\sum_{i:x_i^A=1 \wedge x_i^B=1}$, and similarly for $\sum_{10}, \sum_{01}, \sum_{00}$ for different alleles of SNPs A and B. We can calculate $\bar{Y}_{AB}$ and $\bar{Y}_{\neg AB}$ as

$$\bar{Y}_{AB} = \frac{1}{N_{AB}} \sum_{11} y_i \sim N(\mu + \beta_{AB}, \frac{\sigma^2}{N p_A p_B}),$$

$$\bar{Y}_{\neg AB} = \frac{1}{N_{\neg AB}} \sum_{00,01,10} y_i \sim N(\mu, \frac{\sigma^2}{N(1 - p_A p_B)}) \tag{6}$$

We normalize the difference between $\bar{Y}_{AB}$ and $\bar{Y}_{\neg AB}$ to obtain the following statistic S_{AB}, which is normally distributed with mean $\lambda_{AB}\sqrt{N}$ (the non-centrality parameter) and unit variance.

$$S_{AB} = \frac{\bar{Y}_{AB} - \bar{Y}_{\neg AB}}{\sqrt{\frac{\sigma^2}{N p_A p_B (1 - p_A p_B)}}} \sim N(\lambda_{AB}\sqrt{N}, 1), \text{ where } \lambda_{AB} = \frac{\beta_{AB}\sqrt{p_A p_B(1 - p_A p_B)}}{\sigma} \tag{7}$$

According to [15], we set the per-SNP-pair significance level $\alpha = 10^{-12}$. The per-SNP-pair statistic threshold is then $T_2 = -\Phi^{-1}(\alpha/2) = 7.13$. The per-causal-SNP-pair power of a putative causal SNP pair AB can be estimated as

$$P_{BF}(\text{AB}) = \Phi\left(-T_2 + \lambda_{AB}\sqrt{N}\right) + 1 - \Phi\left(T_2 + \lambda_{AB}\sqrt{N}\right) \tag{8}$$

The average power $\overline{P_{BF}}$ is obtained by averaging per-causal-SNP-pair powers over all putative causal SNP pairs.

Assuming the total number of SNPs is M, we define the cost of brute-force method to be the total number of SNP pairs needed for association analysis, that is, $C_{BF}(M) = \binom{M}{2}$.

3.3   Two Stage Model

In the brute force approach, the total number of SNP pairs to be considered is $\binom{M}{2}$ and we need to compute the statistic S_{AB} for all these pairs. Considering the number of SNPs involved in current GWAS, the computational burden makes this strategy infeasible.

We propose a two-stage model to reduce the number of tests needed while maintaining similar power with the brute force approach. In the first stage, we propose two statistic thresholds T_a and T_b and perform the single marker test on all SNPs. In the second stage, we pair all SNPs that are significant under threshold T_a with those significant SNPs under threshold T_b. Then we perform a pairwise association test between traits and all those pairs. The SNP pairs which pass the per-SNP-pair statistic threshold T_2 are considered to be statistically associated with the trait.

The analysis of single marker test in the first stage is quite similar to that of the one SNP association test in Section 3.1. We derive the similar equations with (1), (2) and (3) except that the effect size of SNP A becomes $p_B\beta_{AB}$, when the pair of SNP A and SNP B is the causal SNP pair. So the statistic S_A of SNP A becomes

$$S_A = \frac{\bar{Y}_A - \bar{Y}_{\neg A}}{\sqrt{\frac{\sigma^2}{N \cdot p_A \cdot (1-p_A)}}} \sim N(\lambda_A \sqrt{N}, 1), \ where \ \lambda_A = \frac{p_B \beta_{AB} \sqrt{p_A(1-p_A)}}{\sigma} \quad (9)$$

The analysis of SNP B is the same except that we switch p_A and p_B in the equations.

Assume a pair of SNPs A and B are putatively associated with a trait. The underlying effect size β_{AB} could either be positive or negative. Here we first analyze the case where the true effect size is positive. To find such positive pairwise association in our model, S_A must be no less than T_a, S_B must be no less than T_b (or vice versa, but here we only analyze one case since we will show in Section 3.5 that the other case is not necessary) and S_{AB} must be at least T_2. Hence, we need to consider three statistics and three thresholds to compute the analytical power of the two-stage model. Under the assumption that we are aware the effect size is positive, the per-causal-SNP-pair power of a putative causal SNP pair AB can be denoted as

$$P_2^+(AB) = P(S_A \geq T_a, S_B \geq T_b \ and \ S_{AB} \geq T_2) \quad (10)$$

However, considering the fact that whether the effect size is positive or negative is hidden from us, we also need to calculate the probability where S_{AB} is less than $-T_2$, that is,

$$P_2^-(AB) = P(S_A \leq -T_a, S_B \leq -T_b \ and \ S_{AB} \leq -T_2) \quad (11)$$

So, the per-causal-SNP-pair power of a putative causal SNP pair AB is

$$P_2(AB) = P_2^+(AB) + P_2^-(AB) \quad (12)$$

The analysis for the case where the true effect size is negative is exactly the same except that the non-centrality parameters for S_A, S_B and S_{AB} are negative.

To calculate the value of $P_2(AB)$, we need to take into account correlations between statistics. The two statistics S_A and S_{AB} are correlated because both involve SNP A. Similarly, we have a correlation between S_B and S_{AB}. We assume SNPs are independent, and hence there is no correlation between S_A and S_B. The average power $\overline{P_2}$ is obtained by averaging per-causal-SNP-pair powers over all putative causal SNP pairs. Computing the analytical power of the two-stage model is complicated as a result of the correlations between statistics. We estimate the power using a multivariate normal distribution (MVN) framework as in Section 3.4.

Denote the per-SNP significance level corresponding to the statistic thresholds T_a and T_b in the first stage to be α_A and α_B, respectively. Then we have $\alpha_A = 2\Phi(-T_a)$ and $\alpha_B = 2\Phi(-T_b)$. The cost of the two stage model can be computed as $C_{TS}(M, \alpha_A, \alpha_B) \approx M^2 \alpha_A \alpha_B / 2$.

Let's measure the cost saving by the ratio between cost of brute-force method (C_{BF}) and that of the two-stage model (C_{TS}):

$$\frac{C_{BF}(M)}{C_{TS}(M, \alpha_A, \alpha_B)} = \frac{\binom{M}{2}}{M^2 \alpha_A \alpha_B / 2} \approx \frac{1}{\alpha_A \alpha_B} \tag{13}$$

And we define the power loss to be

$$1 - \frac{\overline{P_2}}{\overline{P_{BF}}} \tag{14}$$

For a given dataset, there exists a trade-off between the power loss and cost saving. The trade off is controlled by the two thresholds T_a and T_b. We carefully design the thresholds to achieve high cost saving while maintaining low power loss. The details of the algorithm is summarized in Section 3.5.

3.4   Estimating the Two Stage Power Using the MVN

In this section, we provide an approach to compute the power of the two stage model in Equation 12. The distribution of association statistics S_A, S_B and S_{AB} has been derived in Section 3.2 and 3.3. We aim to compute the power in Equation 12 for any given thresholds T_a, T_b and T_2.

For many widely used statistical tests, the statistics over multiple markers asymptotically follow a Multivariate Normal Distribution(MVN) [11, 18]. To derive the analytical power of our two stage model, we use the framework of MVN proposed by [8]. This method creates a MVN using the non-centrality parameters (NCPs) of statistics as a mean vector in the MVN. The NCPs of S_A, S_B, and S_{AB} are already derived in Equations (7) and (9). So the mean vector is $(\lambda_A \sqrt{N}, \lambda_B \sqrt{N}, \lambda_{AB} \sqrt{N})$. The covariance matrix in the MVN will be the correlations among statistics. We assume SNPs are independent of each other, so the correlation between S_A and S_A is 1, and the correlation between S_A and S_B is 0. The covariance matrix is as follows:

$$\begin{pmatrix} 1 & 0 & Cor(S_A, S_{AB}) \\ 0 & 1 & Cor(S_B, S_{AB}) \\ Cor(S_A, S_{AB}) & Cor(S_B, S_{AB}) & 1 \end{pmatrix}$$

We only need to compute the correlation between S_A (or S_B) and S_{AB} to derive the complete MVN. To find a correlation between two statistics, S_A and S_{AB}, we use the following formula where $\text{Var}(X)$ denotes the variance of X and $\text{Cov}(X, Y)$ denotes the covariance between X and Y,

$$\text{Var}(X + Y) = \text{Var}(X) + \text{Var}(Y) + 2\text{Cov}(X, Y) \tag{15}$$

In our model, $X = S_A$ and $Y = S_{AB}$, and $\text{Var}(S_A) = \text{Var}(S_{AB}) = 1$. Then we can compute $\text{Cov}(S_A, S_{AB})$ as

$$\text{Cov}(S_A, S_{AB}) = (1/2)\text{Var}(S_A + S_{AB}) - 1 \tag{16}$$

Hence, we need to derive $\text{Var}(S_A + S_{AB})$ to find the covariance or the correlation between statistics. The covariance and the correlation are equivalent in this case because variances of statistics are 1.

Using Equations (7) and (9), we can write $S_A + S_{AB}$ as

$$S_A + S_{AB} = \sqrt{N/\sigma^2} \left(\theta_A \left(\bar{Y}_A - \bar{Y}_{\neg A} \right) + \theta_{AB} \left(\bar{Y}_{AB} - \bar{Y}_{\neg AB} \right) \right) \tag{17}$$

where $\theta_A = \sqrt{p_A(1 - p_A)}$ and $\theta_{AB} = \sqrt{p_A p_B(1 - p_A p_B)}$.

We then decompose $\bar{Y}_A$, $\bar{Y}_{\neg A}$ and $\bar{Y}_{AB}$ in Equation (17) in terms of alleles of SNPs A and B (x_i^A and x_i^B). Substituting Equations (1), (2), (5) and (6) into Equation (17) and rearranging common terms, we have

$$S_A + S_{AB} = \sqrt{\frac{N}{\sigma^2}} \left[P \sum_{11} y_i + Q \sum_{10} y_i - R \sum_{01} y_i - S \sum_{00} y_i \right] \tag{18}$$

where

$$P = \frac{\theta_A}{Np_A} + \frac{\theta_{AB}}{Np_A p_B}, \quad Q = \frac{\theta_A}{Np_A} - \frac{\theta_{AB}}{N(1 - p_A p_B)}$$

$$R = \frac{\theta_A}{N(1 - p_A)} + \frac{\theta_{AB}}{N(1 - p_A p_B)}, \quad S = \frac{\theta_A}{N(1 - p_A)} + \frac{\theta_{AB}}{N(1 - p_A p_B)}$$

Note that Equation (18) consists of independent terms: each $\sum_{ab} y_i$ term represents a sum of trait values of disjoint individuals, where $ab = 11, 10, 01$ and 00, respectively. Hence, if we take the variance of $S_A + S_{AB}$, covariances among all terms are 0, and $\text{Var}(S_A + S_{AB})$ is a sum of variances of $\sum_{ab} y_i$ terms. Also, note that $\text{Var}(y_i) = \sigma^2$, and hence $\text{Var}\left(\sum_{11} y_i\right)$ is a sum of σ^2 over individuals who have minor alleles at both SNPs A and B. We can then compute the variance of $S_A + S_{AB}$ as

$$\frac{N}{\sigma^2} \left[P^2 \text{Var}\left(\sum_{11} y_i\right) + Q^2 \text{Var}\left(\sum_{10} y_i\right) + R^2 \text{Var}\left(\sum_{01} y_i\right) + S^2 \text{Var}\left(\sum_{00} y_i\right) \right]$$

$$= N \left[P^2 Np_A p_B + Q^2 Np_A(1 - p_B) + R^2 N(1 - p_A)p_B + S^2 N(1 - p_A)(1 - p_B) \right] \tag{19}$$

We can also compute $\text{Var}(S_B + S_{AB})$ similarly using Equation (19) by exchanging p_A and p_B.

Up to now we obtained all parameters for the MVN framework. Then, we can compute the power as the area outside of the significance threshold under the

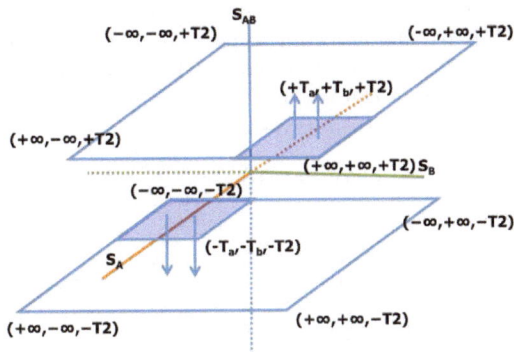

Fig. 2. The volume of the two cubes under the MVN is the power of our two stage model

MVN we created. Fig. 2 helps to illustrate the ideas. We can see that in the three dimension space of the MVN framework for statistics S_A, S_B and S_{AB}, the two cubes on the corners correspond to the significance region. Using the MVN, we can compute the power of our two stage model for any given thresholds T_a, T_b and T_{AB} by summing up the volume of these two cubes under the MVN. This method yields a very accurate estimate of power when there exist correlations among statistics, and hence it provides an appropriate framework to compute the analytical power of our model.

3.5 Efficient Pairwise Association Test Using TEPAA

In previous sections, we have illustrated how to calculate the power and cost savings of our two stage model for any given threshold. In this section, we provide a framework, TEPAA, to determine the first thresholds which generate a relatively small number of SNP pairs for pairwise association test in the second stage while losing a small amount of power compared to the brute force approach.

From Equation 12 and Section 3.4, we can see that the joint distribution between the association statistics of single SNPs and the association statistic of a pair of SNPs depends on the MAFs of the pair of SNPs. MAFs are observable values, so we can categorize all SNP pairs based on the combination of their MAFs. Since MAFs are continuous value, we can discretize the MAFs into bins to have a small number of combinations. After removing rare variants, we can categorize all SNPs into 9 bins, with step size 0.05. In order to detect the pairwise association for all SNP pairs, we break all combinations of SNP pairs into two cases. First we pair SNPs within different bins and this results in $\binom{9}{2}$ categories. The second case is to combine SNPs within one bin. So totally we have $\binom{9}{2}+9$ categories of SNP pairs.

Assuming the power of brute force approach is 50%, we can calculate the effect size β_{AB} from Equation 8. Then for each category of SNP pairs, we can compute the power loss and cost savings from Equations 13 and 14 with the

Table 1. The threshold for SNP A/SNP B and cost savings in various combination of MAFs to achieve power loss of 1%. Here we assume the MAF of SNP A is smaller than that of SNP B in each pair. The first and second number in each cell is the threshold for SNP A (α_A) and SNP B (α_B), respectively. These two thresholds are scaled by 10^{-2}. The third number in each cell is the cost saving, which is the ratio between cost of brute-force method and that of the two-stage model.

MAF of SNP A \ MAF of SNP B	0.1	0.15	0.2	0.25	0.3	0.35	0.4	0.45	0.5
0.1	34/34 /8	8/50 /25	7/58 /25	5/62 /32	2/76 /66	0.82/84 /145	0.26/79 /487	0.10/84 /1190	0.02/90 /5555
0.15	-	14/14 /51	3/24 /139	3/31 /107	2/46 /108	1/58 /172	0.35/54 /529	0.13/62 /1241	0.03/69 /4830
0.2	-	-	5/5 /400	2/9 /556	2/16 /312	1/21 /476	0.47/31 /686	0.19/58 /907	0.05/69 /2899
0.25	-	-	-	3/3 /1100	2/5 /1000	1/7 /1429	1/16 /625	0.26/21 /1831	0.10/42 /2380
0.3	-	-	-	-	1/1 /1e5	1/3 /3333	1/4 /2500	0.62/12 /1344	0.13/16 /4807
0.35	-	-	-	-	-	0.6/0.6 /2.7e4	0.5/1 /2e4	0.1/2 /5e4	0.03/8 /4e4
0.4	-	-	-	-	-	-	0.3/0.3 /1.1e5	0.1/0.6 /1.6e5	0.1/1 /1e5
0.45	-	-	-	-	-	-	-	0.2/0.2 /2.5e5	0.1/0.5 /2e5
0.5	-	-	-	-	-	-	-	-	0.1/0.1 /1e6

MVN, given two first stage significance levels α_A and α_B. We do an exhaustive search over the space $[0, 1)$ with a small step size to find the optimal values of α_A and α_B to achieve best cost saving while maintaining power loss 1%. The values of α_A and α_B are shown in Table 1 when there are 5, 326 samples in the dataset.

For SNPs in each bin, we carry out the single marker test and sort the association statistics of single SNP. Then for each category of SNP pairs, we do a binary search in each involved bin to find all significant SNPs under the precomputed significance level. The selected SNPs are then paired for the second stage pairwise association test. Based on the pre-computed values of α_A and α_B, we can estimate the cost savings for each category of SNP pairs as in Table 1. We propose a threshold for each bin for each category of SNP pairs, and the bins are disjoint. So, in the calculation of Equation 10, we only need to consider the case where $S_A > T_a$ and $S_B > T_b$ and it is not necessary to consider the case $S_A > T_b$ and $S_B > T_a$. We have the same conclusion in the calculation of Equation 11.

Although the calculation is based on the assumption that the brute force approach has power 50%, our approach is robust to the effect size. We did simulations for different effect sizes, which generate different power for the brute force approach. The cost saving of TEPAA is stable when achieving 1% power loss under various effect size.

4   Conclusion

In this paper, we proposed a two-stage model to detect SNP pairs associated with trait. The key idea behind our method is that we model the joint distribution between association statistics at single SNPs and association statistics at pairs of SNPs to allow us to apply a two-stage model that provides guarantees that we detect associations of pairs of SNPs with small number of tests while losing very little power. We rapidly eliminate from consideration pairs of SNPs which with high probability are not associated with the trait. Using extensive simulations, we show that our approach can reduce the computational time by a factor of 60 while only loosing approximately 1% of the power compared to the brute-force approach.

Acknowledgement. Z.W., J.H.S. and E.E. are supported by National Science Foundation grants 0513612, 0731455, 0729049, 0916676, 1065276, 1302448 and 1320589, and National Institutes of Health grants K25-HL080079, U01-DA024417, P01-HL30568, P01-HL28481, R01-GM083198, R01-MH101782 and R01-ES022282. We acknowledge the support of the NINDS Informatics Center for Neurogenetics and Neurogenomics (P30 NS062691). S. S. is supported by USA-Israel Binational Science Foundation and Israel Science Foundation. J. A. L. has been partially supported by the Saiotek and IT-609-13 programs (Basque Government), TIN2010-14931 (Spanish Ministry of Science and Innovation), COMBIOMED network in computational biomedicine (Carlos III Health Institute).

References

1. Altshuler, D., Hirschhorn, J.N., Klannemark, M., Lindgren, C.M., Vohl, M.C., Nemesh, J., Lane, C.R., Schaffner, S.F., Bolk, S., Brewer, C., et al.: The common pparγ pro12ala polymorphism is associated with decreased risk of type 2 diabetes. Nature Genetics 26(1), 76–80 (2000)
2. Bertina, R.M., Koeleman, B.P.C., Koster, T., Rosendaal, F.R., Dirven, R.J., de Ronde, H., Van Der Velden, P.A., Reitsma, P.H., et al.: Mutation in blood coagulation factor v associated with resistance to activated protein c. Nature 369(6475), 64–67 (1994)
3. Brem, R.B., Storey, J.D., Whittle, J., Kruglyak, L.: Genetic interactions between polymorphisms that affect gene expression in yeast. Nature 436(7051), 701–703 (2005)
4. Brinza, D., Schultz, M., Tesler, G., Bafna, V.: Rapid detection of gene-gene interactions in genome-wide association studies. Bioinformatics (2010)
5. Wellcome Trust Case Control Consortium. Genome-wide association study of 14,000 cases of seven common diseases and 3,000 shared controls. Nature 447(7145), 661–678 (2007)
6. Corder, E.H., Saunders, A.M., Strittmatter, W.J., Schmechel, D.E., Gaskell, P.C., Small, G.W., Roses, A.D., Haines, J.L., Pericak-Vance, M.A.: Gene dose of apolipoprotein e type 4 allele and the risk of alzheimer's disease in late onset families. Science 261(5123), 921–923 (1993)

7. Evans, D.M., Marchini, J., Morris, A.P., Cardon, L.R.: Two-stage two-locus models in genome-wide association. PLoS Genet. 2(9), e157 (2006)
8. Han, B., Kang, H.M., Eskin, E.: Rapid and accurate multiple testing correction and power estimation for millions of correlated markers. PLoS Genet. 5, e1000456 (2009)
9. Kostem, E., Eskin, E.: Efficiently identifying significant associations in genome-wide association studies. J. Comput. Biol. 9 (2013)
10. Kostem, E., Lozano, J.A., Eskin, E.: Increasing power of genome-wide association studies by collecting additional snps. Genetics 188(2), 449–460 (2011)
11. Lin, D.Y.: An efficient monte carlo approach to assessing statistical significance in genomic studies. Bioinformatics 21(6), 781–787 (2005)
12. Listgarten, J., Lippert, C., Kang, E.Y., Xiang, J., Kadie, C.M., Heckerman, D.: A powerful and efficient set test for genetic markers that handles confounders. Bioinformatics 4 (2013)
13. Ljungberg, K., Holmgren, S., Carlborg, O.: Simultaneous search for multiple qtl using the global optimization algorithm direct. Bioinformatics 20(12), 1887–1895 (2004)
14. Millstein, J., Conti, D.V., Gilliland, F.D., Gauderman, W.J.: A testing framework for identifying susceptibility genes in the presence of epistasis. The American Journal of Human Genetics 78(1), 15–27 (2006)
15. Prabhu, S., Pe'er, I.: Ultrafast genome-wide scan for snp-snp interactions in common complex disease. Genome Research 22(11), 2230–2240 (2012)
16. Saxena, R., Voight, B.F., Lyssenko, V., Burtt, N.P., de Bakker, P.I.W., Chen, H., Roix, J.J., Kathiresan, S., Hirschhorn, J.N., Daly, M.J., Hughes, T.E., Groop, L., Altshuler, D., Almgren, P., Florez, J.C., Meyer, J., Ardlie, K., Boström, K.B., Isomaa, B., Lettre, G., Lindblad, U., Lyon, H.N., Melander, O., Newton-Cheh, C., Nilsson, P., Orho-Melander, M., Rastam, L., Speliotes, E.K., Taskinen, M.-R.R., Tuomi, T., Guiducci, C., Berglund, A., Carlson, J., Gianniny, L., Hackett, R., Hall, L., Holmkvist, J., Laurila, E., Sjögren, M., Sterner, M., Surti, A., Svensson, M., Svensson, M., Tewhey, R., Blumenstiel, B., Parkin, M., Defelice, M., Barry, R., Brodeur, W., Camarata, J., Chia, N., Fava, M., Gibbons, J., Handsaker, B., Healy, C., Nguyen, K., Gates, C., Sougnez, C., Gage, D., Nizzari, M., Gabriel, S.B., Chirn, G.-W.W., Ma, Q., Parikh, H., Richardson, D., Ricke, D., Purcell, S.: Genome-wide association analysis identifies loci for type 2 diabetes and triglyceride levels. Science 316(5829), 1331–1336 (2007)
17. Schpbach, T., Xenarios, I., Bergmann, S., Kapur, K.: Fastepistasis: a high performance computing solution for quantitative trait epistasis. Bioinformatics 26(11), 1468–1469 (2010)
18. Seaman, S.R., Muller-Myhsok, B.: Rapid simulation of p values for product methods and multiple-testing adjustment in association studies. American Journal of Human Genetics 76(3), 399–408 (2005)
19. Williams, S.M., Addy, J.H., Phillips, J.A., Dai, M., Kpodonu, J., Afful, J., Jackson, H., Joseph, K., Eason, F., Murray, M.M., Epperson, P., Aduonum, A., Wong, L.J., Jose, P.A., Felder, R.A.: Combinations of variations in multiple genes are associated with hypertension. Hypertension 36(1), 2–6 (2000)
20. Wu, M.C., Kraft, P., Epstein, M.P., Taylor, D.M., Chanock, S.J., Hunter, D.J., Lin, X.: Powerful snp-set analysis for case-control genome-wide association studies. Am. J. Hum. Genet. 86(6), 929–942 (2010)
21. Wu, M.C., Lee, S., Cai, T., Li, Y., Boehnke, M., Lin, X.: Rare-variant association testing for sequencing data with the sequence kernel association test. Am. J. Hum. Genet. 89(1), 82–93 (2011)

22. Xiang, W., Can, Y., Qiang, Y., Hong, X., Xiaodan, F., Nelson, T., Weichuan, Y.: Boost: A fast approach to detecting gene-gene interactions in genome-wide case-control studies. The American Journal of Human Genetics 87, 325–340 (2010)
23. Yanchina, E.D., Ivchik, T.V., Shvarts, E.I., Kokosov, A.N., Khodzhayantz, N.E.: Gene-gene interactions between glutathione-s transferase m1 and matrix metallo-proteinase 9 in the formation of hereditary predisposition to chronic obstructive pulmonary disease. Bulletin of Experimental Biology and Medicine 137(1), 64–66 (2004)
24. Yang, C., He, Z., Wan, X., Yang, Q., Xue, H., Yu, W.: Snpharvester: a filtering-based approach for detecting epistatic interactions in genome-wide association studies. Bioinformatics 25(4), 504–511 (2009)
25. Zhang, X., Huang, S., Zou, F., Wang, W.: Team: efficient two-locus epistasis tests in human genome-wide association study. Bioinformatics 227, i217–i227 (2010)
26. Zhang, X., Pan, F., Xie, Y., Zou, F., Wang, W.: COE: A General Approach for Efficient Genome-Wide Two-Locus Epistasis Test in Disease Association Study. In: Batzoglou, S. (ed.) RECOMB 2009. LNCS, vol. 5541, pp. 253–269. Springer, Heidelberg (2009)

A Geometric Clustering Algorithm and Its Applications to Structural Data

Shutan Xu, Shuxue Zou, and Lincong Wang*

The College of Computer Science and Technology, Jilin University,
Changchun, Jilin, China
{xushutan,sandrorzou,wlincong}@gmail.com

Abstract. An important feature of structural data especially those from structural determination and protein-ligand docking programs is that their distribution could be both uniform and non-uniform. Traditional clustering algorithms developed specifically for non-uniformly distributed data may not be adequate for their classification. Here we present a geometric partitional algorithm that could be applied to both uniformly and non-uniformly distributed data. The algorithm is a top-down approach that recursively selects the outliers as the seeds to form new clusters until all the structures within a cluster satisfy certain requirements. The applications of the algorithm to a diverse set of data from NMR structure determination, protein-ligand docking and simulation show that it is superior to the previous clustering algorithms for the identification of the correct but minor clusters. The algorithm should be useful for the identification of correct docking poses and for speeding up an iterative process widely used in NMR structure determination.

Keywords: clustering algorithm, structure determination, protein-ligand docking, structure classification.

1 Introduction

Recently, we have witnessed a rapid growth of not only DNA sequencing data but also three-dimensional (3D)[1] structural data such as those from biomolecular nuclear magnetic resonance (NMR) spectroscopy and protein-ligand docking as well as molecular dynamics (MD) simulation and protein structure prediction. These techniques output not a single but an ensemble of structures. A variety of traditional clustering algorithms both of hierarchical and partitional [1, 2], being able to first assign the data points to groups (clusters) and then identify a representative for each cluster, have been applied to their analysis and visualization in order to discover common structural features such as protein fold, active

---

* Corresponding author.
[1] Abbreviations used: 3D, three dimensional; NMR, nuclear magnetic resonance; PDB, protein data bank; RMSD, root-mean-square deviation; NOE, nuclear Overhauser effect; VDW, van der Waals; MD, molecular dynamics; GA, genetic algorithm; MC, Monte-Carlo; SA, simulated annealing; SAA, solvent accessible surface area.

R. Sharan (Ed.): RECOMB 2014, LNBI 8394, pp. 356–370, 2014.
© Springer International Publishing Switzerland 2014

site and correct pose [3–10]. However, it remains unclear which algorithm is the most suitable for the clustering of 3D structural data because of the inherent difficulty associated with high dimensionality [2]. For example, a past study concluded that there was no perfect "one size fits all" algorithm for the clustering of MD trajectories [4], and May [3] had questioned whether a hierarchical approach is appropriate for the clustering of structural data by forcing them into a dendrogram. An important feature of structural data especially those from NMR structural determination and protein-ligand docking is that their distribution could be rather uniform in a few large regions but non-uniform in others, and thus may not be properly described by a Gaussian mixture model. Traditional clustering algorithms developed specifically for non-uniformly distributed data may not be adequate for their classification. In this paper, we present a novel geometric partitional algorithm that could be applied to both uniformly and non-uniformly distributed data. The algorithm is a top-down approach that recursively partitions all the data points of a previously-generated cluster into c new clusters where c is a user-specified number. It stops and then outputs a final set of clusters that satisfy the classification requirement that no metric distances between any pair of data points in any cluster are larger than a user-specified value. Compared with the previous clustering algorithms, the salient features of our geometric partitional algorithm are (a) it uses the global information in the beginning, (b) is deterministic, and (c) could handle both uniformly and non-uniformly distributed data. We have applied the algorithm to the classification of a diverse set of data: the intermediate structures from an NMR structure determination project, poses from protein-ligand docking and MD trajectories from an ab-initio protein folding simulation. A comparison with other partitional algorithms such as k-medoids shows that our algorithm could classify the data with higher accuracy. Though the final set of clusters from our algorithm may be similar to those from complete-link or average-link hierarchical algorithms, the clusters from our algorithm are more uniform in terms of their structural and physical properties. More importantly, our algorithm outperforms the previous algorithms in identifying the minor clusters with "good" properties (the correct clusters) that are often to be discarded by other criteria used for the selection of representative structures. The rest of the paper is organized as follows. In section 2 we first present the algorithm and then describe the data sets. In section 3 we present and compare the results of applying both our algorithm and three other clustering algorithms to identify the clusters with good scores, and discuss the significance of the geometric algorithm for speeding up the iterative NMR structure determination process and for the selection of accurate docking poses.

2   The Algorithm and Data Set

In this section, we first present our novel geometric partitional algorithm for the clustering of structural data. Then we describe the data sets used for assessing the performance of the new and the previous three clustering algorithms.

---

[2] A structure with N_a atoms has dimension $d = 3N_a$.

2.1   The Geometric Partitional Algorithm

The similarity metric. Our algorithm employs a recursive top-down procedure that clusters a set of structures (data points) $\mathbf{S}$ using a pairwise root-mean square distance (RMSD) d_{ij} between two structures i, j as a similarity metric though other metrics could also be used. All the pairwise d_{ij}s in $\mathbf{S}$ are pre-computed and stored in set $\mathbf{D}$.

The algorithm. The algorithm itself proceeds as follows. Let $\mathbb{C}_s$ denote the set of clusters at recursive step s that are themselves generated at step $s - 1$. At the initial step $s = 1$, $\mathbb{C}_1$ has only a single cluster $\mathbf{S}$ to which all the data belong. At step s, for each cluster $\mathbf{C} \in \mathbb{C}_s$, the algorithm first computes m points, $\mathbf{c}_\mu \in \mathbf{C}, \mu = 1, \ldots, m$, as the seeds for m new clusters, $\mathbf{C}_\mu \in \mathbb{C}_{s+1}, \mu = 1, \ldots, m$, and then assigns all the remaining points in $\mathbf{C}$ to $\mathbf{C}_\mu \in \mathbb{C}_{s+1}, \mu = 1, \ldots, m$ where $3 \leq m \leq N_c$ while N_c is a user-specified number. The above m seed points are defined and computed as follows. The first two points, $\mathbf{c}_1$ and $\mathbf{c}_2$, whose RMSD is the largest among all the pairwise d_{ij} in cluster $\mathbf{C} \in \mathbb{C}_s$, seed the first two clusters, $\mathbf{C}_1 \in \mathbb{C}_{s+1}$ and $\mathbf{C}_2 \in \mathbb{C}_{s+1}$. A point $\mathbf{c}_3$ in $\mathbf{C} - \{\mathbf{c}_1, \mathbf{c}_2\}$ that may seed a new cluster, the third cluster $\mathbf{C}_3 \in \mathbb{C}_{s+1}$, is the point that together with the above two points $\mathbf{c}_1, \mathbf{c}_2$ form a triangle with the largest area among all the triangles in $\mathbf{C} - \{\mathbf{c}_1, \mathbf{c}_2\}$. Similarly, a point $\mathbf{c}_4$ in $\mathbf{C} - \{\mathbf{c}_1, \mathbf{c}_2, \mathbf{c}_3\}$ that may seed a new cluster, the fourth cluster $\mathbf{C}_4 \in \mathbb{C}_{s+1}$, is the point that together with $\mathbf{c}_1, \mathbf{c}_2, \mathbf{c}_3$ form a tetrahedron with the largest volume among the tetrahedrons formed by all the quadruples in $\mathbf{C} - \{\mathbf{c}_1, \mathbf{c}_2, \mathbf{c}_3\}$. Finally, a point $\mathbf{c}_m$ in $\mathbf{C} - \{\mathbf{c}_1, \mathbf{c}_2, \ldots, \mathbf{c}_{m-1}\}$ may seed the last cluster $\mathbf{C}_m \in \mathbb{C}_{s+1}$ that together with $\{\mathbf{c}_1, \ldots, \mathbf{c}_{m-1}\}$ form a polyhedron that has the largest Cayley-Menger determinant [11] among the polyhedra formed by all the n-tuples. For each point $p \in \mathbf{C} - \{\mathbf{c}_1, \mathbf{c}_2, \ldots, \mathbf{c}_m\}$, the algorithm assigns it to the kth cluster $\mathbf{C}_k \in \mathbb{C}_{s+1}$ where k is determined by

$$\arg\min_k d_{p\mathbf{C}_k}, \quad k = 1, \ldots, m \tag{1}$$

where $d_{p\mathbf{C}_k}$ is the RMSD between p and the seed $\mathbf{c}_k$.

In the following we present the key steps of the algorithm at recursive step s with as an input one of clusters $\mathbf{C} \in \mathbb{C}_s$ generated at step $s - 1$ and assume that $N_c = 4$.

1. SEARCH for the first two seed points, $\mathbf{c}_1$ and $\mathbf{c}_2$, whose metric $d_{12} \in \mathbf{D}$, is the largest among all the pairs of structures in $\mathbf{C}$
2. IF $d_{12} \leq d_{max}$
    STOP {*no new clusters*}
3. INITIALIZE two new clusters $\mathbf{C}_1$ and $\mathbf{C}_2$ with $\mathbf{c}_1$ and $\mathbf{c}_2$ as their respective seeds
4. SEARCH for the third seed point $\mathbf{c}_3$ in $\mathbf{C} - \{\mathbf{c}_1, \mathbf{c}_2\}$ that together with $\mathbf{c}_1, \mathbf{c}_2$ forms a triangle with the largest area among all the possible triangles
5. IF any of d_{13} and d_{23} is smaller than d_{max}
    (a) FOR each point p in $\mathbf{C} - \{\mathbf{c}_1, \mathbf{c}_2\}$
        ASSIGN it to $\mathbf{C}_1$ if $d_{p\mathbf{c}_1} \leq d_{p\mathbf{c}_2}$, otherwise to $\mathbf{C}_2$
    (b) FOR both clusters $\mathbf{C}_1$ and $\mathbf{C}_2$
        Go to Step 1

6. SEED a third cluster $\mathbf{C}_3$ with $\mathbf{c}_3$
7. SEARCH for the fourth seed point $\mathbf{c}_4$ in $\mathbf{C} - \{\mathbf{c}_1, \mathbf{c}_2, \mathbf{c}_3\}$ that together with $\mathbf{c}_1, \mathbf{c}_2, \mathbf{c}_3$ forms a tetrahedron with the largest volume among all the possible tetrahedrons
8. IF any of d_{14}, d_{24}, d_{34} is smaller than d_{max}
   (a) ASSIGN each point p in $\mathbf{C} - \{\mathbf{c}_1, \mathbf{c}_2, \mathbf{c}_3\}$ to either $\mathbf{C}_1, \mathbf{C}_2, \mathbf{C}_3$ according to equation (1)
   (b) FOR each cluster $\mathbf{C}_j, j = 1, 2, 3$.
       Go to Step 1
9. ELSE
   (a) SEED a cluster $\mathbf{C}_4$ with $\mathbf{c}_4$
   (b) ASSIGN each point p in $\mathbf{C} - \{\mathbf{c}_1, \mathbf{c}_2, \mathbf{c}_3, \mathbf{c}_4\}$ to one of $\mathbf{C}_j, j = 1, 2, 3, 4$
   (c) FOR each cluster $\mathbf{C}_j, j = 1, 2, 3, 4$
       Go to Step 1

where d_{max} is a user-defined maximum RMSD such that all the structures in the same cluster must have their pairwise RMSDs less than d_{max}. This condition will be called *the cluster restraint requirement*. In step 2 if the largest pairwise RMSD among all the points in a cluster is less than d_{max}, no more partition are required, thus stops the recursive procedure.

The mathematical background. Our algorithm is based on the following two propositions. Let $d_{\mathbf{c}_i \mathbf{c}_m}, i = 1, ..., m - 1$, denote the $m - 1$ RMSDs between the last seed $\mathbf{c}_m$ and the previous $m - 1$ seeds $\mathbf{c}_i, i = 1, ..., m - 1$.

Proposition 1. *If all the $d_{\mathbf{c}_i \mathbf{c}_m}$s are larger than d_{max}, there must exist at least an mth cluster seeded with the point $\mathbf{c}_m$ such that the polyhedron formed by points $\mathbf{c}_i, i = 1, ..., m$ has the largest Cayley-Menger determinant.*

Proposition 2. *If at least one of the $m-1$ RMSDs $d_{\mathbf{c}_i \mathbf{c}_m}$ is less than d_{max}, then there exists no new clusters at the current recursive step but further recursive partition is required for the previous cluster.*

Please see the Supplementary Materials for their proofs.

Running time. Let the number of structures be n. It takes $O(n^2)$ to populate the set $\mathbf{D}$ of all pairwise RMSDs and $|\mathbf{D}| = O(n^2)$ time to find the minimum value in $\mathbf{D}$. The worst-case time complexity occurs when d_{max} is so small that each structure forms a separate cluster. In this case, the depth of recursive search is $O(\log_{N_c} n^2)$ while at each recursive step, it takes $O(n^2)$ to find all the seeds assuming that it takes a constant time to compute the area of a triangle, the volume of a tetrahedron and a Cayley-Menger determinant. Thus the worst-case time complexity is $O(n^2) \log_{N_c} n$ where $N_c \geq 4$.

2.2 Structural Data Set

The structural data to which we have applied our algorithm as well as hierarchical (complete-link and average-link) and k-medoids algorithms includes (a) two sets of intermediate structures from an NMR structure determination project,

(b) 22 sets of poses from protein-ligand docking, and (c) a set of trajectories from an ab-initio MD simulation. Due to the space limitation, we will not present the application to the MD data here. In the following we describe both the data and the computational processes through which they are generated.

NMR Data Set. The two sets of intermediate structures chosen for the comparison of clustering algorithms are from the structure determination project of the protein SiR5 with 101 residues. Its NMR structure was determined by one of the authors using an iterative procedure of automated/manual nuclear Overhasuer effect (NOE) restraint assignment followed by structure computation using CYANA/Xplor with conformational sampling achieved by simulated annealing (SA). A large number of intermediates need to be generated during the iterative process in order to properly sample the huge conformational space defined as the set of all the structures that satisfy the experimentally-derived restraints to the same extent. In contrast to the final set of 20 structures deposited in the PDB (2OA4), the intermediates especially those from an early stage of the iterative process are less uniform in terms of structural similarity, molecular mechanics energy and restraint satisfaction. The pairwise RMSDs are computed only for C_α atoms of residues 20-70 since almost no long-range NOEs were observed for the rest. The d_{max} for both geometric and complete-link hierarchical clustering algorithms are either 1.0Å or 1.5Å. Each cluster is assessed by its average van der Waals (VDW) energy, NOE restraint violation defined as the number of NOE restraints with violation ≥ 0.5 Å, and the average pairwise RMSD d_a between all the pairs of structures within a cluster, and the average RMSD d_f between the structures in the cluster and the centroid of the 20 structures in 2OA4.

The Set of Poses from Protein-Ligand Docking. Structural clustering plays an increasingly important role in both protein-ligand docking and virtual screening [10] since a large amount of poses or library hits are typically generated during a docking or virtual screening process. To demonstrate the importance of clustering to protein-ligand docking, we have performed rescoring experiments on 22 sets of poses[3] generated using GOLD software suit (version 1.2.1) [14]. Several rounds of docking are performed using a binding site specified by a manually picked center with a 20.0Å radius. GOLD requires a user to pick a point that together with a user-specified radius define a sphere inside which poses are searched for using a genetic algorithm (GA). We use the default parameters as provided by GOLD except the requirement that no two poses having their pairwise RMSDs <1.5Å are generated. All the ligand heavy atoms are included in the pairwise RMSD computation. The 3D starting conformation for each ligand is generated by Corina [15]. A set of 500 poses are saved for each complex.

A well-known difficulty with the current scoring functions for protein-ligand docking is that they often fail to rank in the top positions the poses that are most

---

[3] The PDBIDs of the corresponding 22 protein-ligand complexes are 1AAQ, 1A9U, 1ACJ, 1BAF, 1CBS, 1CTR, 1DI8, 1EAP, 1FKG, 1TNI, 1V48, 1GPK, 1Q41, 1Q1G, 1P62, AOYT, ˙1NAV, 1N2J, 1GMY, 1VSN, 1MS6, 2XU5.

similar to the experimentally-determined one. To investigate whether clustering could provide the guarantee that the top-ranked clusters have high probability to be composed of the poses that are most similar to the experimental one, we first perform a series of clustering experiments with decreasing d_{max} values using our geometric partitional algorithm. We then rank the most populated clusters whose combined number of poses either exceed 90% of the total number of poses for larger d_{max} or 75-50% for smaller d_{max}. The ranking is based on their cluster-wide average values of both the GOLD scoring function S_g that consists of three items: ligand internal energy G_i, intermolecular VDW energy G_w and intermolecular hydrogen bond energy G_{hb} and our newly-developed scoring function S_t that also has three items: G_i, E_e the electrostatic energy computed using the partial charge from Corina and the electrostatic potential from APBS [16], and S_{aa} the change in solvent accessible surface area (SAA) of the ligand before and after the binding.

$$S_g = G_i + g_e G_w + g_s G_{hb} \tag{2}$$
$$S_t = G_i + k_e E_e + k_s S_{aa} \tag{3}$$

where g_e, g_s, k_e, k_s are weighting factors. The details of our scoring function, its rational and practical performance will be described elsewhere.

3   Results and Discussion

To evaluate the performance of our algorithm, to compare it with the previous algorithms for structural data classification and to demonstrate the importance of clustering to structural analysis, we have applied them to a diverse set of data including two sets of intermediate structures from an NMR structure determination project and 22 sets of poses from protein-ligand docking. In the following, we first present the results in detail and then discuss their significance for the selection of correct representative structures in the iterative NMR structure determination process and the identification of the correct poses from protein-ligand docking.

3.1   NMR Structural Ensemble

In theory the computation of structures using sparse and inexact geometric restraints derived from NMR experiments is an $\mathcal{N}P$-hard problem [17] because of restraint sparseness and measurement errors. At present, mainly heuristics such as SA and Monte-Carlo (MC) have been employed to search for a small subset of the set of all the structures that satisfy the restraints to the same extent, the conformational space. In practice, due to possible assignment errors and the difficulty of obtaining unambiguous assignment for many restraints especially in the beginning, NMR structure determination is an iterative process in which either a structural biologist or an automated program initializes the computation with a small number of restraints that have unique assignments, then uses the computed structures to assign additional possibly ambiguous restraints that are to become the input for the next cycle of computation. The process stops when the computed structures converge according to certain criteria. During the iterative process, a large number of intermediate structures are generated in order to properly sample the conformational space. However, all but a small subset of intermediates must be discarded in the next cycle due to time and space limitation. There exists no well-established criteria for such a selection though it is typically achieved using a user-specified threshold for a scoring function used in the structure determination. Such a selection assumes that there exists only a single or a few large clusters of structures that satisfy the restraints, a condition that may be difficult to meet especially in the early stages when only a small number of restraints per residue are available. A different selection of representative structures in the iterative process may lead to different ensembles of structures in the PDB as demonstrated by an investigation into two ensembles of NMR-derived structures of the protein Sox-5 HMG-box reported by two groups [7]. In this paper, we have applied four algorithms to two sets of intermediates to assess how the distribution of intermediates could affect the selection of representative structures and which algorithms are most suitable for such a task. The first set has 301 intermediates from an early stage of the SiR5 project while the second has 159 intermediates from a late stage. The clusters are analyzed in terms of the number of structures N_s per cluster, d_a, d_f, VDW energy and NOE violation. In the following we only present the clusters obtained with d_{max}=1.5 Å. Similar but larger numbers of clusters are generated with d_{max}=1.0 Å.

Geometric clustering. The first set of 301 structures are classified into 18 clusters with d_{max}=1.5Å, of which half are singletons. The five most populated clusters have 283 structures in total accounting for 94% of all the structures (Table 1). Their d_a and d_f values vary widely and they also have large VDW energy and NOE violation. The largest cluster has 253 structures and these intermediates differ largely from the final 20 structures with d_f=2.57Å. Among the top five clusters, the third cluster with only 9 structures have the smallest d_f. By comparison, the 159 structures in the second set (Table S1, Supplementary Materials) distribute more uniformly in terms of both d_a and d_f, and their corresponding VDW and NOE values are smaller and have narrower ranges. They are classified into 35 clusters with d_{max}=1.5Å, of which about half (17) are singletons. The seven most populated clusters have 122 structures in total accounting for 75% of all the structures. The largest cluster has only 25 structures with d_a=1.02Å and a range from 0.45Å to 1.49Å, and d_f=1.17 and a range from 1.02Å to 1.56Å. The largest cluster has the second smallest d_f and differs from the smallest d_f by only 0.1Å. In contrast, the largest cluster in the first set has the second largest d_f among the top five clusters. For the second set, the more populated clusters tend to have smaller d_a and d_f with narrower ranges, smaller VDW energy and less NOE violation. This is in contrast with the clusters from the first set whose corresponding values are not only larger but also have much bigger variation.

Complete-link. The first set is classified into 15 clusters with $d_{max}=1.5\text{Å}$, of which six are singletons. The first three largest clusters have 278 structures in total accounting for 92% of all the structures (Table 1). They have d_a, d_f, VDW energy and NOE violation similar to those from the geometric clustering. In particular, the largest cluster is identical to that from geometric clustering. For the second set, complete-link generates 34 clusters with 18 of them are singletons. The first six most populated clusters have 118 structures accounting for 74% of the total structures. These six clusters also have d_a, d_f, VDW and NOE values similar to those from the geometric clustering for the second set. However, its largest cluster has 66 structures that is more than the combined number of structures in the top three clusters from the geometric algorithm.

Average-link. The clusters from the first set are almost identical to those from complete-link except that d_a has larger range as expected (Table 1). For the second set, it outputs 26 clusters with 15 of them being singletons and the first two most populated clusters have 126 structures in total. Of the two clusters d_a, d_f, VDW and NOE values are similar to those from both geometric and complete-link clustering. However, the largest cluster from the second set has 120 structures: that is close to the combined number of structures of all the non-singletons from either geometric or complete-link algorithm.

k-medoids. It classifies the first set into six clusters with one singleton cluster and the largest cluster is almost identical to that from the other algorithms. However, d_a has must wider range, e.g., from 0.12–3.09Å. For the second set, the k-medoids only produces a single non-singleton with 126 structures. It basically merges all the non-singletons from any of the above three algorithms into a single cluster.

The importance of clustering to the correct selection of representative structures. The first set of 301 structures are from an early stage of the iterative process for protein SiR5. The largest clusters generated by the four algorithms are similar to each other, and include about 84% of the total structures (Table 1). However, each of them has rather large d_f though their d_a, VDW and NOE values are likely to be small. The selection of the largest but not representative cluster based solely on molecular mechanics energy and NOE violation had led astray of the iterative process that was only corrected late through manual intervention. Had we applied any of the four algorithms, the correct clusters (the third cluster from the geometric and the second from the complete-link, average-link and k-medoids algorithms) might have not been discarded in the early stage and the time-consuming manual intervention might have been avoided. Among the correct clusters from the four algorithms, the geometric algorithm produces the most accurate one. By the contrast, for the second set that is from a late stage (the refinement stage) of the iterative process, almost any of the most populated clusters from any of the four algorithms could be used to assign additional NOEs (Table S1, Supplementary Materials). Of the four algorithms the geometric algorithm tends to generate the largest number of evenly sized clusters while both the k-medoids and average-link output only one or two large clusters.

Table 1. A list of the clusters on the set of 301 structures by four clustering algorithms. The listed are the most populated clusters with the number of structures $N_S \geq 3$ from geometric and complete-link algorithms generated with a $d_{max}=1.5$ Å, and the non-singletons from average-link and k-medoids algorithms. The cluster shown with the boldfaced font has the smallest d_f among all the clusters. The three numbers are respectively the range and average. For k-medoids the number of initial clusters is 10 with the initial centers to be selected randomly. Please refer to the Supplementary Materials for the implementation of the complete-link, average-link and k-medoids algorithms.

Geometric:

Cluster	N_S	d_a	d_f	NOE viol	VDW energy
1	253	0.12–1.44, 0.52	2.17–2.80, 2.57	91–123, 109	307.3–970.6, 468.6
2	10	0.38–1.39, 0.96	1.76–2.44, 2.01	112–164, 133	644.6–854.7, 727.7
3	**9**	**0.44–1.39, 0.85**	**1.54–1.87, 1.73**	**112–134, 121**	**591.6–753.7, 690.4**
4	7	0.59–1.39, 1.10	3.24–3.45, 3.37	146–172, 158	670.9–937.8, 811.6
5	4	0.71–1.47, 1.17	1.93–2.66, 2.31	115–158, 137	743.1–822.0, 780.8

Complete-link:

1	253	0.12–1.44, 0.52	2.17–2.80, 2.57	91–123, 109	307.3–970.6, 468.6
2	**18**	**0.39–1.49, 0.91**	**1.64–2.44, 1.91**	**112–164, 128**	**591.6–854.7, 712.9**
3	6	0.59–1.39, 1.11	3.24–3.45, 3.36	153–172, 160	670.9–937.8, 812.6

Average-link:

1	253	0.12–1.44, 0.52	2.17–2.80, 2.57	91–123, 109	307.3–970.6, 468.6
2	**23**	**0.38–2.47, 1.12**	**1.54–2.66, 1.95**	**112–164, 129**	**591.6–854.7, 722.4**
3	9	0.59–1.81, 1.21	3.18–3.50, 3.37	146–172, 159	670.9–937.8, 792.5

k-medoids:

1	255	0.12–3.09, 0.64	2.17–3.45, 2.60	91–172, 111	307.3–937.8, 469.0
2	**19**	**0.38–1.97, 0.99**	**1.54–2.27, 1.85**	**112–164, 125**	**591.6–854.7, 709.9**
3	10	1.07–3.29, 2.29	3.50–4.44, 3.98	116–286, 257	225.9–797.8, 320.3
4	10	0.30–2.12, 1.18	2.27–2.79, 2.48	101–158, 116	743.1–970.6, 844.5
5	6	1.25– 2.53, 1.74	2.96–3.73, 3.41	207–267, 242	268.4–351.2, 298.1

In conclusion, the geometric partitional algorithm is most suitable for the selection of minor but correct representatives from the ensemble of intermediates.

3.2   Protein-Ligand Docking

A well-known difficulty with the current scoring functions for protein-ligand docking is that they often fail to rank the docked poses correctly [18] (Figs 1, 2). Because both the correct and incorrect poses are similarly ranked, it greatly reduces the value of the computational results to the practitioners such as medicinal chemists for either lead identification or optimization. One reason for improper ranking is that the scoring functions themselves have errors. From an algorithmic viewpoint, the failure originates also from the formulation of the docking problem as a global optimization problem that seeks to find the minimum in a scoring function with many variables. The complexity of the scoring functions forces the current docking programs to rely on heuristics such as GA or MC to search for the minimum. However, such a formulation is not consistent with the statistical mechanics conclusion that an experimentally-measured pose corresponds to the ensemble average, not necessary the global minimum of a scoring function [19]. Assuming that a cluster represents a statistical ensemble, a good scoring function should be able to identify the best (or correct) cluster with high probability though it may fail to assign the best score to the pose that is the closest to the experimental one. Here a best cluster means the cluster whose average RMSD, d_f, to the experimental pose is the smallest among all the clusters. Using our geometric partitional algorithm, we have applied both GOLD and a newly-developed scoring functions (Eqn. 2, 3) to 22 sets of poses to determine which one is better suitable for the identification of the best clusters. In the following we describe in detail the results on two sets of poses that represent the extreme cases among the 22 sets: our scoring function works well for the first but no 100% guarantee is provided for the second.

The first example is human CRABP2 complexed with an RA analog (1CBS). We first generate three sets of clusters with decreasing d_{max} values (d_{max}=5.0, 4.0, 3.0Å), the average scores are then computed for each cluster (Table 2). Smaller d_{max} generates smaller but more accurate clusters. With d_{max}=5.0Å, there are four major clusters while the poses in each of them distribute rather uniformly (Figs 1a, 1c). Both GOLD and our scores could select correctly the most populated cluster as the best cluster. However, with d_{max}=4.0Å, GOLD picks wrongly the third cluster as the best one while our score identifies correctly the second one. With d_{max}=3.0, GOLD still selects the wrong cluster (the third cluster with 91 poses) (Figs 1b) while our score identifies correctly the 6th cluster (15 poses) as the best one with d_f=2.2Å (Fig. 1d). The main reason for the failure of the GOLD scoring function is that it does not include any term that accounts for the contribution of the intermolecular electrostatic interactions to the binding affinity. For CRABP2 it is well-known that the electrostatic interaction between the carboxylic group of the RA analog and two arginine residues (R111 and R132) contributes greatly to the binding [20].

The second example is an HIV protease complexed with a peptide analog (1AAQ). We first generate three sets of clusters with decreasing d_{max} values (d_{max}=4.5, 3.5, 3.0Å), the average scores are then computed for each cluster (Table 3). We starts with d_{max}=4.5Å since only a single large cluster is generated with d_{max}=5.0Å. With d_{max}=4.5Å there are four major clusters while the poses in each of them distribute very uniformly (Figs 2a, 2c). GOLD picks wrongly the third cluster as the best one while our score identifies the second as the best though the most populated one has slightly smaller d_f. With d_{max}=3.5Å GOLD still picks wrongly the third (76 poses) as the best (Fig. 2b) while our score identifies the 4th (47 poses) (Fig. 2d), 5th (33 poses) and 7th (11 poses) clusters as the best ones with respective d_f of 2.7Å, 3.5Å and 11.5Å. With d_{max}=3.0Å, GOLD selects the wrong cluster (the 15th cluster with 5 poses) as the best while our score identifies correctly the 14th cluster (6 poses) as the best with d_f=2.1Å. For the HIV protease, the exclusion of electrostatic interaction in the GOLD scoring function may still contribute to its failure though the latter likely plays

Table 2. Gold score vs our score of the most populated clusters for 1CBS poses. The clusters are generated using three decreasing d_{max}s. The listed clusters include more than 90% of the total poses. N_s, S_T, S_G and d_f are respectively the number of structures in a cluster, the average score computed using our and GOLD scoring functions, and the average RMSD between the GOLD generated poses and the experimental pose. The lower a score, the better. The three columns with the boldfaced numbers have the lowest average score as computed by our scoring function.

$d_{max} = 5.0$Å:

N_s	**186**	160	51	45
S_T	**-10.0**	-9.5	-6.5	-5.3
S_G	**-44.1**	-43.0	-35.5	-31.3
d_f	**6.3**	9.6	9.5	12.2

$d_{max} = 4.0$Å:

N_s	160	**126**	91	45	25	18	6
S_T	-9.5	**-10.5**	-10.0	-5.3	-6.5	-6.7	-5.7
S_G	-43.0	**-41.9**	-45.4	-31.3	-34.7	-37.7	-33.7
d_f	9.6	**5.3**	6.2	12.2	9.5	9.5	9.5

$d_{max} = 3.0$Å:

N_s	158	106	91	20	17	**15**	10	8	7	6
S_T	-9.4	-10.2	-10.0	-6.5	-5.0	**-12.5**	-6.8	-6.5	-5.3	-6.1
S_G	-43.1	-42.9	-45.4	-34.8	-30.6	**-36.5**	-37.6	-37.9	-31.6	-28.1
d_f	9.6	5.8	6.2	9.5	12.4	**2.2**	9.3	9.7	12.3	12.4

a small role. Though our scoring function outperforms the GOLD function in all the 22 cases tested it remains challenging for our function to select the correct cluster with 100% confidence. In this case, a dozen of outliers with very low electrostatic energy or ligand internal energy must be removed, otherwise, with small d_{max}, both our and GOLD score may mistake the wrong clusters as the best ones. A systematic approach for such outlier detection and for minimizing their ill-effects are under the development.

Table 3. Gold score vs our score of the most populated clusters for 1AAQ poses. The clusters are generated using three decreasing d_{max}s. With d_{max}=3.0Å the clusters whose number of poses is ≤1.0% of the total number of poses are not shown. The listed clusters include more than 85% of the total. The symbols have the same meanings as those in Table 2.

$d_{max} = 4.5$Å:

N_s	179	**131**	80	52
S_T	-13.3	**-13.6**	-12.8	-13.5
S_G	-59.1	**-58.8**	-61.7	58.5
d_f (Å)	3.8	**4.3**	11.2	11.2

$d_{max} = 3.5$Å:

N_s	127	91	76	**49**	33	31	11	10
S_T	-12.8	-13.1	-12.7	**-14.7**	-14.7	-13.1	-14.7	-14.3
S_G	-58.7	-58.8	-61.8	**-59.7**	-59.3	-58.6	-59.9	-59.0
d_f	4.2	4.6	11.2	**2.7**	3.5	11.2	11.3	11.5

$d_{max} = 3.0$Å:

N_s	115	81	20	19	15	14	11	10	8	7	6	6	6	**6**
S_T	-12.7	-13.1	-12.7	-13.0	-13.8	-11.8	-14.5	-11.9	-13.7	-12.9	-14.2	-14.9	-15.1	**-15.5**
S_G	-59.3	-58.9	-62.5	-62.9	-57.6	-63.1	-55.9	-61.1	-60.7	-58.6	-53.6	-53.4	-62.2	**-62.4**
d_f	4.2	4.6	11.2	11.4	3.7	11.0	11.0	11.1	2.5	11.1	11.0	2.4	11.2	**2.1**

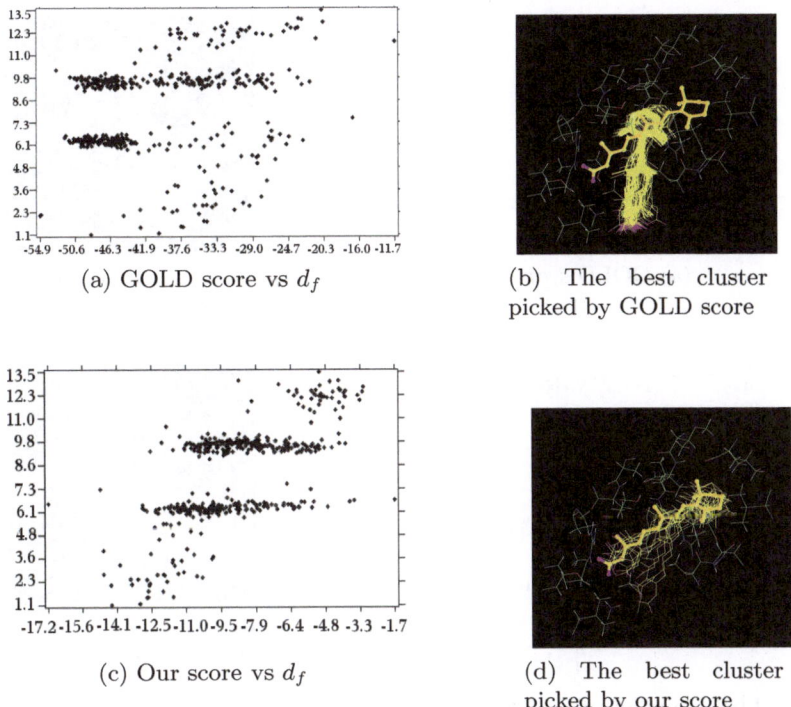

(a) GOLD score vs d_f

(b) The best cluster picked by GOLD score

(c) Our score vs d_f

(d) The best cluster picked by our score

Fig. 1. A comparison of GOLD and our scoring functions for best cluster selection for 1CBS. The x-axis and y-axis in (a, c) are respectively the score and d_f, the RMSD between the docked poses and experimental pose. GOLD ranks 82th the pose with the smallest d_f while our score ranks it to the fifth. The lower a score, the better. The clusters in (b, d) are generated using the geometric algorithm with a d_{max}=3.0Å. The protein atoms C, O, N and H in the binding site are colored respectively in green, red, blue and white while the C and O atoms of the ligand are colored in yellow and magenta. The experimental pose is depicted in a stick-and-ball model. The figures are prepared using our own molecule visualization program.

The complexity of the scoring functions force almost all of the current docking programs to rely on heuristics for optimization. However, a heuristic search may not cover the pose space adequately as being demonstrated in the above two examples: the poses with small d_f to the experimental one are the minority: less than 5% of the total poses. Another noticeable feature of the set of poses generated by GOLD is that the poses inside a few large clusters have similar GOLD scores though their d_f values differ greatly. Working together with our scoring function the geometric algorithm, capable of classifying both uniformly and non-uniformly distributed data, is ideally suitable for the identification of these minor clusters populated with the correct poses.

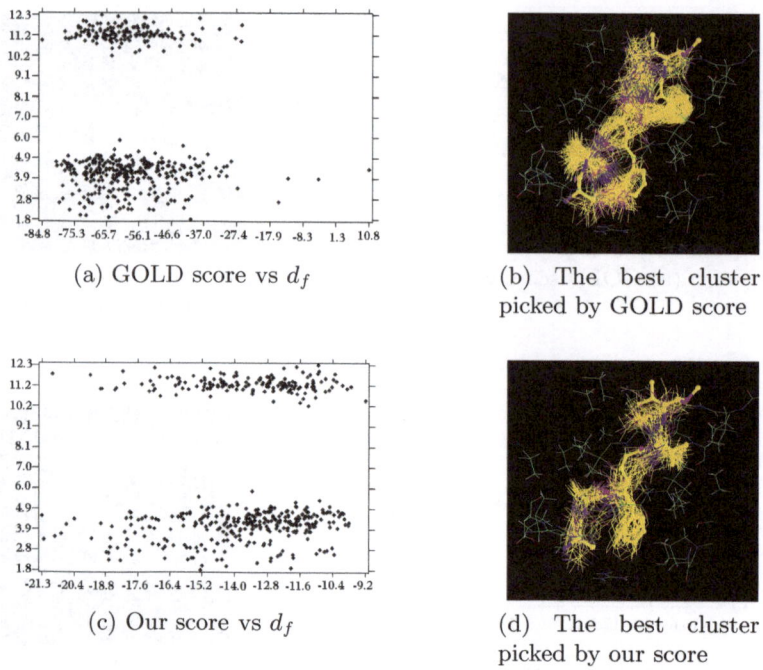

(a) GOLD score vs d_f

(b) The best cluster picked by GOLD score

(c) Our score vs d_f

(d) The best cluster picked by our score

Fig. 2. The comparison of GOLD vs our scoring function for best cluster selection for 1AAQ. The x-axis and y-axis in (a, c) are respectively the score and d_f. GOLD ranks 461th the pose with the smallest RMSD while our score ranks 188th. The protein, ligand and the poses are depicted and their atoms are colored in the same manner as in Fig.1. The figures are prepared using our molecule visualization program.

3.3   Algorithmic Comparison

Data classification as achieved by a clustering algorithm is a natural exploratory process for discovery and thus such algorithms have found wide applications in many different areas. However, as shown by Kleinberg [21] there exists no best or universal clustering algorithm. The classification of structural data especially those computed using restraints must take into consideration their unique features such as the distribution of data may be both uniform and non-uniform or both regular and irregular because of the sparseness of the input restraints, the large error in the current scoring functions, the limited sampling provided by heuristics and the extreme energy level degeneracy of biomolecules in solution [19]. In the following we discuss the unique features of the geometric partitional algorithm and compare it with the previous algorithms to show that it is both efficient and more suitable than the previous algorithms for the analysis of structural data.

In our algorithm the seeds for new clusters are the data points that form the largest polyhedron and the points of a previous cluster are divided into the new clusters according to their minimum distances to the seeds. These seed points are likely to be labeled as "outliers" by previous algorithms but our algorithm initializes the clustering with them and thus ensures them and together with their neighborhoods to be in different clusters. This is in contrast with the average-link and k-medoids algorithms and their variants that try to assign the data into the smallest number of clusters. Consequently, the representatives of the clusters from our algorithm sample the data space more uniformly than hierarchical and much more uniformly than k-medoids algorithms do. In additional, unlike hierarchical algorithms that only optimize an objective function locally, our algorithm takes into consideration the global information at the very beginning. The time complexity of our algorithm is $O(n^2 \log n)$ that is the same as the agglomerative hierarchical algorithm implemented with a priority queue [12]. Furthermore, the implementation suggests that our algorithm is faster than the hierarchical algorithms likely because of the base in the logarithmic function is ≥ 4 rather than 2 as in a typical hierarchical algorithm. The geometric algorithm is somewhat similar to the minimum-diameter divisive hierarchical algorithm by Guenoche, Hansen, and Jaumard [22]. Their key difference lies in how a previous cluster is divided into new clusters: in the minimum-diameter hierarchical algorithm, two new clusters are generated by an expensive search for the two balls with the minimum diameters while in our algorithm up to four new clusters are initialized with the seeds computed in linear time in terms of the number of data points in the previous cluster.

The geometric algorithm differs largely from k-medoids algorithms and thus have no problems associated with them such as (a) the tendency to find hyperspherical clusters, (b) the danger of falling into local minimal, and (c) the variability in results that depends on the choice of the initial seeds. Because our algorithm classifies the data by iteratively separating them into smaller clusters according to their distances to the seeds, it is not to be affected by irregular or non-uniform distributions as it is for a density-related clustering algorithm such as the k-medoids. The results from the applications to the clustering of both intermediate structures and poses suggest that the k-medoids algorithms are not suitable for the classification of structural data.

A possible drawback of our algorithm is that a prior knowledge is required to specify a d_{max} value and several d_{max} values may need to be tried in order to find the best classification for a data set. As far as the structural data is concerned, it is not difficult for the practitioners to find a reasonable value for d_{max} based on the quality of the data or the required precision in the final clusters.

References

1. Jain, A.K., Murty, M.N., Flynn, P.J.: Data clustering: a review. ACM Comput. Surv. 31(3), 264–323 (1999)
2. Jain, A.K.: Data clustering: 50 years beyond k-means. Pattern Recognition Letters 31(8), 651–666 (2010)
3. May, A.C.W.: Toward more meaningful hierarchical classification of protein three-dimensional structures. PROTEINS 37(1), 20–29 (1999)

4. Shao, J., Tanner, S.W., Thompson, N., Cheatham, T.E.: Clustering molecular dynamics trajectories: 1. characterizing the performance of different clustering algorithms. J. Chem. Theory Comput. 3(6), 2312–2334 (2007)
5. Keller, B., Daura, X., van Gunsteren, W.F.: Comparing geometric and kinetic cluster algorithms for molecular simulation data. J. Chem. Phys. 132(7), 074110 (2010)
6. Bottegoni, G., Rocchia, W., Cavalli, A.: Application of conformational clustering in protein–ligand docking. In: Computational Drug Discovery and Design, pp. 169–186. Springer (2012)
7. Adzhubei, A.A., Laughton, C.A., Neidle, S.: An approach to protein homology modelling based on an ensemble of NMR structures: application to the Sox-5 HMG-box protein. Protein Engineering 8(7), 615–625 (1995)
8. Domingues, F.S., Rahnenführer, J., Lengauer, T.: Automated clustering of ensembles of alternative models in protein structure databases. Protein Eng. Des. Sel. 17(6), 537–543 (2004)
9. Sutcliffe, M.J.: Representing an ensemble of NMR-derived protein structures by a single structure. Protein Sci. 2(6), 936–944 (1993)
10. Downs, G.M., Barnard, J.M.: Clustering Methods and Their Uses in Computational Chemistry, pp. 1–40. John Wiley & Sons, Inc. (2003)
11. Blumenthal, L.: Theory and applications of distance geometry, 2nd edn. Chelsea Publishing Company (1970)
12. Day, W.H.E., Edelsbrunner, H.: Efficient algorithms for agglomerative hierarchical clustering methods. Journal of Classification 1(1), 7–24 (1984)
13. Lloyd, S.P.: Least squares quantization in PCM. IEEE Trans. Inf. Theory 28(2), 129–136 (1982)
14. Jones, G., Willett, P., Glen, R.C.: Molecular recognition of receptor sites using a genetic algorithm with a description of desolvation. J. Mol. Biol. 245, 43–53 (1995)
15. Sadowski, J., Gasteiger, J., Klebe, G.: Comparison of automatic three-dimensional model builders using 639 x-ray structures. J. Chem. Inf. Comput. Sci. 34(4), 1000–1008 (1994)
16. Baker, N.A., Sept, D., Joseph, S., Holst, M.J., McCammon, J.A.: Electrostatics of nanosystems: Application to microtubules and the ribosome. PNAS 98(18), 10037–10041 (2001)
17. Wang, L., Mettu, R., Donald, B.R.: A polynomial-time algorithm for *de novo* protein backbone structure determination from NMR data. J. Comput. Biol. 13(7), 1276–1288 (2006)
18. Warren, B.L., Andrews, C.W., Capelli, A.M., Clarke, B., Lalonde, J., Lambert, M.H., Lindvall, M., Nevins, N., Semus, S.F., Senger, S., Tedesco, G., Wall, I.D., Woolven, J.M., Peishoff, C.E., Head, M.S.: A critical assessment of docking programs and scoring functions. J. Med. Chem. 49(20), 5912–5931 (2006)
19. Landau, L.D., Lifshitz, E.M.: Statistical Physics, vol. 5. Pergamon Press, Oxford (1980)
20. Wang, L., Li, Y., Yan, H.: Structure-function relationships of cellular retinoic acid-binding proteins: Quantitative analysis of the ligand binding properties of the wild-type proteins and site-directed mutants. J. Biol. Chem. 272(3), 1541–1547 (1997)
21. Kleinberg, J.: An impossibility theorem for clustering. In: Proc. 2002 Conf. Advances in Neural Information Processing Systems, Vancouver, Canada, vol. 15, pp. 463–470. International Institute of Informatics and Systemics (2002)
22. Gunoche, A., Hansen, P., Jaumard, B.: Efficient algorithms for divisive hierarchical clustering with the diameter criterion. Journal of Classification 8(1), 5–30 (1991)

A Spatial-Aware Haplotype Copying Model with Applications to Genotype Imputation

Wen-Yun Yang[1], Farhad Hormozdiari[1], Eleazar Eskin[1,2],
and Bogdan Pasaniuc[2,3,*]

[1] Department of Computer Science, University of California Los Angeles
[2] Department of Human Genetics, University of California Los Angeles
[3] Department of Pathology and Laboratory Medicine,
University of California Los Angeles
bpasaniuc@mednet.ucla.edu

Abstract. Ever since its introduction, the haplotype copy model has proven to be one of the most successful approaches for modeling genetic variation in human populations with applications ranging from ancestry inference to genotype phasing and imputation. Motivated by coalescent theory, this approach assumes that any chromosome (haplotype) can be modeled as a mosaic of segments copied from a set of chromosomes sampled from the same population. At the core of the model is the assumption that any chromosome from the sample is equally likely to contribute a priori to the copying process. Motivated by recent works that model genetic variation in a geographic continuum, we propose a new spatial-aware haplotype copy model that jointly models geography and the haplotype copying process. We extend hidden Markov models of haplotype diversity such that at any given location, haplotypes that are closest in the genetic-geographic continuum map are a priori more likely to contribute to the copying process than distant ones. Through simulations starting from the 1000 Genomes data, we show that our model achieves superior accuracy in genotype imputation over the standard spatial-unaware haplotype copy model. In addition, we show the utility of our model in selecting a small personalized reference panel for imputation that leads to both improved accuracy as well as to a lower computational runtime than the standard approach. Finally, we show our proposed model can be used to localize individuals on the genetic-geographical map on the basis of their genotype data.

1 Introduction

Complex population demography coupled with the presence of recombination hotspots have shaped genetic variation in the human genome into blocks of markers with similar recent ancestry [1–3]. This recent ancestry sharing induces dependencies among variants in the form of linkage disequilibrium (LD), i.e. the non-random association of alleles at two or more loci [4]. Therefore, the

---

* Corresponding author.

R. Sharan (Ed.): RECOMB 2014, LNBI 8394, pp. 371–384, 2014.
© Springer International Publishing Switzerland 2014

observed LD patterns across the genome are the result of a population's demo-graphic history and are modeled in a wide-range of problems from population genetic inferences [5, 6] to medical population genetics [7, 8]. Most notably, LD has enabled the era of genome-wide association studies that use a small number of variants (as compared to all variation in the genome) to assay variation across the entire human genome [9]. Thus, modeling population LD is a fundamental problem in computational genetics with applications ranging from genotype im-putation and haplotype inference to locus-specific and genome-wide ancestry inference [7, 10–15].

Although many approaches for modeling LD have been proposed [3][16], one of the most successful framework has been introduced by Li and Stephens (widely referred to as the *haplotype copy model* [16]). Drawing on coalescent theory, in this model, a haplotype sampled from a population is viewed as a mosaic of seg-ments of previously sampled haplotypes. This mosaic structure can be efficiently modeled within a hidden Markov model to achieve very accurate solutions to many genetic problems such as genotype imputation [7, 10, 11], ancestry infer-ence [14, 15], quality control in genome-wide association studies [17], detection of identity by descent (IBD) segments [18, 19], estimating recombination rates [20], haplotype phasing [21], migration rates [22] and calling of genotypes at low cov-erage sequencing [23, 24].

At the core of the Li and Stephens [16] model lies a hidden Markov model (HMM) that emits haplotypes through a series of segmental copies from the pool of previously observed haplotypes. The hidden states in the HMM indicate which haplotype from the reference panel to copy from while emission probabil-ities allow for potential mutation events observed since the most recent common ancestor of the target and the reference copy haplotype. Recombination events are modeled through the transition probabilities; the probability of copying from the same reference haplotype at successive loci is much higher than switching to another haplotype, based on the idea the probability of having a recombi-nation between two neighboring loci is low. Motivated by coalescent theory in randomly mating populations, the a priori probability of switching the copy process to another haplotype is equally likely among all the previously observed haplotypes. However, since human populations show a tremendous amount of structure across geography [25–27] (inline with isolation-by-distance models), it is likely that haplotypes physically closer in geography to the target haplo-type contribute significantly more to the copy process. Furthermore, with the emergence of high-throughput sequencing that is generating massive amount of data [28–30], existing methods are increasingly computationally intensive due to the ever larger samples of haplotypes that can be used as reference. Although a commonly used approach for reducing computational burden is to downsample the reference panels [31–33] (often in an ad-hoc manner) a principled approach for selection of a reference panel for optimizing performance is currently lacking.

In this paper, we propose a new approach to modeling genetic variation in structured populations that incorporates ideas from both the haplotype copying model [16] and the spatial structure framework that models genetic variation as function of geography [26, 27]. That is, we propose a haplotype copy model that a priorly up weights the contribution of haplotypes closer in geographical distance to the copying process. We accomplish this by jointly modeling geography and the copying process. Each haplotype is associated with a geographical position; when copying into a new haplotype with known location, we instantiate an HMM that has switching transition probabilities up weighted for haplotypes closer in geographical space to the target haplotype.

We use real data from the 1000 Genomes project [2] to show that the our spatial-aware approach fits the data significantly better than the standard model. Through a masking procedure followed by a leave-one-out experiment we show that our spatial-aware method significantly increases imputation accuracy especially for lower frequency variation (e.g. an improvement of 6% (2%) for low-frequency (common) variation in Asian data). We also show that our approach can be used to select a small personalized reference panel for imputation that increases imputation accuracy while significantly reducing imputation runtime (up to 10-fold). Finally, we show how our model can be used in a supervised manner to infer locations on the genetic-geographic map for individuals based on their genetic data.

2 Methods

2.1 The Standard Haplotype Copying Model

We start by briefly introducing the standard haplotype copying model [16] for modeling LD in a population. Let $H \in \{0,1\}_{N \times L}$ be a matrix of haplotypes (which we will refer to as the *reference panel*), where $h_{ij} \in \{0,1\}$ indicates if the i-th individual at the j-th position (SNP) contains the reference or the alternate allele. N denotes the number of haplotypes in the reference panel and L the number of SNPs in the data. Let $h \in \{0,1\}_{1 \times L}$ be a multi-locus haplotype which we will refer to as the *target haplotype* where $h_i \in \{0,1\}$ indicates the i-th SNP. The haplotype copy model views the target haplotype as being composed of a mosaic of segments from haplotypes of the reference panel.

Formally, we define a hidden Markov model (HMM) specified by a triple (S, τ, ω), where S is the set of states, τ is the transition probability, and ω is the emission probability function. The set S contains state variables $\{s_1, \ldots, s_L\}$ where $s_k = \{1, 2, \cdots N\}$ indicates from what reference haplotype is the k-th allele in the target haplotype copied from. The transition probability τ is non-zero only between pairs of states in consecutive sets of states S, which can be defined between SNP k and SNP $k + 1$ as follows

$$\tau_k(i,j) = \begin{cases} \theta_k + (1 - \theta_k)/N & i = j \\ (1 - \theta_k)/N & i \neq j \end{cases} \quad , \text{ where } \quad \theta_k = \exp(-\rho d_k).$$

Here d_k is the physical distance between SNP k and SNP $k + 1$ and $\rho = 4N_e c$ where N_e is the effective population size, c is the average rate of crossover per unit physical distance per meiosis (e.g. 10^{-8}). This can be easily extended to use recombination maps with varying recombination events at different loci in the genome. The emission probability mimics the mutation process and can be defined as follows

$$\omega(h_k, s_k; H) = \begin{cases} 1 - \epsilon & h_k = H_{s_k,k} \\ \epsilon & \text{otherwise} \end{cases}, \quad \text{where} \quad \epsilon = \frac{N}{N + \left(\sum_{m=1}^{N-1} 1/m\right)^{-1}}.$$

where N denotes the number of reference haplotypes. Intuitively the copying process is more accurate as the reference sample size grows and it is more likely to find in the reference a haplotype closely matching the target one.

The likelihood of the target haplotype h is defined as:

$$P(h|S, H; \lambda) = P(S) \prod_k P(h_k|s_k, H) = \prod_k \tau_k(s_{k-1}, s_k) \left(\prod_k \omega(h_k, s_k; H)\right) \quad (1)$$

and can be efficiently estimated using the forward/backward algorithm. Inference in this model is performed using standard HMM approaches such as Viterbi or posterior decoding. For example, if the target haplotype has any of the alleles missing, posterior decoding can be employed to estimate the most likely values conditional on the model and the rest of the target haplotype.

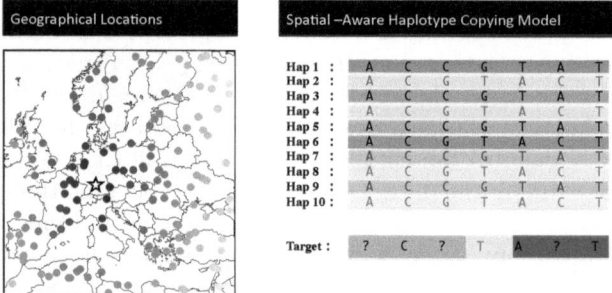

Fig. 1. An illustration of spatial haplotype copying model. In the left panel, the location for target haplotype is shown using the star. All haplotypes in the data are color coded using the distance to the target location (light more distant, darker are closer). We enforce the transition rates (that encode the copy switching) to give higher weight to haplotypes closer to the target haplotype. A haplotype at the target location is more likely to contain mosaic segments from haplotypes that are closer to the target location.

2.2   A Spatial-Aware Haplotype Copying Model

A drawback of the standard haplotype copying model comes from the equal treatment of reference haplotypes; that is, a priori all haplotypes from the reference panel are equally likely to contribute to the target haplotype. This effect motivates us to propose the following approach to take spatial effect into account in the model. Let $X = \{x_1, \ldots, x_N\}$ indicate the locations for all N reference haplotypes and x indicate the location for target haplotype. In a scenario where the location of the individuals are not known, we estimate their locations from genotype data using methods such as PCA [25], SPA [26] or LOCO-LD [27]. Then, instead of using uniform switching probability across all reference haplotypes, we assign higher probability to haplotypes located closer to the target haplotype. Formally, we redefine the transition rate τ between SNP k and SNP $k+1$ as:

$$\tau_k(i,j) = \begin{cases} \theta_k + (1 - \theta_k)p_j & i = j \\ (1 - \theta_k)p_j & i \neq j \end{cases} \quad \text{where} \quad p_j = \frac{\exp(-\lambda\psi(x, x_j))}{Z}.$$

The function $\psi(x, x_j)$ denotes a distance function between x and x_j (e.g. Euclidean distance) and Z is a normalization factor to ensure the probability definition. The parameter λ specifies the effect of geographical distance. It is worth mentioning that this spatial-aware model can be reduced to standard haplotype copying model by setting $\lambda = 0$, such that $p_j = 1/N$; therefore our approach can be viewed as a generalization of the standard Li and Stephens model. An illustration of our model is shown in Figure 1. Intuitively a large value for λ indicates a more pronounced spatial effect (less probability to copy from distant haplotypes), while $\lambda = 0$ reverts to assigning equal a priori probability.

The likelihood of the target haplotype is defined as before by summing on all paths in the model (Eq 1). Inference in this model can be performed as in the standard haplotype copy model using a combination of Viterbi and posterior decoding as function of the particular application.

2.3   Estimation of Spatial Effect Parameter λ

A pre-requisite step in applying our model is the specification of λ. It is necessary to estimate the λ before using the model for various applications, as the value of λ could vary significantly across individuals or populations. We estimate λ through maximum likelihood estimation (MLE). Starting from the likelihood of the target haplotype h (Eq 1), we marginalize over all possible values of hidden variables S to obtain likelihood as function of λ:

$$L(h; \lambda) = \sum_S P(h|S, H) \tag{2}$$

However, this overall likelihood function is infeasible to optimize directly, as the number of all possible values of S is very large L^N. Although the likelihood computation can be reduced by forward-backward algorithm to $O(NL)$, the gradient is still very expensive to compute, as the calculation would involve a forward-backward in $O(NL)$ and a summation of $O(N^2L)$ terms. When the number of reference haplotypes is large, this gradient would be infeasible to compute. Fortunately, the gradient for the Q function in EM algorithm is much simpler to compute than the gradient of likelihood function in (2). It is also

---

Algorithm 1. Learning Algorithm for Parameter λ Estimation

1: Setting optimization parameters R and C (e.g., $R = 1 \times 10^3$ and $C = 20$)
2: Pre-computing $\psi(x, x_j)$ for all reference haplotype j, and θ_k for all k.
3: Randomly initialize $\lambda^{(0)} > 0$
4: **for** t from 0 to T **do**
5:     Perform forward-backward algorithm to get the forward/backward probability
6:     Compute stochastic gradient $g(\lambda^{(t)})$ by sampling R pairs of i and j in (4)
7:     Setting $\lambda^{(t+1)} = \lambda^{(t)} + \dfrac{1}{t + C} \cdot g(\lambda^{(t)})$
8: **end for**
9: Output $\lambda^{(T+1)}$

---

guaranteed that the gradient of the Q function will be an increasing direction for the original likelihood function, which is a theoretical property of the EM algorithm. Thus, we resort to compute the gradient of the Q function instead of the gradient of original likelihood function.

First, the Q function in EM algorithm can be written as follows

$$Q(\lambda, \lambda^{(t)}) = \sum_S P(S) \ln P(h, S; \lambda)$$

$$\propto \sum_{kij} P(s_{k-1} = i, s_k = j; \lambda^{(t)}) \ln \tau_k(i, j; \lambda) \tag{3}$$

The gradient for this Q function can be calculated as follows

$$\frac{\partial Q}{\partial \lambda} = -\sum_{kij} P(s_{k-1} = i, s_k = j; \lambda^{(t)}) \left(\frac{\psi(x, x_j) - \sum_l \psi(x, x_l) p_l}{1 + I(i = j)\left(\frac{\theta_k}{(1 - \theta_k)p_j}\right)} \right) \tag{4}$$

where the identity function $I(i = j)$ is equal to 1 when $i = j$ and 0 otherwise. However, simply calculation of this gradient will also be inefficient with the complexity $O(N^2 L)$, which is still expensive for thousands of reference haplotypes and millions of SNPs. We resort to computing a stochastic gradient for the Q function, and apply it to the original likelihood function as a searching direction. We estimate the gradient by sampling over N haplotypes, instead of enumerating all of them. In practice, between each pair of SNP k and SNP $k + 1$, we randomly sample 1000 pairs of $s_{k-1} = i$ and $s_k = j$, instead of all N^2 pairs. The overall algorithm for efficient optimization of the spatial effect parameter λ is described in Algorithm 1.

2.4   Localization of Individuals Based on Their Genetic Data

Another appealing application for spatial-aware haplotype copying model is to localize individuals on the map. That is, given locations X for all reference panel haplotypes, we seek to find the best location x for the target haplotype to maximize the likelihood of the data. The algorithm follows similar procedure as above section 2.3. The difference mainly comes from a different Q function as follows

$$Q(x, x^{(t)}) = \sum_S P(S) \ln P(h, S; x)$$

$$\propto \sum_{kij} P(s_{k-1} = i, s_k = j; x^{(t)}) \ln \tau_k(i, j; x) \tag{5}$$

which is parameterized by x instead of λ as in Equation (3). However, this change leads to non-concavity of the function in general. But since there is only one parameter to estimate, and the function is well behaved in practice, we can still compute the gradient for the Q function and apply it to the stochastic gradient descent method. The gradient for the Q function in Equation (5) can be calculated as follows

$$\frac{\partial Q}{\partial x} = -\sum_{kij} P(s_{k-1} = i, s_k = j; x^{(t)}) \lambda \left(\frac{\frac{\partial \psi(x, X_j)}{\partial x} - \sum_l p_l \cdot \frac{\partial \psi(x, X_j)}{\partial x}}{1 + I(i = j) \left(\frac{\theta_k}{(1 - \theta_k)p_j} \right)} \right) \tag{6}$$

we can use Euclidean distance $\psi(x, X_j) = ||x - X_j||_2$ as a sufficient estimation of spatial distance. Thus, the gradient of the distance metric becomes

$$\frac{\partial \psi(x, X_j)}{\partial x} = \frac{x - X_j}{||x - X_j||_2}$$

The overall algorithm is similar as Algorithm 1 for optimizing λ, except for replacement of λ by x and the gradients correspondingly.

3   Experimental Results

3.1   Estimation of Spatial Copying Effect in the 1000 Genomes Data

We applied our methods to data generated part of the 1000 Genomes project [2]. A total of 1092 individuals were collected from 14 populations across the European, Asian, African and American continents. For all of our simulations we used $157,827$ SNPs on chromosome 22, where 79.5% of SNPs are rare SNPs (allele frequency < 0.05), and the rest 20.5% are common SNPs; although the original data contained $473,481$ SNPs, for computational efficiency we down sampled to every third SNP. Among the considered SNPs, we assumed that only $2,931$ SNPs present on the Affymetrix 6.0 SNP array are collected and the remaining SNPs will be imputed using our model. This amounts to using 1.86% SNPs to impute the rest 98.14% SNPs. We apply PCA [25] to assign a geographical location to each individual in the dataset. Although we note that the imputation performance can be further improved if denser SNPs are assumed to be typed, we expect the general trends reported below to maintain.

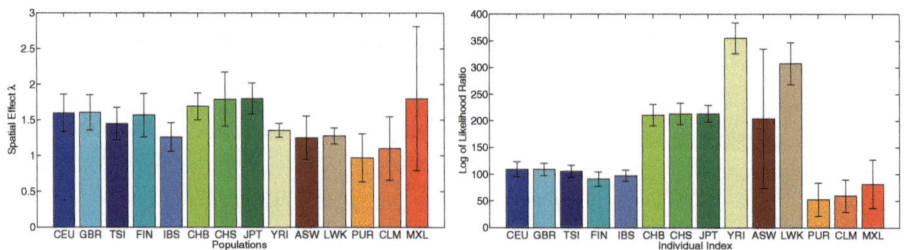

Fig. 2. Estimated spatial copying effects λ^* across different populations in 1000 Genomes data. Left shows the average λ^* across all individuals in a given population while right displays the log likelihood ratio of the model with λ^* as compared to $\lambda = 0$. The error bars indicate the standard deviations for each population.

Starting from the 2,931 SNPs, we estimated the spatial effect parameter λ for each of the 2,184 haplotypes in the dataset. The average λ values are 1.54, 1.76, 1.30 and 1.32 for European, Asian, African and American populations, respectively (Figure 2). Generally, the higher value of λ corresponds to stronger spatial copying effect, which leads to more segments copied from nearby haplotypes. To test the significance of spatial effect, we compared the likelihoods of the data (the 2,184 haplotypes) within the model assuming no spatial effect ($\lambda = 0$) versus the model with spatial effect (λ^* estimated from the data). The log likelihood ratio between spatial haplotype copying model and standard haplotype model is given in Figure 2. The likelihood is computed for each haplotype being emitted from the rest of haplotypes. Across all populations we observe that the model with a spatial effect fits the data much better than the model with no spatial assignment. This is expected since we use haplotypes across all continents (except the target) in the reference panel, and it is expected that haplotypes share more continental-specific segments.

3.2 Spatial-Aware Model Improves Imputation Accuracy

Having established that the model with spatial effect fits the data much better than the standard model with no spatial effect, we focused next on haplotype imputation (a standard approach in genome-wide association studies through pre-phasing [34]). We carry out a leave-one-out procedure to perform the evaluation. In each round, we select one haplotype as a target and use the rest as the reference panel. To remove potential bias, instead of using all haplotypes, we randomly select one haplotype from each individual to use a total of 1,092 haplotypes (i.e. each round imputes one haplotype from the remaining 1,091). The imputation results are evaluated using the average per-SNP r^2 correlation coefficients averaged across all leave-one-out rounds for either all haplotypes, or for data within each population.

We first demonstrate the effect of the lambda parameter on imputation accuracy by applying our model using a wide range of lambda parameter values.

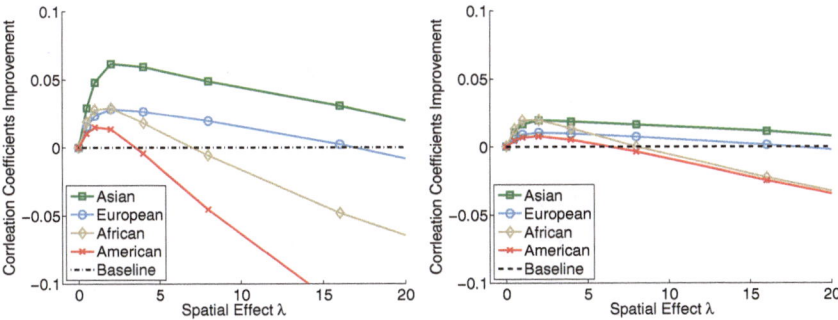

Fig. 3. Effect of spatial copying parameter λ on imputation accuracy. Left shows results for low-frequency (1−5%) while right displays results for common variants (> 5%). The maximum accuracy is attained at a $\lambda \approx 2$, close to the maximum likelihood estimate for λ (1.3 to 1.7, see Section 3.1).

Compared with the baseline method ($\lambda = 0$), we observe that a clear improvement is obtained for a value of λ around 2, especially for European and Asian populations (see Figure 3). This is consistent with the spatial model fitting those populations (see Figure 2). We also observe that the spatial model improves the imputation of rare variants more significantly than common variants, which is expected as the rare variants are more clustered geographically [35]. Moreover, the improvement for Asian and European populations is larger than for African and American populations.

Although we have shown that spatial model improves accuracy, in practice the value of λ is unknown and needs to be estimated from the data. We reassessed the accuracy of our approach by not setting λ to pre-specified values but by estimating it from the data. The performance of the model using the maximum likelihood λ^* over baseline method is given in Table 1. As before, we observe a larger improvements for rare variants than common variants. A plausible explanation for this effect is that that rare variants are more clustered in geography [35] than common variants. Overall for all populations, the improvement is highly correlated with allele frequency. The trend is shown in Figure 4, where we can see that the improvement is higher for SNPs with lower allele frequency.

3.3 Selection of a Personalized Reference Panel for Imputation to Increase Performance

Inspired by the significant spatial haplotype copying effect in experiments, we hypothesized that imputation efficiency can be improved by only using a personalized reference panel composed only from geographically close haplotypes [32, 31]. First, we expect that most of the reference haplotypes are not contributing haplotype segments to target haplotype. In Figure 5, we observe that the number of copied haplotypes decreases with higher λ (e.g. an average of 100 haplotypes

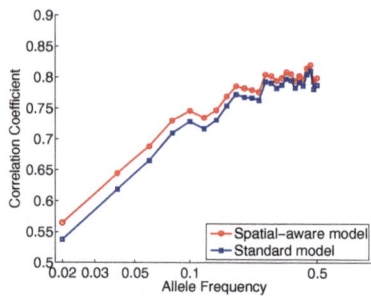

Fig. 4. Absolute imputation improvement across all spectrum of allele frequencies. Spatial-aware model uses λ^* inferred from the data.

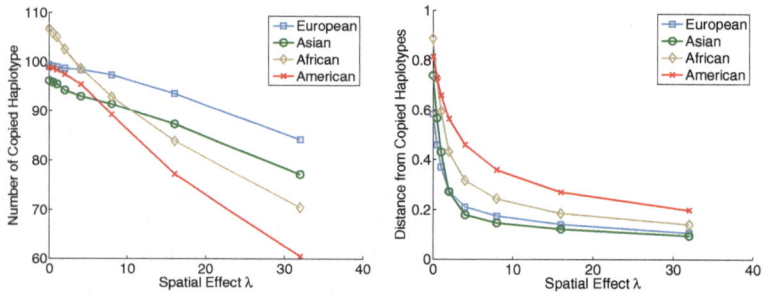

Fig. 5. Spatial effect on copied haplotypes from reference. Left shows that the number of copied haplotypes decreases while the spatial effect parameter is larger. Right shows that the averaged distance from copied haplotype decreases while the spatial effect parameter is larger.

Table 1. Performance of spatial model compared to the standard model

	Methods	European	Asian	African	American
	Baseline ($\lambda = 0$)	0.5560	0.4115	0.4833	0.5549
Low Frequency Variants	Spatial model with λ^*	0.5834	0.4364	0.4912	0.5654
	Relative Improvement	4.92 %	6.05 %	1.63 %	1.89 %
	Baseline ($\lambda = 0$)	0.7790	0.7189	0.6498	0.7701
Common Variants	Spatial model with λ^*	0.7939	0.7326	0.6605	0.7765
	Relative Improvement	1.90 %	1.91 %	1.64 %	0.84 %

are used in the copy process of a new target among 1091 reference haplotypes). On the other hand, in Figure 5, we plot the distance of those useful reference haplotypes from the target haplotype, weighted by the posterior. We observe there is a significant decrease of haplotype copying distance for higher λ value. It strongly suggests that the haplotype copying model can be significantly sped up by only keeping a small number of nearby haplotypes as reference panel. To assess this scenario, we re-imputed the target data using gradual decreasing

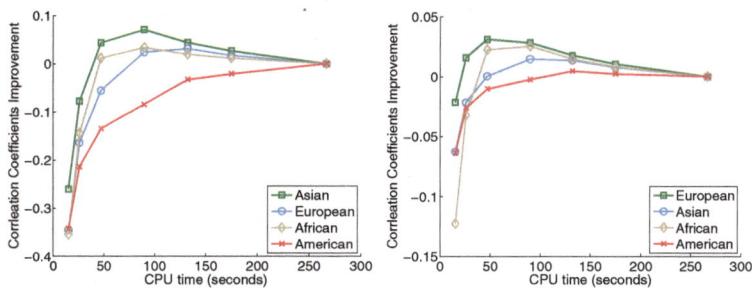

Fig. 6. Imputation accuracy versus computational time. Left shows low-frequency variants (1-5%) while right shows results over common variants (> 5%).

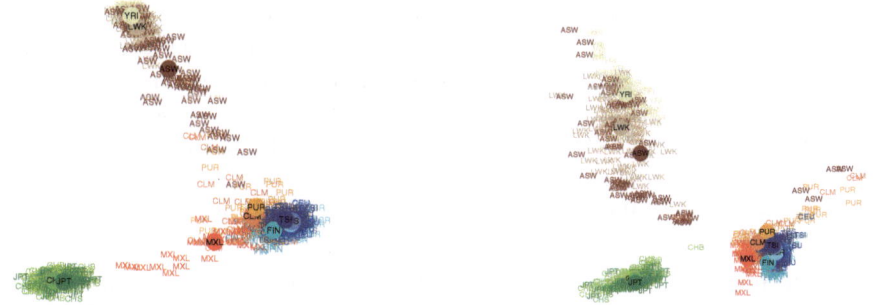

Fig. 7. Left shows results of PCA on chromosome 22 of the 1000 Genomes data while right shows results of our leave one out procedure to localize 1000 Genomes individuals

sizes for the reference panel (1091, 800, 600, 400, 200, 100 and 50) where we only keep the most nearby haplotypes in geographical space. The relation between imputation correlation coefficients and computational CPU time is shown in Figure 6. We observe that the computational time can be improved linearly in the size of reference panel but the imputation performance is also improved even using less number of reference haplotypes. For rare variants, the best imputation performance is obtained at 400 haplotypes and for common variants, the best imputation performance is obtained at 200 haplotypes.

3.4   Localization of Individuals on a Map

Finally, we explored whether we can use our approach to infer the location on the map of a new individual given data of individuals with known locations. We localized individual haplotypes using spatial-aware copying model with optimal λ value estimated before assuming known locations for the rest of the haplotype data. That is, in each round, we apply spatial-aware model to infer the optimal x^* for one individual using all other other individuals as reference panel (PCA was used to infer locations for the reference panel). We observe that spatial-aware model is able to well identify individual locations, in terms of the clear separating of different continents (see Figure 7). We observe a high correlation coefficient between the PCA and our inferred geographical ($r = 0.87$), thus showing that our approach can potentially be used to localize individuals on a map given training data with known locations (see Figure 7) .

4   Conclusions

The haplotype copying model plays an important role in a wide variety of genetic applications. A major drawback is that the model assumes that all haplotypes in the reference panel equally contribute a priori to the observed haplotype. In this paper, we have proposed a spatial-aware haplotype copying model that takes the spatial effects into account. We have also presented a highly efficient algorithm to estimate the spatial effect parameter before using the proposed model. We applied the proposed model to the 1000 genomes data set for several applications. First, we estimate the likelihood ratio between the spatial-aware model and spatial-unaware model, and a significant improvement is observed. Second, we test the application of imputation using spatial-aware model and obtain significant improvement over standard model. Finally, we apply this model to localize individuals and the results indicate high accuracy can be obtained.

Acknowledgement. W.Y., F.H. and E.E. are supported by National Science Foundation grants 0513612, 0731455, 0729049, 0916676, 1065276, 1302448 and 1320589, and National Institutes of Health grants K25-HL080079, U01-DA024417, P01-HL30568, P01-HL28481, R01-GM083198, R01-MH101782 and R01-ES022282. W.Y. and B.P. are partially supported by National Institutes of Health awards R03-CA162200 and R01-GM053275. We acknowledge the support of the NINDS Informatics Center for Neurogenetics and Neurogenomics (P30 NS062691).

References

1. Gibbs, R.A., Belmont, J.W., Hardenbol, P., Willis, T.D., Yu, F., et al.: The international hapmap project. Nature 426(6968), 789–796 (2003)
2. Consortium, G.P., Abecasis, G.R., Altshuler, D., Auton, A., Brooks, L.D., Durbin, R.M., Gibbs, R.A., Hurles, M.E., McVean, G.A.: A map of human genome variation from population-scale sequencing. Nature 467(7319), 1061–1073 (2010)
3. Daly, M.J., Rioux, J.D., Schaffner, S.F., Hudson, T.J., Lande, E.S.: High-resolution haplotype structure in the human genome. Nature Genetics 29(6), 229–232 (2001)
4. Kruglyak, L.: Prospects for whole-genome linkage disequilibrium mapping of common disease genes. Nature Genetics 22(2), 139–144 (1999)
5. Lohmueller, K.E., Bustamante, C.D., Clark, A.G.: Methods for human demographic inference using haplotype patterns from genomewide single-nucleotide polymorphism data. Genetics 182(1), 217–231 (2009)
6. Pool, J.E., Hellmann, I., Jensen, J.D., Nielsen, R.: Population genetic inference from genomic sequence variation. Genome Res. 20(3), 291–300 (2010)

7. Marchini, J., Howie, B., Myers, S., McVean, G., Donnelly, P.: A new multipoint method for genome-wide association studies by imputation of genotypes. Nature Genetics 39(7), 906–913 (2007)
8. Li, Y., Willer, C.J., Ding, J., Scheet, P., Abecasis, G.R.: Mach: using sequence and genotype data to estimate haplotypes and unobserved genotypes. Genet. Epidemiol. 34(8), 816–834 (2010)
9. de Bakker, P.I.W., Yelensky, R., Pe'er, I., Gabriel, S.B., Daly, M.J., Altshuler, D.: Efficiency and power in genetic association studies. Nat. Genet. 37(11), 1217–1223 (2005)
10. Howie, B.N., Donnelly, P., Marchini, J.: A flexible and accurate genotype imputation method for the next generation of genome-wide association studies. PLoS Genetics 5(6), e1000529 (2009)
11. Howie, B., Fuchsberger, C., Stephens, M., Marchini, J., Abecasis, G.R.: Fast and accurate genotype imputation in genome-wide association studies through prephasing. Nature Genetics 44(8), 955–959 (2012)
12. Chung, C.C., Kanetsky, P.A., Wang, Z., Hildebrandt, M.A.T., Koster, R., Skotheim, R.I., Kratz, C.P., Turnbull, C., Cortessis, V.K., Bakken, A.C., Bishop, D.T., Cook, M.B., Erickson, R.L., Foss, S.D., Jacobs, K.B., Korde, L.A., Kraggerud, S.M., Lothe, R.A., Loud, J.T., Rahman, N., Skinner, E.C., Thomas, D.C., Wu, X., Yeager, M., Schumacher, F.R., Greene, M.H., Schwartz, S.M., McGlynn, K.A., Chanock, S.J., Nathanson, K.L.: Meta-analysis identifies four new loci associated with testicular germ cell tumor. Nature Genetics 45(6), 680–685 (2013)
13. Savage, S.A., Mirabello, L., Wang, Z., Gastier-Foster, J.M., Gorlick, R., Khanna, C., Flanagan, A.M., Tirabosco, R., Andrulis, I.L., Wunder, J.S., Gokgoz, N., Patio-Garcia, A., Sierrasesmaga, L., Lecanda, F., Kurucu, N., Ilhan, I.E., Sari, N., Serra, M., Hattinger, C., Picci, P., Spector, L.G., Barkauskas, D.A., Marina, N., de Toledo, S.R.C., Petrilli, A.S., Amary, M.F., Halai, D., Thomas, D.M., Douglass, C., Meltzer, P.S., Jacobs, K., Chung, C.C., Berndt, S.I., Purdue, M.P., Caporaso, N.E., Tucker, M., Rothman, N., Landi, M.T., Silverman, D.T., Kraft, P., Hunter, D.J., Malats, N., Kogevinas, M., Wacholder, S., Troisi, R., Helman, L., Fraumeni, J.F., Yeager, M., Hoover, R.N., Chanock, S.J.: Genome-wide association study identifies two susceptibility loci for osteosarcoma. Nature Genetics 45(7), 799–803 (2013)
14. Pasaniuc, B., Sankararaman, S., Kimmel, G., Halperin, E.: Inference of locus-specific ancestry in closely related populations. Bioinformatics 25(12), i213–i221 (2009)
15. Price, A.L., Tandon, A., Patterson, N., Barnes, K.C., Rafaels, N., Ruczinski, I., Beaty, T.H., Mathias, R., Reich, D., Myers, S.: Sensitive detection of chromosomal segments of distinct ancestry in admixed populations. PLoS Genetics 5(6), e1000519 (2009)
16. Li, N., Stephens, M.: Modeling linkage disequilibrium and identifying recombination hotspots using single-nucleotide polymorphism data. Genetics 165(4), 2213–2233 (2003)
17. Han, B., Kang, H.M., Eskin, E.: Rapid and Accurate Multiple Testing Correction and Power Estimation for Millions of Correlated Markers. PLoS Genet. 5(4), e1000456+ (2009)
18. Browning, S.R.: Multilocus association mapping using variable-length markov chains. Am. J. Hum. Genet. 78(6), 903–913 (2006)
19. Browning, S.R., Browning, B.L.: High-resolution detection of identity by descent in unrelated individuals. Am. J. Hum. Genet. 86(4), 526–539 (2010)

20. Wegmann, D., Kessner, D.E., Veeramah, K.R., Mathias, R.A., Nicolae, D.L., Yanek, L.R., Sun, Y.V., Torgerson, D.G., Rafaels, N., Mosley, T., Becker, L.C., Ruczinski, I., Beaty, T.H., Kardia, S.L.R., Meyers, D.A., Barnes, K.C., Becker, D.M., Freimer, N.B., Novembre, J.: Recombination rates in admixed individuals identified by ancestry-based inference. Nature Genetics 43(9), 847–853 (2011)
21. Delaneau, O., Marchini, J., Zagury, J.F.: A linear complexity phasing method for thousands of genomes. Nature Methods 9(2), 179–181 (2012)
22. Roychoudhury, A., Stephens, M.: Fast and accurate estimation of the population-scaled mutation rate, theta, from microsatellite genotype data. Genetics 176(2), 1363–1366 (2007)
23. Pasaniuc, B., Rohland, N., McLaren, P.J., Garimella, K., Zaitlen, N., Li, H., Gupta, N., Neale, B.M., Daly, M.J., Sklar, P., Sullivan, P.F., Bergen, S., Moran, J.L., Hultman, C.M., Lichtenstein, P., Magnusson, P., Purcell, S.M., Haas, D.W., Liang, L., Sunyaev, S., Patterson, N., de Bakker, P.I.W., Reich, D., Price, A.L.: Extremely low-coverage sequencing and imputation increases power for genome-wide association studies. Nat. Genet. 44(6), 631–635 (2012)
24. Li, Y., Sidore, C., Kang, H.M., Boehnke, M., Abecasis, G.R.: Low-coverage sequencing: implications for design of complex trait association studies. Genome Res. 21(6), 940–951 (2011)
25. Novembre, J., Johnson, T., Bryc, K., Kutalik, Z., Boyko, A.R., Auton, A., Indap, A., King, K.S., Bergmann, S., Nelson, M.R., Stephens, M., Bustamante, C.D.: Genes mirror geography within europe. Nature 456(7218), 98–101 (2008)
26. Yang, W.Y., Novembre, J., Eskin, E., Halperin, E.: A model-based approach for analysis of spatial structure in genetic data. Nature Genetics 44(6), 725–731 (2012)
27. Baran, Y., Quintela, I., Carracedo, A., Pasaniuc, B., Halperin, E.: Enhanced localization of genetic samples through linkage-disequilibrium correction. Am. J. Hum. Genet. (May 2013)
28. Mardis, E.R.: The impact of next-generation sequencing technology on genetics. Trends Genet. 24(3), 133–141 (2008)
29. Schuster, S.C.: Next-generation sequencing transforms today's biology. Nature Methods 5(1), 16–18 (2008)
30. Shendure, J., Mitra, R.D., Varma, C., Church, G.M.: Advanced sequencing technologies: methods and goals. Nat. Rev. Genet. 5(5), 335–344 (2004)
31. Howie, B., Marchini, J., Stephens, M.: Genotype imputation with thousands of genomes. G3: Genes, Genomes, Genetics 1(6), 457–470 (2011)
32. Paaniuc, B., Avinery, R., Gur, T., Skibola, C.F., Bracci, P.M., Halperin, E.: A generic coalescent-based framework for the selection of a reference panel for imputation. Genetic Epidemiology 34(8), 773–782 (2010)
33. Liu, E.Y., Li, M., Wang, W., Li, Y.: Mach-admix: Genotype imputation for admixed populations. Genetic Epidemiology 37(1), 25–37 (2013)
34. Howie, B., Fuchsberger, C., Stephens, M., Marchini, J., Abecasis, G.R.: Fast and accurate genotype imputation in genome-wide association studies through pre-phasing. Nature Genetics (2012)
35. Nelson, M.R., Wegmann, D., Ehm, M.G., Kessner, D., St. Jean, P., Verzilli, C., Shen, J., Tang, Z., Bacanu, S.A., Fraser, D., Warren, L., Aponte, J., Zawistowski, M., Liu, X., Zhang, H., Zhang, Y., Li, J., Li, Y., Li, L., Woollard, P., Topp, S., Hall, M.D., Nangle, K., Wang, J., Abecasis, G., Cardon, L.R., Zöllner, S., Whittaker, J.C., Chissoe, S.L., Novembre, J., Mooser, V.: An Abundance of Rare Functional Variants in 202 Drug Target Genes Sequenced in 14,002 People. Science 337(6090), 100–104 (2012)

Traversing the *k*-mer Landscape of NGS Read Datasets for Quality Score Sparsification

Y. William Yu, Deniz Yorukoglu, and Bonnie Berger

Massachusetts Institute of Technology, Cambridge MA 02139, USA
bab@mit.edu
http://people.csail.mit.edu/bab/

Abstract. It is becoming increasingly impractical to indefinitely store raw sequencing data for later processing in an uncompressed state. In this paper, we describe a scalable compressive framework, Read-Quality-Sparsifier (RQS), which substantially outperforms the compression ratio and speed of other de novo quality score compression methods while maintaining SNP-calling accuracy. Surprisingly, RQS also improves the SNP-calling accuracy on a gold-standard, real-life sequencing dataset (NA12878) using a *k*-mer density profile constructed from 77 other individuals from the 1000 Genomes Project. This improvement in downstream accuracy emerges from the observation that quality score values within NGS datasets are inherently encoded in the *k*-mer landscape of the genomic sequences. To our knowledge, RQS is the first scalable sequence-based quality compression method that can efficiently compress quality scores of terabyte-sized and larger sequencing datasets.

Availability: An implementation of our method, RQS, is available for download at: http://rqs.csail.mit.edu/.

Keywords: RQS, quality score, sparsification, compression, accuracy, variant calling.

1 Introduction

In the past two decades, genomic sequencing capabilities have increased exponentially, outstripping advances in computing power and storage [1, 2]. Capitalizing on this deluge of data is heavily dependent on the ability to efficiently store, process and extract meaningful biological insights from sequencing datasets, which is becoming correspondingly more difficult as more data is generated.

Early studies on compressing NGS datasets have mainly focused on compressing sequence data itself, aiming to leverage the inherent redundancy present in read sequences to reduce the space needed for storing 'raw' reads [3–7]. Furthermore, Loh et al. [8] demonstrated that it is possible to further exploit this redundancy through the use of succinct data structures that allow us to operate directly on the compressed data, saving CPU time as well as space.

That said, the Phred quality scores encoding the "base-calling confidence" take up more than twice the space on disk as the read sequence itself

R. Sharan (Ed.): RECOMB 2014, LNBI 8394, pp. 385–399, 2014.
© Springer International Publishing Switzerland 2014

($\sim 2.3x - 2.8x$ for Illumina reads). Furthermore, it is more challenging to compress the quality scores [9], as they not only have a larger alphabet size (ranging from 63 to 94 characters depending on the sequencing technology), but also have limited repetitive patterns as well as little direct correlation with the bases sequenced.

Computational methods for NGS quality compression fall into two main camps: lossless compression methods that include general-purpose text-compressors, such as GZIP, BZIP2 or 7zip, as well as methods specialized to exploit the local similarity of quality values for further compression [7, 10, 11]; and lossy compression methods that aim to achieve further compression by sacrificing the ability to reconstruct the original quality values [9, 12, 13].

A recent area of investigation in quality score compression is exploiting sequence read information within NGS datasets in order to boost the compression of quality scores. Most of these methods need to compute expensive whole-genome alignments of the NGS read dataset, then use additional position and coverage information obtained from the alignments to compress quality values [4, 6, 14]. However, it has been shown that alignment-agnostic methods can also utilize sequence data to achieve better compression of quality values [15], but again require costly operations (e.g. BWT) to be run on the entire read dataset. Thus far neither approach has been able to truly address the scalability problem of quality score compression for terabyte-sized or larger NGS datasets.

In this paper, we introduce a highly efficient, scalable, and alignment-free k-mer based algorithm, "Read-Quality-Sparsifier" (RQS), which sparsifies quality score values by smoothing a large fraction of quality score values based on the k-mer neighborhood of their corresponding positions in the read sequences. In particular, RQS constructs a comprehensive database, or dictionary, of commonly occurring k-mers throughout a population-sized read dataset. Once this database is constructed it can be used to compress any given read dataset by identifying k-mers within each read that have a small Hamming distance from the database; assuming that any divergent base in a k-mer likely corresponds to a SNP or machine error, we preserve quality scores for probable variant locations and discard the rest.

Our "coarse" representation of quality scores leads to great savings in storage. Throwing away this much information significantly improves the compression ratios, allowing us on average to store quality scores at roughly 0.4 bits per value (from the original size of $6 - 7$ bits). The scalability of our method arises from the fact that the k-mer database needs to be constructed only once for any given species, and the quality sparsification stage is very efficient. As a surprising result, our quality sparsification method not only significantly outperforms other de novo quality compression methods based on efficiency and compression ratio, but is also able to improve downstream variant calling accuracy. The improvement in downstream SNP-calling accuracy of the compressed dataset emerges from the fact that base calling confidences within NGS datasets are inherently encoded in the k-mer landscape of genomic sequences. Notably, supported by experimental results on annotated real data, **our study demonstrates that**

k-mer density profiles of read sequences are more informative on average than ~95% of the quality score information.

We validate RQS on real NGS exome and genome read datasets taken from 1000 Genomes Project Phase 1 [16]. Specifically, we demonstrate the superior compression ratio and efficiency of our method on Illumina read sequences of 77 British individuals (see Appendix B), comparing the variant call accuracy to other de novo quality compression methods. We also give preliminary results showing that, compared to a ground truth genotype annotation of NA12878, RQS compression achieves better downstream SNP-call accuracy compared to the uncompressed quality values.

While lossy compression of quality scores has not been widely adopted by biologists due to loss of precision [6], our RQS method remediates this effect by improving the accuracy of downstream genotyping applications. It does so by capitalizing on the k-mer landscape of a read dataset. The usefulness of k-mer frequencies for inferring knowledge about the error content of a read sequence has been studied [17–19]—in fact, many sequence-correction and assembly methods directly or indirectly make use of this phenomenon [20–23]; however, to our knowledge, RQS is the first such method to traverse the k-mer landscape for quality score compression, thereby improving efficiency, compression-ratio, and accuracy.

2 Methods

At a high level, RQS is divided into two separate stages (Figure 1). In the first preprocessing stage, we generate a dictionary, D, of all k-mers that appear with high multiplicity in a representative collection (corpus) of reads. In the second sparsification stage, we look at the k-mers in a read. k-mers that are close to our dictionary (as measured by Hamming distance) have nearly all of their quality scores discarded. RQS keeps only the low quality scores for bases where the k-mer differs from the dictionary.

More precisely, with NGS read data, we are given a corpus C of reads with depth coverage t of some consensus sequence G. We will assume an independent accuracy rate of p for each base call, and let $q = 1 - p$ be the variation rate (whether caused by machine error or by a SNP). As usual, we identify reverse complements together. Let γ_k be the multiset of all k-mers that appear in a read γ, counting multiplicity. Similarly, let G_k and C_k be respectively the multiset of all k-mers that appear in G and the reads of C.

Let $\Delta(x, y)$ be the Hamming distance between two k-mers x and y, and let $\Delta(x, D)$ be the minimum Hamming distance from x to any k-mer in D. Then we generate a dictionary of all k-mers that appear at least r times in C_k, which approximates G_k for r, k and p sufficiently large. These "good" k-mers are then used to identify high confidence base calls in reads: if a k-mer is within Hamming distance 1 from D, we assign high confidence to all locations where there is concordance among it and all its Hamming neighbors in D. Each read can be covered by overlapping k-mers, allowing us to identify high confidence base calls in that case as well.

The sparsification procedure then consists of two separate steps. First, we discard quality values for all high confidence base calls. Then, we discard quality values for all base calls above some threshold value $\mathcal{Q}$. In our implementation, for later downstream analysis, we replace all discarded values with $\mathcal{Q}$.

Below we present pseudocode for a simplified version of RQS (an efficient implementation is described in appendix A). DICT(C, k, r) (alg. 1) takes the corpus of reads C and returns a list D of all k-mers that appear at least r times. Using Hamming distance from D, MARK_KMER(x, D) (alg. 2) generates a boolean vector marking each position in a k-mer x that corresponds to a high-confidence call. MARK_READ(γ, D) (alg. 3) then repeatedly calls MARK_KMER to generate the vector of high-confidence calls in read γ. SPARSIFY_RQ$(\gamma, Q, D, \mathcal{Q})$ (alg. 4) calls MARK_READ to locate high-confidence calls and then discards both the corresponding quality scores and quality scores above a cut-off threshold $\mathcal{Q}$.

Input: C, k, r
Output: D
  $D = \{\,\}$
  $A = [0, \ldots, 0] \in \mathbb{N}^{4^k}$
  for $x \in C_k$ **do**
    $A[x]+\,=1$
  for $x \in [4^k]$ **do**
    if $A[x] \geq r$ **then**
      D.append(x)
  return D

Algorithm 1. DICT(C, k, r): Generates a dictionary of all k-mers that appear at least r times in a corpus C of reads

Input: x, D
Output: M
  if $\Delta(x, D) > 1$ **then**
    $M = [F, \ldots, F] \in \{T, F\}^k$
  else
    $M = [T, \ldots, T] \in \{T, F\}^k$
    for $y \in D$ s.t. $\Delta(x, y) = 1$ **do**
      for $i \in [k]$ **do**
        if $x_i \neq y_i$, **then** $M[i] = F$
  return M

Algorithm 2. MARK_KMER(x, D): Marks high confidence locations in a k-mer x using a dictionary D

Input: γ, D
Output: $\mathcal{M}$
  Let x^a be the k-mer in γ starting at a.

  Cover γ by k-mers $\{x^{a_1}, \ldots, x^{a_n}\}$.
  for $i \in [n]$ **do**
    $M^i = $ MARK_KMER(x^{a_i}, D)
    $\overline{M}^i = [F, \ldots, F] \in \{T, F\}^{length(\gamma)}$
    for $j \in [k]$ **do** $\overline{M}^i_{j+a_i-1} = M^i_j$
  $\mathcal{M} = \overline{M}^1$ OR $\cdots$ OR $\overline{M}^n$
  return $\mathcal{M}$

Algorithm 3. MARK_READ(γ, D): Marks high confidence calls in read γ using dictionary D

Input: $\gamma, Q, D, \mathcal{Q}$
Output: Q'
  $Q' = Q$
  $\mathcal{M} = $ MARK_READ(γ, D)
  for $i \in length(\gamma)$ **do**
    if $(Q_i > \mathcal{Q})$ OR $(\mathcal{M}_i = T)$ **then**
      $Q'_i = \mathcal{Q}$
  return Q'

Algorithm 4. SPARSIFY_RQ $(\gamma, Q, D, \mathcal{Q})$: Sparsifies the quality vector Q associated with read γ using a dictionary D; discarded qualities replaced with $\mathcal{Q}$

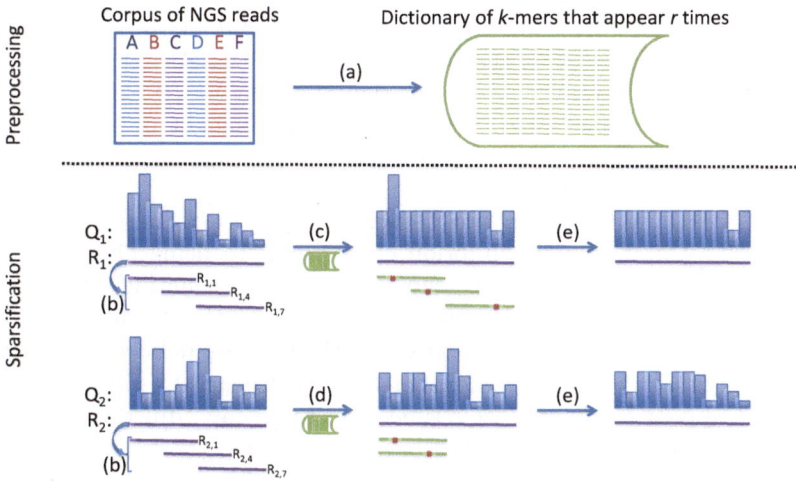

Fig. 1. Preprocessing. (a) A dictionary of k-mers that appear at least r times in a corpus of NGS reads is generated. **Sparsification.** R is a read sequence and Q is a corresponding quality score string. $R_{i,j}$ is a k-mer within R_i starting at position j. (b) We choose multiple k-mers to cover the read sequence. If a particular k-mer is within Hamming distance 1 of a k-mer in the dictionary, then we can sparsify the quality scores that correspond to the positions in the k-mer. (c) $R_{1,1}, R_{1,4}$, and $R_{1,7}$ are exactly distance 1 from the dictionary. We mark the locations where they do not match the dictionary k-mer and smooth all the other quality scores. Note that although $R_{1,4}$ has a mismatch, that location is still smoothed because the location is covered by $R_{1,1}$, which does not have a corresponding mismatch at the same position in the read. (d) Only $R_{2,1}$ is within Hamming distance 1 of the dictionary. However, it has two Hamming neighbors in the dictionary. We only smooth the quality scores where there is concord among all Hamming neighbors and $R_{2,1}$. Neither $R_{2,4}$ nor $R_{2,7}$ contribute because they are too far away from the dictionary. (e) Last, we smooth all quality scores above a threshold.

2.1   Theoretical Guarantees

For completeness, we present an analysis of the composition of D, showing that under certain conditions, $D \approx G_k$. Let $\Sigma = \{A, C, G, T\} \cong \mathbb{Z}/4\mathbb{Z}$ be the alphabet, and let $G \in \Sigma^N$ be the consensus sequence—note, we will assume that we are not given G. Let the multiset counting multiplicity of all k-length substrings of G be denoted by G_k.

We construct an idealized variation model combining both machine error in the base call and the presence of SNPs. Let $p \in [0,1]$ and $q = 1 - p$. Let the random variable $\sigma : [0,1] \to \Sigma$ be defined by

$$\sigma(\omega) = \begin{cases} 0, & \text{if } \omega \in [0, p) \\ 1, & \text{if } \omega \in [p, p + q/3) \\ 2, & \text{if } \omega \in [p + q/3, 1 - q/3) \\ 3, & \text{if } \omega \in [1 - q/3, 1] \end{cases} \tag{1}$$

Thus $\forall l \in \Sigma$, $l + \sigma = l$ with probability p. This is to say that a base is read correctly with probability p and incorrectly with probability q.

Given $x \in \Sigma^k$, define x_i as the ith letter (base) of x. For all $x \in \Sigma^k$, define independently the Σ^k-valued random variables R^x by $\forall i, R_i^x = x_i + \sigma_i$, where $\sigma_i, \ldots, \sigma_k$ are i.i.d. copies of σ. Thus, R^x can be thought of as a read of x, including machine errors and SNPs. Let $\hat{G}_k \equiv \{R^x | x \in G_k\}$. $\hat{G}_k$ thus corresponds to a version of G_k with noise.

We are given t independent noisy copies $\hat{G}_{k,1}, \ldots, \hat{G}_{k,t}$ of G_k, but with a low error rate $q > 0$. This assumption corresponds to being given reasonably accurate reads covering the target genome t times and counting all k-mers. We want to recover a dictionary D_r approximating G_k (without multiplicity) from the collection $\bar{G}_k \equiv \hat{G}_{k,1}, \ldots, \hat{G}_{k,t}$. Let C_k be the multiset defined as the disjoint union of $\hat{G}_{k,1}, \ldots, \hat{G}_{k,t}$. We will construct the dictionary by simply taking all k-mers that appear in C_k at least r times, where r is an adjustable parameter. Intuitively, this process should work because provided the variation rate q is small, $R^x = x$ often so there will be many exact copies of x in C_k if $x \in G_k$; however, then $R^x = y$ only rarely for $x \in G_k$ and $y \notin G_k$, so there will not be many copies of y in C_k.

Let $\Delta(x, y)$ be the Hamming distance between $x, y \in \Sigma^k$. Let $\Delta(x, G_k) = \min_{y \in G_k} \Delta(x, y)$. Let $\alpha(x) = |\{i : x \in \hat{G}_{k,i}\}|$, the number of times x appears in C_k. Let us denote the i.i.d. copies of R^x in $\hat{G}_{k,1}, \ldots, \hat{G}_{k,t}$ by $R^{x,1}, \ldots, R^{x,t}$. Then for $x \in G_k$,

$$\mathbb{P}(x \in \hat{G}_{k,j}) \geq \mathbb{P}(x = R^{x,j}) \geq p^k, \tag{2}$$

which is just the chance that a noise-free version of x was stored in $\hat{G}_{k,j}$. Let $\mathbb{1}_{x \in \hat{G}_{k,j}}$ be an indicator variable for the event $\{x \in \hat{G}_{k,j}\}$. Then

$$(\alpha(x)|x \in G_k) = \sum_{j=1}^{t} \mathbb{1}_{x \in \hat{G}_{k,j}} \implies \mathbb{E}(\alpha(x)|x \in G_k) \geq tp^k. \tag{3}$$

Furthermore, by applying a Chernoff bound, for any $\delta_1 > 0$,

$$\mathbb{P}\left((\alpha(x)|x \in G_k) < (1 - \delta_1)tp^k\right) \leq e^{\frac{-\delta_1^2 tp^k}{2 + \delta_1}}. \tag{4}$$

Recall that we defined our dictionary $D_r \equiv \{x \in C_k | \alpha(x) \geq r\}$, which consists of all members of C_k with multiplicity at least r. Then,

$$\mathbb{E}\left|\{x \in D_{tp^k(1-\delta_1)} | x \in G_k\}\right| \geq |unique(G_k)| \left(1 - \exp\left(\frac{-\delta_1^2 tp^k}{2 + \delta_1}\right)\right), \tag{5}$$

where $unique(G_k)$ is the k-mer set obtained by discarding multiplicity of k-mers in G_k. So long as tp^k is reasonably large and $(1 - \delta_1)$ is not very big, most of G_k is expected to fall in $D_{tp^k(1-\delta_1)}$.

We also want to be able to say that most elements not in G_k do not fall in D_r. For simplicity of analysis, let us assume that all G_k is well-separated and sparse in Σ^k (NB: this assumption does not hold for repetitive regions in the genome) so that for $x, y \in G_k$, $\mathbb{P}(R^x = R^y | x \neq y)$ is negligible. Then we can separately consider for each $x \in G_k$, the number of collisions among $R^{x,1}, \ldots, R^{x,t}$. $\mathbb{P}(R^x = y | \Delta(x,y) = d) = \binom{k}{d} p^{k-d} q^d \leq (kq)^d$. If $kq < 1$, the probability mass decreases and is spread thinner for higher d, so it is sufficient to bound collisions conditional upon all the probability mass staying within Hamming distance 1. Note that this assumption also implies that $|unique(G_k)| = |G_k|$.

By symmetry, under these conditions, $\mathbb{P}(R^x = y | \Delta(x,y) = 1) = \frac{1}{3k}$ since there are $3k$ possible positions for R^x to go. Let $\mathbb{1}_{R^x = y}$ be an indicator variable. Then for every y,

$$(\alpha(y) | \Delta(x,y) = 1) = \sum_{j=1}^{t} \mathbb{1}_{R^{x,j} = y} \implies \mathbb{E}(\alpha(y) | \Delta(x,y) = 1) = \frac{t}{3k}. \tag{6}$$

By again applying Chernoff, for any $\delta_2 > 0$,

$$\mathbb{P}\left((\alpha(y) | \Delta(x,y) = 1) > (1 + \delta_2) \frac{t}{3k} \right) \leq e^{\frac{-\delta_2^2 t}{3k(2 + \delta_2)}}. \tag{7}$$

Putting it all together,

$$\mathbb{E}\left| \{x \in D_{\frac{t}{3k}(1 + \delta_2)} | x \notin G_k \} \right| \leq |G_k| 3k \exp\left(\frac{-\delta_2^2 t}{3k(2 + \delta_2)} \right). \tag{8}$$

Thus, equation 8 shows that as read multiplicity r increases D_r contains exponentially fewer k-mers that are not in G_k. Additionally, equation 5 shows that if r is small compared tp^k, D_r contains nearly all of G_k. Whether or not there exists a value of r that makes D_r sufficiently close to G_k for our purposes is of course dependent on the exact parameters k, q, and t. However, because the simplifying assumptions we made do not perfectly reflect real data, instead of attempting to compute r, we swept over different values of r in our results section. These bounds do however show that as coverage t grows, there are parameters for which D_r asymptotically approaches G_k. As we demonstrate in the results, we are close enough to that regime for accurate and effective compression.

3   Results and Discussion

RQS performs impressively in terms of both compression rates and effects on downstream variant calling. In the first experiments, we demonstrate that RQS' performance is superior to existing methods in a more careful analysis of effects on downstream variant calling on chromosome 21 using the gold standard of NA12878. Most interestingly, RQS improves downstream variant calling, despite throwing away most of the quality scores. The second experiment demonstrates RQS' ability to successfully scale to the whole human genome by using a sampling algorithm for generating the dictionary.

Datasets. We generated our dictionary from a subset of the reads from the genomes of 77 British individuals with data taken from the 1000 Genomes Project, Phase 1 (see Appendix B for details). Read lengths ranged from 50-110bp, and there was a total depth coverage of 460 across all 77 individuals. Variant calling was performed using samtools [24], and BZIP2 was used to further compress the sparsified quality scores. For chromosome 21 analyses, we first filtered the reads in our corpus by those mapping to chromosome 21 using BWA [25] in combination with GATK [26]. For the whole human genome, we again used the same 77 British individuals and considered all reads mapping to any chromosome using a sampling approach.

Parameters. We chose $k = 32$ for two reasons. Importantly, 32-mers can be stored efficiently in 64-bit numbers, facilitating both ease of implementation and runtime. Additionally, our theoretical results were dependent on k-mers from our corpus being sparse; thus, k needs to be sufficiently large. We chose $Q = 40$ as 40 was close to the average quality score in the corpus.

Choice of read multiplicity r for inclusion in the dictionary is highly dependent on read depth of the corpus. Additionally, our theoretical guarantees assume random distribution of k-mers in the corpus, which is not necessarily true for repetitive regions of the human genome. Thus, we used several different values of r in our experiments. For our last analysis though, we sweep over k-mer multiplicity $r = 25, 50, 100, 200, 350, 550$ to provide some guidance as to the trade-offs involved.

3.1   RQS Out-Performs Existing Compression Methods

Here we show that our method offers a reduction in size over state-of-the-art lossless algorithms currently available and performs at least as well as lossy algorithms we have encountered in the literature. We measured the performance of several different compressors on the quality scores of HG02215, chromosome 21. General purpose lossless text compressors naturally have perfect fidelity for downstream variant calling, but also are unable on their own to achieve compression levels better than roughly 3.6 bits per quality score, which is only just over a 50% reduction in space. Of the three general purpose compressors we used, 7zip (PPMd) outperformed both BZIP2 and GZIP.

Read-Quality-Sparsifier, as implemented in this paper, only makes use of redundancy by replacing scores with the constant threshold value Q, and so must be paired with one of the general purpose compressors to compress the modified scores. Surprisingly, the relative efficacy of compression after preprocessing with RQS with $r = 50$ (see Section 3.4 for choice of read multiplicity parameter) did not match that of the original scores; indeed, RQS + BZIP2 was by far the most effective combination, using only 0.2540 bits per quality value for storage.

In addition to the usual general purpose compressors, we also compared our compressive framework to QualComp [27], which features a tuning parameter to specify the number of bits needed for quality scores per read. We chose to sweep the QualComp parameter, bits per read, to match RQS' compression level and accuracy. As displayed in Table 1, RQS performs considerably better on accuracy than QualComp when a comparable compression level was chosen. Indeed, for QualComp to match the F-measure of RQS, QualComp needed over 6 times the space.

Lastly, we compared against the method of Janin et al, 2013 [15], which uses a computation-intensive Burrows-Wheeler transform and least-common-prefixes table to smooth high-confidence reads. To ensure comparability, we display in Table 1 best-in-class parameters taken from those used in their paper. Although one set of parameters did have slightly higher accuracy than RQS, it came at the cost of needing over 10 times the disk space. Even at 5 times less compression, their method was not as accurate at variant calling, as measured by the balanced F-measure.

Table 1. Relative compression rates of different compressors on the HG02215, chromosome 21 dataset. For each method, best results with respect to F-measure are bolded. Note that QualComp has its own quality storage format, whereas for Janin et al, we used 7zip (PPMd) to postprocess the smoothed quality scores (as this gave the best results for their method).

Method	Size	Bits/Q	Precision	Recall	F-measure
Uncompressed	273 MiB	8.0000	1	1	1
GZIP	143 MiB	4.1923	1	1	1
BZIP2	133 MiB	3.8791	1	1	1
7zip (PPMd)	**124 MiB**	**3.6269**	**1**	**1**	**1**
RQS ($r = 50$) + GZIP	14 MiB	0.3825	0.9867	0.9963	0.9914
RQS ($r = 50$) + BZIP2	**8.7 MiB**	**0.2540**	**0.9867**	**0.9963**	**0.9914**
RQS ($r = 50$) + 7zip (PPMd)	11 MiB	0.2935	0.9867	0.9963	0.9914
QualComp (25 bits/read)	9.4 MiB	0.2747	0.8988	0.9934	0.9436
QualComp (50 bits/read)	19 MiB	0.5494	0.9329	0.9943	0.9625
QualComp (100 bits/read)	38 MiB	1.0988	0.9746	0.9949	0.9846
QualComp (150 bits/read)	**57 MiB**	**1.6482**	**0.9874**	**0.9957**	**0.9914**
Janin et al (2013) ($c = 0, s = 1, r = 29$)	3.4 MiB	0.0987	0.9279	0.9754	0.9509
Janin et al (2013) ($c = 1, s = 1, r = 29$)	4.4 MiB	0.1284	0.9283	0.9751	0.9510
Janin et al (2013) ($c = 1, s = 5, r = 29$)	43 MiB	1.2402	0.9903	0.9887	0.9894
Janin et al (2013) ($c = 0, s = 5, r = 29$)	43 MiB	1.2416	0.9902	0.9887	0.9894
Janin et al (2013) ($c = 0, s = 10, r = 29$)	**90 MiB**	**2.6336**	**0.9953**	**0.9944**	**0.9948**

3.2   Comparison of Effects on Genotyping Accuracy by RQS, QualComp, and Janin et al

In the previous section, we considered any differences from the genotype information obtained from the original uncompressed data to be errors. In this section we show preliminary results suggesting that the differences between the SNP-calls of the RQS-compressed data and uncompressed data do not necessarily indicate errors, but actually some of them are corrections of the false positive and false negative SNP-calls obtained using the uncompressed data, improving the overall area-under-curve (AUC) of the SNP-calls. To demonstrate this, we tested our compressive framework on NA12878 genome, an extensively studied individual within the 1000 Genomes Project, for which a high-quality trio-validated genotype annotation is available [28].

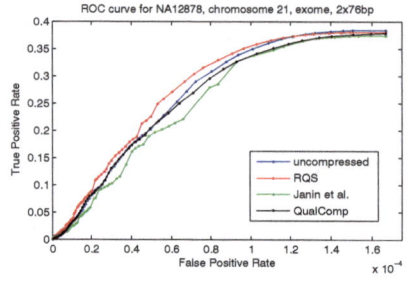

Method	BitsQ	Time (s)	AUC ($\times 10^{-05}$)
Uncompressed	8	0	4.4733
RQS, $r = 50$	0.2264	19	4.6239
QualComp†	0.2631	130	4.3499
Janin et al.*	0.7216	251	4.1903

† QualComp was run with parameters set so that 20 bits per read were used.

* RQS was run with parameters $c = 0$, $s = 2$, $r = 29$.

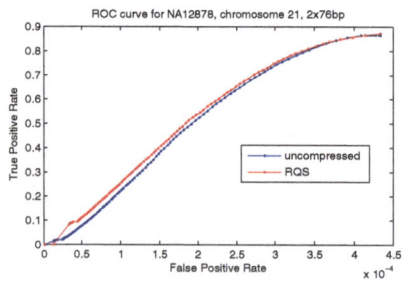

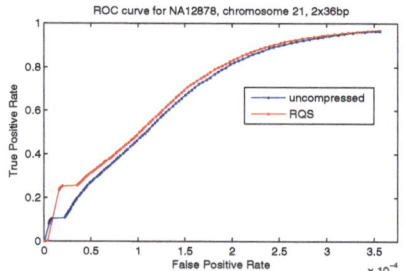

Fig. 2. ROC curves of genotype calls made from the NA12878 dataset, both before and after compression, for exome reads of length 76bp, genome reads of length 36bp and 76bp. For the exome dataset (top), we also compared genotype call accuracies to QualComp and Janin et al. AUC values are included, integrated up to the largest of the maximum false positive rates to ensure comparability. To our knowledge, RQS is unique in improving accuracy, as measured by AUC, through compression. Note that the maximum true positive rate in the exome dataset is limited by the total fraction of variants covered by exome reads.

Table 2. Depth-of-coverage and read length information for datasets used in Figure 2

Dataset	Read depth-of-coverage	Read length
NA12878, chromosome 21 (76bp)	16.9	2x76bp
NA12878, chromosome 21 (36bp)	22.5	2x36bp
NA12878, exome, chromosome 21	5.15	2x76bp

In order to compress the quality scores of NGS reads of NA12878 chromosome 21, we reused the dictionary generated from the set of 77 British individuals, of which NA12878 is not a member. Table 2 shows the read length, coverage and sequence type of the NGS datasets used for this experiment.

For each read set, we used samtools to call SNPs over the uncompressed dataset and computed the ROC curve using annotated variant locations in chromosome 21. Then the same test was performed after compressing the quality scores. As it is demonstrated in Figure 2, even though the ROC curves occasionally cross, AUC values favor the RQS-compressed data. Because the exome dataset was much smaller, it was more tractable for multiple comparisons with other compression tools. Thus, for the exome dataset specifically, we also compare the ROC curves of Qualcomp and Janin et al. at comparable compression levels. Unlike the other methods, RQS does not decrease AUC when compared to that of the uncompressed SNP calls. Furthermore, it is also demonstrably faster.

3.3 Compressing Large-Scale Whole-Genome Sequencing Datasets with RQS

We demonstrate that RQS scales to the whole genome using a probabilistic dictionary construction algorithm, as counting the exact number of appearances in the corpus of k-mers of a large whole-genome NGS dataset of 77 individuals is computation and memory intensive. Our sampling-based dictionary construction method identifies 32-mers that in expectation appear at least 500 times in the original corpus—by taking only 32-mers that appear at least 5 times in a randomly-chosen 1% of the reads.

Compressing the sampled reads using this method resulted in an average compression of 1.934 bits per quality score. Compressing all the reads—both sampled and unsampled—of HG02215 (one representative genome from the corpus) resulted in an average compression of 1.8406 bits/score. As is shown in the next section, these compression levels are better than the rates achieved with $r = 350$ or $r = 550$ when analyzing just chromosome 21 (Figure 3). These preliminary results indicate that we can effectively perform dictionary generation with a probabilistic sampling scheme and also larger (and potentially more redundant) genome sequences can facilitate better compression.

3.4 Effect of Read Multiplicity on Compression and Accuracy

We have used several different values for read multiplicity r. Here, we apply RQS to the chromosome 21 reads in the corpus itself while sweeping over r to demonstrate the effect on compression and accuracy of variant calling; F-measure, precision, and recall are measured against the variant calls generated from the uncompressed data. Note that it is the ratio of read multiplicity to total depth coverage (here, 460) that matters, rather than absolute read multiplicity. However, since each read is independently compressed, once the dictionary is generated from the corpus, read coverage no longer affects the compression ratio.

As depicted in Figure 3, across multiple r values, RQS is able to maintain high fidelity with respect to the variants called in the uncompressed dataset. Furthermore, note that the loss of fidelity at $r < 200$ is not necessarily negative, as the improvements in accuracy shown in Figure 2 naturally require that the SNP-calls after compression differ from those for the uncompressed quality scores. However, we observe that compression rates become dramatically worse with r values higher than 200. This indicates that, for $r > 200$, our dictionary becomes too sparse to achieve high compression ratios.

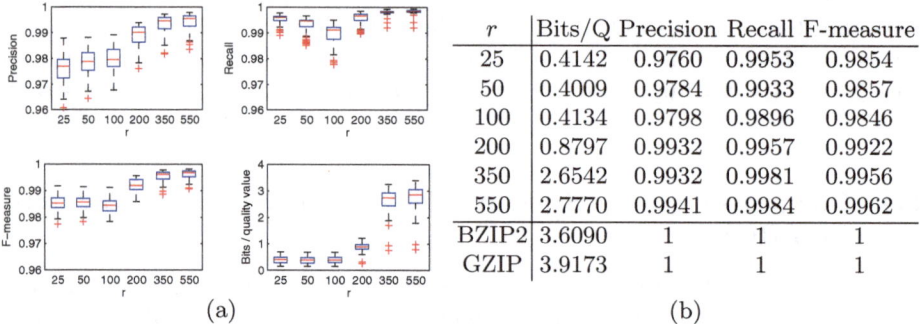

r	Bits/Q	Precision	Recall	F-measure
25	0.4142	0.9760	0.9953	0.9854
50	0.4009	0.9784	0.9933	0.9857
100	0.4134	0.9798	0.9896	0.9846
200	0.8797	0.9932	0.9957	0.9922
350	2.6542	0.9932	0.9981	0.9956
550	2.7770	0.9941	0.9984	0.9962
BZIP2	3.6090	1	1	1
GZIP	3.9173	1	1	1

(a)   (b)

Fig. 3. (a) Box plots of precision, recall, F-measure, and compression ratios across multiple values of read multiplicity r; each data point corresponds to one of the 77 genomes. Compression rates shown here include post-processing with BZIP2. (b) Table of averages for the same data shown in (a), along with compression levels of BZIP2 and GZIP.

4   Conclusions

Here we have shown that our RQS compressive framework capitalizes on the redundancies in the k-mer landscape of NGS read data to discard and sparsify nearly all quality scores, while at the same time enhancing compression ratio, speed, and downstream genotyping accuracy.

Acknowledgements. Y.W.Y. gratefully acknowledges support from the Fannie and John Hertz Foundation. This research was partially supported by the National Institutes of Health (NIH) R01GM108348. This content is solely the responsibility of the authors and does not reflect the official views of the NIH.

References

1. Berger, B., Peng, J., Singh, M.: Computational solutions for omics data. Nature Reviews Genetics 14, 333–346 (2013)
2. Kahn, S.D.: On the future of genomic data. Science 331(6018), 728–729 (2011)
3. Apostolico, A., Lonardi, S.: Compression of biological sequences by greedy off-line textual substitution. In: Proceedings of the Data Compression Conference, DCC 2000, pp. 143–152. IEEE (2000)
4. Kozanitis, C., Saunders, C., Kruglyak, S., Bafna, V., Varghese, G.: Compressing genomic sequence fragments using SlimGene. Journal of Computational Biology 18(3), 401–413 (2011)
5. Jones, D.C., Ruzzo, W.L., Peng, X., Katze, M.G.: Compression of next-generation sequencing reads aided by highly efficient de novo assembly. Nucleic Acids Research 40(22), e171 (2012)
6. Fritz, M.H.Y., Leinonen, R., Cochrane, G., Birney, E.: Efficient storage of high throughput DNA sequencing data using reference-based compression. Genome Research 21, 734–740 (2011)
7. Deorowicz, S., Grabowski, S.: Compression of DNA sequence reads in FASTQ format. Bioinformatics 27(6), 860–862 (2011)

8. Loh, P.R., Baym, M., Berger, B.: Compressive genomics. Nature Biotechnology 30, 627–630 (2012)
9. Bonfield, J.K., Mahoney, M.V.: Compression of FASTQ and SAM format sequencing data. PloS one 8(3), e59190 (2013)
10. Hach, F., Numanagic, I., Alkan, C., Sahinalp, S.C.: SCALCE: boosting sequence compression algorithms using locally consistent encoding. Bioinformatics 28(23), 3051–3057 (2012)
11. Tembe, W., Lowey, J., Suh, E.: G-SQZ: compact encoding of genomic sequence and quality data. Bioinformatics 26(17), 2192–2194 (2010)
12. Popitsch, N., von Haeseler, A.: NGC: lossless and lossy compression of aligned high-throughput sequencing data. Nucleic Acids Research 41(1), e27 (2013)
13. Wan, R., Anh, V.N., Asai, K.: Transformations for the compression of FASTQ quality scores of next-generation sequencing data. Bioinformatics 28(5), 628–635 (2012)
14. Christley, S., Lu, Y., Li, C., Xie, X.: Human genomes as email attachments. Bioinformatics 25(2), 274–275 (2009)
15. Janin, L., Rosone, G., Cox, A.J.: Adaptive reference-free compression of sequence quality scores. Bioinformatics (2013)
16. Consortium, T.G.P.: An integrated map of genetic variation from 1,092 human genomes. Nature 491, 1 (2012)
17. Yang, X., Chockalingam, S.P., Aluru, S.: A survey of error-correction methods for next-generation sequencing. Briefings in Bioinformatics 14(1), 56–66 (2013)
18. Melsted, P., Pritchard, J.K.: Efficient counting of k-mers in DNA sequences using a bloom filter. BMC Bioinformatics 12(1), 333 (2011)
19. Marçais, G., Kingsford, C.: A fast, lock-free approach for efficient parallel counting of occurrences of k-mers. Bioinformatics 27(6), 764–770 (2011)
20. Kelley, D.R., Schatz, M.C., Salzberg, S.L., et al.: Quake: quality-aware detection and correction of sequencing errors. Genome. Biol. 11(11), 116 (2010)
21. Liu, Y., Schröder, J., Schmidt, B.: Musket: a multistage k-mer spectrum-based error corrector for Illumina sequence data. Bioinformatics 29(3), 308–315 (2013)
22. Ilie, L., Molnar, M.: RACER: Rapid and accurate correction of errors in reads. Bioinformatics 29(19), 2490–2493 (2013)
23. Grabherr, M.G., Haas, B.J., Yassour, M., Levin, J.Z., Thompson, D.A., Amit, I., Adiconis, X., Fan, L., Raychowdhury, R., Zeng, Q., et al.: Full-length transcriptome assembly from RNA-Seq data without a reference genome. Nature Biotechnology 29(7), 644–652 (2011)
24. Li, H., Handsaker, B., Wysoker, A., Fennell, T., Ruan, J., Homer, N., Marth, G., Abecasis, G., Durbin, R., et al.: The sequence alignment/map format and SAMtools. Bioinformatics 25(16), 2078–2079 (2009)
25. Li, H., Durbin, R.: Fast and accurate long-read alignment with Burrows–Wheeler transform. Bioinformatics 26(5), 589–595 (2010)
26. DePristo, M.A., Banks, E., Poplin, R., Garimella, K.V., Maguire, J.R., Hartl, C., Philippakis, A.A., del Angel, G., Rivas, M.A., Hanna, M., et al.: A framework for variation discovery and genotyping using next-generation DNA sequencing data. Nature Genetics 43(5), 491–498 (2011)
27. Ochoa, I., Asnani, H., Bharadia, D., Chowdhury, M., Weissman, T., Yona, G.: QualComp: a new lossy compressor for quality scores based on rate distortion theory. BMC Bioinformatics 14, 187 (2013)
28. Consortium, T.G.P.: A map of human genome variation from population-scale sequencing. Nature 467, 1061–1073 (2010)

A   Experimental Setup and Implementation of RQS

In the Results section, in order to perform an in-depth comparison between the performance of RQS and other compression methods on NGS reads from a large number of individuals, we reduced the domain of the genome to chromosome 21 ($\sim 1.5\%$ of whole genome in size).

In sections 3.1 and 3.2, read alignments were generated using BWA [25] in combination with GATK (Genome Analysis Toolkit) [26]. After chromosome 21 alignments were extracted, positional alignment information within mappings were removed before running RQS and other compression schemes.

In section 3.3, we used the reads aligned to the whole genome (instead of only chromosome 21), again removing all positional information from the alignments. Due to the large number of possible k-mers, it was not possible to fit the entire hash table into memory during dictionary generation. Though it is possible to compute exact k-mer frequencies using a parallelization approach, here we designed a less computation and memory intensive sampling-based dictionary construction method and demonstrated that the quality of the constructed dictionary is not affected.

In the RQS implementation, the dictionary generation from the corpus of reads was implemented using an unordered set data type in C++. For detecting all reads within Hamming distance 1 to the dictionary D; we stored D in a Boost multi-index hash table with 4 keys. Each key covers 24 out of 32 bases and each base of the 32-mer is covered exactly by 3 keys. Defining the key in this way allows us to aggressively prune most 32-mers within the dictionary, guaranteeing that each matching key implies a close match $\Delta(x, D) \leq 8$. If the total number of matches is < 96, we check the remaining nucleotides to verify that the matching 32-mers are within Hamming distance of 1. Otherwise, we enumerate all 96 neighboring 32-mers of the query 32-mer and check whether any of these exist in the dictionary.

During the sparsification step of our experiments, we chose $\mathcal{Q}$, the replacement quality value for discarded and smoothed positions, to be "40". Although this is a user-defined parameter within our implementation, we selected this value as it was close to the average quality score value in our dataset.

After the sparsification step was completed, as a rough measure of entropy, we ran the quality scores through BZIP2 and recorded the number of bits per quality score required for storage. Although a production implementation would probably use a custom file format storing only a subset of the quality scores, using a modification of the standard SAM file format allowed for easy input into downstream analysis tools.

As a lossy compression method, there are two separate criteria upon which we can compare different methods: (1) the compression ratio and (2) the accuracy of downstream analysis. For the comparison between compression ratios, we fixed an F-measure threshold of 0.99 and searched for the best compression ratio of different tools across a range of input parameters, satisfying the minimum accuracy threshold. For fixed accuracy-level, RQS displayed superior compression rates compared to other methods (see Table 1). For the comparison between downstream variant call accuracies, we fixed a compression rate of 25x and searched for the best downstream variant call accuracy (with respect to the uncompressed data) of different tools across a range of parameters, satisfying minimum compression rate. For each fixed compression-level, RQS gave the most accurate F-measure values.

B   Data Sources

Corpus from 1000 Genomes Project Phase 1:

```
HG00096 HG00100 HG00103 HG00106 HG00108 HG00111 HG00112 HG00114 HG00115
HG00116 HG00117 HG00118 HG00119 HG00120 HG00122 HG00123 HG00124 HG00125
HG00126 HG00127 HG00131 HG00133 HG00136 HG00137 HG00138 HG00139 HG00140
HG00141 HG00142 HG00143 HG00145 HG00146 HG00148 HG00149 HG00150 HG00151
HG00152 HG00154 HG00155 HG00156 HG00157 HG00158 HG00159 HG00160 HG00231
HG00232 HG00233 HG00236 HG00237 HG00239 HG00242 HG00243 HG00244 HG00245
HG00246 HG00247 HG00249 HG00250 HG00251 HG00252 HG00253 HG00254 HG00256
HG00257 HG00258 HG00259 HG00260 HG00261 HG00262 HG00263 HG00264 HG00265
HG01334 HG01789 HG01790 HG01791 HG02215
```

Exome reads for NA12878: `ftp://ftp-trace.ncbi.nih.gov/1000genomes/ftp/phase1/data/NA12878/exome_alignment/NA12878.mapped.illumina.mosaik.CEU.exome.20110411.bam`

Human Genome 18: `ftp://ftp-trace.ncbi.nih.gov/1000genomes/ftp/technical/working/20101201_cg_NA12878/Homo_sapiens_assembly18.fasta`

Reconstructing Breakage Fusion Bridge Architectures Using Noisy Copy Numbers

Shay Zakov* and Vineet Bafna

Department of Computer Science and Engineering,
University of California, San Diego, CA, USA
szakov@eng.ucsd.edu

Abstract. The *Breakage Fusion Bridge* (*BFB*) process is a key marker for genomic instability, producing highly rearranged genomes in relatively small number of cell cycles. While the process itself was observed during the late 1930's, little is known about the extent of BFB in tumor genome evolution. This is partly due to methodological requiring the rare observation of a spontaneous BFB occurence, or rigorous assays for identifying BFB-modified genomes after the process has ceased. Moreover, BFB can dramatically increase copy numbers of chromosomal segments, which in turn hardens the tasks of both reference assisted and *ab initio* genome assembly.

Based on available data such as *Next Generation Sequencing* (NGS) and *Array Comparative Genomic Hybridization* (aCGH) data, we show here how BFB evidence may be identified, and how to predict all possible evolutions of the process with respect to observed data. Specifically, we describe practical algorithms that, given a chromosomal arm segmentation and noisy segment copy number estimates, produce all segment count vectors supported by the data that can be produced by BFB, and all corresponding BFB architectures. This extends the scope of analyses described in our previous work, which produced a single count vector and architecture per instance.

We apply these analyses to a comprehensive human cancer dataset, demonstrate the effectiveness and efficiency of the computation, and suggest methods for further assertions of candidate BFB samples. An online Appendix, the source code of our tool, and analyses results, are available at http://cseweb.ucsd.edu/~vbafna/bfb.

1 Introduction

The origin of a tumor cell is marked by genomic instability [9]. Spontaneous, viral, or other kinds of mechanisms may cause genomic segment deletions, duplications, translocations, inversions, etc., producing rearranged genomes with a possibly malignant nature. Thus, decoding mechanisms that generate rearranged genomes is critical to understanding cancer. Numerous mechanisms were proposed, including the faulty repair of double-stranded DNA breaks by recombination or end-joining and polymerase hopping caused by replication fork

---

* Corresponding author.

R. Sharan (Ed.): RECOMB 2014, LNBI 8394, pp. 400–417, 2014.
© Springer International Publishing Switzerland 2014

collapse [5,10]. These mechanisms are generally not directly observable, so their elucidation requires the deciphering of often subtle clues after genomic instability has ceased. An important source of information in this respect is the *architecture* of the rearranged genome, i.e. the description of its chromosomes in terms of concatenations of segments from the original genome.

Breakage Fusion Bridge (BFB) is one model of a genome rearrangement process, which was first proposed by Barbara McClintock in the 1930's [13,14]. Recently, it has seen renewed interest as a possible mechanisms in tumor genome evolution [3,4]. BFB begins with a telomeric loss on a chromosome, including a loss of a sequential pattern that signals the location of chromosome termination. During cell division the telomere-lacking chromosome replicates, and its two sister chromatids fuse together (possibly due to some DNA repair mechanism falsely induced by the cell). This fusion produces a dicentric chromosome of palindromic structure, which is later torn apart at some random point as the centromeres of the dicentric chromosome migrate to opposite poles of the cell. One part of the torn chromosome includes the fusion region and some tandemly inverted chromosomal suffix duplication, and the other part lacks the corresponding suffix. The two daughter cells receive these rearranged chromosomes, both are missing the telomeric region, and the cycle can repeat again (Fig. 1).

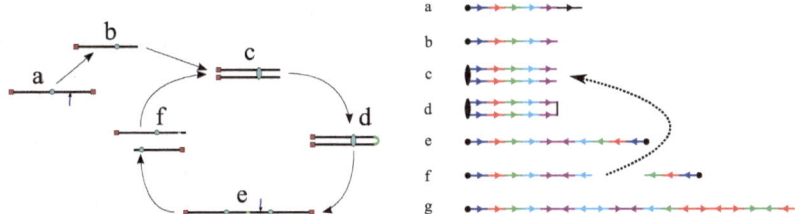

Fig. 1. The BFB process. (a) A normal chromosome. (b) The chromosome looses its telomere. (c) The chromosome is duplicated during cell devision. (d) Sister chromatids are fused together. (e) Centromeres migrate to opposite poles of the cell. (f) The fused chromosome is torn apart at some random position, causing one copy to have an inverted suffix duplication, while the other copy has a trimmed suffix. Both copies lack a telomere, and therefore may undergo additional BFB cycles. (g) After several BFB cycles, the chromosome architecture exhibits significant increases in segment copy numbers, as well as fold-back patterns.

In contrast to other mechanisms, BFB can actually be observed in progress using methods that have been available for decades [14]. Cytogenetic techniques can reveal the anaphase bridges, dicentric chromosomes, and homogeneously staining regions that have long been the canonical evidence for BFB. However, these techniques are useful only in cases where the BFB cycles are ongoing. While useful in understanding the mechanism, they do not address the question of whether BFB occurs extensively in evolving tumor genomes.

Recently, researchers (including us) have started looking at modern available data in order to demonstrate BFB occurrence after the process has ceased, including *Fluorescent In Situ Hybridization* (FISH), *Array Comparative Genomic Hybridization* (aCGH), and *Next Generation Sequencing* (NGS) data. These methods take advantage of distinctive BFB features exposed by such data, including the abundance of fold-back inversions (i.e. duplicated chromosomal segments arranged in a head-to-head orientation) [3,4], patterns of interleaving segments of alternating orientations [12,17], and combinatorial properties of segment counts when copy number variations are due to BFB [11,20]. In fact, if the architecture of the rearranged genome is known, it is possible to decide if this architecture can be produced by BFB [11].

Partial knowledge regarding the architecture can be reveled by FISH analyses [12], which uses fluorescence markers to identify the physical locations of predetermined sequences on the rearranged genome. However, such experiments are relatively expensive, and can only be performed in a small number of cases. A more common measurement is NGS data, which contain a big set of short sequenced reads extracted from a donor genome. Such data is typically used for predicting the entire donor genomic sequence by computationally assembling the reads, sometimes facilitated by consulting a similar pre-sequenced reference genome. Unfortunately, BFB and other mechanisms can produce massively rearranged and highly repetitive genomes. This hardens the task of assembly-based sequencing due to the multiple ambiguous manners the repetitive reads may be assembled, and the lack of a relevant reference template. Nevertheless, NGS data can still be analyzed in order to infer some indirect information regarding the donor genome architecture [1,6,15,19]. After aligning the reads against a reference genome, their genomic location distribution can be used in order to identify segments on the reference genome of coherent read coverage, and to estimate the number of times each such segment repeats in the donor genome. We will refer to the output of the latter kind of analysis as *copy number data*. Other methods to obtain copy number data are based on analyzing aCGH data [7,8,16,18] (Fig. 2). Due to the noisy nature of both NGS and aCGH data, count estimates may be inaccurate, and the true segment count is likely to fall within some interval of integers around the estimated value. We use the term *noisy copy number data* when referring to information regarding such intervals of possible count values. In addition to copy number data, NGS data can be used in order to produce contigs (chromosomal segments which may be assembled unambiguously), and aberrant segment adjacencies can be exposed by discordant reads, restricting the set of possible contig-based architectures.

In previous work [11,20], we showed how to analyze noisy copy number data in order to decide is it likely to observe the input data under the assumption the underlying rearrangement process is BFB. Specifically, we designed algorithms that produce a single BFB architecture over the given segments in which segment counts are supported by the data, if such an architecture exists. We applied these algorithm in order to analyze a comprehensive aCGH dataset of cancer cell lines [2], as well as sequence data from primary tumors [4], and

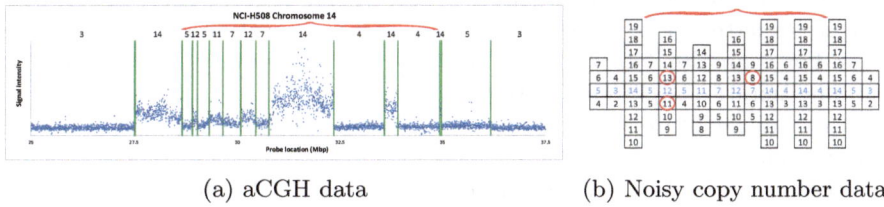

(a) aCGH data   (b) Noisy copy number data

Fig. 2. (a) aCGH data for a part of the q-arm of human chromosome 14 in the NCI-H508 cell line. Each data point corresponds to a probe on the array, where its x-coordinate gives the probe's sequence chromosomal position, and y-coordinate gives its measured intensity (log-ratio). The data points are clustered into segments, and an estimated segment copy number appears above each segment. (b) Possible count intervals around the estimated counts. The counts in the region under the red curly bracket are supported by a BFB architecture, if changing the count estimate of the second segment in this region from 12 to 13 or to 11, and of the seventh segment from 7 to 8. Data is taken from [2] (segmentation and copy number analysis were computed using the PICNIC software [8]).

identified a small subset of candidate samples exhibiting BFB hallmarks. Here, we extend the scope of the analysis, and describe algorithms that report *all* count settings supported by the data which can be explained by BFB, and all corresponding BFB architectures. Although the theoretical time bounds for these new algorithms may be exponential, we show that in practice they are efficient, and apply an Informed Search optimization that further improves their practical efficiency.

Our proposed algorithms satisfy an important need, therefore. While our work postulates the existence of BFB using statistical arguments, additional physical assertions can be obtained with FISH and aberrant read analyses. Starting with noisy copy number data, our tool can be used to enumerate all possible BFB architectures. These candidate architectures can then be used towards a small set of FISH experiments (with a limited number of fluorescence markers) to validate and refine the genomic architecture.

2   Formalism and Previous Results

Computational BFB-related problems were previously formulated in [11,20]. For completeness, we give here the main definitions from these works.

A *DNA segment* σ is a string over the DNA nucleotide alphabet A, C, G, T. The *reversed segment* of a segment σ, denoted here by $\overline{\sigma}$, is the string obtained by reading σ backwards, and replacing each nucleotide with its complementary nucleotide ($A \leftrightarrow T, C \leftrightarrow G$). For example, the reverse of a segment $\sigma = CGGAT$ is the segment $\overline{\sigma} = ATCCG$. In the rest of this paper, it is assumed we operate on a given chromosomal arm with a fixed segmentation, and denote its list of k segments by $\Sigma = \{\sigma_1, \sigma_2, \ldots, \sigma_k\}$, ordered from the centromeric segment σ_1 to the telomeric segment σ_k. The term "string" refers to a genomic architecture over these segments, i.e. a concatenation of segments from Σ and their reversed forms. Greek letters $\alpha, \beta, \gamma, \rho$ denote strings, and bar notation indicates reversed strings. For example, if $\alpha = \sigma_1\sigma_3\overline{\sigma}_2$, $\overline{\alpha} = \sigma_2\overline{\sigma}_3\overline{\sigma}_1$. An empty string is denoted by ε. The notation $\alpha_{l,t}$ represents the string $\sigma_l\sigma_{l+1}\ldots\sigma_t$ (thus when $t < l$, $\alpha_{l,t} = \varepsilon$). To facilitate reading, $\sigma_1, \sigma_2, \sigma_3, \ldots$ are replaced by A, B, C, $\ldots$ in concrete examples.

A *BFB cycle* applied over a chromosomal arm can be viewed as a special rearrangement procedure, in which some telomeric suffix of the arm is duplicated, inverted, and concatenated tandemly at the telomeric end of the arm. A *BFB process* is a consecutive application of BFB cycles. This notion is formally captured by the following definition.

Definition 1. *For two strings α, β, say that $\alpha \xrightarrow{BFB} \beta$ if $\alpha = \beta$, or $\alpha = \rho\gamma$ for some strings ρ, γ such that $\gamma \neq \varepsilon$, and $\rho\gamma\overline{\gamma} \xrightarrow{BFB} \beta$. Say that α is an l-BFB string if $\alpha_{l,t} \xrightarrow{BFB} \alpha$ for some t, and say that α is a BFB string if it is an l-BFB string for some l.*

Note that by definition $\varepsilon = \alpha_{l,l-1}$ is an l-BFB string for every $l \geq 1$. The *count vector* $\vec{n}(\alpha) = [n_1, n_2, \ldots, n_k]$ of a string α is a vector of integers, where for every $1 \leq l \leq k$, n_l is the total number of occurrences of σ_l and $\overline{\sigma}_l$ in α. For example, for $\alpha = \text{ABCD}\overline{\text{D}}\overline{\text{C}}\text{C}$, $\vec{n}(\alpha) = [1,1,3,2]$. Say that a vector $\vec{n}$ is a *BFB vector* if there exists some BFB string α such that $\vec{n} = \vec{n}(\alpha)$. In the previous example $\vec{n}(\alpha)$ is a BFB vector, due to the BFB process $\alpha_{1,4} = \text{ABCD} \xrightarrow{BFB} \text{ABCD}\overline{\text{D}}\overline{\text{C}} \xrightarrow{BFB} \text{ABCD}\overline{\text{D}}\overline{\text{C}}\text{C} = \alpha$.

The computational analyses presented in this paper aim to detect evidence for BFB, given a pre-analyzed segmentation of the genome and corresponding copy number data. We assume that noisy copy number data is represented by a *weight function* $W = \{w_{l,n} \mid 1 \leq l \leq k, n = 0,1,2,\ldots\}$, where $w_{l,n}$ is a nonnegative *weight* of the count n with respect to the l-th segment. It may be assumed w.l.o.g. that all weights $w_{l,n}$ satisfy $0 \leq w_{l,n} \leq 1$. The *weight* of a count vector $\vec{n} = [n_1, n_2, \ldots, n_k]$ is given by $W(\vec{n}) = \prod\limits_{1 \leq i \leq k} w_{i,n_i}$, and by assumption $0 \leq W(\vec{n}) \leq 1$. In some cases, we refer to prefixes $\vec{n}_{1,l-1} = [n_1, n_2, \ldots, n_{l-1}]$ and suffixes $\vec{n}_{l,k} = [n_l, n_{l+1}, \ldots, n_k]$ of $\vec{n}$, which may be empty if $l = 1$ or $l = k+1$, respectively. Define the weights of such sub-vectors accordingly, i.e. $W(\vec{n}_{1,l-1}) = \prod\limits_{1 \leq i < l} w_{i,n_i}$ and $W(\vec{n}_{l,k}) = \prod\limits_{l \leq i \leq k} w_{i,n_i}$, where the weight of an empty vector is 1 by definition. Thus, for every $1 \leq l \leq k+1$, $W(\vec{n}) = W(\vec{n}_{1,l-1}) \cdot W(\vec{n}_{l,k})$.

If some data analysis produces segment count probabilities $\Pr(n_l = n)$ for every segment σ_l and every count $n = 0,1,2,\ldots$, weights can be set to these probabilities choosing $w_{l,n} = \Pr(n_l = n)$. This way, the weight of a count vector is the probability this vector reflects the true segment counts, given the observed data. Another way to set weights given such probabilities would be to choose weights by setting $w_{l,n} = \frac{\Pr(n_l=n)}{\Pr(n_l=n_l^*)}$, where n_l^* is the most likely count for the l-th segment. Here, the weight of a count vector gives the ratio between its probability and the probability of a most likely vector. Nevertheless weights are more general than probabilities, and can be used as a heuristic count error modeling even when no probabilistic model is available.

In [20], several variants of BFB problems where formulated. Below we restate these problems, and add two new variants addressed in the current work:

BFB Problem Variants
Input: *a count vector $\vec{n}$, or a weight function W and a weight $0 < \eta \leq 1$.*

1. **The decision variant** [20]: *given $\vec{n}$, decide if $\vec{n}$ is a BFB vector.*
2. **The string search variant** [20]: *if $\vec{n}$ is a BFB vector, find a BFB string α such that $\vec{n} = \vec{n}(\alpha)$.*
3. **The vector search variant** (or **the distance variant** in [20]): *given W and η, report a maximum weight BFB vector $\vec{n}$ in case there exists such a vector with $W(\vec{n}) \geq \eta$, and otherwise report "FAILED".*
4. **The exhaustive vector search variant:** *given W and η, report all BFB vectors $\vec{n}$ with $W(\vec{n}) \geq \eta$.*
5. **The exhaustive string search variant:** *given W and η, report all BFB strings α such that $W(\vec{n}(\alpha)) \geq \eta$.*

For a count vector $\vec{n}$, define $N(\vec{n}) = \sum_{1 \leq l \leq k} n_l$ and $\tilde{N}(\vec{n}) = \sum_{1 \leq l \leq k} \log(n_l)$.

Note that $N(\vec{n})$ is the total length of a string admitting $\vec{n}$, and $\tilde{N}(\vec{n})$ is proportional to the number of bits needed for representing $\vec{n}$. For a weight function W and a weight η, define $N(W, \eta) = \max \{N(\vec{n}) : W(\vec{n}) \geq \eta\}$, and $\tilde{N}(W, \eta) = \max \left\{ \tilde{N}(\vec{n}) : W(\vec{n}) \geq \eta \right\}$. In [20], it was shown that the BFB decision variant can be solved using $O(\tilde{N}(\vec{n}))$ bit operations (i.e. linear time in the input length), the string search variant can be solved in $O(N(\vec{n}))$ operations (i.e. linear time in the output length), and that the vector search variant can be solved using at most a sub-exponential number of operations $2^{O(\log^2 N(W,\eta))}$. Here, we give algorithms for the two new exhaustive search variants. While theoretically the output of these algorithms can be exponential with respect to $N(W, \eta)$, we show that for realistic inputs this output is manageable. In addition, we describe an *Informed Search* (IS) approach that significantly reduces the running time in practice by eliminating irrelevant search paths and traversing only paths which are guaranteed to produce valid solutions. Next, we describe some ideas taken from [20], upon which the algorithms presented here are built.

An *l-BFB palindrome* is an *l*-BFB string of the form $\beta = \alpha\overline{\alpha}$. It can be shown that $\beta = \alpha\overline{\alpha}$ is an *l*-BFB palindrome if and only if α is an *l*-BFB string. By definition, $\varepsilon = \varepsilon\overline{\varepsilon}$ is an *l*-BFB palindrome for every $l \geq 1$. In addition, observe that when $\beta = \alpha\overline{\alpha}$ we have that $\vec{n}(\beta) = 2\vec{n}(\alpha)$. This allows to replace the question "is there a BFB string admitting the count vector $\vec{n}$" by the equivalent question "is there a BFB palindrome admitting the count vector $2\vec{n}$".

An *l-block* is a string of the form $\beta = \sigma_l\beta'\overline{\sigma}_l$, where β' is an $(l + 1)$-BFB palindrome. It can be shown that an *l*-block is a special form of an *l*-BFB palindrome, and that every *l*-BFB palindrome is some palindromic concatenation of

l-blocks (though not every palindromic concatenation of l-blocks is a valid l-BFB palindrome). These observations allow to adopt a "layered" view of BFB palindromes, as follows (Fig. 3). Let $\beta = \alpha\bar{\alpha}$ be a 1-BFB palindrome, where $\vec{n}(\beta) = 2\vec{n}(\alpha) = [2n_1, 2n_2, \ldots, 2n_k]$. Therefore, β is a palindromic concatenation of 1-blocks, and denote by B^1 the collection of all these blocks. Every 1-block in B^1 is a string of the form $A\beta'\bar{A}$, where β' is some 2-BFB palindrome. As there are $2n_1$ occurrences of A and $\bar{A}$ in β, and each block in B^1 contains exactly two such occurrences, the total number of blocks in B^1 is exactly n_1. Masking the letters A and $\bar{A}$ from all blocks in B^1, the collection becomes a 2-BFB palindrome collection of size n_1. The 2-BFB palindromes in this collection can be further decomposed into 2-blocks, yielding a collection B^2 of 2-blocks. Similarly as above, B^2 contains exactly n_2 blocks. This process can continue inductively, yielding for every $1 \le l \le k$ a corresponding collection B^l of l-blocks, whose size is n_l. One may also imagine an additional collection in this series B^{k+1}, containing zero $(k+1)$-blocks.

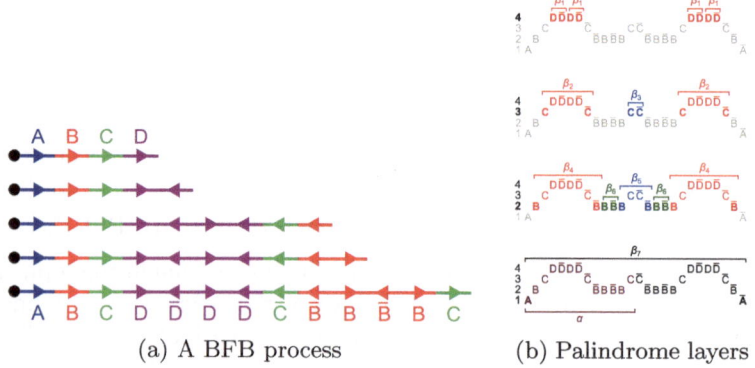

(a) A BFB process   (b) Palindrome layers

Fig. 3. (a) A BFB process generating a string α: ABCD $\xrightarrow{\text{BFB}}$ ABCDD̄ $\xrightarrow{\text{BFB}}$ ABCDD̄D̄D̄C̄B̄ $\xrightarrow{\text{BFB}}$ ABCDD̄D̄D̄C̄B̄B $\xrightarrow{\text{BFB}}$ ABCDD̄D̄D̄C̄B̄B̄B̄BC. (b) The layers of the BFB palindrome $\beta = \alpha\bar{\alpha}$. The blocks in each layer are marked with annotations of the form β_i.

This layered view is exploited in a reversed order by the algorithms in [20], developing a BFB palindrome given an input count vector $\vec{n} = [n_1, n_2, \ldots, n_k]$: Starting with an empty collection B^{k+1} of $(k+1)$-blocks, the algorithm computes iteratively a sequence of collections $B^k, B^{k-1}, \ldots, B^1$, each collection B^l is an l-block collection of size n_l. In order to generate B^l, the algorithm first concatenates $(l+1)$-blocks from B^{l+1}, forming a collection B of $(l+1)$-BFB palindromes of size n_l (this procedure is called *folding*). Then, each $(l+1)$-BFB palindrome $\beta' \in B$ is wrapped with a pair of σ_l segments, rendering it into an l-block $\beta = \sigma_l\beta'\bar{\sigma}_l$, and B^l is set to be the collection containing all these l-blocks. The final collection of 1-blocks B^1 is folded one more time into a single 1-BFB palindrome $\beta = \alpha\bar{\alpha}$, and the algorithm returns the half-length prefix α of this palindrome as a BFB string admitting the input count vector $\vec{n}$.

Fig. 3b illustrates a possible run of the algorithm over the input count vector $\vec{n} = [1, 5, 3, 4]$. First, the algorithm initializes an empty collection of 5-blocks B^5. In the first iteration, there is a need to perform concatenations of blocks in B^5, and produce $n_4 = 4$ 5-BFB palindromes. Such 5-BFB palindromes may only be obtained by concatenating zero elements (as there are no elements in B^5), and so four empty strings are generated in this folding process, yielding the 5-BFB palindrome collection $\{4\varepsilon\}$. Next, each 5-BFB palindrome in this collection is wrapped by $\sigma_4 = \mathrm{D}$ and $\bar{\sigma}_4 = \bar{\mathrm{D}}$, producing the collection of 4-blocks $B^4 = \{4\mathrm{D}\varepsilon\bar{\mathrm{D}}\} = \{4\beta_1\}$. In the next iteration, the collection B^4 needs to reduce its size from $n_4 = 4$ into $n_3 = 3$ by concatenating its elements to produce 3-BFB palindromes. In this example, there are two concatenations of two elements the form $\beta_1\beta_1$, and one concatenation of zero elements that produces an empty string ε. The 4-BFB palindromes in the resulting folded collection $\{2\beta_1\beta_1, \varepsilon\}$ are wrapped by $\sigma_3 = \mathrm{C}$ and $\bar{\sigma}_3 = \bar{\mathrm{C}}$, yielding the 3-block collection $B^3 = \{2\mathrm{C}\beta_1\beta_1\bar{\mathrm{C}}, \mathrm{C}\varepsilon\bar{\mathrm{C}}\} = \{2\beta_2, \beta_3\}$. This process continues for two more iterations, generating similarly the collections $B^2 = \{2\beta_4, \beta_5, 2\beta_6\}$ and $B^1 = \{\beta_7\}$. All elements in the last collection B^1 are then concatenated into a single 1-BFB palindrome β (in this example B^1 contains a single element β_7, and so $\beta = \beta_7$), and the returned string α is the half-length prefix of this palindrome.

The ability of the schematic algorithm above to process the entire input vector $\vec{n}$ and produce a corresponding BFB string depends on its ability to fold intermediate collections B^l computed along its run. In cases where it cannot fold some intermediate block collection, it returns a fail message, implying no BFB string admits the input vector $\vec{n}$. A case where folding cannot be applied is for example the case where $n_2 = 2$, $B^2 = \{\mathrm{BC}\bar{\mathrm{C}}\bar{\mathrm{B}}, \mathrm{B}\bar{\mathrm{B}}\}$, and $n_1 = 1$. In this case, since both possible concatenations $\mathrm{BC}\bar{\mathrm{C}}\bar{\mathrm{B}}\mathrm{B}\bar{\mathrm{B}}$ and $\mathrm{B}\bar{\mathrm{B}}\mathrm{BC}\bar{\mathrm{C}}\bar{\mathrm{B}}$ of the two elements in B^2 are non-palindromic, the folding procedure must fail at this stage. Another example of a fail folding is the case where $n_2 = 3$, $B^2 = \{\mathrm{BC}\bar{\mathrm{C}}\bar{\mathrm{B}}, 2\mathrm{B}\bar{\mathrm{B}}\}$, and $n_1 = 1$. In this case, though there exists a palindromic concatenation $\mathrm{B}\bar{\mathrm{B}}\mathrm{BC}\bar{\mathrm{C}}\bar{\mathrm{B}}\mathrm{B}\bar{\mathrm{B}}$ of all three elements in B^2, this concatenation is not a valid BFB palindrome (since any 2-BFB string containing the segment C must start with the prefix BC), and so the collection may not be folded.

In [20], it was shown that the ability to fold a block collection depends on a property called the *signature* of the collection. A signature of an l-BFB palindrome collection is an infinite sequence of integers $\vec{s} = [s_0, s_1, s_2, \ldots]$ with the following properties: (1) the first nonzero element in $\vec{s}$ (if there is such an element) must be positive, (2) the *cardinality* of the signature, defined by

$$\|\vec{s}\| = \sum_{d=0}^{\infty} 2^d \mathrm{abs}(s_d)$$ (where $\mathrm{abs}(s_d)$ is the absolute value of s_d), equals to the

size of the collection to which the signature corresponds, and (3) wrapping the collection (i.e. replacing each l-BFB palindrome β in the collection with an $(l-1)$-block $\sigma_{l-1}\beta\bar{\sigma}_{l-1}$) does not change its signature. In this sense, a signature can be thought of as a generalization of a binary representation of an integer, in which the coefficients may be other integers besides 0 and 1 (with the additional restriction of a positive first nonzero element, and the fact the absolute coefficient value is taken when computing the corresponding summation). The prefix of a signature $\vec{s}$ up to its d-th element is denoted by $\vec{s}_d = [s_0, s_1, \ldots, s_d]$. Due to being relatively technical, we omit here the formal signature definition and refer intrigue readers to [20] for a full explanation on how to derive collection signatures.

From the signature cardinality definition, it follows that for a signature $\vec{s}$ such that $\|\vec{s}\| = n$, all signature elements s_i for $i > \log n$ are zeros, thus signatures can be explicitly represented by a (small) finite number of nonzero elements. In addition, it follows that the only signature of an empty collection is $\vec{s} = [0, 0, \ldots]$, and the only signature of a collection containing a single element is $\vec{s} = [1, 0, \ldots]$ (the "$\ldots$" notation implies that the remaining signature elements are zeros). Otherwise, two collections of the same size may have different signatures. Signatures can be ranked according to their *lexicographic order*. That is, say that $\vec{s} < \vec{s}'$ if there exists an index d such that $\vec{s}_{d-1} = \vec{s}'_{d-1}$ and $s_d < s'_d$, and say that $\vec{s} \leq \vec{s}'$ if $\vec{s} < \vec{s}'$ or $\vec{s} = \vec{s}'$.

Lemma 1. *Let B be an l-block collection with a signature $\vec{s}$. For any folding B' of B and its corresponding signature $\vec{s}'$, $\vec{s} \leq \vec{s}'$. In addition, for any signature $\vec{s}'$ such that (1) $\vec{s} \leq \vec{s}'$ and (2) $\vec{s}'$ is the lexicographically minimal signature among all signatures of cardinality $\|\vec{s}'\|$ that meet (1), there exists a folding B' of B whose signature is $\vec{s}'$.*

The proof of Lemma 1 follows from Claims 14 and 28 in [20] (Supporting Information). The signatures corresponding to the 4 block collections implied by the BFB palindrome presented in Fig. 3b are $\vec{s}^4 = [0, 0, 1, 0, \ldots]$, $\vec{s}^3 = [1, -1, 0, \ldots]$, $\vec{s}^2 = [1, 0, -1, 0, \ldots]$, and $\vec{s}^1 = [1, 0, \ldots]$, respectively. Observe that the cardinality of each signature equals to the size of the corresponding collection (or the corresponding count in $\vec{n} = [1, 5, 3, 4]$), and that $\vec{s}^{l+1} \leq \vec{s}^l$ for every $1 \leq l < 4$.

It follows from Lemma 1 that a vector $\vec{n} = [n_1, n_2, \ldots, n_k]$ is a BFB count vector if and only if there exists a series of lexicographically non-increasing signatures $\vec{s}^1, \vec{s}^2, \ldots, \vec{s}^k$ such that $n_l = \|\vec{s}^l\|$ for every $1 \leq l \leq k$, and the first signature in this series satisfies $\vec{s}^1 \leq [1, 0, \ldots]$ (the signature of a collection with one element, due to the last concatenation of all 1-blocks in B^1 into a single palindrome). Call such a signature series a *valid signature series* for $\vec{n}$, and so we get the following conclusion:

Conclusion 1. *A vector $\vec{n}$ is a BFB vector if and only if it has a valid signature series. Moreover, any sub-sequence of a BFB vector is also a BFB vector, evident by the corresponding sub-series of a valid signature series for the full vector.*

For example, the vector $\vec{n} = [3, 4]$ is a BFB vector, due to the valid signature series $\vec{s}^1 = [1, -1, 0, \ldots]$, $\vec{s}^2 = [0, 0, 1, 0, \ldots]$. A corresponding BFB string may be obtained by AB $\xrightarrow{\text{BFB}}$ AB$\bar{\text{B}}$ $\xrightarrow{\text{BFB}}$ AB$\bar{\text{B}}\text{B}\bar{\text{A}}$ $\xrightarrow{\text{BFB}}$ AB$\bar{\text{B}}\text{B}\bar{\text{A}}\text{A}$. An example for a vector that does not have a valid signature series is the vector

$\vec{n} = [4, 3]$: the only signatures with cardinality 4 are the signatures $[0, 0, 1, 0, \ldots]$, $[0, 2, 0, \ldots]$, $[2, 1, 0, \ldots]$, $[2, -1, 0, \ldots]$, and $[4, 0, \ldots]$. Among these signatures, the only ones who lexicographically precede the signature $[1, 0, \ldots]$ are the signatures $[0, 0, 1, 0, \ldots]$ and $[0, 2, 0, \ldots]$. Nevertheless, the only signatures of cardinality 3 are $[1, -1, 0, \ldots]$, $[1, 1, 0, \ldots]$, and $[3, 0, \ldots]$, and none of them precedes the two possible 4-cardinality signatures.

As a matter of fact, restating Algorithm DECISION-BFB in [20] (Supporting Information), one can describe it as follows. Setting $\vec{s}^{k+1}$ to be the the the signature $[0, 0, \ldots]$ of an empty collection (which is also the lexicographically minimal among all signatures), the algorithm produces iteratively the signatures $\vec{s}^k, \ldots, \vec{s}^1$ in a valid signature series for the input vector $\vec{n} = [n_1, n_2, \ldots, n_k]$. Each signature $\vec{s}^l$ is obtained by applying the minimal lexicographic increment to $\vec{s}^{l+1}$ so that it would admit the cardinality $\|\vec{s}^l\| = n_l$. The algorithm returns true if and only if all increments are successful.

3 Algorithms

In this section we develop algorithms for the two exhaustive search variants of the BFB problem. To do so, we first describe some ideas and subroutines that would allow efficient implementations of these algorithms.

Let $\vec{n} = [n_1, n_2, \ldots, n_k]$ be a BFB vector, and let $1 \leq l \leq k + 1$. Define the *right-maximal signature* $R(\vec{n}_{1,l-1})$ of the prefix $\vec{n}_{1,l-1} = [n_1, n_2, \ldots, n_{l-1}]$ of $\vec{n}$ to be $[1, 0, \ldots]$ if $l = 1$, and otherwise to be the lexicographically maximal signature $\vec{s}^{l-1}$ in some valid signature series $\vec{s}^1, \ldots, \vec{s}^{l-1}$ for $\vec{n}_{1,l-1}$. Similarly, define the *left-minimal signature* $L(\vec{n}_{l,k})$ of the suffix $\vec{n}_{l,k} = [n_l, n_{l+1}, \ldots, n_k]$ of $\vec{n}$ to be $[0, 0, \ldots]$ if $l = k+1$, and otherwise to be the lexicographically minimal signature $\vec{s}^l$ in some valid signature series $\vec{s}^l, \ldots, \vec{s}^k$ for $\vec{n}_{l,k}$.

Lemma 2. *Let $\vec{n} = [n_1, n_2, \ldots, n_k]$ be a BFB vector. For every $1 \leq l' \leq l \leq k + 1$, $L(\vec{n}_{l,k}) \leq R(\vec{n}_{1,l-1})$, $R(\vec{n}_{1,l-1}) \leq R(\vec{n}_{1,l'-1})$, and $L(\vec{n}_{l,k}) \leq L(\vec{n}_{l',k})$.*

Proof. We start by showing the first inequality in the lemma. If $l = 1$ or $l = k+1$, $L(\vec{n}_{l,k}) \leq R(\vec{n}_{1,l-1})$ follows immediately. Otherwise, consider a valid signature series $\vec{s}^1, \vec{s}^2, \ldots, \vec{s}^k$ for $\vec{n}$. Note that its prefix $\vec{s}^1, \vec{s}^2, \ldots, \vec{s}^{l-1}$ is a valid signature series for $\vec{n}_{1,l-1}$, and its suffix $\vec{s}^l, \vec{s}^{l+1}, \ldots, \vec{s}^k$ is a valid signature series for $\vec{n}_{l,k}$. Thus, by definition, $L(\vec{n}_{l,k}) \leq \vec{s}^l \leq \vec{s}^{l-1} \leq R(\vec{n}_{1,l-1})$.

To show the second inequality in the lemma, let $\vec{s}^1, \vec{s}^2, \ldots, \vec{s}^{l-1}$ be a valid signature series for $\vec{n}_{1,l-1}$ such that $\vec{s}^{l-1} = R(\vec{n}_{1,l-1})$. Observe similarly as above that $R(\vec{n}_{1,l-1}) = \vec{s}^{l-1} \leq \vec{s}^{l'-1} \leq R(\vec{n}_{1,l'-1})$. The last inequality in the lemma is shown symmetrically. □

The MIN-DECREMENT procedure (Algorithm 1) gets as an input a signature $\vec{s}$ and an integer $n \geq 0$, and returns the lexicographically maximal signature $\vec{s}'$ such that $\vec{s}' \leq \vec{s}$ and $\|\vec{s}'\| = n$ if such a signature exists, and otherwise returns a fail message. Here, for an integer $m \neq 0$, the notation d_m represents the *parity degree* of m, which is defined to be the maximum integer d_m such that m divides by 2^{d_m}. Thus, for example, $d_{13} = d_{13 \cdot 2^0} = 0$, and $d_{-12} = d_{-3 \cdot 2^2} = 2$. The correctness of this computation is shown in the online Appendix. Symmetrically, the MIN-INCREMENT procedure gets as an input a signature $\vec{s}$ and an integer $n \geq 0$, and returns the lexicographically minimal signature $\vec{s}'$ such that $\vec{s} \leq \vec{s}'$ and $\|\vec{s}'\| = n$ if such a signature exists, and otherwise returns a fail message. The pseudo-code for this procedure is given in the online Appendix, and its proof is symmetric to that of the MIN-DECREMENT procedure.

Algorithm 1. MIN-DECREMENT$(\vec{s}, n)$

Input: A signature $\vec{s}$ and an integer $n \geq 0$.
Output: The lexicographically maximal signature $\vec{s}' \leq \vec{s}$ such that $\|\vec{s}'\| = n$, or the message "FAILED" if there is no such signature.

1 Let $m = \|\vec{s}\| - n$. **If** $m = 0$ **then return** $\vec{s}$.
2 **Else if** *there is an integer* $0 \leq d \leq d_m$ *such that* $n \geq \|\vec{s}_{d-1}\| + 2^d \max\{-s_d + 1, 0\}$ **then**
3   Let d be the maximum integer meeting the condition above. Initialize $\vec{s}'$ so that
       $\vec{s}'_{d-1} = \vec{s}_{d-1}$, and $s'_d = s_d - 2$ if $d < d_m$, or $s'_d = s_d - 1$ if $d = d_m$.
4   **If** $n \geq \|\vec{s}'_d\|$ **then** set $s'_{d+1} \leftarrow \frac{n - \|\vec{s}'_d\|}{2^{d+1}}$.
5   **Else** set $s'_d \leftarrow \frac{n - \|\vec{s}'_{d-1}\|}{2^d}$.
6   **Return** $\vec{s}'$.
7 **Else return** "FAILED".

Lemma 3. *If $\vec{n}_{1,l-1} = [n_1, \ldots, n_{l-1}]$ is a BFB vector, $\vec{s} = R(\vec{n}_{1,l-1})$, and $\vec{s}' = MIN\text{-}DECREMENT(\vec{s}, n_l)$, then $\vec{s}'$ is the right-maximal signature for the BFB vector $\vec{n}_{1,l} = [n_1, \ldots, n_{l-1}, n_l]$. Symmetrically, if $\vec{n}_{l+1,k} = [n_{l+1}, \ldots, n_k]$ is a BFB vector, $\vec{s} = L(\vec{n}_{l+1,k})$, and $\vec{s}' = MIN\text{-}INCREMENT(\vec{s}, n_l)$, then $\vec{s}'$ is the left-minimal signature for the BFB vector $\vec{n}_{l,k} = [n_l, n_{l+1}, \ldots, n_k]$.*

Proof. We show the first part of the lemma, where the second part is shown symmetrically. First, note that the constructed vector $\vec{n}_{1,l}$ is indeed a BFB vector, due to the corresponding valid signature series obtained by adding $\vec{s}'$ to a valid signature series for $\vec{n}_{l-1}$ whose last signature is $\vec{s}$. Note that $\|R(\vec{n}_{1,l})\| = \|\vec{s}'\| = n_l$. From Lemma 2 $R(\vec{n}_{1,l}) \leq \vec{s}$, and since $\vec{s}' = \text{MIN-DECREMENT}(\vec{s}, n_l)$ it follows that $R(\vec{n}_{1,l}) \leq \vec{s}'$. From the maximality of $R(\vec{n}_{1,l})$, $R(\vec{n}_{1,l}) = \vec{s}'$. $\qquad\square$

In the rest of this section, let W be a weight function, and $0 < \eta \leq 1$ some weight threshold. Let $0 \leq l \leq k$, and consider the set of all signature-weight pairs of the form $\langle R(\vec{n}_{1,l}), W(\vec{n}_{1,l}) \rangle$ such that $\vec{n}_l = [n_1, n_2, \ldots, n_l]$ is a BFB vector and $W(\vec{n}_{1,l}) \geq \eta$. Say that the pair $\langle \vec{s}, w \rangle$ within this set *dominates* the pair $\langle \vec{s}', w' \rangle$ if $\vec{s}' \leq \vec{s}$ and $w' \leq w$. Define the *l-th boundary curve* C^l with respect to W and η as the maximal subset of these pairs satisfying that no pair in C^l dominates another pair in C^l, and note that C^l is unique. Traversing the pairs in C^l from lowest to highest lexicographic signature rank, the series of signature values strictly increases, while the series of weight values strictly decreases, yielding a steps-like curve (Fig. 4). Algorithm 2 generates all boundary curves for W and η, which will later be exploited by algorithms for the BFB exhaustive vector and string search variants.

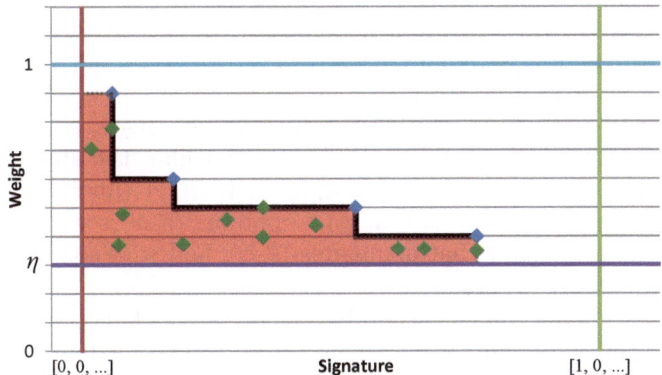

Fig. 4. A boundary curve. Points correspond to pairs of the form $\langle \vec{s}, w \rangle$, with x-coordinate reflecting the lexicographic rank of $\vec{s}$ and y-coordinate equals to w. Blue points belong to the boundary curve, and green points are dominated by points on the curve.

Algorithm 2. BOUNDARY-CURVES (W, η)

Input: A weight function W and a weight η.
Output: All boundary curves for W and η.

1 Set $C^0 \leftarrow \{\langle [1, 0, \ldots], 1 \rangle\}$.
2 **For** $l \leftarrow 1$ **to** k **do**
3   Set $C^l \leftarrow \emptyset$.
4   **For each** n and $\langle \vec{s}', w' \rangle \in C^{l-1}$ *s.t.* $w' \cdot w_{l,n} \geq \eta$ *and MIN-DECREMENT*$(\vec{s}', n)$
    does not fail **do**
5   Let $\vec{s}$ be the output of MIN-DECREMENT$(\vec{s}', n)$, and let $w = w' \cdot w_{l,n}$.
6   **If** $\langle \vec{s}, w \rangle$ *is not dominated by any pair in* C^l **then**
7   Add $\langle \vec{s}, w \rangle$ into C^l, and remove from C^l all pairs dominated by $\langle \vec{s}, w \rangle$.

8 **Return** $\left\{ C^0, C^1, \ldots, C^k \right\}$.

Proof (Algorithm 2). Note that a pair in C^0 corresponds to a right-maximal signature and a weight of an empty vector. By definition, the only such pair is the pair $\langle [1, 0, \ldots], 1 \rangle$, and the algorithm correctly sets C^0 to contain this single pair (line 1). Now, assuming inductively the algorithm has computed correctly the curve C^{l-1}, we prove it also computes correctly C^l. It is clear from lines 6 and 7 of the algorithm that no pair in the set C^l computed by the algorithm dominates another pair in this set. It is therefore remains to show that after the l-th loop iteration was executed (1) for every BFB vector $\vec{n}_{1,l} = [n_1, \ldots, n_l]$ with $W(\vec{n}_{1,l}) \geq \eta$ there exists a pair $\langle \vec{s}, w \rangle \in C^l$ which dominates $\langle R(\vec{n}_{1,l}), W(\vec{n}_{1,l}) \rangle$, and (2) for every pair $\langle \vec{s}, w \rangle \in C^l$ there exists some BFB vector $\vec{n}_{1,l} = [n_1, \ldots, n_l]$ such that $\vec{s} = R(\vec{n}_{1,l})$ and $w = W(\vec{n}_{1,l})$.

We start by showing (1). Let $\vec{n}_{1,l} = [n_1, \ldots, n_{l-1}, n_l]$ be a BFB vector with $W(\vec{n}_{1,l}) \geq \eta$, and consider its prefix $\vec{n}_{1,l-1} = [n_1, \ldots, n_{l-1}]$. Observe that $W(\vec{n}_{n,l-1}) = \frac{W(\vec{n}_{1,l-1})}{w_{l,n_l}} \geq \eta$. As $\vec{n}_{1,l-1}$ is also a BFB vector, the inductive assumption implies that C^{l-1} contains a pair $\langle \vec{s}', w' \rangle$ that dominates $\langle R(\vec{n}_{1,l-1}), W(\vec{n}_{1,l-1}) \rangle$. From Lemma 2, $R(\vec{n}_{1,l}) \leq R(\vec{n}_{1,l-1}) \leq \vec{s}'$. Since $\|R(\vec{n}_{1,l})\| = n_l$, running MIN-DECREMENT $(\vec{s}', n_l)$ does not fail, and returns a signature $\vec{s}$ such that $R(\vec{n}_{1,l}) \leq \vec{s} \leq \vec{s}'$ and $\|\vec{s}\| = n_l$. As $w' \cdot w_{l,n_l} \geq W(\vec{n}_{1,l-1}) \cdot w_{l,n_l} = W(\vec{n}_{1,l}) \geq \eta$, it follows that the algorithm runs the code in lines 5-7 with respect to n_l and $\langle \vec{s}', w' \rangle$. In particular, the algorithm updates C^l with the pair $\langle \vec{s}, w \rangle$ for $w = w' \cdot w_{l,n_l} \geq W(\vec{n}_{1,l})$ (lines 6-7). Therefore, at the end of the l-th iteration, either C^l contains $\langle \vec{s}, w \rangle$, or it contains some other signature-weight pair that dominates $\langle \vec{s}, w \rangle$, and so it contains a pair that dominates $\langle R(\vec{n}_{1,l}), W(\vec{n}_{1,l}) \rangle$.

To show (2), assume that C^l contains a pair $\langle \vec{s}, w \rangle$. This pair was added to C^l in line 7 of the algorithm, which means there exists some pair $\langle \vec{s}', w' \rangle \in C^{l-1}$ such that for $n_l = \|\vec{s}\|$, $\vec{s} = $ MIN-DECREMENT$(\vec{s}', n_l)$, and $w = w' \cdot w_{l,n_l} \geq \eta$. From the inductive assumption, there is BFB vector $\vec{n}_{1,l-1} = [n_1, \ldots, n_{l-1}]$ such that $\vec{s}' = R(\vec{n}_{1,l-1})$ and $w' = W(\vec{n}_{1,l-1})$. For the vector $\vec{n}_{1,l} = [n_1, \ldots, n_{l-1}, n_l]$, lemma 3 implies that $\vec{s} = R(\vec{n}_{1,l})$. In addition $W(\vec{n}_{1,l}) = w$, and the lemma follows. $\qquad\square$

Finally, we present Algorithm 3 for the BFB exhaustive vector search variant. The algorithm processes the segments of the input one by one, starting from the k-th segment down to the first segment. The notation $[n, \vec{n}]$ is used for denoting a vector whose first element is the integer n, and its remaining suffix is the vector $\vec{n}$.

Algorithm 3. EXHAUSTIVE-VECTOR-SEARCH (W, η)

Input: A weight function W and a weight $0 < \eta \leq 1$.
Output: All BFB vectors $\vec{n} = [n_1, n_2, \ldots, n_k]$ satisfying $W(\vec{n}) \geq \eta$.

1 Generate all boundary curves $C^0, C^1, \ldots, C^k$ with respect to W and η using Algorithm 2. If C^k is empty, return the message "NO SOLUTION" and halt.
2 Set Q^{k+1} to be the collection containing a single empty vector.
3 **For** $l \leftarrow k$ **down to** 1 **do**
4 $\quad$ Set $Q^l \leftarrow \emptyset$.
5 $\quad$ **For each** $\vec{n}_{l+1,k} \in Q^{l+1}$ *and count* n *such that* $W([n, \vec{n}_{l+1,k}]) \geq \eta$ *and* MIN-INCREMENT $(L(\vec{n}_{l+1,k}), n)$ *does not fail* **do**
6 $\quad\quad$ Let $\vec{n}_{l,k} = [n, \vec{n}_{l+1,k}]$, and let $\vec{s} = $ MIN-INCREMENT$(L(\vec{n}_{l+1,k}), n)$.
7 $\quad\quad$ **If** *there exists a pair* $\langle \vec{s}', w' \rangle \in C^{l-1}$ *such that* $\vec{s} \leq \vec{s}'$ *and* $w' \cdot W(\vec{n}_{l,k}) \geq \eta$ **then**
8 $\quad\quad\quad$ Add $\vec{n}_{l,k}$ to Q^l.
9 **Return** Q^1.

Proof (Algorithm 3). By definition, if the boundary curve C^k is empty, it implies there is no BFB vector $\vec{n} = [n_1, \ldots, n_k]$ with $W(\vec{n}) \geq \eta$. In this case, the algorithm correctly reports there is no solution to the input (line 1).

Otherwise, we show for every $1 \leq l \leq k+1$ that the following invariant holds: *After Q^l is fully computed, Q^l contains $\vec{n}_{l,k} = [n_l, \ldots, n_k]$ if and only if $\vec{n}_{l,k}$ is a suffix of some BFB vector $\vec{n} = [n_1, \ldots, n_k]$ of weight $W(\vec{n}) \geq \eta$.* In particular, this invariant proves that the returned value Q^1 (line 9) is indeed the solution for the BFB exhaustive vector search variant, and so it only remains to establish the correctness of the invariant.

For $l = k+1$, the fact that Q^{k+1} contains a single empty suffix (line 2) derives the invariant in a straightforward manner. Otherwise, assuming inductively the invariant holds with respect to Q^{l+1}, we prove it also holds with respect to Q^l.

Let $\vec{n} = [n_1, \ldots, n_k]$ be a BFB vector of weight $W(\vec{n}) \geq \eta$, and consider its two suffixes $\vec{n}_{l,k} = [n_l, n_{l+1} \ldots, n_k]$ and $\vec{n}_{l+1,k} = [n_{l+1} \ldots, n_k]$. From the inductive assumption, $\vec{n}_{l+1,k} \in Q^{l+1}$. From Lemma 3, $\vec{s} = L(\vec{n}_{l,k})$ satisfies that $\vec{s} =$ MIN-INCREMENT$(L(\vec{n}_{l+1,k}), n_l)$. Since $W(\vec{n}_{l,k}) \geq W(\vec{n}) \geq \eta$, the condition in line 5 holds, and lines 6-8 are executed with respect to $\vec{n}_{l,k}$ and $\vec{s}$. Note that the prefix $\vec{n}_{1,l-1} = [n_1, \ldots, n_{l-1}]$ of $\vec{n}$ is a BFB vector with $W(\vec{n}_{1,l-1}) \geq W(\vec{n}) \geq \eta$. From the definition of C^{l-1}, there exists a pair $\langle \vec{s}', w' \rangle \in C^{l-1}$ that dominates the pair $\langle R(\vec{n}_{1,l-1}), W(\vec{n}_{1,l-1}) \rangle$. From Lemma 2, $L(\vec{n}_{l,k}) \leq R(\vec{n}_{1,l-1}) \leq \vec{s}'$. In addition, $w' \cdot W(\vec{n}_{l,k}) \geq W(\vec{n}_{1,l-1}) \cdot W(\vec{n}_{l,k}) = W(\vec{n}) \geq \eta$, and so the condition in line 7 holds, and the algorithm adds $\vec{n}_{l,k}$ into Q^l in line 8.

For the other direction of the invariant, let $\vec{n}_{l,k} = [n_l, n_{l+1}, \ldots, n_k] \in Q^l$. Due to the manner it was constructed (lines 5-6), its suffix $\vec{n}_{l+1,k} = [n_{l+1}, \ldots, n_k]$ is in Q^{l+1}, and from Lemma 3, $\vec{n}_{l,k}$ is a BFB vector with $L(\vec{n}_{l,k}) = \vec{s}$. From line 7, there exists a pair $\langle \vec{s}', w' \rangle \in C^{l-1}$ such that $\vec{s} \leq \vec{s}'$ and $w' \cdot W(\vec{n}_{l,k}) \geq \eta$, and so from the definition of C^{l-1} there exists a BFB vector $\vec{n}_{1,l-1} = [n_1, \ldots, n_{l-1}]$ for which $R(\vec{n}_{1,l-1}) = \vec{s}'$ and $W(\vec{n}_{1,l-1}) = w'$. The concatenation of $\vec{n}_{1,l-1}$ and $\vec{n}_{l,k}$ gives the vector $\vec{n} = [n_1, \ldots, n_{l-1}, n_l, \ldots, n_k]$, whose weight satisfies $W(\vec{n}) = W(\vec{n}_{1,l-1}) \cdot W(\vec{n}_{l,k}) = w' \cdot W(\vec{n}_{l,k}) \geq \eta$. In addition, $\vec{n}$ is a BFB vector, due to the corresponding valid signature series obtained by concatenating a valid signature series for $\vec{n}_{1,l-1}$ that ends with $\vec{s}'$ and a valid signature series for $\vec{n}_{l,k}$ that starts with $\vec{s}$, concluding this direction of the proof.   □

The algorithm for the exhaustive BFB string search variant applies a similar approach in order to produce all BFB strings whose count vector weights are at least η. It starts by generating signature curves exactly as done by Algorithm 3. Then, in each iteration l, instead of computing a set Q^l of count vectors, the algorithm computes a set P^l of l-block collections. At the end of the iteration, P^l contains all l-block collections B^l such that there exists some 1-BFB palindrome β in which the l-th layer's block collection is B^l, and the weight of the vector $\vec{n}$ such that $\vec{n}(\beta) = 2\vec{n}$ satisfies $W(\vec{n}) \geq \eta$. The initial collection P^{k+1} contains a single empty $(k+1)$-block collection. In the l-th iteration, for each $(l+1)$-block collection $B^{l+1} \in P^{l+1}$, all possible foldings of B^{l+1} are enumerated. For each such folding, its signature and weight are examined against C^{l-1} similarly as done in line 7 of Algorithm 3, and if meeting the condition all elements in the collection are wrapped, and the resulting l-block collection B^l is added into P^l. Due to space limits, we omit the details for the process of enumerating all foldings of B^{l+1}, which will be described in an extended version of this manuscript.

4   Results

In order to test our algorithms we have used cancer data taken from the Cancer Genome Project dataset [2]. This data covers aCGH samples (Affymetrix Genome-Wide Human SNP Array 6.0) from 746 human cancer cell lines. Segmentation and segment copy numbers are as reported by [2], who used the PICNIC software [8] for this analysis. In total, the dataset contains about 35 thousands chromosomal arms (746 samples, 23 or 24 chromosomes per sample, two arms per chromosome), each arm is segmented, and each segment is assigned an estimated copy number based on the observed aCGH data. As shown in [20], short BFB-like count vectors have a high probability to emerge even when the genome was rearranged with mechanisms different from BFB. Thus, in order to detect significant BFB evidence we have filtered the set of chromosomal arms to include only arms with at least eight consecutive segments such that no adjacent segments share the same copy number estimation. After this filtration, the remaining subset included 6589 chromosomal arms. As the estimated counts reflect the expected segment copy numbers in all copies of the chromosome in the sample, we have corrected the counts by reducing $p - 1$ from each count, where p is the ploidy (i.e. the number of copies) of the chromosome in the sample. Typically $p = 2$, but since these are heavily rearranged cancer genomes, chromosomal losses and whole chromosomal duplications are not rare. We therefore allowed the value of p to vary between 1 and 5, and run the BFB analyses for each value.

As currently no analysis tool available produces count weights, we have derived such weights from the expected counts reported by PICNIC (after correcting for ploidy). Specifically, for a segment whose observed count is n, the weight of a count n' was defined by $\frac{\Pr(n|n')}{\Pr(n|n)}$, where $\Pr(x|\lambda) = \frac{\lambda^x e^{-\lambda}}{x!}$ is the probability to observe the value x for a random variable distributing according to the *Poisson* distribution with parameter λ. For each of the obtained weight functions, we used the DISTANCE-BFB algorithm from [20] to report all longest BFB sub-vectors with weight at least $\eta = 0.7$. Out of the 6589 samples, 54 samples had for at least one ploidy value $1 \leq p \leq 5$ a BFB sub-vector of length at least 8. Some samples had long BFB sub-vectors with respect to more than one ploidy value, and the total number of obtained BFB vectors was 86.

Then, we considered the segment coordinates and weight functions corresponding to the obtained sub-vectors, and run Algorithm 3 in order to find all BFB vectors of weights at least $\eta = 0.7$ with respect to these weight functions. For these 86 instances, a total number of 19154 heavy BFB vectors were found, with an average of 222 solutions per-instance. This reviles an interesting property of the problem when applied over this data: the vast majority of samples, 6535 out of 6589, cannot be explained by any BFB count vector (and thus are unlikely to be obtained from BFB), yet each one of those 54 samples who can be explained by BFB has about several tens or hundreds of corresponding count vectors.

The above analysis was run by two variants of our algorithm - the IS variant described by Algorithm 3, and a variant that runs a similar procedure without applying the IS optimization (essentially, it runs the same code as Algorithm 3, with the exception it does not generate the signature curves in line 1, and does not apply the condition in line 7 before adding new elements to collections Q^l). The disadvantage of the non-IS variant is in that sets of the form Q^l maintain BFB vectors $\vec{n}_{l,k} = [n_l, \ldots, n_k]$ which may not be suffixes of some BFB vectors $\vec{n} = [n_1, \ldots, n_k]$ of weight at least η. To measure the gain of the IS algorithm, we count the number of signature increment attempts the algorithms perform (line 5). On average, the IS variant performed 57-fold less increments, with a total number of 5672346 incrementation attempt over all 86 vectors, versus 325343441 for the non-IS algorithm. While the IS variant has a clear efficiency advantage over the non-IS variant, this advantage might be considered more modest than expected. A possible reason for that is that maximum copy number values reported in [2] were limited to 14, even when the data suggests higher copy numbers. In general, higher copy numbers usually imply a higher number of alternative heavy counts, which in turn induce a higher number of possible heavy count vectors. For example, when comparing the two algorithms over the synthetic count vector $\vec{n} = [3, 8, 111, 8, 5, 150, 11, 170, 4, 53, 100, 75, 49, 10, 42, 18]$, using the same Poisson-based weights as described above and requiring that output vectors weigh at least $\eta = 0.85$, the non-IS algorithm runs 218 second[1] and performs over 20 million signature increments, whereas the IS algorithm runs 120 milliseconds and performs 635 signature increments. Both algorithms return exactly the same output - a set of 18 BFB vectors. Other simulated inputs can cause memory explosion for the non-IS variant, while handled efficiently by the IS variant.

5   Discussion and Conclusions

The problem of detecting breakage fusion bridge is challenging, but significant progress has been made in the last few years. Our work suggests that while rare, BFB does occur in tumor derived cell lines and also in primary tumors. In this work, we describe algorithms that can be used to enumerate all possible BFB architectures given uncertain copy number data.

The results of our analyses heavily depend on the input weights, which in turn depend on separated analyses applied to biological data. While we used here a simple Poisson-based model in order to render fixed available count estimations into weight functions, it is clear that more realistic weighing can be applied. Examining Fig. 2 for example, one can observe that different segments demonstrate different variance in signal intensities, implying that some count estimates are more reliable than others. Incorporating segment lengths and signal variance information when choosing count weights is likely to produce more meaningful weights and improve the quality of the analyses output.

---

[1] Running time was measured for an intel Core i7 processor with Microsoft Windows 7 operating system, code is implemented in Java.

Different measurements can yield other types of BFB evidence. For example, deep sequencing experiments can sequence reads spanning genomic breakpoints. In a BFB modified genome, it is expected that many of these breakpoints reflect fold-back inversions (i.e. concatenations between reference segments and their inverted form), while such fold-back patterns are less common in other rearrangement mechanisms [4]. Thus, identification of high or low fold-back pattern frequencies can support or weaken the conjecture BFB has occurred, respectively. Such evidence is less frequent in currently available data, as reliable breakpoint information requires sequencing to a relatively high depth of coverage (while copy number data can be obtained also from sequencing with a lower depth of coverage or from aCGH experiments). When given though, such information can be integrated and improve the quality of BFB calling [20].

Acknowledgements. The authors are thankful to the anonymous RECOMB reviewers for their helpful comments. The research was supported by grants from the NIH (RO1-HG004962), and the NSF (CCF-1115206, IIS-1318386).

References

1. Alkan, C., Kidd, J.M., Marques-Bonet, T., Aksay, G., Antonacci, F., Hormozdiari, F., Kitzman, J.O., Baker, C., Malig, M., Mutlu, O., Sahinalp, S.C., Gibbs, R.A., Eichler, E.E.: Personalized copy number and segmental duplication maps using next-generation sequencing. Nat. Genet. 41, 1061–1067 (2009)
2. Bignell, G.R., Greenman, C.D., Davies, H., Butler, A.P., Edkins, S., Andrews, J.M., Buck, G., Chen, L., Beare, D., Latimer, C., Widaa, S., Hinton, J., Fahey, C., Fu, B., Swamy, S., Dalgliesh, G.L., Teh, B.T., Deloukas, P., Yang, F., Campbell, P.J., Futreal, P.A., Stratton, M.R.: Signatures of mutation and selection in the cancer genome. Nature 463(7283), 893–898 (2010)
3. Bignell, G.R., Santarius, T., Pole, J.C., Butler, A.P., Perry, J., Pleasance, E., Greenman, C., Menzies, A., Taylor, S., Edkins, S., Campbell, P., Quail, M., Plumb, B., Matthews, L., McLay, K., Edwards, P.A., Rogers, J., Wooster, R., Futreal, P.A., Stratton, M.R.: Architectures of somatic genomic rearrangement in human cancer amplicons at sequence-level resolution. Genome. Res. 17, 1296–1303 (2007)
4. Campbell, P.J., Yachida, S., Mudie, L.J., Stephens, P.J., Pleasance, E.D., Stebbings, L.A., Morsberger, L.A., Latimer, C., McLaren, S., Lin, M.L., McBride, D.J., Varela, I., Nik-Zainal, S.A., Leroy, C., Jia, M., Menzies, A., Butler, A.P., Teague, J.W., Griffin, C.A., Burton, J., Swerdlow, H., Quail, M.A., Stratton, M.R., Iacobuzio-Donahue, C., Futreal, P.A.: The patterns and dynamics of genomic instability in metastatic pancreatic cancer. Nature 467(7319), 1109–1113 (2010)
5. Carr, A.M., Paek, A.L., Weinert, T.: DNA replication: failures and inverted fusions. Semin. Cell Dev. Biol. 22(8), 866–874 (2011)
6. Chiang, D.Y., Getz, G., Jaffe, D.B., O'Kelly, M.J., Zhao, X., Carter, S.L., Russ, C., Nusbaum, C., Meyerson, M., Lander, E.S.: High-resolution mapping of copy-number alterations with massively parallel sequencing. Nat. Methods 6(1), 99–103 (2009)
7. Eckel-Passow, J.E., Atkinson, E.J., Maharjan, S., Kardia, S.L., de Andrade, M.: Software comparison for evaluating genomic copy number variation for Affymetrix 6.0 SNP array platform. BMC Bioinformatics 12, 220 (2011)

8. Greenman, C.D., Bignell, G., Butler, A., Edkins, S., Hinton, J., Beare, D., Swamy, S., Santarius, T., Chen, L., Widaa, S., Futreal, P.A., Stratton, M.R.: PICNIC: an algorithm to predict absolute allelic copy number variation with microarray cancer data. Biostatistics 11(1), 164–175 (2010)
9. Hanahan, D., Weinberg, R.A.: Hallmarks of cancer: the next generation. Cell 144(5), 646–674 (2011)
10. Hastings, P.J., Lupski, J.R., Rosenberg, S.M., Ira, G.: Mechanisms of change in gene copy number. Nat. Rev. Genet. 10(8), 551–564 (2009)
11. Kinsella, M., Bafna, V.: Combinatorics of the breakage-fusion-bridge mechanism. J. Comput. Biol. 19(6), 662–678 (2012)
12. Kitada, K., Yamasaki, T.: The complicated copy number alterations in chromosome 7 of a lung cancer cell line is explained by a model based on repeated breakage-fusion-bridge cycles. Cancer Genet. Cytogenet. 185, 11–19 (2008)
13. McClintock, B.: The Production of Homozygous Deficient Tissues with Mutant Characteristics by Means of the Aberrant Mitotic Behavior of Ring-Shaped Chromosomes. Genetics 23, 315–376 (1938)
14. McClintock, B.: The Stability of Broken Ends of Chromosomes in Zea Mays. Genetics 26, 234–282 (1941)
15. Medvedev, P., Stanciu, M., Brudno, M.: Computational methods for discovering structural variation with next-generation sequencing. Nat. Methods 6, 13–20 (2009)
16. Olshen, A.B., Venkatraman, E.S., Lucito, R., Wigler, M.: Circular binary segmentation for the analysis of array-based dna copy number data. Biostatistics 5(4), 557–572 (2004)
17. Reshmi, S.C., Roychoudhury, S., Yu, Z., Feingold, E., Potter, D., Saunders, W.S., Gollin, S.M.: Inverted duplication pattern in anaphase bridges confirms the breakage-fusion-bridge (bfb) cycle model for 11q13 amplification. Cytogenetic and Genome Research 116(1-2), 46–52 (2007)
18. Venkatraman, E.S., Olshen, A.B.: A faster circular binary segmentation algorithm for the analysis of array cgh data. Bioinformatics 23(6), 657–663 (2007)
19. Yoon, S., Xuan, Z., Makarov, V., Ye, K., Sebat, J.: Sensitive and accurate detection of copy number variants using read depth of coverage. Genome. Res. 19(9), 1586–1592 (2009)
20. Zakov, S., Kinsella, M., Bafna, V.: An algorithmic approach for breakage-fusion-bridge detection in tumor genomes. Proceedings of the National Academy of Sciences 110(14), 5546–5551 (2013)

Reconciliation with Non-binary
Gene Trees Revisited

Yu Zheng and Louxin Zhang

Department of Mathematics, National University of Singapore,
10 Lower Kent Ridge, Singapore 119076
{matzhyu,matzlx}@nus.edu.sg

Abstract. By reconciling the phylogenetic tree of a gene family with
the corresponding species tree, it is possible to infer lineage-specific duplications and losses with high confidence and hence annotate orthologs
and paralogs. However, the currently available reconciliation methods
for non-binary gene trees are computationally expensive for being ap-
plied on a genomic level. Here, an $O(|G| + |S|)$ algorithm is presented
to reconcile an arbitrary gene tree G with its corresponding species tree
S, where $|\cdot|$ denotes the number of nodes in the corresponding tree.
The improvement is achieved through two innovations: a fast computation of compressed child-image subtrees and efficient reconstruction of
irreducible duplication histories.

1   Introduction

Given the importance of accurately annotated gene relationships in evolutionary and functional studies of biological systems [15,23], significant efforts
have been invested in developing methods to identify orthologs and paralogs
[1,5,7,12,16,21,22]. A pair of genes in different species whose last common ancestor (LCA) corresponds to a speciation event are orthologs [10]. Two genes (in the
same or different species) that descend from a gene duplication event are paralogs. Knowing the orthologs and paralogs of species permits one to reconstruct
the duplication history within a gene family.

In practice, this is often done by reconciling the phylogenetic tree (the gene
tree) of a family with the corresponding species tree, and inferring the lineage-
specific duplication and loss events [11,16]. Although a plethora of reconciliation
methods have been developed over the past two decades (see the review paper
[6]), only recently has this reconciliation process been generalized to non-binary
gene trees (see the survey articles [9,24]). The ability to reconcile non-binary
gene trees substantially expands the application of this method in comparative
genomics. First, it expands the range of tools: many widely-used phylogenetic
programs such as MrBayes [14] produce non-binary gene trees if there is not
enough signal in the data to date the divergences. Moreover, reconciling non-
binary gene trees that have been obtained by contracting weak branches in binary
gene trees produces more accurate duplication events than working directly on
corresponding binary ones (our unpublished data). Second, our method allows

R. Sharan (Ed.): RECOMB 2014, LNBI 8394, pp. 418–432, 2014.
© Springer International Publishing Switzerland 2014

us to design fast heuristic programs for genome-wide mapping of orthologs and paralogs. For example, SYNERGY [23] implicitly assumes that the gene tree of every gene family is a star tree and heuristically reconciles the star gene tree with the input species tree, which relieves the substantial preprocessing burden of building binary gene trees for individual gene families. It inspires us to work on a bottom-up approach for reconciling non-binary gene trees.

For binary gene trees and species trees, there is an accepted reconciliation process which has been proven to produce the unique duplication history with the fewest gene duplications and losses [3,13], and whose computational complexity is linear with respect to the number of nodes in the two trees [4,19,25]. However, no one has yet designed a linear-time reconciliation algorithm for non-binary gene trees that is guaranteed to generate the history with the minimum number of duplication and loss events. Furthermore, the uniqueness of the result is not so clear for non-binary gene trees, where reconciliation may produce different duplication histories for different cost models [26], where a linear-time algorithm was obtained for the duplication cost model. Chang and Eulenstein [2] developed the first algorithm for the problem, but their solution has cubic complexity. The dynamic programming algorithm of Durand *et al.* [8] has the same worst-case time complexity but it can also solve the problem under any affine cost model. Recently, a quadratic algorithm was proposed by Lafond *et al.* [17]. All these methods are computationally intensive when applied on a genomic scale.

In this paper, we present a linear-time algorithm that solves the problem. Our bottom-up approach can incorporate multiple sources of information on gene similarity, including sequence similarity and conserved gene order, and is efficient enough to be used on a genomic level. Hence, it provides a valuable framework for the genome-wide mapping of orthologs and paralogs in any group of species with a known phylogeny, while taking advantage of the rapid increase in fully sequenced genomes.

The rest of this paper is divided into six sections. The reconciliation problem and different cost models are introduced in Section 2. Section 3 presents an algorithm to simultaneously computes all compressed child-image subtrees of the species tree in linear time, which immediately leads to an improved reconciliation method. Section 4 introduces the concept of irreducible duplication history. Section 5 presents a simple algorithm that takes $O(|G| + |S|)$ operations to reconcile a gene tree G and the corresponding species tree S, where $|\cdot|$ denotes the number of nodes in the corresponding tree. In Section 6, we use simulated data to compare the time efficiency of our algorithm with other methods. We conclude with suggestions for future work. All the proofs omitted in this extended abstract can be found in the full version of the work.

2   Concepts and Notions

2.1   Definitions

Let $T = (V(T), E(T))$ be a rooted tree in which one node is designated as the root and the branches are oriented away from the root. $V(T)$ is the set of all nodes, and $E(T)$ is the set of all branches (directed edges). For two nodes $u, v \in V(T)$, v is the parent of u (and, equivalently, u is a child of v) if $(v, u) \in E(T)$. Further, v is an ancestor of u (equivalently, u is a descendant of v), written $v \prec u$, if the unique path from the root to u passes through v. We write $v \preceq u$ if $u = v$ or $v \prec u$. For $U \subseteq V(T)$, lca(U) denotes the most recent common ancestor of the nodes in U. The depth of a node in T is the number of branches in the path from the root to it. In this paper, we also use the following notation:

- $|T|$ denotes the number of nodes in T;
- $p(u)$ denotes the parent of a non-root node $u \in V(T)$;
- $V_{\mathrm{lf}}(T)$ denotes the set of leaves (terminal nodes) of T;
- $V_{\mathrm{it}}(T)$ denotes the set of internal (non-leaf) nodes of T;
- Ch(u) denotes the set of children of $u \in V_{\mathrm{it}}(T)$;
- $T(u)$ denotes the subtree consisting of u and all descendants of u;
- $T|_U$ denotes the subtree induced by a subset $U \subseteq V(T)$: the nodes of $T|_U$ are $V' = \{v \in V(T) \mid \mathrm{lca}(U) \preceq v \preceq u \in U\}$ and the edges of $T|_U$ are $E(T) \cap (V' \times V')$.

A node v is said to be binary if it has two children. T is binary if every internal node is binary. If T is non-binary, a binary tree T' is said to be a binary refinement of T if, for every $u \in V(T)$, $v \in V(T')$ exists such that $V_{\mathrm{lf}}(T(u)) = V_{\mathrm{lf}}(T'(v))$ or, equivalently, if T can be obtained from T' by branch contraction.

2.2   Species Trees

A species tree S is a rooted tree in which each leaf is associated with a unique species. For node $u \in V_{\mathrm{lf}}(S)$, the branch $(p(u), u)$ represents the species that labels u. For $u \in V_{\mathrm{it}}(S)$, $(p(u), u)$ represents the common ancestor of all the species that label the leaves in $S(u)$ and u represents a speciation event.

In the present paper, we assume that a species tree is binary, and that the branch entering the root represents the common ancestor of all the species in the tree, called the *root* branch (Figure 1A).

2.3   Gene Trees and Gene Duplication History

The gene tree reconstructed from the DNA or protein sequences of a gene family represents evolutionary relationships in these genes. However, it may not explicitly represent the duplication history of the gene family. Without knowing the true orthologous and paralogous relationships in the family members, we do not need to distinguish the members that are sampled from the same species. Hence, we label each leaf in the gene tree that represents a gene with the species that hosts the gene today. In the resulting tree, leaves are not uniquely labeled in general. Also, gene trees do not need to be binary.

Consider a family F of genes sampled from a collection X of species with a known phylogenetic tree S. Assume that F evolved from a unique ancestral gene

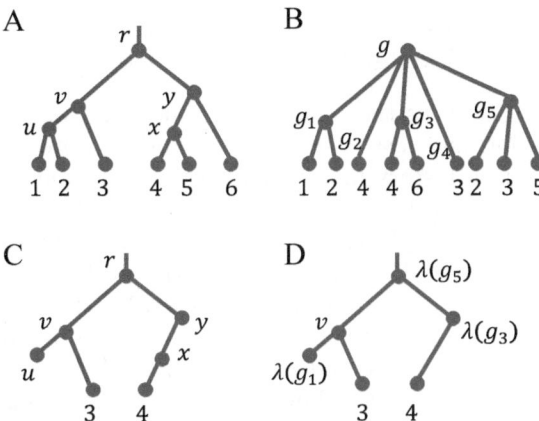

Fig. 1. A. A binary species tree over six species 1–6. **B.** A gene tree of nine genes: two each from species 2, 3 and 4, and one each from species 1, 5 and 6. The gene tree has two non-binary nodes. The child-image subtree of g and its compressed version are shown in panels **C** and **D**, respectively. Here, $\lambda(g_1) = u$, $\lambda(g_2) = 4$, $\lambda(g_3) = y$, $\lambda(g_4) = 3$, and $\lambda(g_5) = r$.

through k gene duplications and m gene losses in the ancestral species (that is, branches) of S (Figure 2A). We further assume that (i) each duplication event gives rise to one new copy of the gene involved; (ii) each copy, as well as the original duplicated gene, has exactly one descendant gene in an species, unless one of the m loss events occurs in the ancestors of that species. The topology H of the duplication history $\mathcal{H}$ of F is a rooted tree whose leaves are labeled with genes. Since S is binary, each degree-2 node $u \in V(H)$ corresponds to a gene loss, and each degree-3 node with children u and v represents a duplication if it does not correspond to a species tree node (Figure 2C). We use such types of trees to represent duplication histories.

The duplication (respectively loss) cost $d_{\mathcal{H}}$ ($l_{\mathcal{H}}$) of $\mathcal{H}$ is defined as the number of duplication (loss) events occurring in it. Its mutation cost is defined as $d_{\mathcal{H}} + l_{\mathcal{H}}$. If we assign the weights w_d and w_l to duplication and loss events respectively, the (w_d, w_l)-affine cost of $\mathcal{H}$ is defined as $w_d d_{\mathcal{H}} + w_l l_{\mathcal{H}}$.

2.4   The Reconciliation Problem

The duplication history $\mathcal{H}$ of a gene family F can be inferred by reconciling its gene tree G and the corresponding species tree S. The symbol g^s denotes a gene $g \in F$ in species s. For $U \subseteq V(G)$, $\lambda(U) \overset{\text{def}}{=} \{\lambda(u) \mid u \in U\}$. The *lca reconciliation* is the map $\lambda : V(G) \to V(S)$ defined as:

$$\lambda(g) = \begin{cases} s & \text{if } g = g^s \in V_{\text{lf}}(G), \\ \text{lca}\left(\lambda(\text{Ch}(g))\right) & \text{if } g \in V_{\text{it}}(G). \end{cases} \tag{1}$$

If G is binary, λ induces the unique duplication history of F that has the minimum duplication and loss costs [3,13]. In other words, it finds the most parsimonious evolution history in a reconciliation cost model. Furthermore, for $g \in V_{\text{it}}(G)$, g is inferred to be a duplication node if $\lambda(g) \in \{\lambda(g') \mid g' \in \text{Ch}(g)\}$. The corresponding gene duplication event occurs in the branch $(p(\lambda(g)), \lambda(g))$ in S, and gene loss occurs in each branch off the path from $\lambda(g)$ to $\lambda(g')$ for each $g' \in \text{Ch}(g)$ in the inferred duplication history. We define the cost of the lca reconciliation of G and S to be the cost of the corresponding duplication history for each of the duplication, loss, and affine cost models.

If G is non-binary, it is not clear how many duplication events can be inferred and where they should occur in the most parsimonious duplication history of F. The problem of reconciling an arbitrary gene tree G and a binary species tree S is formulated as follows [9]:

Instance: A contracted version of the true gene tree G of a family of genes F observed in species with a known binary species tree S and a reconciliation cost model c.

Solution: A duplication history of F whose cost is $\min_{G' \in \mathcal{BR}(G)} c(G', S)$, where $\mathcal{BR}(G)$ is the set of all binary trees that refine G.

Note that $V(G) \subseteq V(T)$ for every $T \in \mathcal{BR}(G)$. The lca reconciliation of T and S maps every node in G to the same node in S for every $T \in \mathcal{BR}(G)$. Therefore, we simply need to infer the duplication history from each ancestral gene g to its children in the subtree $S|_{\lambda(\text{Ch}(g))}$ (called the *child-image subtree*) (Figure 1C), for each $g \in V_{\text{it}}(G)$ separately. In the next section, we discuss our algorithm for non-binary nodes in G, which is identical to the simple rule mentioned above when applied to binary nodes.

3   Compressed Image Subtrees

By definition, $\lambda(g)$ is the root of $S|_{\lambda(\text{Ch}(g))}$. If $S|_{\lambda(\text{Ch}(g))}$ contains non-root degree-2 nodes, its size can be much larger than $|\text{Ch}(g)|$. To design a fast algorithm for reconciling G and S, we need to compress $S|_{\lambda(\text{Ch}(g))}$ by contracting all non-root degree-2 nodes except for those in $\lambda(\text{Ch}(g))$ for each g (Figure 1D). The compressed version of $S|_{\lambda(\text{Ch}(g))}$ is indicated by $I(g)$ and defined as follows.

Let P be a path from p_1 to p_2 in $S|_{\lambda(\text{Ch}(g))}$ such that p_1 and p_2 are either of degree 3 or in $\lambda(\text{Ch}(g))$, and all the middle nodes are of degree 2 and not in $\lambda(\text{Ch}(g))$. Note that any parsimonious duplication history from g to its children can only have gene loss events in the first branch of P, gene duplication events in the last branch of P, or both. $I(g)$ is obtained from $S|_{\lambda(\text{Ch}(g))}$ by replacing each of these paths with a single branch. Note that if the depths of p_1 and p_2 in S are known, when working on $I(g)$, we can compute the gene losses occurring

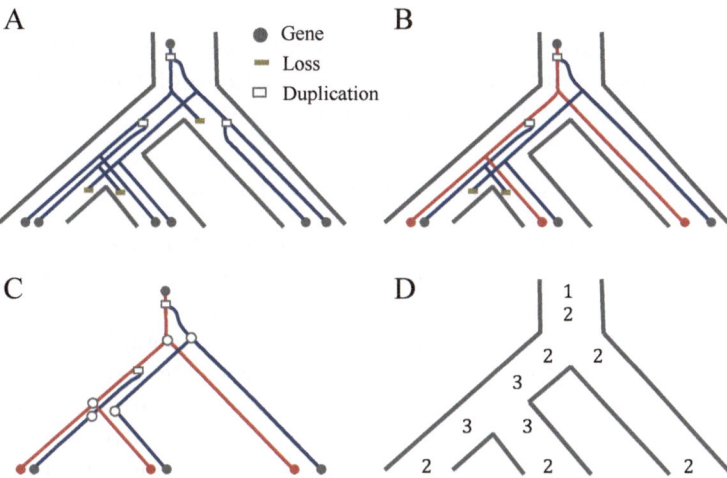

Fig. 2. A. A duplication history that does not have a minimum duplication cost: in the rightmost lineage, a duplication and a loss occur. **B.** An irreducible duplication history equivalent to the duplication history in A. Here, the oldest gene lineage is colored red, the right-handed copy in the first two leaves are the descendants of the gene duplicate produced in the left lineage, and the right-handed copy in the rightmost leaf is the descendant of the duplicate produced in the root branch. **C.** The gene tree that represents the duplication history in B where circle nodes correspond to species tree nodes and square nodes are duplication nodes. **D.** The numbers of genes flowing into (top) and out of (bottom) all the branches for the irreducible duplication history in B.

in the branches leading away from P. Most importantly, $|I(g)| \leq 2|\mathrm{Ch}(g)|$ for each g and hence $\sum_{g \in V_{\mathrm{it}}(G)} |I(g)| \leq 2|G|$. Additionally, we have the following fact.

Theorem 1. *It takes linear time $O(|G|+|S|)$ to construct the compressed child-image subtrees of all the internal nodes of G in S.*

Finally, we assume that for each $s \in I(g)$, the depth $d(s)$ in S is computed and is stored in the data structure along with other information on node s. Note that, for each s, $d(s) - d(p(s))$ is the number of branches in the path from $p(s)$ to s in S and is used to compute the gene loss cost of the duplication history from g to its child genes in S.

Theorem 1 leads immediately to an improved method for tree reconciliation. By implementing a dynamic programming algorithm to resolve different non-binary gene tree nodes in their corresponding $I(g)$, we can compute an optimal reconciliation of G and S in time $O(|S|+d^2 \times |G|)$ in the affine cost model, where d is the maximum degree of a node in G.

4   Irreducible Duplication Histories

To develop a linear time algorithm for reconciling non-binary gene trees, we also need to focus on a special type of duplication histories of gene families. In this section, we introduce this type of duplication histories.

4.1   Equivalence of Gene Duplication Histories

Consider a feasible duplication history $\mathcal{H}$ from g to $\mathrm{Ch}(g)$ in the child-image subtree $S|_{\lambda(\mathrm{Ch}(g))}$. If duplication and loss occur in the same branch (Figure 2A), we can eliminate one duplication and one loss to obtain a new duplication history with fewer events (Figure 2B), because we do not distinguish the elements in $\mathrm{Ch}(g)$. Hence, the duplication history of $\mathrm{Ch}(g)$ with the smallest duplication cost does not allow both duplication and loss to occur in the same branch. Considering a branch e as the population of the representative species, we use $n_{\mathcal{H}}^{\mathrm{in}}(e)$ and $n_{\mathcal{H}}^{\mathrm{out}}(e)$ to denote the numbers of genes flowing into and out of a branch e. For $u \in V\left(S|_{\lambda(\mathrm{Ch}(g))}\right)$, we define:

$$\omega(u) \overset{\mathrm{def}}{=} |\{g' \in \mathrm{Ch}(g) : \lambda(g') = u\}|. \tag{2}$$

The following conditions hold for a duplication history with the minimum duplication cost as shown in Figure 2D:

(C1) For $e \in E\left(S|_{\lambda(\mathrm{Ch}(g))}\right)$, $n_{\mathcal{H}}^{\mathrm{in}}(e) \geq 1$ and $n_{\mathcal{H}}^{\mathrm{out}}(e) \geq 1$. If e is the root branch, $n_{\mathcal{H}}^{\mathrm{in}}(e) = 1$.

(C2) For any leaf u, $n_{\mathcal{H}}^{\mathrm{out}}(e) = \omega(u)$, where $e = (p(u), u)$.

(C3) For $e = (u, v)$ and $e' = (v, w)$ in $S|_{\lambda(\mathrm{Ch}(g))}$, $n_{\mathcal{H}}^{\mathrm{out}}(e) = n_{\mathcal{H}}^{\mathrm{in}}(e') + \omega(v)$.

(C4) In every branch e, k duplications occur iff $n_{\mathcal{H}}^{\mathrm{out}}(e) - n_{\mathcal{H}}^{\mathrm{in}}(e) = k$; similarly, k losses occur in e iff $n_{\mathcal{H}}^{\mathrm{in}}(e) - n_{\mathcal{H}}^{\mathrm{out}}(e) = k$.

Define:

$$\Sigma_{\mathcal{H}} = \left\{ \left(e, n_{\mathcal{H}}^{\mathrm{in}}(e), n_{\mathcal{H}}^{\mathrm{out}}(e)\right) \mid e \in E\left(S|_{\lambda(\mathrm{Ch}(g))}\right) \right\}. \tag{3}$$

Two duplication histories $\mathcal{H}$ and $\mathcal{H}'$ from g to $\mathrm{Ch}(g)$ are said to be equivalent if $\Sigma_{\mathcal{H}} = \Sigma_{\mathcal{H}'}$. Clearly, any given value of $\Sigma_{\mathcal{H}}$ may be achieved by a large number of histories with the same duplication and loss costs. In this work, we infer a duplication history by determining the values of the three arguments defined in (3) for all branches. One benefit of taking this approach is that our method effectively outputs a large set of optimal duplication histories that reconcile the input gene and species trees.

4.2   Irreducible Duplication Histories

A duplication process copies an existing gene, giving rise to two versions of the gene. A duplication history from g to $\mathrm{Ch}(g)$ is *irreducible* if the ancestral gene representing g in the root branch does not experience any loss event, so that it has a descendant in every leaf of $S|_{\lambda(\mathrm{Ch}(g))}$ (the red lineage in Figure 2B), and if every duplication event copies the descendant of this oldest gene in the branch where the event occurs. Note that a history with no duplication is also irreducible. Such limiting cases are called *speciation histories*.

In general, different children of g may be mapped to the same node in $S|_{\lambda(\mathrm{Ch}(g))}$. We consider $\lambda(\mathrm{Ch}(g))$ to be a multiset, meaning that each element has a multiplicity. It is not hard to see that an irreducible duplication history $\mathcal{H}$ from g to $\mathrm{Ch}(g)$ induces the following decomposition of $\lambda(\mathrm{Ch}(g))$ in $S|_{\lambda(\mathrm{Ch}(g))}$:

$$\lambda(\mathrm{Ch}(g)) = D_0 \uplus D_1 \uplus \ldots \uplus D_k, \tag{4}$$

where $\uplus$ is the sum operation for multisets[1], such that the following hold: (i) k equals the number of duplication events in $\mathcal{H}$; (ii) $D_0 = V_{\mathrm{lf}}(S|_{\lambda(\mathrm{Ch}(g))})$, representing the oldest gene lineage; (iii) $D_i \overset{\mathrm{def}}{=} \{x \in \lambda(\mathrm{Ch}(g)) \,|\mathrm{the}$ gene made by E_i has a descendant in $x\}$, where E_i is the i-th duplication event of $\mathcal{H}$ occurring in the branch entering $\mathrm{lca}(D_i)$ for $1 \leq i \leq k$.

For example, there are two duplication events in the irreducible duplication history in Fig. 2B. Assume the species are named S_1, S_2, S_3 from left to right in the species tree, the duplicated gene produced in the first (i.e. top) duplication event has descendants in S_2 and S_3, whereas the duplicated gene produced by the second duplication has only one descendant in S_1. Hence, this irreducible duplication history corresponds to the following decomposition:

$$\{S_1, S_2, S_3\} \uplus \{S_2, S_3\} \uplus \{S_1\}.$$

Conversely, such a decomposition of $\lambda(\mathrm{Ch}(g))$ uniquely defines an irreducible duplication history from g to $\mathrm{Ch}(g)$ in $S|_{\lambda(\mathrm{Ch}(g))}$. Therefore, we have the following theorem.

Theorem 2. *Every duplication history $\mathcal{H}$ from g to $\mathrm{Ch}(g)$ is equivalent to an irreducible duplication history $\mathcal{H}'$ such that $d_{\mathcal{H}} \geq d_{\mathcal{H}'}$ and $l_{\mathcal{H}} \geq l_{\mathcal{H}'}$.*

5   Linear Time Reconciliation Algorithm

In order to infer a duplication history with the minimum mutation cost, by Theorem 2, we need only to find the decomposition $\lambda(\mathrm{Ch}(g))/V_{\mathrm{lf}}\big(S|_{\lambda(\mathrm{Ch}(g))}\big) = D_1 \uplus D_2 \uplus \cdots \uplus D_k$ that minimizes $k + \sum l_i$, where l_i is the loss cost of the speciation history defined by D_i. This is because the number of gene losses in the speciation history defined by $V_{\mathrm{lf}}\big(S|_{\lambda(\mathrm{Ch}(g))}\big)$ is fixed. We refer to this as a minimum decomposition. Note that $D_1 \uplus D_2 \uplus \cdots \uplus D_k$ corresponds to the set of all child genes that are the descendants of ancestral genes produced by duplication. For each leaf in $S|_{\lambda(\mathrm{Ch}(g))}$, all but one of the genes mapped to the leaf are descendants of duplicated ancestral genes; these gene children are called *redundant* gene copies. The descendant of the oldest gene in each leaf is called the *basal* gene copy.

---

[1] The multiplicity of an element equals the sum of the multiplicities in the operands.

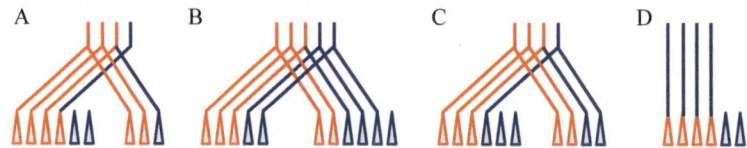

Fig. 3. A schematic view of merging partial decompositions for the three possible cases (**A–C**) where u has two children, and also for the case when u has only one child (**D**). Good trees and defect trees are colored orange and blue respectively in the decompositions $\mathcal{D}_1$ (left) and $\mathcal{D}_2$ (right). The $\omega(u)$ singleton trees added at the current node are not shown in each case.

We now present a linear-time algorithm for finding a minimum decomposition of the redundant gene copies by working on the compressed child-image subtree $I(g)$. For the sake of clarity, we assume that for any $(u, v) \in E(I(g))$, the difference between the depths of v and u in the species tree S is one. (We describe how to generalize this to more general cases later.)

A rooted tree is called a *defective tree* if there is at least one degree-2 node in the middle of every path from the root to a leaf. It is called a *good tree* if there is a root-to-leaf path in which all but the end nodes are of degree 3. Note that a speciation history is a subtree of $I(g)$.

Theorem 3. *Let*
$$\mathcal{D}: \quad D_1 \uplus D_2 \uplus \cdots \uplus D_k$$
be the minimum decomposition of $\lambda(\mathrm{Ch}(g))/V_{\mathrm{lf}}(I(g))$. If $\mathcal{D}$ gives a duplication history such that redundant gene copies have the minimum gene loss cost, compared to all other duplication histories with the same mutation cost, then for each i, the speciation history $I(g)|_{D_i}$ satisfies the following:
(1). The subtree $T(u)$ below any degree-2 node u cannot be a defective tree.
(2). $I(g)|_{D_i}$ must be a good tree.

Theorem 3 motivated us to design a bottom-up recursive algorithm for finding the minimum decomposition of $\lambda(\mathrm{Ch}(g))/V_{\mathrm{lf}}(I(g))$, thereby reconstructing a duplication history from g to its children with the minimum mutation cost. By the theorem, any component in a minimum decomposition of $\lambda(\mathrm{Ch}(g))/V_{\mathrm{lf}}(I(g))$ induces a good subtree that has a special structural property. Hence, for subset $V' \subset V_{\mathrm{lf}}(I(g))$, we use the induced subtree $I(g)|_{V'}$ to represent V'. As such, we use a set of subtrees to represent a partial decomposition obtained at each internal node.

At a leaf $u \in V_{\mathrm{lf}}(I(G))$, the partial decomposition consists of $\omega(u) - 1$ singleton trees, which are considered good trees.

Let u be a node with two children in $I(g)$, u_1 and u_2. Consider a partial decomposition $\mathcal{D}_1$ of $\{x \in \lambda(\mathrm{Ch}(g))/V_{\mathrm{lf}}(I(g)) \mid x \in V_{\mathrm{lf}}(I(g)(u_1))\}$ into $b(u_1)$ trees and a partial decomposition $\mathcal{D}_2$ of $\{x \in \lambda(\mathrm{Ch}(g))/V_{\mathrm{lf}}(I(g)) \mid x \in V_{\mathrm{lf}}(I(g)(u_2))\}$ into $b(u_2)$ trees. We attempt to merge these two partial decompositions to obtain a decomposition of $\{x \in \lambda(\mathrm{Ch}(g))/V_{\mathrm{lf}}(I(g)) \mid x \in V_{\mathrm{lf}}(I(g)(u))\}$. By Theorem 3, each component of a minimum decomposition induces a good subtree. However, for a good subtree X and an internal node y, $X \cap I(g)(y)$ can be a defective tree. Hence, a partial decomposition may contain defective trees. We distinguish between defect trees and good trees.

Assume that $a(u_1)$ out of $b(u_1)$ trees are good in $\mathcal{D}_1$, that $a(u_2)$ out of $b(u_2)$ trees are good in $\mathcal{D}_2$, and that $a(u_2) \leq a(u_1)$. We merge $\mathcal{D}_1$ and $\mathcal{D}_2$ by considering the following two cases (Figure 3).

1. $a(u_2) \leq b(u_2) < a(u_1) \leq b(u_1)$ (Figure 3A). Merge $a(u_2)$ pairs of good trees, $b(u_2) - a(u_2)$ pairs of good and defective trees, extend $a(u_1) - b(u_2)$ good trees from $\mathcal{D}_1$, and discard $b(u_1) - a(u_1)$ defective trees from $\mathcal{D}_1$. Further, add $\omega(u)$ singleton trees, which are good trees.
2. $a(u_2) \leq a(u_1) \leq \min\{b(u_1), b(u_2)\}$ (Figures 3B and 3C). Merge $a(u_2)$ pairs of good trees, $a(u_1)-a(u_2)$ pairs of good and defective trees, $\min\{b(u_1), b(u_2)\}-a(u_1)$ pairs of defective trees, and discard $b(u_2) - b(u_1)$ defective trees from $\mathcal{D}_2$ if $b(u_2) > b(u_1)$ or $b(u_1) - b(u_2)$ defective trees from $\mathcal{D}_1$ otherwise. Add $\omega(u)$ singleton trees.

Proposition 1. *Let $m_1 \leq m_2 \leq m_3 \leq m_4$ be the arrangement of $\{a_1, a_2, b_1, b_2\}$ from smallest to largest. Merging $\mathcal{D}_1$ and $\mathcal{D}_2$ produces $\omega(u) + m_2$ good trees and $m_3 - m_2$ defect trees to merge, and detects $m_4 - m_3$ defect trees to discard.*

At an internal node u with only one child u_1 (Figure 3D), we create $\omega(u)$ singleton trees, extend all good trees, and discard all the defective trees in the decomposition $\mathcal{D}_1$.

Using this bottom-up merging procedure, we obtain a set of good and defective trees at the root of $I(g)$. This set of trees defines a minimal decomposition of $\lambda(\mathrm{Ch}(g))/V_{\mathrm{lf}}(I(g))$. More specifically, each good tree corresponds to a component of the minimal decomposition. However, each defective tree corresponds to $k \geq 2$ components, where k equals the cardinality of the maximum incomparable degree-2 internal nodes in the tree. Similarly, each defective tree discarded at an internal nodes also corresponds to several components of the minimal decomposition.

For m real numbers $i_1, i_2, \cdots, i_m$, we use $\mathrm{median}\{i_1, i_2, \ldots, i_m\}$ to denote their median if m is odd. For each $u \in V(I(g))$, we use $b(u)$ to denote the number of trees in the decomposition obtained at u in which $a(u)$ out of $b(u)$ trees are good trees. For u and $k \geq 0$, we define the following:

$$\mathrm{dist}\,(k, [a(u), b(u)]) = \min_{x \in [a(u), b(u)]} |x - k|,$$

$$k' = \mathrm{median}\{k, a(u), b(u)\} - \omega(u), \tag{5}$$

$$f(u, k') = \begin{cases} 0 & \text{if } u \text{ is a leaf,} \\ C(u_1, k') + k' & \text{if } \mathrm{Ch}(u) = \{u_1\}, \\ C(u_1, k') + C(u_2, k') & \text{if } \mathrm{Ch}(u) = \{u_1, u_2\}, \end{cases}$$

and

$$C(u, k) = \mathrm{dist}\,(k, [a(u), b(u)]) + f(u, k'). \tag{6}$$

```
Input An annotated compressed child-image subtree I(g);
Output The number of genes flowing into and out of branches in I(g).
1.Traversing I(g) in post-order
   Compute a(u) and b(u) at node u:
   if (u is a leaf) {
      a(u) = ω(u) − 1;  b(u) = ω(u) − 1;
   } else if (Ch(u) = {u₁, u₂}) {
      max_a = max(a(u₁), a(u₂));  min_b = min(b(u₁), b(u₂));
      a(u) = ω(u) + min(max_a, min_b);
      b(u) = ω(u) + max(max_a, min_b);
   } else if (Ch(u) = {u₁}) {
      a(u) = ω(u);  b(u) = a(u₁) + ω(u);
   }
2.Traversing I(g) in pre-order
   /* in(u) and out(u) denote the number of genes */
   /* flowing into and out of the branch (p(u), u) */
   Compute in(u) and out(u) at node u:
   if (u is the root) {
      α(u) = 0;  β(u) = a(u);
   } else {
      α(u) = β(p(u)) − ω(p(u));  β(u) = median{α(u), a(u), b(u)};
   }
   /* factor in the basal copy in each branch */
   in(u) = 1 + α(u);  out(u) = 1 + β(u);
```

Fig. 4. A linear-time algorithm for reconstructing evolution from g to its children. Here, we assume that $d(u) = d(p(u)) + 1$ for each u in $I(g)$.

Theorem 4. *Let r be the root of $I(g)$. The decomposition $\mathcal{D}_r$ obtained by this merging procedure determines a duplication history of redundant gene copies with the minimum mutation cost $C(r, 0)$.*

Theorem 4 suggests a two-step algorithm for reconstructing evolution from g to its children in linear time (Figure 4). First, we compute the numbers of good and defective trees obtained at the internal nodes in $I(g)$ by visiting all the nodes in order from leaf to root, which guarantees that we visit all the children of a node before the node itself. We then identify duplications and losses by computing the numbers of genes flowing into and out of the branches in $I(g)$, top down from root to leaf. To take into account the basal gene copies, we add one to the numbers of ancestral gene copies flowing into and out of each branch. Figure 5 gives an example to illustrate this algorithm.

Recall that we assume $d(u) = d(p(u)) + 1$ for each $u \in V(I(g))$ in the algorithm described above. It can be modified for general cases by (i) finding all maximal subtrees of $I(g)$ that do not contain any branch (u, v) such that $d(v) > d(u) + 2$ in S and then (ii) for each subtree T found in (i), replacing every branch (u, v) such that $d(v) = d(u) + 2$ by the two-branch path between u and v in S and then

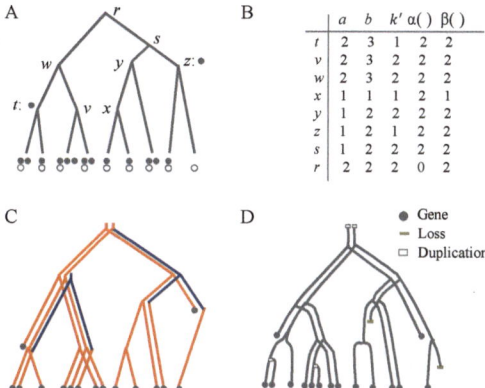

Fig. 5. Illustration of the reconciliation algorithm. **A**. A compressed child-image tree $I(g)$ with basal (empty circles) and redundant (solid circles) child genes of g drawn beside their image nodes. **B**. The values of a, b, α and β at internal nodes. **C**. The subtrees to be merged at each node in $I(g)$. Two good trees are obtained after the merging process terminates at the root. A subtree obtained at each node is good if its root is connected to one blue branch at most. **D**. The optimal duplication history from g to its children (in the compressed child-image tree) obtained from the optimal decomposition in C, whose minimum mutation cost is 6; the basal gene copies are not shown.

applying the algorithm to the resulting subtree T'. The complete version of this algorithm can be found in the full version of this work.

6   Experimental Tests

We compared a naive dynamic programming method (DP) (found in [8]) and a modified dynamic programming method (DP+C) (which applies the dynamic programming technique to the compressed child-image subtrees) to the proposed linear-time method (LT) using simulated data. For fair comparison, we implemented DP. Our version of DP is slightly faster than the dynamic-programming-based program found in NOTUNG [8] but, to be fair, the latter has several other features, such as listing all the inferred optimal solutions.

All three programs were run to reconcile non-binary gene trees with the mutation cost, using the same machine (3.4GHz and 8G RAM). We measured their run times for 100 reconciliations between a non-binary tree containing $1.2n$ genes and its corresponding species trees over n species. For each size n, both species tree and a binary gene tree with $1.2n$ leaves were generated using the Yule model. Finally, a non-binary gene tree was obtained from the binary gene tree by contracting each edge with a fixed rate p. We examined 40 cases by allowing n to take 10 different values in the range from 1,000 to 10,000 and setting the edge contraction rate p to 0.4, 0.6, 0.7 or 0.8 (Figure 6). We also ran LT on 50 different tree sizes in the range from 2,000 to 100,000, which are too large for

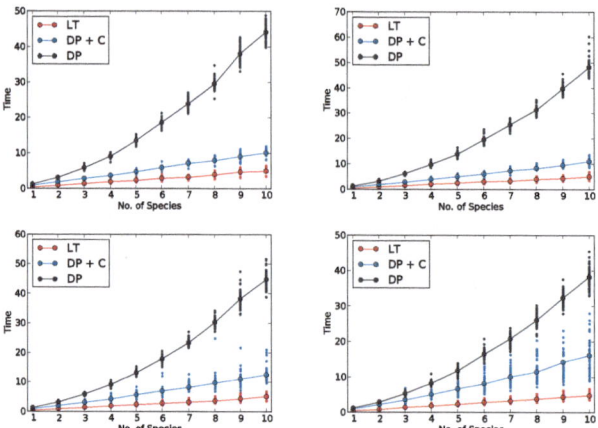

Fig. 6. Comparison of three algorithms: dynamic programming (DP), dynamic programming with compressed child-image subtrees (DP+C) and the proposed linear time (LT) algorithm. Four figures are drawn for the four different edge contraction rates 0.4 (top left), 0.6 (top right), 0.7 (bottom left), and 0.8 (bottom right). The run time is given in microseconds and the number of species is in thousands.

the other two methods (Figure 7). The results summarized in the figure confirm that the run times of LT are linearly proportional to the size of the gene trees. LT is slightly faster than DP+C , and 5 to 20 times faster than DP for gene trees with thousands of genes.

We also ran the program PolytomeSolver, which implements the algorithm presented in [17]. Both LT and DP+C were faster than PolytomeSolver for gene trees with hundreds of leaves. However, PolytomeSolver aborted on some large species trees (those of height >64, for example). We exclude it from comparison analyses in this conference version.

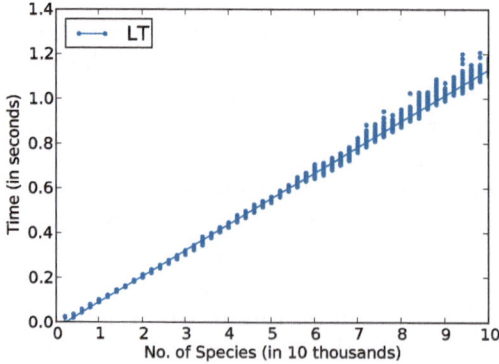

Fig. 7. The run times of LT are linearly proportional to the sizes of gene trees

7   Discussion and Future Work

Here we present a linear-time algorithm to reconcile the non-binary gene tree of a gene family and the corresponding species tree to reconstruct the duplication history of the gene family with the minimum mutation cost. The reconciliation time is an order of magnitude faster than others achieved using compressed child-image trees and working on irreducible duplication histories.

Our approach has several important benefits. First, we do not consider incomplete lineage sorting (ILS) events, which may not be rare and hence cannot be ignored in certain circumstances [18,20,27]. Since the effect of an ILS event on the divergence of gene and species trees is similar to that of a duplication event, the concepts proposed here can easily be extended to take ILS events in account. Second, the output of our program is actually a class of optimal duplication histories, not just one individual history. This is because the program may assign multiple duplications to a branch in the species trees, and these duplications can be arranged in different ways. Third, our linear-time algorithm is fast and hence is ideal for providing an on-line service for tree reconciliation (see our TxT server http://phylotoo2.appspot.com/rgt/, where the source code is also available). Finally, our bottom-up approach can incorporate multiple sources of information on gene similarity, including sequence similarity and conserved gene order, when it is applied to genome-wide studies of the evolution of gene families. This is definitely an interesting future project.

References

1. Bansal, M.S., Alm, E.J., Kellis, M.: Reconciliation revisited: Handling multiple optima when reconciling with duplication, transfer, and loss. In: Deng, M., Jiang, R., Sun, F., Zhang, X. (eds.) RECOMB 2013. LNCS, vol. 7821, pp. 1–13. Springer, Heidelberg (2013)
2. Chang, W.-C., Eulenstein, O.: Reconciling gene trees with apparent polytomies. In: Chen, D.Z., Lee, D.T. (eds.) COCOON 2006. LNCS, vol. 4112, pp. 235–244. Springer, Heidelberg (2006)
3. Chauve, C., El-Mabrouk, N.: New perspectives on gene family evolution: Losses in reconciliation and a link with supertrees. In: Batzoglou, S. (ed.) RECOMB 2009. LNCS, vol. 5541, pp. 46–58. Springer, Heidelberg (2009)
4. Chen, K., Durand, D., Farach-Colton, M.: NOTUNG: a program for dating gene duplications and optimizing gene family trees. J. Comput. Biol. 7, 429–447 (2000)
5. Chen, Z.Z., Deng, F., Wang, L.: Simultaneous identification of duplications, losses, and lateral gene transfers. IEEE-ACM Trans. Comput. Biol. Bioinformatics 9, 1515–1528 (2012)
6. Doyon, J.P., et al.: Models, algorithms and programs for phylogeny reconciliation. Briefings Bioinform. 12, 392–400 (2011)
7. Dufayard, J.-F., et al.: Tree pattern matching in phylogenetic trees: automatic search for orthologs or paralogs in homologous gene sequence databases. Bioinformatics 21, 2596–2603 (2005)
8. Durand, D., Halldorsson, B., Vernot, B.: A hybrid micro-macroevolutionary approach to gene tree reconstruction. J. Comput. Biol. 13, 320–335 (2006)

9. Eulenstein, O., Huzurbazar, S., Liberles, D., Eulenstein, O., et al.: Reconciling phylogenetic trees. In: Dittmar, K., Liberles, D. (eds.) Evolution After Duplication, pp. 185–206. Wiley-Blackwell, USA (2010)

10. Fitch, W.M.: Distinguishing homologous from analogous proteins. Syst. Zool. 19, 99–113 (1970)

11. Goodman, M., et al.: Fitting the gene lineage into its species lineage, a parsimony strategy illustrated by cladograms constructed from globin sequences. Syst. Zool. 28, 132–163 (1979)

12. Goodstadt, L., Ponting, C.: Phylogenetic reconstruction of orthology, paralogy, and conserved synteny for dog and human. PLoS Comput. Biol. 2, e133 (2006)

13. Górecki, P., Tiuryn, J.: DLS-trees: a model of evolutionary scenarios. Theoret. Comput. Sci. 359, 378–399 (2006)

14. Huelsenbeck, J.P., Ronquist, F.: MRBAYES: Bayesian inference of phylogenetic trees. Bioinformatics 17, 754–755 (2001)

15. Kellis, M., et al.: Sequencing and comparison of yeast species to identify genes and regulatory elements. Nature 423, 241–254 (2003)

16. Kristensen, D.M., Wolf, Y.I., Mushegian, A.R., Koonin, E.V.: Computational methods for gene orthology inference. Briefings Bioinform. 12, 379–391 (2011)

17. Lafond, M., Swenson, K.M., El-Mabrouk, N.: An optimal reconciliation algorithm for gene trees with polytomies. In: Raphael, B., Tang, J. (eds.) WABI 2012. LNCS, vol. 7534, pp. 106–122. Springer, Heidelberg (2012)

18. Pollard, et al.: Widespread discordance of gene trees with species tree in *Drosophila*: evidence for incomplete lineage sorting. PLoS Genet. 2(10), e173 (2006)

19. Schieber, B., Vishkin, U.: On finding lowest common ancestors: simplification and parallelization. SIAM J. Comput. 17, 1253–1262 (1988)

20. Stolzer, M., et al.: Inferring duplications, losses, transfers and incomplete lineage sorting with non-binary species trees. Bioinformatics 28(18), i409–i415 (2012)

21. Storm, C., Sonnhammer, E.: Automated ortholog inference from phylogenetic trees and calculation of orthology reliability. Bioinformatics 18, 92–99 (2002)

22. Tatusov, R.L., et al.: A genomic perspective on protein families. Science 278, 631–637 (1997)

23. Wapinski, I., et al.: Natural history and evolutionary principles of gene duplication in fungi. Nature 449, 54–61 (2007)

24. Warnow, T.: Large-scale multiple sequence alignment and phylogeny estimation. In: Models and Algorithms for Genome Evolution, pp. 85–146. Springer, UK (2013)

25. Zhang, L.X.: On a Mirkin-Muchnik-Smith conjecture for comparing molecular phylogenies. J. Comput. Biol. 4, 177–187 (1997)

26. Zheng, Y., Wu, T., Zhang, L.: A linear-time algorithm for reconciliation of non-binary gene tree and binary species tree. In: Widmayer, P., Xu, Y., Zhu, B. (eds.) COCOA 2013. LNCS, vol. 8287, pp. 190–201. Springer, Heidelberg (2013)

27. Zheng, Y., Zhang, L.: Effect of incomplete lineage sorting on tree-reconciliation-based inference of gene duplication. In: Cai, Z., Eulenstein, O., Janies, D., Schwartz, D. (eds.) ISBRA 2013. LNCS, vol. 7875, pp. 261–272. Springer, Heidelberg (2013)

Learning Protein-DNA Interaction Landscapes by Integrating Experimental Data through Computational Models

Jianling Zhong[1], Todd Wasson[2], and Alexander J. Hartemink[1,3,⋆]

[1] Computational Biology & Bioinformatics, Duke University, Durham, NC 27708
[2] Lawrence Livermore National Laboratory, Livermore, CA 94550
[3] Department of Computer Science, Duke University, Durham, NC 27708
{zhong,amink}@cs.duke.edu

Abstract. Transcriptional regulation is directly enacted by the interactions between DNA and many proteins, including transcription factors, nucleosomes, and polymerases. A critical step in deciphering transcriptional regulation is to infer, and eventually predict, the precise locations of these interactions, along with their strength and frequency. While recent datasets yield great insight into these interactions, individual data sources often provide only noisy information regarding one specific aspect of the complete interaction landscape. For example, chromatin immunoprecipitation (ChIP) reveals the precise binding positions of a protein, but only for one protein at a time. In contrast, nucleases like MNase and DNase reveal binding positions for many different proteins at once, but cannot easily determine the identities of those proteins. Here, we develop a novel statistical framework that integrates different sources of experimental information within a thermodynamic model of competitive binding to jointly learn a holistic view of the *in vivo* protein-DNA interaction landscape. We show that our framework learns an interaction landscape with increased accuracy, explaining multiple sets of data in accordance with thermodynamic principles of competitive DNA binding. The resulting model of genomic occupancy provides a precise, mechanistic vantage point from which to explore the role of protein-DNA interactions in transcriptional regulation.

Keywords: protein-DNA interaction landscape, thermodynamic modeling, genomic data integration, competitive binding, COMPETE.

1 Introduction

As an essential component of transcriptional regulation, the interaction between DNA-binding factors (DBFs) and DNA has been studied extensively by experimentalists. To map genome-wide protein-DNA interactions, two basic categories of experimental techniques have been developed: chromatin immunoprecipitation

---

⋆ Corresponding author.

R. Sharan (Ed.): RECOMB 2014, LNBI 8394, pp. 433–447, 2014.
© Springer International Publishing Switzerland 2014

(ChIP) based methods (numerous studies in many organisms, but a few examples for yeast are [9, 18, 19]); and nuclease digestion based methods that profile chromatin with either DNase [11] or MNase [10]. To reveal high-resolution DNA interaction sites for a single antibody-targeted factor, ChIP methods can be used, especially the recently developed ChIP-exo methods [19] that use lambda exonuclease to obtain precise positions of protein binding. Nuclease digestion methods can be used to efficiently assay genome-wide DNA occupancy of all proteins at once, without explicit information about protein identities. These and other experimental efforts over the past decade have generated a large amount of data regarding the chromatin landscape and its role in transcriptional regulation. We now need computational models that can effectively integrate all this data to generate deeper insights into transcriptional regulation.

A popular set of computational models use this data to search for overrepresented DNA sequences bound by certain DBFs; these are often applied in the setting of motif discovery [4, 9, 15, 23]. More recently, models have been applied to DNase-seq data to identify 'digital footprints' of DBFs [3, 11, 14, 16]. However, many of these approaches share certain drawbacks. First, protein binding is typically treated as a binary event amenable to classification: either a protein binds at a particular site on the DNA sequence or it does not. However, both empirical and theoretical work has demonstrated that proteins bind DNA with continuous occupancy levels (as reviewed by Biggin [1]). Second, most computational methods model the binding events for one protein at a time instead of taking into consideration the interactions among different DBFs, especially nucleosomes. Although the work of Segal et al. [22], Kaplan et al. [12], and Teif and Rippe [24] are notable exceptions, these all consider small genomic regions and include only a few transcription factors (TFs); Segal et al. [22] ignored the role of nucleosomes altogether. Third, and most importantly, almost all current methods fail to integrate different kinds of datasets. This is insufficient because data from one kind of experiment only reveals partial information about the *in vivo* protein-DNA interaction landscape. For example, ChIP datasets only contain binding information for one specific protein under a specific condition; nuclease digestion datasets provide binding information for all proteins, but do not reveal the identities of the proteins; and protein binding microarray (PBM) experiments only look at sequence specificity of one isolated protein in an *in vitro* environment.

We previously published a computational model of protein-DNA interactions, termed COMPETE [25], that overcomes the first two drawbacks above by representing the competitive binding of proteins to DNA within a thermodynamic ensemble. Interactions between proteins and DNA are treated as probabilistic events, whose (continuous) probabilities are calculated from a Boltzmann distribution. COMPETE can easily include a large number of DBFs, including nucleosomes, and can efficiently profile entire genomes with single base-pair resolution. However, a major limitation of COMPETE is that it is a purely theoretical model of binding, based on thermodynamic first principles but not guided by data regarding *in vivo* binding events. Indeed, it is possible for COMPETE to predict superfluous binding events that are inconsistent with observed data (see Supplemental Figure S1). It is therefore necessary to develop a new computational framework for jointly interpreting experimentally-derived data regarding genomic occupancy within a model built upon the thermodynamic foundation of COMPETE.

Here, we develop just such a method: a general framework for combining both a thermodynamic model for protein-DNA interactions (along the lines of COMPETE) and a new statistical model for learning from experimental observations regarding those interactions. Information from different experimental observations can be integrated to infer the actual thermodynamic interactions between DBFs and a genome. In this particular study, we demonstrate the use of this framework by integrating paired-end micrococcal nuclease sequencing (MNase-seq) data, which reveals information about the binding occupancy of both nucleosomes and smaller (subnucleosomal) factors. Our framework also integrates protein binding specificity information from PBM data and produces a more accurate and realistic protein-DNA interaction landscape than COMPETE alone, along with a mechanistic explanation of MNase-digested fragments of different sizes. The cross-validated performance of our framework is significantly higher than several baselines to which we compared it. Our framework is flexible and can easily incorporate other data sources as well, and thus represents a general modeling framework for integrating multiple sources of information to produce a more precise view of the interaction landscape undergirding transcriptional regulation.

2   Methods

2.1   Modeling Protein-DNA Interaction

We model the binding of DBFs (e.g., transcription factors and nucleosomes) to DNA along a probabilistic continuum, and we incorporate explicit competition between different DBFs. The ensemble average of the probability with which a particular DBF binds a specific position of the sequence can be derived from thermodynamic principles. To calculate this average probability, consider a specific binding configuration i from the ensemble, where i can be viewed as an instantaneous snapshot of the dynamic competition between DBFs for binding sites along the genome. Following the Boltzmann distribution, the unnormalized probability w_i of configuration i can be shown to be $w_i = \prod_{t=1}^{N_i} X_t \times P(S_t, E_t | DBF_t)$, where t is an index over the N_i DBF binding sites in configuration i. To simplify notation, we have treated each unbound nucleotide as being bound by a special kind of 'empty' DBF. In the above expression, X_t denotes a weight associated with DBF t, while S_t and E_t denote the start and end position of the DBF binding site, respectively. $P(S_t, E_t | DBF_t)$ is the probability of observing the DNA sequence between S_t and E_t, given that DBF t is bound there. If we use p_i to denote the probability of configuration i after normalization by the partition function, we can write the probability that DBF t binds at a specific position j as $\sum_{i \in I(t,j)} p_i$, where $I(t, j)$ is the subset of binding configurations in the ensemble that have DBF t bound at sequence position j.

This model can be formulated analogously to a hidden Markov model (HMM) [17], in which the states correspond to the binding of different DBFs and the observations are the DNA sequence. The various probabilities, along with the partition function, can then be calculated efficiently using the forward-backward algorithm. For transcription factors, we have chosen to represent $P(S_t, E_t | DBF_t)$ using a position weight matrix (PWM), but more sophisticated models can also be used (e.g., relaxing positional independence, or based on energies rather than probabilities [26]). Regardless, binding models from different sources and of different forms can be easily incorporated into our model, generating the appropriate states and sequence emission probabilities. We use the curated PWMs from Gordân et al. [7], derived from *in vitro* PBM experiments, as the input protein binding specificities and consider them fixed (though our framework also could allow them to be updated).

The analogues of HMM transition probabilities in our model are the DBF weights, but these are not constrained to be probabilities. To allow this flexibility, we adopt a more general statistical framework called a Boltzmann chain [21] which can be understood as a HMM that allows the use of any positive real numbers for these weights. Because of the analogy with an HMM, we henceforth refer to these DBF weights as 'transition weights' and denote them collectively as a vector $\boldsymbol{X} = (X_1, X_2, \ldots, X_D)$, where D is the number of different kinds of DBFs. We treat the D elements of $\boldsymbol{X}$ as free parameters, and we will fit them using experimentally-derived genomic data.

We should note that the DBF transition weights in a Boltzmann chain are sometimes called 'concentrations'. However, it is important to point out that these transition weights are not the same as bulk cellular protein concentrations, of the kind that can sometimes be measured experimentally [5]. Bulk cellular protein concentrations are not necessarily indicative of the availability of a DBF to bind DNA, because they do not account for phenomena like sub-cellular localization or extra-nuclear sequestration, protein activation through post-translational modification or ligand or co-factor binding, or the number of DBFs already bound to DNA. In contrast, our transition weights correspond to nuclear concentrations of active proteins that are free and available to bind DNA. In this sense, our weight parameters are more reasonably interpreted not as cellular concentrations but rather as the chemical potentials of the DBFs for interacting with the genome.

2.2 Using Paired-End MNase-Seq Data as a Measure of Genomic Occupancy Level of DNA-Binding Proteins

We used paired-end MNase-seq data from Henikoff et al. [10]. Based on their protocol, the length of the sequencing fragments correspond roughly to the size of the protein protecting that part of the DNA; the number of fragments mapping to the location correlates with the binding strength or occupancy. Therefore, to measure the level of occupancy of different DNA binding proteins, we separate

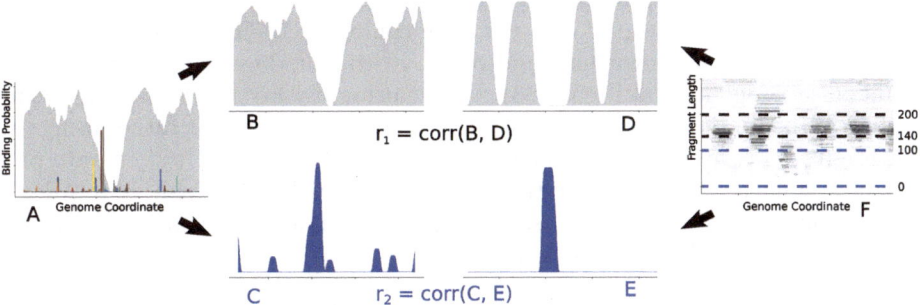

Fig. 1. Overview of objective function evaluation. (A) Predicted probability that each particular DBF binds at a given genome position, as calculated by COMPETE, given current DBF weights. We then separate these probabilities into two profiles: (B) predicted nucleosome binding profile and (C) predicted composite TF binding profile in which protein identities have been removed; the latter is smoothed to make it comparable to a short fragment coverage profile. Similarly, we separate the observed MNase-seq fragments (F) into long (140–200bp) and short (0–100bp) fragments, which are summed to produce measures of coverage. (D) Total long fragment coverage is processed into a large protein binding profile, which is compared to predicted nucleosomal binding, arriving at Pearson correlation r_1. (E) Total short fragment coverage is processed into a small protein binding profile, which is compared to predicted composite TF binding, arriving at Pearson correlation r_2. For this promoter, the quantity h that appears in our objective function (the pseudo-likelihood) is simply the geometric mean of the two correlations, after they are rescaled to lie in the interval $[0, 1]$: $h = \frac{1}{2}\sqrt{(1 + r_1) \times (1 + r_2)}$. The complete pseudo-likelihood over all promoters is then optimized with respect to the DBF weights using the inference method described below.

the fragments into long (140–200bp) and short (0–100bp) fragment groups and count the number of fragments in each group that cover a specific genomic location (called long and short fragment coverage, respectively). The long fragment coverage is used as a measure of the occupancy of large protein complexes, which are mainly nucleosomes, while the short fragment coverage is used as a measure of the occupancy of smaller proteins, which are mainly transcription factors.

To reduce noise in the MNase-seq data, we process the noisy fragment data into binding profiles through thresholding and smoothing. We define two thresholds: a bottom threshold T_b and a top threshold T_t. Coverage values that are below T_b are converted to 0, while those above T_t are converted to 1; coverage values between the two thresholds are normalized linearly to $[0, 1]$. We then smooth the track using a Gaussian kernel of bandwidth B_m. We process long and short fragment coverage data separately to get the large and small protein binding profiles, respectively (Figure 1 D and E). We choose $T_b = 200$ and $T_t = 500$, with $B_m = 10$ for short fragment coverage and $B_m = 30$ for long fragment coverage. These values give satisfying results in terms of reducing noise while retaining clear peaks. Because the choices are a little arbitrary, we performed a sensitivity analysis and observed that our results are largely unaffected across a broad range of these parameters (see Supplemental Figure S2). We also note that MNase is known to prefer to cut A/T compared to G/C. We assessed the severity of this well-known bias and observed that it does not affect our final results (see Supplemental Figure S3). This is primarily because we are not using profiles of the total number of cuts at each genomic position, but rather using the full fragments (available as a result of paired-end sequencing) to generate profiles of fragment coverage; while the former would be highly sensitive to MNase bias, the latter is relatively insensitive to the small fluctuations in fragment end locations introduced by MNase bias.

2.3   Selecting a Subset of TFs and Promoter Regions

Our framework has the capability to include all *S. cerevisiae* transcription factors. However, our choice of transcription factors is limited by available high quality binding preference data. In addition, adding more TFs increases the dimensionality of the parameter space and therefore the computation time required to explore the space. In this study, we chose a set of 42 TFs with available high quality binding preference data. These TFs cover a wide range of cellular functions, including the widely-studied transcriptional regulators Reb1, Rap1, and Abf1 (possessing some chromatin remodeling activity), TFs involved in pheromone response (Ste12 and Tec1), TFs involved in stress response (like Msn4), and TFs involved in cell cycle regulation (Fkh1, Mbp1, and so forth). We also included some TFs, like Pho2 and Phd1, that regulate a large number of genes according to MacIsaac et al. [15]. While these 42 do not represent all yeast TFs, they are collectively responsible for 66% of the genome-wide protein-DNA interactions reported by MacIsaac et al. [15] (at p-value < 0.005 and conservation level 3).

Having selected our 42 TFs, we next chose a set of promoter regions that, according to MacIsaac et al. [15] (at p-value < 0.005 and conservation level 3), seem to be bound exclusively by those TFs. For this study, we focus on 81 such promoter regions, and extracted MNase-seq data for these loci as follows. If the promoter is divergently transcribed, we extracted the MNase-seq data between the two TATA elements, plus 200bp downstream of each TATA element. For the other (non-divergent) promoters, we extracted MNase-seq data 500bp upstream of the TATA element (or 100bp upstream of the end of the upstream gene, whichever is smaller), and 200bp downstream of the TATA element. Locations of TATA elements were taken from Rhee and Pugh [20].

2.4   Incorporating MNase-seq Data through an Objective Function

We model MNase-seq data through a pseudo-likelihood function, conditioned on COMPETE outputs. To calculate the pseudo-likelihood function, we process the COMPETE output TF binding probabilities as following: the binding probability of each COMPETE output TF binding event is expanded to a flanking region of C_e bp, and is then dropped linearly to 0 for another C_r bp; we then sum the expanded binding probability of all TFs (truncating values larger than 1) and smooth the track using a Gaussian kernel of bandwidth B_c to get a composite TF binding profile (Figure 1C). We process the occupancy profile in such a way for two reasons: (a) the resolution of the short fragment coverage does not distinguish protection from adjacent proteins, and (b) MNase does not completely digest all unprotected DNA, leaving some additional nucleotides flanking any TF's actual binding site. We choose $C_e = C_r = B_c = 10$, though, as with the threshold and bandwidth parameters discussed above, varying the specific values tends to have only small effects on the model predictions. We do not process the nucleosome profile predicted by COMPETE since the model already takes nucleosome padding into consideration.

For promoter region m, we calculate two correlations: the Pearson correlation $r_{1,m}$ between the nucleosome binding profile and the MNase-seq long fragment coverage profile, and the Pearson correlation $r_{2,m}$ between the composite TF binding profile and the MNase-seq small protein coverage profile. The complete pseudo-likelihood function we seek to maximize is defined as:

$$L(\boldsymbol{X}) = \prod_{m=1}^{M} h_m(\boldsymbol{X}) \quad \text{where} \quad h_m(\boldsymbol{X}) = \frac{1}{2}\sqrt{(1 + r_{1,m}) \times (1 + r_{2,m})}.$$

Note that $h_m(\boldsymbol{X})$, which depends on the vector of DBF weights $\boldsymbol{X}$, is the geometric mean of the two rescaled correlations for promoter region m (an example is shown in Figure 1). In this study, $M = 81$.

2.5   Inference Method

We use Markov chain Monte Carlo (MCMC) to explore a posterior distribution based on the pseudo-likelihood function. However, since correlation measures the overall goodness of fit for many genomic locations at once, our pseudo-likelihood function is much flatter than typical likelihood functions. This property can be useful in preventing overfitting, but it also imposes some difficulty for parameter inference. To alleviate this concern, and allow for more efficient MCMC exploration, we apply a temperature parameter τ to each dimension of the search space in order to concentrate the mass of $L(\boldsymbol{X})$ around its modes. We apply a possibly different temperature to each dimension (i.e., each element of the vector $\boldsymbol{X}$) because the pseudo-likelihood in one dimension may be more or less flat than in others. We base our choice of temperature parameter on the MCMC acceptance rate, and empirically set τ for each dimension to be one of $\{0.1, 0.05, 0.01, 0.002\}$. Note that none of these choices change the local maxima of our objective function in any way; they simply may make convergence more efficient.

As for the prior over $\boldsymbol{X}$, a nice feature of our framework is that we can use non-uniform priors if there is reason to do so; later, we explore the possibility of including mildly informative priors for certain TFs where measurements of cellular concentrations in *S. cerevisiae* are available [5]. However, when no relevant information is available, a uniform prior distribution is a natural choice. In what follows, we use a uniform prior over $[-10, 2]$ for log transition weights of

TFs and a uniform prior over $[0, 3]$ for the log transition weight of nucleosomes. Such values are chosen based on the range of TF dissociation constants at their respective optimal binding sites (K_d, as defined and computed by Granek and Clarke [8]). Sig1 has the highest log K_d value of -2.5 and Asg1 has the lowest log K_d value of -7.6. Empirical observations also show that MCMC never produced samples outside these ranges.

In our Gibbs-style MCMC, each iteration consists of an update for each of the transition weight parameters in the model. On a commodity computer cluster, we could compute roughly 25 such iterations per hour.

2.6   Incorporating Pre-initiation Complexes

The pre-initiation complex (PIC) assembles at nucleosome-free promoter regions and facilitates transcription initiation and regulation. PICs compete with other DBFs for binding sites when they are assembled around TATA or TATA-like elements (henceforth referred to as TATA boxes, for simplicity). To account for this competition, we calculate the TATA-binding protein (TBP) binding probability in our model using the DNA binding specificity derived from Rhee and Pugh [20]. Because of the degenerate nature of the TBP binding motif, we amend our model to allow this competition to occur only at TATA boxes (essentially, we set the transition weight for TBP to be 0 at all sequence locations except TATA boxes).

Rhee and Pugh [20] report that core PICs (TBP-associated factors and general transcription factors) assemble approximately 40bp downstream of TATA boxes. The MNase digestion data used here also show an enrichment of short fragments coverage at the same location. Therefore, we approximate the PIC protection by adding the same MNase short fragment coverage shape (scaled by the probability of TBP binding) to the predicted small protein binding probability downstream of the TATA box (see Supplemental Figure S4 for details).

3   Results

3.1   Overall Inference Performance Evaluated by Cross Validation

We randomly split our 81 promoter regions into nine equal sets and performed a standard nine-fold cross validation: parameters were trained on 72 promoter regions using MCMC and we used the average of MCMC samples as trained DBF weights $\hat{X}$; we then calculated $h(\hat{X})$ values for the held out nine promoter regions. Figure 2 shows boxplots of $h(\hat{X})$ values of all the training and testing promoter regions from all the folds of cross validation. We compare the performance to five baselines: (a) average performance when log transition weights are drawn 1000 times uniformly under the prior; or setting the nucleosome transition weight to 35 and TF transition weights to either (b) 8 K_d, (c) 16 K_d, (d) 32 K_d, or (e) 64 K_d.

As Figure 2 shows, our learned model outperforms all five baselines by a significant amount. Note that $h(X) = 0.5$ indicates no correlation on average between

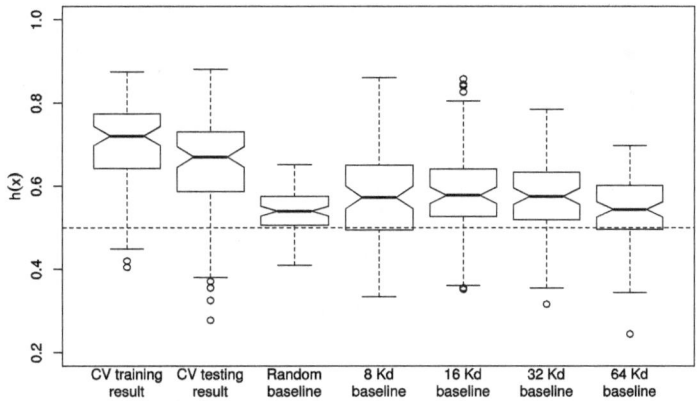

Fig. 2. Comparison of cross validated inference performance to various baselines. Data from the 81 promoter regions were split into nine equal parts. A standard nine-fold cross validation procedure was applied: 72 promoter regions were used as training data to obtain trained DBF weights $\boldsymbol{X}$; we then calculated $h(\hat{\boldsymbol{X}})$ values of the held out nine promoter regions (testing results). 'CV training result' considers the $h(\hat{\boldsymbol{X}})$ values for each promoter when used as training data. 'CV testing result' shows the $h(\hat{\boldsymbol{X}})$ values for each promoter when used as testing data. Uniformly drawn TF transition weights and different multiples of K_d are used as baseline comparisons. Variance is reduced in the random baseline case because each result is the average of 1000 random samples.

the model predictions and observed data. We observe that median performance for the random baseline is still larger than 0.5 even though the TF transition weights are uninformed guesses; this is because the model's emission parameters (derived from *in vitro* experimental data regarding TF and nucleosome binding specificity) are highly informed.

3.2   A Mechanistic Explanation for Paired-End MNase-seq Data

Owing to *in vitro* experiments, our model has knowledge about inherent DBF sequence specificities. The thermodynamic interaction and competition between these DBFs are accounted for by COMPETE. By adding information about *in vivo* DBF binding occupancy levels present in MNase-seq data, our framework can now infer a DBF binding landscape that provides a mechanistic explanation for the observed data.

Figure 3 illustrates examples of predicted binding profiles for each DBF in six promoter regions in the test sets of the nine-fold cross validation, in comparison with the corresponding MNase-seq binding profile tracks (see Supplemental Figure S3 for raw coverage and Supplemental Figure S5 for additional comparisons between composite predicted profiles and processed MNase-seq fragment coverages). These examples span the full spectrum of our framework performance, from strong performance to weak performance. In all cases, our predictions for the TF binding profiles provide a good or fair explanation for the MNase-seq

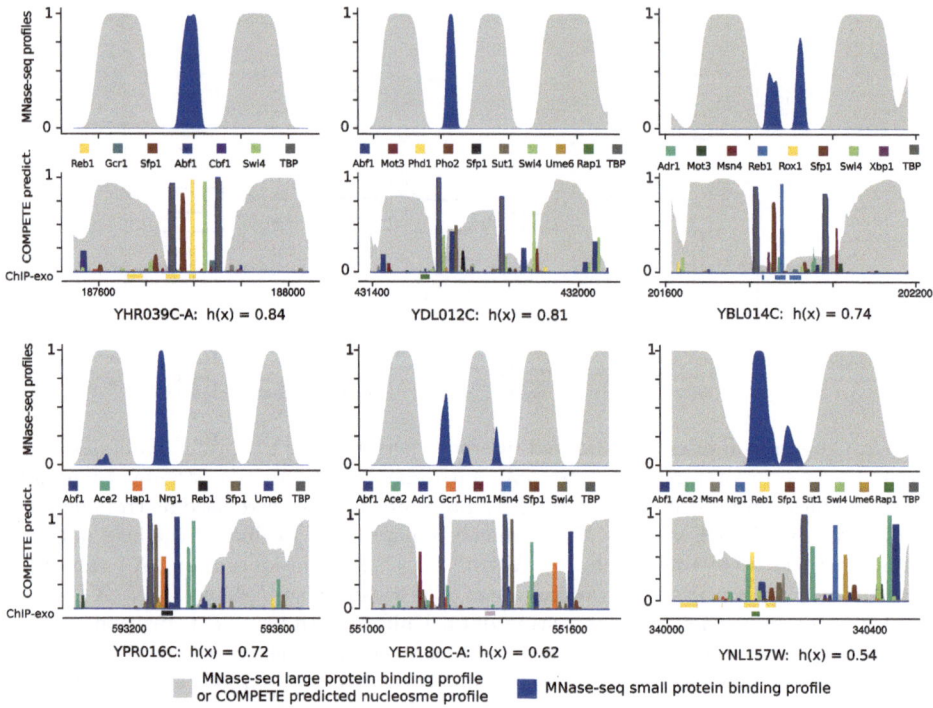

Fig. 3. Predicted binding profiles versus MNase-seq binding profiles. For six promoter regions in our 81 promoter set, we plot the predicted binding profiles when they were evaluated as testing data. We also indicate reported binding sites from ChIP-exo [19] underneath the predicted binding profiles; these have the same color as the corresponding TF's binding probability. No binding event is reported by MacIsaac et al. [15, p-value < 0.001 and conservation level 3] for these promoter regions.

data and are much more consistent with the data compared to random baseline predictions (see Supplemental Figure S1), considering the simplicity of our approach and the complexity of the problem.

One difficulty in interpreting high-throughput nuclease digestion data is identifying the binding proteins at read-enriched regions. Traditional motif matching is not satisfactory when there are multiple potentially overlapping motifs, nor can it assess the strength of protein binding. In contrast, our framework provides a principled interpretation for the data in terms of distinct binding events, each with its own probability of occurrence based on evaluating the probability of every possible binding configuration in the ensemble. This is demonstrated, for example, in the YDL012C and YPR016C promoter regions. Our approach can also capture weak binding events, such as the Reb1 binding events in the YPR016C and YNL157W promoter regions, which are missed in ChIP-chip experiments [15] but are captured in ChIP-exo experiments [19] (Figure 3; reported ChIP-exo binding sites are indicated underneath the predicted TF binding landscape; no binding event is reported by MacIsaac et al. [15, p-value < 0.001 and conservation level 3] for these promoter regions).

Our predictions of nucleosome binding profiles match the data well in spite of the fact that nucleosome positioning is less precise than TF positioning. The predictions reflect the intrinsic uncertainty about nucleosome positioning related to their mobility and only mildly sequence preferences, especially when the MNase-seq large protein binding profile is more noisy, as in the promoter regions of YBL014C and YNL157W (Figure 3; see Supplemental Figure S3 for raw coverage).

3.3   Incorporating Measurements of Protein Concentration through Prior Distributions

We have demonstrated that our framework can achieve good performance using non-informative priors. However, the framework could potentially perform better by incorporating prior information when it is available. For instance, Ghaemmaghami et al. [5] measured cellular protein concentrations using Western blots in *S. cerevisiae* during log phase growth. As discussed above, although cellular protein concentrations are not precisely equivalent to the transition weights we are estimating, the two still might be expected to loosely correlate with one another. We can therefore use these measurements to construct weak prior distributions for the corresponding DBF transition weights. To account for the loose correlation between the two, as well as experimental measurement error, we use a truncated normal prior for log transition weights with a large standard deviation of 2 (so a standard deviation in each direction corresponds to multiplying the weight by 1/100 or 100, respectively). We calculate the mean for this normal prior by converting measurements from Ghaemmaghami et al. [5] to molar concentration using a yeast cell volume of 5×10^{-14}L [2]. The resulting prior means are in the range of -8 to -6 in log scale. Note that nine of the 42 TFs in our model do not have measurements available, and thus their priors remain uniform, as described above.

When we utilize this prior information, we observe no change in training performance and a marginal increase in testing performance (median $h(\hat{X})$ increases by 0.013; Figure 4). Such an insignificant result could arise for multiple reasons: (a) the aforementioned difference between cellular concentration and the model's transition weights means that the information provided by the measured concentrations might not even be relevant; (b) the noisy physiological measurements of both cellular concentration and cell volume means that the measurements we used might not be quite accurate; or (c) the weak prior we utilized in the model because the measured concentrations are not trusted to be very precise means that the objective function landscape might change only slightly.

4   Discussion

We show that integrating information from experimental data within a general framework built on a thermodynamic ensemble model of competitive factor

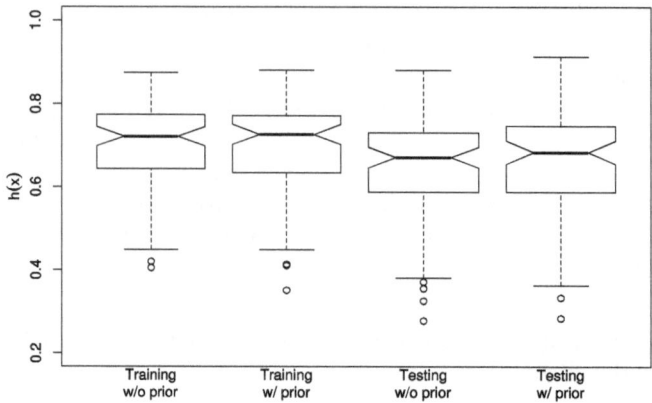

Fig. 4. Comparison of cross validation performance with and without prior information regarding measured cellular protein concentration. Performance for each promoter is measured by the geometric mean $(h(\hat{\boldsymbol{X}}))$ of the two Pearson correlations defined in Figure 1. Each boxplot shows the performance summary of the 81 promoter regions across all the cross validation trials.

binding can improve the accuracy of inferred protein-DNA interactions, providing a more biologically plausible view of the protein-DNA interaction landscape. Such a landscape gives a mechanistic explanation for observed paired-end MNase-seq fragments through various protein binding events, each with its own probability of occurrence. Many of those binding events are weak binding events that are typically missed in other modeling methods, but are captured in our framework; these weaker binding events are also supported by higher resolution experimental data where available [20]. These weak binding events are important: It has been reported that low affinity protein-DNA interactions may be involved in fine-tuning transcriptional regulation and are common along the genome [1, 22, 23]. Our framework's predictions agree with this viewpoint: 72% of the binding events in our predicted profiles have a probability lower than 0.5. Our framework could thus form an important basis for future computational work that connects transcriptional activity with the protein-DNA interaction landscape.

Our framework does not successfully predict a few TF binding events reported by high resolution ChIP-exo experiments [19], most notably some of the binding sites for Phd1 and Reb1. We believe the primary reason is occasional mismatches between our input TF PWMs and these proteins' actual *in vivo* DNA-binding specificities. For Phd1, Rhee and Pugh [20] report several distinct *in vivo* motifs. However, the Phd1 PWM we used in our framework comes from *in vitro* data [27] and does not match the *in vivo* DNA-binding specificity of Phd1 reported by Rhee and Pugh [20]. Similarly, for Reb1, Rhee and Pugh [20] report that 40% of Reb1 binding sites are so-called 'secondary binding sites', with motifs that deviate from the TTAGGC consensus of the *in vitro* PWM we are using.

This mismatch in DNA binding specificity may account for much of the discrepancy between our predicted profiles and reported binding sites. However, some caution should be taken when interpreting *in vivo* ChIP data, since the assay cannot distinguish between direct protein-DNA interaction and indirect interaction [6]. We also note that our current framework only includes a subset of all yeast TFs. Some unexplained short fragment coverage peaks, such as those in the YBL014C promoter region, could indicate the binding of DBFs that are not in our set. These and other discrepancies may have an impact on our overall inference, resulting in missing binding events (or possibly even superfluous binding events, because of the competition that is inherent in our model).

In the promoters of YNL157W and YDL012C, our predictions do not include Rap1 binding events even though they are reported in ChIP-exo experiments. However, we believe this results from the nature of Rap1 binding: Lickwar et al. [13] report that Rap1 binding on non-ribosomal protein promoters, like the two mentioned above, is highly dynamic and involves fast turnover. Such binding events are possibly captured in ChIP experiments because of cross-linking, but may be difficult to observe in an MNase-based digestion experiment if the latter does not involve a cross-linking step. Incidentally, the two ChIP-determined Rap1 binding events are not close to MNase-seq small fragment coverage peaks. One possible use of our framework for extending the results shown here would be to incorporate data from ChIP-based experiments and use the framework to estimate parameters that reflect information from both kinds of data.

We designed our model to take advantage of published data on PIC positions [20]. Such data provides additional protein occupancy information that is likely the result of mechanisms beyond TBP-related protein sequence specificity and protein competition. We observed that adding PICs allowed the nucleosome free regions to agree better with the MNase fragment data, because the PICs both enhance the exclusion of nucleosomes and explain some of the small MNase fragments downstream of the TATA-like element (so that TFs are not needed to provide that explanation).

We also demonstrate the use of prior information in our framework through incorporating measured bulk cellular protein concentration. The model performance improved marginally, which can be interpreted two ways. On the one hand, it is reassuring that one need not have measured cellular protein concentrations in order to perform effective inference. The fact that our uniform priors work as well as having priors informed by measured concentrations means that the measured concentrations available currently are not critical for good performance. However, that said, it is also reassuring that our framework has the ability to incorporate this sort of prior information when available because we anticipate such data will only improve. As measurement technologies enable us to move from bulk cellular concentrations toward nuclear concentrations of active TFs, we anticipate that the ability to incorporate prior information will become more useful, if not for achieving better results then perhaps at least for more rapid convergence toward optima when we move to higher-dimensional inference (e.g., more TFs).

With adequately fitted parameters, our framework has the potential to perform *in silico* simulation for various environmental conditions by changing the protein concentrations. For example, we could simulate *in silico* heat shock by increasing the concentration of heat shock response factors in our model. We could also investigate how certain single nucleotide polymorphisms (SNP) affect the overall protein-DNA interaction landscape, not just at the site of the SNP but propagating to the surrounding region due to altered competition.

This work represents a first step toward a more general framework. By specifying probabilistic distributions appropriate for other kinds of experiments—like ChIP-seq, FAIRE-seq, or DNase-seq—the framework can integrate other sources of data through a joint likelihood. As more and larger-scale sequencing projects are carried out, such a framework will prove extremely valuable for integrating different pieces of information to infer a more precise view of the protein-DNA interactions that govern transcriptional regulation.

Supplemental information is available from http://www.cs.duke.edu/~amink/

Acknowledgments. We would like to thank Jason Belsky, Kaixuan Luo, Yezhou Huang, and Michael Mayhew for helpful discussions and comments.

References

[1] Biggin, M.: Animal transcription networks as highly connected, quantitative continua. Developmental Cell 21(4), 611–626 (2011)

[2] Bryan, A.K., Goranov, A., Amon, A., et al.: Measurement of mass, density, and volume during the cell cycle of yeast. Proceedings of the National Academy of Sciences 107(3), 999–1004 (2010)

[3] Chen, X., Hoffman, M., Bilmes, J., et al.: A dynamic Bayesian network for identifying protein-binding footprints from single molecule-based sequencing data. Bioinformatics 26(12), i334–i342 (2010)

[4] Foat, B., Morozov, A., Bussemaker, H.: Statistical mechanical modeling of genome-wide transcription factor occupancy data by MatrixREDUCE. Bioinformatics 22(14), e141–e149 (2006)

[5] Ghaemmaghami, S., Huh, W.K., Bower, K., et al.: Global analysis of protein expression in yeast. Nature 425(6959), 737–741 (2003)

[6] Gordân, R., Hartemink, A.J., Bulyk, M.: Distinguishing direct versus indirect transcription factor-DNA interactions. Genome Research 19(11), 2090–2100 (2009)

[7] Gordân, R., Murphy, K., McCord, R., et al.: Curated collection of yeast transcription factor DNA binding specificity data reveals novel structural and gene regulatory insights. Genome Biology 12(12), R125 (2011)

[8] Granek, J., Clarke, N.: Explicit equilibrium modeling of transcription-factor binding and gene regulation. Genome Biology 6(10), R87 (2005)

[9] Harbison, C., Gordon, D., Lee, T., et al.: Transcriptional regulatory code of a eukaryotic genome. Nature 431(7004), 99–104 (2004)

[10] Henikoff, J., Belsky, J., Krassovsky, K., et al.: Epigenome characterization at single base-pair resolution. Proceedings of the National Academy of Sciences 108(45), 18318–18323 (2011)

[11] Hesselberth, J., Chen, X., Zhang, Z., et al.: Global mapping of protein-DNA interactions in vivo by digital genomic footprinting. Nature Methods 6(4), 283–289 (2009)

[12] Kaplan, T., Li, X.Y., Sabo, P., et al.: Quantitative models of the mechanisms that control genome-wide patterns of transcription factor binding during early Drosophila development. PLoS Genetics 7(2), e1001290 (2011)

[13] Lickwar, C.R., Mueller, F., Hanlon, S.E., et al.: Genome-wide protein-DNA binding dynamics suggest a molecular clutch for transcription factor function. Nature 484(7393), 251–255 (2012)

[14] Luo, K., Hartemink, A.J.: Using DNase digestion data to accurately identify transcription factor binding sites. In: Pacific Symposium on Biocomputing, pp. 80–91. World Scientific (2013)

[15] MacIsaac, K., Wang, T., Gordon, D., et al.: An improved map of conserved regulatory sites for Saccharomyces cerevisiae. BMC Bioinformatics 7, 113 (2006)

[16] Pique-Regi, R., Degner, J.F., Pai, A.A., et al.: Accurate inference of transcription factor binding from DNA sequence and chromatin accessibility data. Genome Research 21(3), 447–455 (2011)

[17] Rabiner, L.: A tutorial on hidden Markov models and selected applications in speech recognition. Proceedings of the IEEE 77(2), 257–286 (1989)

[18] Ren, B., Robert, F., Wyrick, J., et al.: Genome-wide location and function of DNA binding proteins. Science 290(5500), 2306–2309 (2000)

[19] Rhee, H., Pugh, B.: Comprehensive genome-wide protein-DNA interactions detected at single-nucleotide resolution. Cell 147(6), 1408–1419 (2011)

[20] Rhee, H., Pugh, B.: Genome-wide structure and organization of eukaryotic preinitiation complexes. Nature 483(7389), 295–301 (2012)

[21] Saul, L., Jordan, M.: Boltzmann chains and hidden Markov models. Advances in Neural Information Processing Systems, pp. 435–442. MIT Press (1995)

[22] Segal, E., Raveh-Sadka, T., Schroeder, M., et al.: Predicting expression patterns from regulatory sequence in Drosophila segmentation. Nature 451(7178), 535–540 (2008)

[23] Tanay, A.: Extensive low-affinity transcriptional interactions in the yeast genome. Genome Research 16(8), 962–972 (2006)

[24] Teif, V., Rippe, K.: Calculating transcription factor binding maps for chromatin. Briefings in Bioinformatics 13(2), 187–201 (2012)

[25] Wasson, T., Hartemink, A.J.: An ensemble model of competitive multi-factor binding of the genome. Genome Research 19(11), 2101–2112 (2009)

[26] Weirauch, M.T., Cote, A., Norel, R., et al.: Evaluation of methods for modeling transcription factor sequence specificity. Nature Biotechnology 31(2), 126–134 (2013)

[27] Zhu, C., Byers, K., McCord, R., et al.: High-resolution DNA-binding specificity analysis of yeast transcription factors. Genome Research 19(4), 556–566 (2009)

Imputation of Quantitative Genetic Interactions in Epistatic MAPs by Interaction Propagation Matrix Completion

Marinka Žitnik[1] and Blaž Zupan[1,2]

[1] Faculty of Computer and Information Science, University of Ljubljana, Tržaška 25, 1000, Slovenia
[2] Department of Molecular and Human Genetics, Baylor College of Medicine, Houston, TX-77030, USA
{marinka.zitnik,blaz.zupan}@fri.uni-lj.si

Abstract. A popular large-scale gene interaction discovery platform is the Epistatic Miniarray Profile (E-MAP). E-MAPs benefit from quantitative output, which makes it possible to detect subtle interactions. However, due to the limits of biotechnology, E-MAP studies fail to measure genetic interactions for up to 40% of gene pairs in an assay. Missing measurements can be recovered by computational techniques for data imputation, thus completing the interaction profiles and enabling downstream analysis algorithms that could otherwise be sensitive to largely incomplete data sets. We introduce a new interaction data imputation method called interaction propagation matrix completion (IP-MC). The core part of IP-MC is a low-rank (latent) probabilistic matrix completion approach that considers additional knowledge presented through a gene network. IP-MC assumes that interactions are transitive, such that latent gene interaction profiles depend on the profiles of their direct neighbors in a given gene network. As the IP-MC inference algorithm progresses, the latent interaction profiles propagate through the branches of the network. In a study with three different E-MAP data assays and the considered protein-protein interaction and Gene Ontology similarity networks, IP-MC significantly surpassed existing alternative techniques. Inclusion of information from gene networks also allows IP-MC to predict interactions for genes that were not included in original E-MAP assays, a task that could not be considered by current imputation approaches.

Keywords: genetic interaction, missing value imputation, epistatic miniarray profile, matrix completion, interaction propagation.

1 Introduction

The epistatic miniarray profile (E-MAP) technology [1–4] is based on a synthetic genetic array (SGA) approach [5,6] and generates quantitative measures of both positive and negative genetic interactions (GIs) between gene pairs. E-MAP was developed to study the phenomenon of epistasis, wherein the presence of

R. Sharan (Ed.): RECOMB 2014, LNBI 8394, pp. 448–462, 2014.
© Springer International Publishing Switzerland 2014

one mutation modulates the effect of another mutation. The power of epistasis analysis is greatly enhanced by quantitative GI scores [2]. E-MAP has provided high-throughput measurements of hundreds of thousands of GIs in yeast [1,4,7] and has been shown to significantly improve gene function prediction [7]. However, E-MAP data suffer from a large number of missing values, which can be as high as ~40% for a given assay (see also Table 1). Missing values correspond to pairs of genes for which the strength of the interaction could not be measured during the experimental procedure or that were subsequently removed due to low reliability. A high proportion of missing values can adversely affect analysis algorithms or even prevent their use. For instance, missing data might introduce instability in clustering results [8] or bias the inference of prediction models [9]. Accurate imputation of quantitative GIs is therefore an appealing option to improve downstream data analysis and correspondence between genetic and functional similarity [7, 10–13]. Imputed quantitative GIs can be a powerful source for understanding both the functions of individual genes and the relationships between pathways in the cell.

The missing value problem in E-MAPs resembles that from gene expression data, where imputation has been well studied [9, 14, 15]. The objective in both tasks is to estimate the values of missing entries given the incomplete data matrix. Both types of data may exhibit a correlation between gene or mutant profiles, which is indicative of co-regulation in the case of gene expression data and pathway membership in the case of E-MAP data [16]. E-MAP data sets are therefore often analyzed with tools originally developed for gene expression data analysis [17]. However, there are important differences between E-MAP and gene expression data that limit the direct application of gene expression imputation techniques to E-MAPs [16]. E-MAP data are pairwise, symmetric and have substantially different dimensionality than gene expression data sets. They contain considerably more missing values than gene expression data sets (the latter have up to a 5% missing data rate, see [9, 18]). These differences, coupled with the biological significance of E-MAP studies, have spurred the development of specialized computational techniques for recovering missing data in E-MAP-like data sets [16].

In this paper, we propose IP-MC ("interaction propagation matrix completion"), a *hybrid* and *knowledge assisted* method for imputing missing values in E-MAP-like data sets. IP-MC builds upon two concepts, matrix completion and propagation of interaction. Matrix completion uses information on global correlation between entries in the E-MAP score matrix. The interaction propagation serves to exploit the local similarity of genes in a gene network. The use of background knowledge in the form of gene networks gives IP-MC the potential to improve imputation accuracy beyond purely data-driven approaches. This could be especially important for data sets with a small number of genes and a high missing data rate, such as E-MAPs. In the following, we derive a mathematical formulation of the proposed approach and, in a comparative study that includes several state-of-the-art imputation techniques, demonstrate its accuracy across several E-MAP data sets.

2 Related Work

Imputation algorithms for gene expression data sets are reviewed in Liew *et al.* (2011) [9], who categorized them into four classes based on how they utilize or combine local and global information from within the data (*local, global* and *hybrid* algorithms) and their use of domain knowledge during imputation (*knowledge-assisted* algorithms). Local methods, such as k-nearest neighbors (KNNimpute) [14], local least squares (LLS) [19] and adaptive least squares (LSimpute) [18], rely on the local similarity of genes to recover the missing values. Global methods are based on matrix decompositions, such as the singular value decomposition (e.g. SVDimpute [14]), the singular value thresholding algorithm for matrix completion (SVT) [20] and Bayesian principal component analysis (BPCA) [21]. A hybrid imputation approach for gene expression data by Jörnsten *et al.* (2005) [22] estimates missing values by combining estimates from three local and two global imputation methods.

Only a handful of missing data imputation algorithms directly address E-MAP-like data sets. Ulitsky *et al.* (2009) [23] experimented with a variety of genomic features, such as the existence of physical interaction or co-expression between gene pairs, that were used as input to a classification algorithm. The IP-MC differs from this approach as it directly uses the matrix of measured GI scores and does not require data-specific feature engineering. Ryan *et al.* (2010, 2011) [16, 24] considered four general strategies for imputing missing values – three local methods and one global method – and adapted these strategies to address E-MAPs. They modified unweighted and weighted k-nearest neighbors imputation methods (uKNN and wNN, respectively). They also adapted LLS and BPCA algorithms to handle symmetric data. We refer the reader to Ryan *et al.* (2010) [16] for details on the algorithm modifications. We compare their imputation approaches with the IP-MC (see Sec. 5). Pan *et al.* (2011) [25] proposed an ensemble approach to combine the outputs of two global and four local imputation methods based on diversity of estimates of individual algorithms. In this paper we focus on the development of a single algorithm that if necessary could be used in an ensemble, and therefore compare it only with ensemble-free algorithms.

Another avenue of research focuses on predicting qualitative, i.e. binary, instead of quantitative interactions. Qualitative predictions estimate the presence or absence of certain types of interaction rather than their strength [26–29]. A major distinction between these techniques and the method proposed in the paper is that we aim at accurate imputation of quantitative genetic interactions using the scale of GI scores. Individual GI may by itself already provide valuable biological insight, as each interaction provides evidence for a functional relationship between a gene pair. Prediction of synthetic sick and lethal interaction types in *S. cerevisiae* was pioneered by Wong *et al.* (2004) [26], who applied probabilistic decision trees to diverse genomic data. Wong *et al.* [26] introduced *2-hop features* for capturing the relationship between a gene pair and a third gene. They showed that, for example, if protein g_1 physically interacts with protein g_2, and gene g_3 is synthetic lethal with the encoding gene of g_2, then this increases the likelihood of a synthetic lethal interaction between the encoding gene of g_1 and gene g_3. Two-hop features were shown to be crucial when predicting GIs [11, 23, 26] and are the rationale behind our concept of interaction propagation.

3   Methods

We first introduce a probabilistic model of matrix completion for missing value imputation in E-MAP-like data sets. The model predicts scores for missing interaction measurements by employing only the E-MAP score matrix. We then extend it with the notion of interaction propagation. The resulting method, IP-MC, is able to exploit the transitivity of interactions, that is, the relationship between a gene pair and a third gene (see Sec. 2). IP-MC predicts missing values from both E-MAP data and the associated gene network that encodes domain knowledge. Any type of knowledge that can be expressed in the form of a network can be passed to IP-MC. In this paper, we use the Gene Ontology [30] semantic similarity network and protein-protein interaction network.

3.1   Problem Definition and Preliminaries

In the E-MAP study we have a set of genes $(g_1, g_2, \ldots, g_n)$. The genetic interaction between a pair of genes is scored according to the fitness of the corresponding double mutant and reported through an S-score that reflects the magnitude and sign of the observed GI [2]. Scored GIs are reported in the form of a partially observed matrix $\mathbf{G} \in \mathbf{R}^{n \times n}$. In this matrix, $\mathbf{G}_{i,j}$ contains a GI measurement between g_i and g_j. Here, $\mathbf{G}$ is symmetric, $\mathbf{G}_{i,j} = \mathbf{G}_{j,i}$. Without loss of generality, we map GIs to the [0,1]-interval by normalizing $\mathbf{G}$ (step 1 in Fig. 2). Following the imputation, we re-scale the completed (imputed) matrix $\widehat{\mathbf{G}}$ to the original scale of S-scores (step 5 in Fig. 2).

In a gene network every gene g_i has a set of N_{g_i} neighbors, and $\mathbf{P}_{i,j}$ denotes the value of influence that gene $g_j \in N_{g_i}$ has on g_i. These values are given in matrix $\mathbf{P} \in \mathbb{R}^{n \times n}$. We normalize each row of $\mathbf{P}$ such that $\sum_{j=1}^{n} \mathbf{P}_{i,j} = 1$. A non-zero entry $\mathbf{P}_{i,j}$ denotes dependence of the g_i-th latent feature vector to the g_j-th latent feature vector. Using this idea, latent features of genes that are indirectly connected in the network become dependent after a certain number of algorithm steps, the number of steps being determined by the path distance between genes. Hence, information about gene latent representation propagates through the network.

The model inference task is defined as follows: given a pair of genes, g_i and g_j, for which $\mathbf{G}_{i,j}$ (and $\mathbf{G}_{j,i}$) is unknown, predict the quantitative GI between g_i and g_j using $\mathbf{G}$ and $\mathbf{P}$. We employ a probabilistic view of matrix completion to learn gene latent feature vectors. Let $\mathbf{F} \in \mathbb{R}^{k \times n}$ and $\mathbf{H} \in \mathbb{R}^{k \times n}$ be gene latent feature matrices with column vectors $\mathbf{F}_i$ and $\mathbf{H}_j$ representing k-dimensional gene-specific latent feature vectors of g_i and g_j, respectively. The goal is to learn these latent feature matrices and utilize them for missing value imputation in E-MAP-like data sets.

3.2   Matrix Completion Model

We start our derivation by formulating basic matrix completion approach for recovering missing values in $\mathbf{G}$ without considering the additional gene network. Throughout the paper, this approach is denoted by MC. In order to learn low-dimensional gene latent feature matrices $\mathbf{F}$ and $\mathbf{H}$, we factorize observed values in $\mathbf{G}$. The conditional probability of observed GIs is defined as:

$$p(\mathbf{G}|\mathbf{F}, \mathbf{H}, \sigma_{\mathbf{G}}^2) = \prod_{i=1}^{n} \prod_{j=1}^{n} \mathcal{N}(\mathbf{G}_{i,j}|g(\mathbf{F}_i^T \mathbf{H}_j), \sigma_{\mathbf{G}}^2)^{I_{i,j}^{\mathbf{G}}}, \tag{1}$$

where $\mathcal{N}(x|\mu, \sigma^2)$ is a normal distribution with mean μ and variance σ^2 and $I_{i,j}^{\mathbf{G}}$ is an indicator function that is equal to 1 if a GI score between g_i and g_j is available and is 0 otherwise. Notice that Eq. (1) deals only with observed entries in matrix $\mathbf{G}$. Thus, predictions are not biased by setting missing entries in $\mathbf{G}$ to some fixed value, which is otherwise common in matrix factorization algorithms. The function g is a logistic function, $g(x) = 1/(1+e^{-0.5x})$, which bounds the range of $g(\mathbf{F}_i^T \mathbf{H}_j)$ within interval $(0, 1)$. We assume a zero-mean Gaussian prior for gene latent feature vectors in $\mathbf{F}$ as $p(\mathbf{F}|\sigma_{\mathbf{F}}^2) = \prod_{i=1}^{n} \mathcal{N}(\mathbf{F}_i|0, \sigma_{\mathbf{F}}^2 \mathbf{I})$ and similarly, the prior probability distribution for $\mathbf{H}$ is given by $p(\mathbf{H}|\sigma_{\mathbf{H}}^2) = \prod_{i=1}^{n} \mathcal{N}(\mathbf{H}_i|0, \sigma_{\mathbf{H}}^2 \mathbf{I})$.

Through Bayesian inference we obtain the following equation for the log-posterior probability of latent feature matrices $\mathbf{F}$ and $\mathbf{H}$ given the interaction measurements in $\mathbf{G}$:

$$\ln p(\mathbf{F}, \mathbf{H}|\mathbf{G}, \sigma_{\mathbf{G}}^2, \sigma_{\mathbf{F}}^2, \sigma_{\mathbf{H}}^2) = -\frac{1}{2\sigma_{\mathbf{G}}^2} \sum_{i=1}^{n} \sum_{j=1}^{n} I_{i,j}^{\mathbf{G}} (\mathbf{G}_{i,j} - g(\mathbf{F}_i^T \mathbf{H}_j))^2 - \frac{1}{2\sigma_{\mathbf{F}}^2} \sum_{i=1}^{n} \mathbf{F}_i^T \mathbf{F}_i$$

$$- \frac{1}{2\sigma_{\mathbf{H}}^2} \sum_{j=1}^{n} \mathbf{H}_j^T \mathbf{H}_j - \frac{1}{2} (\sum_{i=1}^{n} \sum_{j=1}^{n} I_{i,j}^{\mathbf{G}}) \ln \sigma_{\mathbf{G}}^2 - \frac{1}{2} nk(\ln \sigma_{\mathbf{F}}^2 + \ln \sigma_{\mathbf{H}}^2) + \mathcal{C}. \tag{2}$$

We select the factorized model by finding the maximum *a posteriori* (MAP) estimate. This is equivalent to solving a minimization problem with the objective:

$$\mathcal{L}(\mathbf{G}, \mathbf{F}, \mathbf{H}) = \frac{1}{2} \sum_{i=1}^{n} \sum_{j=1}^{n} I_{i,j}^{\mathbf{G}} (\mathbf{G}_{i,j} - g(\mathbf{F}_i^T \mathbf{H}_j))^2 + \frac{\lambda_{\mathbf{F}}}{2} \sum_{i=1}^{N} \mathbf{F}_i^T \mathbf{F}_i + \frac{\lambda_{\mathbf{H}}}{2} \sum_{j=1}^{N} \mathbf{H}_j^T \mathbf{H}_j, \tag{3}$$

where $\lambda_{\mathbf{F}} = \sigma_{\mathbf{G}}^2 / \sigma_{\mathbf{F}}^2$ and $\lambda_{\mathbf{H}} = \sigma_{\mathbf{G}}^2 / \sigma_{\mathbf{H}}^2$. Interactions in $\mathbf{G}$ are normalized before numerical optimization such that they are between 0 and 1 because their estimates $g(\mathbf{F}^T \mathbf{H})$ are also bounded. We keep the observation noise variance $\sigma_{\mathbf{G}}^2$ and prior variances $\sigma_{\mathbf{F}}^2$ and $\sigma_{\mathbf{H}}^2$ fixed and use a gradient descent algorithm to find the local minimum of $\mathcal{L}(\mathbf{G}, \mathbf{F}, \mathbf{H})$ to infer gene latent feature matrices.

3.3   Interaction Propagation Matrix Completion Model

Interaction propagation matrix completion (IP-MC) extends the basic matrix completion model MC by borrowing latent feature information from neighboring genes in the network $\mathbf{P}$. A graphical example of IP-MC is shown in Fig. 1.

The biological motivation for the propagation of interactions stems from the transitive relationship between a gene pair and a third gene (see Sec. 2) and indicates that the behavior of a gene is affected by its direct and indirect neighbors in the underlying gene network $\mathbf{P}$. In other words, the latent feature vector of gene g, $\mathbf{F}_g$, is in each iteration dependent on the latent feature vectors of its direct neighbors $h \in N_g$ in $\mathbf{P}$. The influence is formulated as $\widehat{\mathbf{F}}_g = \sum_{h \in N_g} \mathbf{P}_{g,h} \mathbf{F}_h$, where $\widehat{\mathbf{F}}_g$ is the estimated latent feature vector of g given feature vectors of its direct neighbors. Thus, the latent feature vectors in $\mathbf{F}$ of genes that are indirectly connected in network $\mathbf{P}$ are dependent and thus, information about their latent representation propagates as the algorithm progresses according to the connectivity of network $\mathbf{P}$.

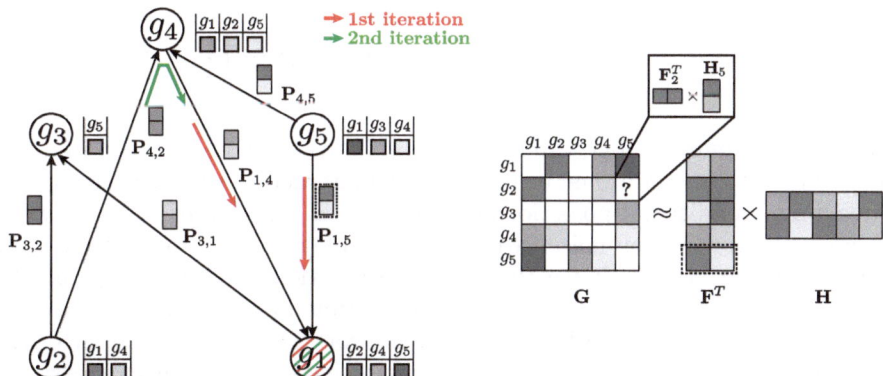

Fig. 1. An example application of the interaction propagation matrix completion algorithm (IP-MC). A hypothetical E-MAP data set with five genes $(g_1, \ldots, g_5)$ is given. Their measured GI profiles are listed next to corresponding nodes in gene network $\mathbf{P}$ (left) and are shown in the sparse and symmetric matrix $\mathbf{G}$ (right). Different shades of grey quantify interaction strength, while white matrix entries in $\mathbf{G}$ denote missing values. Matrices $\mathbf{F}$ and $\mathbf{H}$ are gene latent feature matrices. Gene latent feature vector $\mathbf{F}_{g_i}$ depends in each iteration of IP-MC on the latent feature vectors of g_i's direct neighbors in $\mathbf{P}$. For instance, the latent vector of gene g_1 in $\mathbf{F}$ depends in the first iteration of the IP-MC update (in red) only on its direct neighbors, the latent vectors of g_4 and g_5 ($\mathbf{F}_{g_4}$ and $\mathbf{F}_{g_5}$ are shown on input edges of g_1), whose level of influence is determined by $\mathbf{P}_{1,4}$ and $\mathbf{P}_{1,5}$, respectively. In the second iteration, the update of $\mathbf{F}_{g_1}$ (in green) also depends indirectly on the latent vector of g_2, $\mathbf{F}_{g_2}$. Thus, the influence of gene latent feature vectors propagates in $\mathbf{P}$. Gene latent feature matrix $\mathbf{H}$ is not influenced by the gene neighborhood in $\mathbf{P}$.

Notice that considering gene network $\mathbf{P}$ does not change the conditional probability of observed measurements (Eq. (1)). It only affects gene latent feature vectors in $\mathbf{F}$. We describe them with two factors: the zero-mean Gaussian prior to avoid overfitting and the conditional distribution of gene latent feature vectors given the latent feature vectors of their direct neighbors:

$$p(\mathbf{F}|\mathbf{P}, \sigma_{\mathbf{F}}^2, \sigma_{\mathbf{P}}^2) \propto \prod_{i=1}^{n} \mathcal{N}(\mathbf{F}_i|0, \sigma_{\mathbf{F}}^2 \mathbf{I}) \times \prod_{i=1}^{n} \mathcal{N}(\mathbf{F}_i| \sum_{j \in N_i} \mathbf{P}_{i,j} \mathbf{F}_j, \sigma_{\mathbf{P}}^2 \mathbf{I}). \tag{4}$$

Notice that such formulation of gene latent matrix $\mathbf{F}$ keeps gene feature vectors $\mathbf{F}_i$ both small and close to the latent feature vectors of their direct neighbors. Much like the previous section, we get the following equation through Bayesian inference for the posterior probability of gene latent feature matrices $\mathbf{F}$ and $\mathbf{H}$ given observed GI scores $\mathbf{G}$ and gene network $\mathbf{P}$:

$$p(\mathbf{F}, \mathbf{H}|\mathbf{G}, \mathbf{P}, \sigma_{\mathbf{G}}^2, \sigma_{\mathbf{P}}^2, \sigma_{\mathbf{F}}^2, \sigma_{\mathbf{H}}^2) \propto \prod_{i=1}^{n} \prod_{j=1}^{n} \mathcal{N}(\mathbf{G}_{i,j}|g(\mathbf{F}_i^T \mathbf{H}_j), \sigma_{\mathbf{G}}^2)^{I_{i,j}^{\mathbf{G}}}$$

$$\times \prod_{i=1}^{n} \mathcal{N}(\mathbf{F}_i| \sum_{j \in N_i} \mathbf{P}_{i,j} \mathbf{F}_j, \sigma_P^2 \mathbf{I}) \times \prod_{i=1}^{n} \mathcal{N}(\mathbf{F}_i|0, \sigma_{\mathbf{F}}^2 \mathbf{I}) \times \prod_{j=1}^{n} \mathcal{N}(\mathbf{H}_j|0, \sigma_{\mathbf{H}}^2 \mathbf{I}). \tag{5}$$

We then compute the log-posterior probability to obtain an equation similar to Eq. (2) but with an additional term due to the interaction propagation concept. To maximize conditional posterior probability over gene latent features, we fix the prior and observation noise variance and employ gradient descent on $\mathbf{F}$ and $\mathbf{H}$. In particular, we minimize the objective function similar to Eq. (3) that has an additional term to account for the conditional probability of gene latent features given their neighborhoods in gene network $\mathbf{P}$. The complete algorithm of IP-MC is presented in Fig. 2. In each iteration, gene latent feature matrices $\mathbf{F}$ and $\mathbf{H}$ are updated based on the latent feature vectors from the previous iteration and network neighborhood in $\mathbf{P}$. Successive updates of $\mathbf{F}_i$ and $\mathbf{H}_j$ converge to a maximum *a posteriori* (MAP) estimate of the posterior probability in Eq. (5).

4   Experimental Setup

In the experiments we consider an existing incomplete E-MAP matrix and artificially introduce an additional 1% of missing values for a set of arbitrarily selected gene pairs [16, 25]. These gene pairs and their data constitute a test set on which we evaluate the performance of imputation algorithms. Because of E-MAP symmetry, for a given test gene pair and its corresponding entry $\mathbf{G}_{i,j}$, we also hide the value of $\mathbf{G}_{j,i}$. We repeat this process 30 times and report on the averaged imputation performance.

Notice that the standard performance evaluation procedure of missing value imputation methods for gene expression data is not directly applicable to E-MAPs for the several reasons discussed in [16]. This approach constructs a complete gene expression data matrix by removing genes with missing data and then artificially introduces missing values for evaluation. In gene expression data, a substantially lower fraction of data is missing than in E-MAPs (Table 1)

Input: Sparse matrix $\mathbf{G} \in \mathbb{R}^{n \times n}$ containing S-scores of measured E-MAP interactions, gene network $\mathbf{P} \in \mathbb{R}^{n \times n}$ and parameters $\lambda_{\mathbf{F}} = \lambda_{\mathbf{H}}$, $\lambda_{\mathbf{P}}$, rank k and learning rate α.
Output: Completed E-MAP matrix $\widehat{\mathbf{G}}$.

1. Normalize $\tilde{\mathbf{G}} = (\mathbf{G} - \min_{i,j} \mathbf{G}_{i,j})/\max_{i,j} \mathbf{G}_{i,j}$.
2. Normalize each row of $\mathbf{P}$ such that $\sum_{j=1}^{n} \mathbf{P}_{i,j} = 1$.
3. Sample $\mathbf{F} \sim \mathcal{U}[0,1]^{k \times n}$ and $\mathbf{H} \sim \mathcal{U}[0,1]^{k \times n}$.
4. Repeat until convergence:
   - For $i, j \in 1, 2, \ldots, n$:

$$\frac{\partial \mathcal{L}}{\partial \mathbf{F}_i} = \sum_{j=1}^{n} I_{i,j}^{\tilde{\mathbf{G}}} \mathbf{H}_j g'(\mathbf{F}_i^T \mathbf{H}_j)(g(\mathbf{F}_i^T \mathbf{H}_j) - \tilde{\mathbf{G}}_{i,j}) + \lambda_{\mathbf{F}} \mathbf{F}_i +$$

$$+ \lambda_{\mathbf{P}}(\mathbf{F}_i - \sum_{j \in N_i} \mathbf{P}_{i,j} \mathbf{F}_j) - \lambda_{\mathbf{P}} \sum_{\{j | i \in N_j\}} \mathbf{P}_{j,i}(\mathbf{F}_j - \sum_{l \in N_j} \mathbf{P}_{j,l} \mathbf{F}_l),$$

$$\frac{\partial \mathcal{L}}{\partial \mathbf{H}_j} = \sum_{i=1}^{n} I_{i,j}^{\tilde{\mathbf{G}}} \mathbf{F}_i g'(\mathbf{F}_i^T \mathbf{H}_j)(g(\mathbf{F}_i^T \mathbf{H}_j) - \tilde{\mathbf{G}}_{i,j}) + \lambda_{\mathbf{H}} \mathbf{H}_j.$$

   - Set $\mathbf{F}_i \leftarrow \mathbf{F}_i - \alpha \frac{\partial \mathcal{L}}{\partial \mathbf{F}_i}$ for $i = 1, 2, \ldots, n$.
   - Set $\mathbf{H}_j \leftarrow \mathbf{H}_j - \alpha \frac{\partial \mathcal{L}}{\partial \mathbf{H}_j}$ for $j = 1, 2, \ldots, n$.
5. Compute $\widehat{\mathbf{G}} = g(\mathbf{F}^T \mathbf{H}) \cdot \max_{i,j} \mathbf{G}_{i,j} + \min_{i,j} \mathbf{G}_{i,j}$. Impute missing entry (i, j) as $(\widehat{\mathbf{G}}_{i,j} + \widehat{\mathbf{G}}_{j,i})/2$.

Fig. 2. Interaction propagation matrix completion (IP-MC) algorithm. We observed that parameter values $\lambda_{\mathbf{H}} = \lambda_{\mathbf{F}} = 0.01$ and $\alpha = 0.1$ gave accurate results across a number of different data sets. Parameter $\lambda_{\mathbf{P}}$, which controls the influence of gene network $\mathbf{P}$ on gene latent feature vectors $\mathbf{F}_i$, depended on data set complexity [15]. In data sets with higher complexity, we used a larger $\lambda_{\mathbf{P}}$ ($\lambda_{\mathbf{P}} = 1$).

and removing a small number of genes and experimental conditions does not significantly reduce the size of the data set.

In our experiments we select the number of latent dimensions k and regularization parameters $\lambda_{\mathbf{F}}$ and $\lambda_{\mathbf{P}}$ of IP-MC with the following procedure: For each data set and before the performance evaluation, we leave out 1% of randomly selected known values and attempt to impute them with varying values of parameters in a grid search fashion. Parameter values that result in the best estimation of the left-out values are then used in all experiments involving the data set. Notice that the left-out values are determined before the performance evaluation and are therefore not included in the test data set.

We consider two measures of imputation accuracy. These are the Pearson correlation (CC) between the imputed and the true values, and the normalized root mean squared error (NRMSE) [21] given as $\text{NRMSE} = \sqrt{\mathbb{E}((\widehat{\mathbf{y}} - \mathbf{y})^2)/\text{Var}(\mathbf{y})}$, where $\mathbf{y}$ and $\widehat{\mathbf{y}}$ denote vectors of true and imputed values, respectively. More accurate imputations give a higher correlation score and a lower NRMSE.

To test if the differences in performance between imputation methods are significant, we use the Wilcoxon signed-rank test, a non-parametric equivalent of a paired t-test. Its advantage is that it does not require a normal distribution or homogeneity of variance, but it has less statistical power, so there is the risk that some differences are not recognized as significant.

5   Results and Discussion

We considered three E-MAP data sets and compared IP-MC to five state-of-the-art methods for imputing missing values in E-MAP-like data sets [16]. We set the parameters of these methods to values as proposed in [16] (wNN, LLS, BPCA) or optimized the parameter selection through a grid search (SVT, MC, IP-MC). The evaluated data sets are from the budding yeast *S. cerevisiae*; they differ in their size, the subset of genes that are studied and the proportion of missing values (Table 1). We used GI S-scores reported in original publications:

- Chromosome Biology [7]: This is the largest of the E-MAPs, encompassing interactions between 743 genes involved in various aspects of chromosome biology, such as chromatid segregation, DNA replication and transcriptional regulation.
- RNA processing (RNA) [4]: It focuses on the relationships between and within RNA processing pathways involving 552 mutations, 166 of which are hypomorphic alleles of essential genes.
- The Early Secretory Pathway (ESP) [1]: It generates genetic interaction maps of genes acting in the yeast early secretory pathway to identify pathway organization and components of physical complexes.

Table 1. Overview of the E-MAPs considered

Data set	Genes	Missing Interactions	Measured Interactions
Chromosome Biology [7]	743	34.0%	187,000
Early Secretory Pathway [1]	424	7.5%	83,000
RNA [4]	552	29.6%	107,000

IP-MC considered two different data sources for gene network **P**. The first network was constructed from Gene Ontology [30] (GO) annotation data as a weighted network of genes included in the E-MAP study in which edge weights corresponded to the number of shared GO terms between connected genes, excluding annotations inferred from GI studies (i.e. those with the IGI evidence code). The second network represented physical interaction data from BioGRID 3.2 [31]. The physical interaction network was a binary network in which two genes were connected if their gene products physically interact. Both networks were normalized as described in Sec. 3.1. Depending on a network, we denote their corresponding IP-MC models by IP-MC-GO and IP-MC-PPI, respectively.

5.1   Imputation Performance

Table 2 shows the CC and NRMSE scores of imputation algorithms along with the baseline method of filling-in with zeros. IP-MC-PPI and IP-MC-GO demonstrated the best accuracy on all considered data sets. We compared their scores with the performance of the second-best method (i.e. LLS on Chromosome Biology data set, SVT on ESP data set and MC on RNA data set) and found that improvements were significant in all data sets.

We did not observe any apparent connection between the proportion of missing values in a data set and the performance of any of the imputation methods. The performance was better on smaller ESP and RNA data sets, although differences were small and further investigation appears to be worthwhile.

The baseline method of filling-in with zeros had the worst performance for all data sets. While this approach seems naïve, it is justified by the expectation that most genes do not interact. We observed that BPCA failed to match the performance of weighted neighbor-based and local least squares methods, wNN and LLS, respectively, despite BPCA being an improvement of the KNN algorithm. Both local imputation methods, wNN and LLS, demonstrated good performance across all three data sets. The good performance of neighbor-based methods on larger data sets could be explained by a larger number of neighbors to choose from when imputing missing values, which resulted in more reliable missing value estimates.

Global methods, BPCA, SVT and MC, performed well on the ESP data set but poorly on the much larger Chromosome Biology data set. These methods assume the existence of a global covariance structure among all genes in the E-MAP score matrix. When this assumption is not appropriate, i.e. when the genes exhibit dominant local similarity structures, their imputation becomes less accurate. Notice that the comparable performance of SVT and MC across data sets was expected. Both methods solve related optimization problems and operate under the assumption that the underlying matrix of E-MAP scores is low-rank.

The superior performance of IP-MC models over other imputation methods can be explained by their ability to include circumstantial evidence. As a hybrid imputation approach, IP-MC can benefit from both global information present in E-MAP data and local similarity structure between genes. One could vary the level of influence of global and local imputation aspects on the inferred IP-MC model through the $\lambda_\mathbf{P}$ parameter, where a higher value of $\lambda_\mathbf{P}$ indicates more emphasis on locality. In this way, our approach can adequately address the data of varying underlying complexity [15], where the complexity denotes the difficulty with which the data can be mapped to a lower dimensional subspace. Brock *et al.* (2008) [15] devised an entropy-based imputation algorithm selection scheme based on their observation that global imputation methods performed better on gene expression data with lower complexity and that local methods performed better on data with higher complexity. Thus, their selection scheme could be adapted to work with E-MAP-like data sets and be used to set $\lambda_\mathbf{P}$ in an informed way, which is left for our future work. In additional experiments (results not shown), we found that the performance of IP-MC is robust for a broad range of $\lambda_\mathbf{P}$ parameter values.

Table 2. Accuracy as measured by the Pearson correlation coefficient (CC) and normalized root mean squared error (NRMSE) across three E-MAP data sets and eight imputation methods. MC denotes the matrix completion model (Sec. 3.2). The IP-MC-GO and IP-MC-PPI models are interaction propagation matrix completion models (Sec. 3.3) that utilize annotation and physical interaction data, respectively. For descriptions of other methods see Related Work. Highlighted results are significantly better than the best non-IP-MC method according to the Wilcoxon signed-rank test at 0.05 significance level.

Approach	Chromosome Biology		ESP		RNA	
	CC	NRMSE	CC	NRMSE	CC	NRMSE
Filling with zeros	0.000	1.021	0.000	1.011	0.000	1.000
BPCA ($k = 300$)	0.539	0.834	0.619	0.796	0.589	0.804
wNN ($k = 50$)	0.657	0.744	0.625	0.776	0.626	0.787
LLS ($k = 20$)	0.678	0.736	0.626	0.764	0.626	0.776
SVT ($k = 40$)	0.631	0.753	0.672	0.719	0.649	0.765
MC ($k = 40$)	0.641	0.742	0.653	0.722	0.651	0.760
IP-MC-GO ($k = 60$)	0.691	0.693	**0.732**	**0.648**	**0.727**	**0.641**
IP-MC-PPI ($k = 60$)	**0.722**	**0.668**	**0.742**	0.667	0.701	**0.652**

5.2   Missing Value Abundance and Distribution

Ulitsky *et al.* (2009) [23] described three different scenarios of missing values in E-MAP experiments (Fig. 3). The simplest and the most studied scenario is the *Random* model, for which we assume that missing measurements are generated independently and uniformly by some random process. The *Submatrix* model corresponds to the case when all interactions between a subset of genes (e.g. essential genes) are missing. The *Cross* model arises when all interactions between two disjoint subsets of genes are missing. This model concurs with the situation when two E-MAP data sets that share a subset of genes are combined into a single larger data set. We identify another missing value configuration, which we call the *Prediction* scenario (Fig. 4d). It occurs when GI profiles of a subset of genes are completely missing. Learning in such a setting is substantially harder as these genes do not have any associated measurements. In the previous section, we compared the imputation methods on the Random configuration, and study other configurations in this section. This time we were interested in the effect these configurations have on IP-MC, and we compared the algorithm to its variant MC that does not use additional knowledge (e.g. the gene network).

Fig. 4 reports the predictive performance of our matrix completion approach obtained by varying the fraction of missing values in the four missing data scenarios from Fig. 3. For $x = 5, 10, 20, \ldots, 90$ we hid $x\%$ of E-MAP measurements in ESP data and inferred prediction models. Our results are reasonably accurate (CC > 0.4) when up to 60% of the E-MAP values were hidden for the Random and Submatrix model. Notice that when we hid 60% of the ESP E-MAP measurements, the E-MAP scores were present in less than 40% of the matrix because the original ESP data set already had ~8% missing values (Table 1).

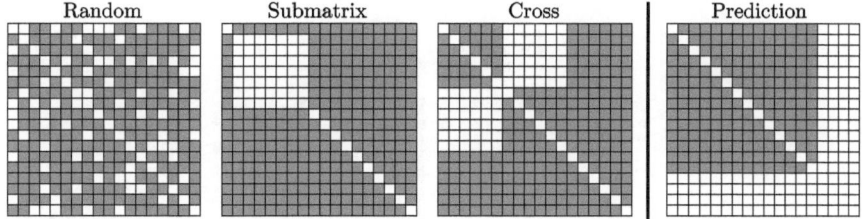

Fig. 3. The four configurations producing missing values in E-MAP data. In Random configuration, a random subset of GIs is hidden. In Submatrix or Cross configurations all interactions between a random subset of genes or two random disjoint subsets of genes, respectively, are hidden. In the Prediction scenario, complete profiles of GIs of a random subset of genes are removed.

When more than 80% of the data were removed, the three considered models still achieved higher accuracy (CC ≈ 0.2) than filling in with zeros. As expected, predictions were more accurate for the Random model than for the Submatrix model for almost all fractions of hidden data (cf. Fig. 4a and Fig. 4b). However, the difference in performance between the Random and the Submatrix models tended to be small when less than 30% or more than 70% of the measurements were hidden. We observed that inclusion of additional genomic data is more useful in structured missing value scenarios, i.e. the Submatrix and Cross models (Fig. 4b–4c).

Imputation accuracy improved (Fig. 4) when E-MAP data were combined with gene annotation (IP-MC-GO) or protein-protein interaction (IP-MC-PPI) networks. These results are not surprising as several studies [6,7,32] showed that if two proteins act together to carry out a common function, deletions of the two encoding genes have similar GI profiles and that gene annotations from the GO and synthetic lethality are correlated, with 12 and 27% of genetic interactions having an identical or similar GO annotation, respectively [6]. Thus, our IP-MC-GO and IP-MC-PPI models could exploit the strong links between functionally similar genes, physically interacting proteins and GIs. The performance of our two integrated models indicates the importance of combining interaction and functional networks for predicting missing values in E-MAP data sets.

Imputation accuracy deteriorated when complete profiles of GIs were removed and IP-MC could only utilize circumstantial evidence (Fig. 4d). This suggests that measured gene pairs in the E-MAP are the best source of information for predicting missing pairs. However, as the percentage of missing GIs increases, the inclusion of other genomic data is more helpful. With the exception of the Prediction model, for which we observed the opposite behavior, the performance difference between MC and IP-MC was small (∼10%) as long as <50% of the data were removed, but rose to above 20% when ≥60% of the data were removed (Fig. 4).

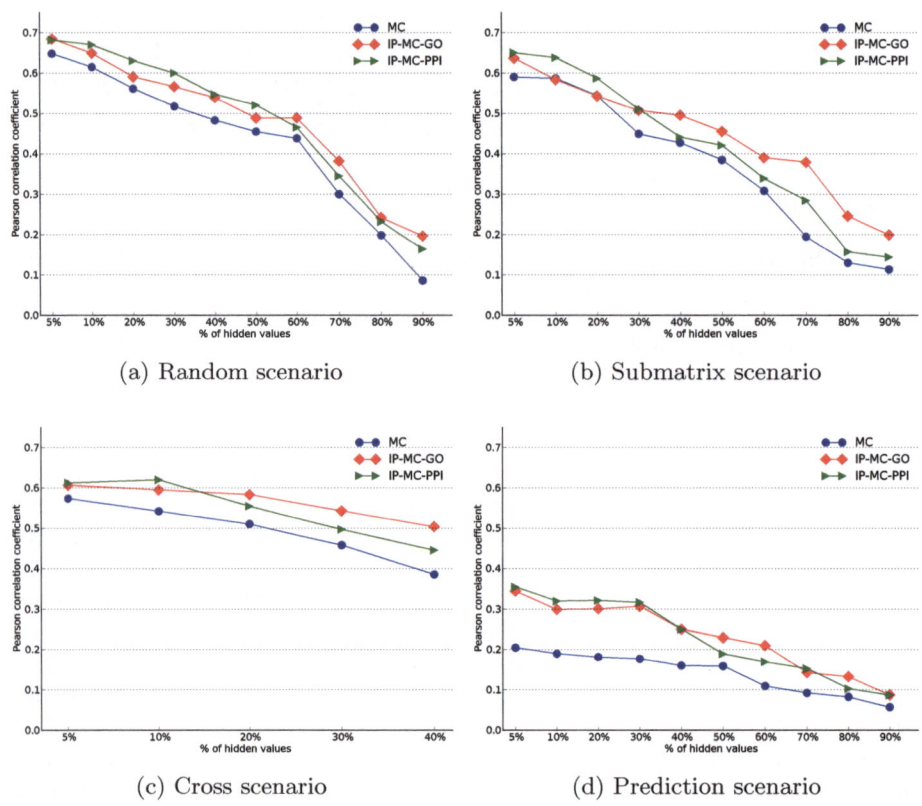

(a) Random scenario   (b) Submatrix scenario

(c) Cross scenario   (d) Prediction scenario

Fig. 4. Performance of imputation methods (Pearson correlation coefficient, CC) proposed in this paper for different fractions of missing values and scenarios of missing value distribution. Refer to the main text and Fig. 3 for descriptions of the missing value scenarios. MC denotes the matrix completion approach (Sec. 3.2). The integrated approaches are represented by IP-MC-GO and IP-MC-PPI (Sec. 3.3). Performance was assessed for the ESP E-MAP data set because it contains the least missing values. The 'Cross' configuration is not applicable when more than 50% of values are missing.

6   Conclusion

We have proposed a new missing value imputation method IP-MC that targets gene interaction data sets. The approach is unique in combining gene interaction and network data through inference of a single probabilistic model. Experiments on epistatic MAP interaction data sets show that the inclusion of additional knowledge is crucial and helps IP-MC to perform better than a number of state-of-the-art algorithms we have included in our study. The results are encouraging, have a potentially high practical value, and were intuitively expected. Gene interaction studies use double-mutant phenotypes to uncover functional dependencies, and additional knowledge that could provide any information on relations between genes should help. Driven by this intuition, the principal novelty of the paper is thus a new knowledge-based missing value imputation approach and the demonstration of its successful application on E-MAP data sets.

Acknowledgments. This work was supported by the Slovenian Research Agency (P2-0209, J2-5480), the National Institute of Health (P01-HD39691) and the European Commission (Health-F5-2010-242038).

References

1. Schuldiner, M., et al.: Exploration of the function and organization of the yeast early secretory pathway through an epistatic miniarray profile. Cell 123(3), 507–519 (2005)
2. Collins, S.R., et al.: A strategy for extracting and analyzing large-scale quantitative epistatic interaction data. Genome Biology 7, R63 (2006)
3. Roguev, A., et al.: Conservation and rewiring of functional modules revealed by an epistasis map in fission yeast. Science 322(5900), 405–410 (2008)
4. Wilmes, G.M., et al.: A genetic interaction map of RNA-processing factors reveals links between Sem1/Dss1-containing complexes and mRNA export and splicing. Molecular Cell 32(5), 735–746 (2008)
5. Tong, A.H.Y., et al.: Systematic genetic analysis with ordered arrays of yeast deletion mutants. Science 294(5550), 2364–2368 (2001)
6. Tong, A.H.Y., et al.: Global mapping of the yeast genetic interaction network. Science 303(5659), 808–813 (2004)
7. Collins, S.R., et al.: Functional dissection of protein complexes involved in yeast chromosome biology using a genetic interaction map. Nature 446(7137), 806–810 (2007)
8. de Brevern, A.G., et al.: Influence of microarrays experiments missing values on the stability of gene groups by hierarchical clustering. BMC Bioinformatics 5(1), 114 (2004)
9. Liew, A.W.C., et al.: Missing value imputation for gene expression data: computational techniques to recover missing data from available information. Briefings in Bioinformatics 12(5), 498–513 (2011)
10. Pu, S., et al.: Local coherence in genetic interaction patterns reveals prevalent functional versatility. Bioinformatics 24(20), 2376–2383 (2008)
11. Bandyopadhyay, S., et al.: Functional maps of protein complexes from quantitative genetic interaction data. PLoS Computational Biology 4(4), e1000065 (2008)
12. Ulitsky, I., et al.: From E-MAPs to module maps: dissecting quantitative genetic interactions using physical interactions. Molecular Systems Biology 4(1) (2008)
13. Järvinen, A.P., et al.: Predicting quantitative genetic interactions by means of sequential matrix approximation. PLoS One 3(9), e3284 (2008)
14. Troyanskaya, O., et al.: Missing value estimation methods for DNA microarrays. Bioinformatics 17(6), 520–525 (2001)
15. Brock, G.N., et al.: Which missing value imputation method to use in expression profiles: a comparative study and two selection schemes. BMC Bioinformatics 9(1), 12 (2008)
16. Ryan, C., et al.: Missing value imputation for epistatic MAPs. BMC Bioinformatics 11(1), 197 (2010)

17. Zheng, J., et al.: Epistatic relationships reveal the functional organization of yeast transcription factors. Molecular Systems Biology 6(1) (2010)
18. Bø, T.H., et al.: LSimpute: accurate estimation of missing values in microarray data with least squares methods. Nucleic Acids Research 32(3), e34 (2004)
19. Kim, H., et al.: Missing value estimation for DNA microarray gene expression data: local least squares imputation. Bioinformatics 21(2), 187–198 (2005)
20. Cai, J.F., et al.: A singular value thresholding algorithm for matrix completion. SIAM Journal on Optimization 20(4), 1956–1982 (2010)
21. Oba, S., et al.: A Bayesian missing value estimation method for gene expression profile data. Bioinformatics 19(16), 2088–2096 (2003)
22. Jörnsten, R., et al.: DNA microarray data imputation and significance analysis of differential expression. Bioinformatics 21(22), 4155–4161 (2005)
23. Ulitsky, I., et al.: Towards accurate imputation of quantitative genetic interactions. Genome Biology 10(12), R140 (2009)
24. Ryan, C., et al.: Imputing and predicting quantitative genetic interactions in epistatic MAPs. In: Network Biology, pp. 353–361 (2011)
25. Pan, X.Y., Tian, Y., Huang, Y., Shen, H.B.: Towards better accuracy for missing value estimation of epistatic miniarray profiling data by a novel ensemble approach. Genomics 97(5), 257–264 (2011)
26. Wong, S.L., et al.: Combining biological networks to predict genetic interactions. PNAS 101(44), 15682–15687 (2004)
27. Kelley, R., Ideker, T.: Systematic interpretation of genetic interactions using protein networks. Nature Biotechnology 23(5), 561–566 (2005)
28. Qi, Y., et al.: Finding friends and enemies in an enemies-only network: a graph diffusion kernel for predicting novel genetic interactions and co-complex membership from yeast genetic interactions. Genome Research 18(12), 1991–2004 (2008)
29. Pandey, G., et al.: An integrative multi-network and multi-classifier approach to predict genetic interactions. PLoS Computational Biology 6(9), e1000928 (2010)
30. Ashburner, M., et al.: Gene Ontology: tool for the unification of biology. Nature Genetics 25(1), 25–29 (2000)
31. Stark, C., et al.: BioGRID: a general repository for interaction datasets. Nucleic Acids Research 34(suppl. 1), D535–D539 (2006)
32. Costanzo, M., et al.: The genetic landscape of a cell. Science 327(5964), 425–431 (2010)

Author Index